Knowledge and Culture in the Earl

Knowledge and Culture in the Early Dutch Republic

Isaac Beeckman in Context

Edited by
Klaas van Berkel,
Albert Clement, and
Arjan van Dixhoorn

Routledge
Taylor & Francis Group
LONDON AND NEW YORK

First published in 2022 by Amsterdam University Press Ltd.

Published 2025 by Routledge
4 Park Square, Milton Park, Abingdon, Oxon OX14 4RN
605 Third Avenue, New York, NY 10158

Routledge is an imprint of the Taylor & Francis Group, an informa business

ISBN: 9789463722537 (hbk)
ISBN: 9781041181828 (pbk)
ISBN: 9781003698531 (ebk)
NUR 685

Cover illustration: Map of the city of Middelburg by Cornelis Goliath, 1657; Beeckman's drawing of two thermoscopes, from his *Journal*

Cover design: Coördesign, Leiden

DOI: 10.5117/9789463722537

Table of Contents

List of Illustrations

Colour illustrations

A Note on Abbreviations

AT	*Oeuvres de Descartes*, publiées par Charles Adam et Paul Tannery, 12 vols. (Paris: L. Cerf, 1897-1910). Nouvelle présentation in 13 vols., Paris: Vrin, 1974-1986
GW	Johannes Kepler, *Gesammelte Werke*, ed. Walther von Dyck and Max Caspar, 22 vols. (Munich: Beck, 1937-2017)
JIB	*Journal tenu par Isaac Beeckman de 1604 à 1634*, publié avec une introduction et des notes par C. de Waard, 4 vols. (The Hague: Martinus Nijhoff, 1939-1953)
ZAM	Zeeuws Archief Middelburg

Preface

In this volume we have brought together more than a dozen new studies about the Dutch natural philosopher Isaac Beeckman (1588-1637). Today, his important role in the initial stages of the Scientific Revolution of the seventeenth century is contested by no one, if only because of his decisive influence on the young René Descartes. Yet, the origins of Beeckman's innovative ideas about the constitution of the natural world and the mechanisms that lay behind natural phenomena deserve more historical investigation. Also, the social and cultural context in which he operated and which partly shaped his ideas and practices awaits further scrutiny. Moreover, his notebook and his particular way of philosophizing shed new light on the cultures of knowledge in the early seventeenth century, especially in the Dutch Republic. By exploring all these different issues, by extending the research into areas that were previously underexplored, and by re-thinking categories of thought that have been taken for granted for too long, we hope that this volume will contribute to a better and richer understanding of the early modern history of knowledge.

Klaas van Berkel, Albert Clement, and Arjan van Dixhoorn

1 Introduction

Klaas van Berkel, Albert Clement, and Arjan van Dixhoorn

In 1905, the discovery of the so-called *Journal* of Isaac Beeckman was a major event in the small community of historians of science in Europe.[1] The manuscript not only contained precious information about Beeckman's meeting with René Descartes in 1618 and their collaboration in deriving the law of falling bodies, but also copies of some unknown letters by Descartes to Beeckman, and an abundance of notes concerning various topics that were of interest to historians of the early modern period, such as the invention of the telescope, the principle of the conservation of movement, the refraction of light, the concept of air pressure and the corpuscular theory of matter in general. Although Beeckman had not been completely unknown before, from this point on his name became firmly entrenched in the grand narrative of what was soon to be called the Scientific Revolution of the seventeenth century. In his famous book *The Origins of Modern Science, 1300-1800* (first published in 1949), Herbert Butterfield refers to Beeckman as 'a man who stimulated others to take an interest in important problems and initiated a number of ideas', though without specifying what these ideas were. In *The Mechanization of the World Picture* (English translation 1961), E.J. Dijksterhuis devoted no less than five pages to Beeckman's work, focusing on his work, with Descartes, on the law of free-falling bodies. In the same vein, John Henry in his slim volume *The Scientific Revolution and the Origins of Modern Science* (second edition, 2002) pointed to Beeckman in the context of the mathematization of natural philosophy. Isaac Beeckman, he says, 'set an impressive example of how to use mathematics in physics'. In his more

1 Throughout this volume, we differentiate between the 'manuscript' or the 'notebook', which has as its title *Loci communes*, but certainly is not an example of that genre, and the *Journal*, that is Cornelis de Waard's edition of the manuscript: *Journal tenu par Isaac Beeckman de 1604 à 1634*, publié avec une introduction et des notes par C. de Waard, 4 vols. (The Hague: Martinus Nijhoff, 1939-1953) [henceforth *JIB*]. See the contribution by Van Berkel to this volume, 'Framing Beeckman'. The original notebook is now preserved in the Zeeuwse Bibliotheek, Middelburg, ms. nr. 6471.

Van Berkel, Klaas, Albert Clement, and Arjan van Dixhoorn (eds.), *Knowledge and Culture in the Early Dutch Republic: Isaac Beeckman in Context*. Taylor & Francis Group 2022
DOI: 10.5117/9789463722537_CH01

recent *The Invention of Science: A New History of the Scientific Revolution* (2015), David Wooton also mentions Beeckman regularly.[2]

Nonetheless, even though it would be incorrect to say that Beeckman has been neglected since his notebook was discovered more than a century ago, it is true that to this day the enormous richness of the *Journal* has not been fully exploited. Like Butterfield, Dijksterhuis only hints at the wealth of interesting topics discussed in Beeckman's notes by saying that although he did not publish his findings, Beeckman's ideas in his *Journal* are to be valued because they give the reader 'some notion of the scientific thought of a gifted man of the early seventeenth century'.[3] The way in which historians of science looked at Beeckman and his *Journal* was mostly determined by the somewhat narrow scope of the historians of science in the greater part of the twentieth century. They were mainly interested in the development of ideas that could be linked to modern science as we know it. Present-day historians of science have a much broader horizon than previous generations and take into account many more aspects of the early modern philosophers' and scholars' occupation with nature. This has resulted in the rise of the social history of science and the history of knowledge in the 1980s through 2000s. Our current understanding of Beeckman has profited greatly from this development. Reading and re-reading the *Journal* constantly offers new perspectives and brings to light new aspects of his life, his thinking and knowledge-making, as well as that of his immediate social surroundings.

Isaac Beeckman: A Brief Outline of His Life

Ever since the publication of Beeckman's *Journal*, the basic facts about his life have been well established.[4] He was born on 8 December 1588 in the city of Middelburg, capital of the province of Zeeland, one of the seven provinces

2 Herbert Butterfield, *The Origins of Modern Science, 1300-1800* (London: Bell and Sons, 1949), p. 71; E.J. Dijksterhuis, *The Mechanization of the World Picture*, trans. by C. Dikshoorn (Oxford: Oxford University Press, 1961), esp. pp. 329-333; John Henry, *The Scientific Revolution and the Origins of Modern Science* (Basingstoke: Houndsmill, 1997), p. 27; David Wootton, *The Invention of Science: A New History of the Scientific Revolution* (London: Allan Lane, 2015).

3 Dijksterhuis, *The Mechanization of the World Picture*, p. 330.

4 In the first volume of his edition of the *Journal*, published in 1939, De Waard included a 'Vie de l'auteur' that is still the basis of our knowledge of Beeckman's life. *JIB*, I, pp. i-xxiv. For a more extensive treatment of his life: Klaas van Berkel, *Isaac Beeckman on Matter and Motion: Mechanical Philosophy in the Making* (Baltimore: Johns Hopkins University Press, 2013), pp. 8-75.

that constituted the independent Dutch Republic that was just beginning to take shape after the Revolt against Habsburg rule. His father, Abraham Beeckman, was an immigrant from the city of Turnhout in Brabant, in the Southern Netherlands. As a child Abraham had fled his native city for the sake of religion (the Beeckman family had converted to Calvinism in the 1560s). First the family moved to London, but in 1585 Abraham settled in Middelburg, where he set up shop as a candle maker. In early 1588, he married Suzanna van Rhee, the daughter of another immigrant from the South. Isaac Beeckman was their first-born son. Their second child, baptized as Jacob, was born in 1590. Isaac Beeckman always remained very close to his younger brother.

Isaac went to primary school in Middelburg, but had to go to the nearby cities of Arnemuiden and Veere to get his secondary education at the Latin schools in those cities. This was due to a long-standing theological dispute of his father with local ministers in Middelburg, who controlled the local Latin Schools. In May 1607, at the age of eighteen, Isaac matriculated at Leiden University, where he studied theology and mathematics. His mentor in this respect was the well-known Ramist professor of mathematics Rudolph Snellius. During his time in Leiden, Beeckman started to record his ideas on a wide range of topics (mathematical, mechanical, natural philosophical, and in later days also medical) in a notebook that evolved into the *Journal* published by Cornelis de Waard. In August 1610 he left Leiden and returned to Middelburg, without having obtained a specific degree. This was not unusual for those who intended to become a minister in the Dutch Reformed Church, since the Church examined future ministers itself. Beeckman helped his father in his workshop and in 1611 established himself as a candle maker in the city of Zierikzee, also in the province of Zeeland. In order to add to his qualification to serve in the ministry, he went to the Huguenot academy at Saumur in 1612. A year later, he passed his exams for the Church, but then found it difficult to find a congregation, presumably because of his father's difficult relations with the Church in Middelburg. Thus Isaac settled for good as a candle maker in Zierikzee, or so it seemed.

In 1616 Beeckman again made a surprising career switch. He sold his shop, moved back to Middelburg and set out to study medicine, possibly with the help of books lent to him by a family friend, the minister, self-educated physician and astronomer Philippus Lansbergen. After two years of intensive study he travelled to the University of Caen in Normandy, where he took his doctoral degree on 6 September 1618, with a dissertation entitled *Theses de febre intermittente*. The most interesting parts of this thesis were the *corollaria* and *quodlibeta*, including theses about air pressure, vacuum, the

early twenty-first century. In the history of knowledge approach, unlike the older social history of science, well-established disciplinary boundaries are being historicized and identified as the nineteenth-century outcome of the making of the new sciences. As the outcome clearly did not exist in the early modern world, these boundaries cannot be used to generate genealogies of the sciences, which means that the histories that were written of these sciences at the time of their making and the institutionalization of their boundaries can no longer be taken for granted. This is the fundamental belief of the new history of knowledge. It calls into question our current epistemic hierarchies and opens up ever new spaces, communities, and networks of knowledge-making. The new approach disregards and even dismantles these hierarchies and urges historians to trace the trajectories of their making and subsequent interaction.[17]

At the same time, social and cultural historians and sociologists uncovered 'non-scientific' types of knowledge, sometimes in relation to the then existing epistemic hierarchies called subjugated forms of knowledge. It is commonly understood that, in early modern Europe, previously subjugated or geographically distant forms of knowledge (from Asia, Africa, the Americas) were integrated into newly configured socially powerful and productive epistemic communities which eventually gave rise to the new sciences.[18] The history of knowledge aims to study the full range of knowledges and their relationships and hierarchies that humans have produced, with modern science being one of the forms among many, and the heir of many more. Recently, it has been argued (again) that (parts of) the humanities should also be recognized as the ancestors of modern science.[19] A potential problem with that claim, however, is that it seems to ignore older claims and reintroduces modern disciplinary hierarchies into the study of their making. It might also have the effect of re-enforcing the current status of the natural sciences at the top of the epistemic hierarchy. In order to write histories of knowledge, as a rule of method, one should

17 On the social history of knowledge in general, see the introductory texts: Peter Burke, *A Social History of Knowledge: From Gutenberg to Diderot* (Cambridge: Polity Press, 2004), and his *What Is the History of Knowledge?* (Cambridge: Polity Press, 2017).

18 The definition and demarcation of the history of knowledge (and what it is not) is discussed in: Lorraine Daston, 'The History of Science and the History of Knowledge', *KNOW: A Journal on the Formation of Knowledge* 1 (2017), pp. 131-154; Johan Östling, David Larsson Heidenblad, and Anna Nilsson Hammar, eds., *Forms of Knowledge: Developing the History of Knowledge* (Lund: Nordic Academic Press, 2020); see also the contributions in the *Journal for the History of Knowledge* 1:1 (2020).

19 Rens Bod, *A New History of the Humanities: The Search for Principles and Patterns from Antiquity to the Present* (Oxford: Oxford University Press, 2013).

that constituted the independent Dutch Republic that was just beginning to take shape after the Revolt against Habsburg rule. His father, Abraham Beeckman, was an immigrant from the city of Turnhout in Brabant, in the Southern Netherlands. As a child Abraham had fled his native city for the sake of religion (the Beeckman family had converted to Calvinism in the 1560s). First the family moved to London, but in 1585 Abraham settled in Middelburg, where he set up shop as a candle maker. In early 1588, he married Suzanna van Rhee, the daughter of another immigrant from the South. Isaac Beeckman was their first-born son. Their second child, baptized as Jacob, was born in 1590. Isaac Beeckman always remained very close to his younger brother.

Isaac went to primary school in Middelburg, but had to go to the nearby cities of Arnemuiden and Veere to get his secondary education at the Latin schools in those cities. This was due to a long-standing theological dispute of his father with local ministers in Middelburg, who controlled the local Latin Schools. In May 1607, at the age of eighteen, Isaac matriculated at Leiden University, where he studied theology and mathematics. His mentor in this respect was the well-known Ramist professor of mathematics Rudolph Snellius. During his time in Leiden, Beeckman started to record his ideas on a wide range of topics (mathematical, mechanical, natural philosophical, and in later days also medical) in a notebook that evolved into the *Journal* published by Cornelis de Waard. In August 1610 he left Leiden and returned to Middelburg, without having obtained a specific degree. This was not unusual for those who intended to become a minister in the Dutch Reformed Church, since the Church examined future ministers itself. Beeckman helped his father in his workshop and in 1611 established himself as a candle maker in the city of Zierikzee, also in the province of Zeeland. In order to add to his qualification to serve in the ministry, he went to the Huguenot academy at Saumur in 1612. A year later, he passed his exams for the Church, but then found it difficult to find a congregation, presumably because of his father's difficult relations with the Church in Middelburg. Thus Isaac settled for good as a candle maker in Zierikzee, or so it seemed.

In 1616 Beeckman again made a surprising career switch. He sold his shop, moved back to Middelburg and set out to study medicine, possibly with the help of books lent to him by a family friend, the minister, self-educated physician and astronomer Philippus Lansbergen. After two years of intensive study he travelled to the University of Caen in Normandy, where he took his doctoral degree on 6 September 1618, with a dissertation entitled *Theses de febre intermittente.* The most interesting parts of this thesis were the *corollaria* and *quodlibeta*, including theses about air pressure, vacuum, the corporeal nature of light, and the principle of inertia. After this he travelled back to Zeeland, but soon moved to the city of Breda in Brabant, partly, as he said himself, to court a young lady. In Breda, by pure coincidence, he met the young René Descartes, who had gone to Breda in order to enlist himself in the army of Maurits, Prince of Orange, the military leader of the Dutch Republic. The two discussed topics in mathematics and music and became friends. Around 1 January 1619, Beeckman left Breda and found employment as a teacher at the Latin School in Utrecht. In 1620 he married Cateline de Cerf, whose family had also fled the Southern Netherlands and moved to Middelburg. Over the years she bore him seven children, most of who, however, died at a very early age.

At the end of 1620, Beeckman moved to the port city of Rotterdam, where his brother Jacob had become principal of the local Latin School and could use some help from his sibling. In 1624 Isaac's position was formalized when he became the vice-principal. In Rotterdam, Beeckman, together with some artisans and merchants, founded the Collegium Mechanicum, an informal society which discussed all sorts of technical projects but also asserted itself as an advisory body for the city government. Beeckman also got involved in a series of disputes on the attitude the Reformed Church should adopt towards the so-called Remonstrants, a liberal faction within the Reformed Church that had been thrown out in 1619. Although Beeckman was to a certain extent an orthodox believer, he belonged to those church members who favoured a more lenient approach to the Remonstrants. In 1627, however, before the dispute was finally settled, Beeckman moved to the nearby city of Dordrecht, where he became principal of the Latin School. The Collegium Mechanicum was disbanded after his departure.

In Dordrecht, Beeckman resumed contact with Descartes, who visited him in 1628 and 1629, before settling down in Amsterdam and elsewhere in the Dutch Republic. Also, the French philosophers Piere Gassendi and Marin Mersenne paid him a visit. In 1630, however, Beeckman and Descartes fell out with each other, purportedly because Descartes was informed (incorrectly) that Beeckman had claimed to be his master, but more probably because Descartes thought that Beeckman might become a rival in publishing his own account of the mechanical philosophy of which Descartes claimed to be the sole originator. Although they re-established a more or less friendly relationship after a year or two, the old friendship never returned and this quarrel cast a shadow over Beeckman's later years. In these final years, the entries in the notebook stopped, except for extensive reports of his efforts to learn the craft of lens grinding. On 19 May 1637, he died of consumption, the disease his brother Jacob had succumbed to in 1629. With her two surviving daughters, Beeckman's widow returned to Middelburg. In 1644 Beeckman's younger brother Abraham succeeded in publishing a collection of entries in his brother's notebook as the *Mathematico-physicarum meditationum, quaestionum, solutionum centuria*, but this publication hardly caused a ripple in the Dutch community of mathematicians, natural philosophers, and physicians that was by then almost hypnotized by the publications of Descartes. Only when De Waard announced his discovery of the *Journal* centuries later did it begin to dawn upon historians of science and philosophy that Beeckman had not been as insignificant as they thought he had been after all.

Historiography

As sure as we are about the outlines of Beeckman's life, historians of science and philosophy are less certain about the interpretation of his natural philosophy, its origins, and its influence. In the classical period of the history of science, attention was mainly focused on the development of the theories and methods of science in its early stages, in what came to be known as the Scientific Revolution. Eminent representatives of this approach who studied the contribution of Beeckman to the development of the new science were Cornelis de Waard, the editor of the *Journal*, E.J. Dijksterhuis, and Alexandre Koyré.[5] In later years, John Schuster, Floris Cohen, Giancarlo Nonnoi, Benedino Gemelli, Henk Kubbinga, and Richard Arthur continued to explore Beeckman's contribution to the rise of modern science.[6] These scholars were surely sensitive to influences traditionally seen as external to science on the development of Beeckman's ideas, but they nevertheless stressed his contribution to the theory of matter, the science of mechanics,

5 Cornelis de Waard, *L'Expérience barométrique, ses antécédants et ses explications. Étude historique* (Thouars: Imprimerie nouvelle, 1936); Cornelis de Waard, 'Sur les règles du choc des corps d'apres Beeckman', in: *Correspondance du P. Marin Mersenne, religieux minime*, publiée par Mme Paul Tannery, editée et annotée par Cornelis de Waard et al., 17 vols. (Paris: Beauchesne, 1932-1988), II (1936), pp. 632-644; Dijksterhuis, *The Mechanization of the World Picture*; Alexandre Koyré, *Études galiléennes* (second ed., Paris: Hermann, 1966).

6 John A. Schuster, *Descartes and the Scientific Revolution, 1618-1634: An Interpretation* (PhD diss., Princeton University, 1977); H. Floris Cohen, *Quantifying Music: The Science of Music at the First Stage of the Scientific Revolution, 1580-1650* (Dordrecht: Reidel, 1984); Giancarlo Nonnoi, *Il pelago d'aria. Galileo, Baliani, Beeckman* (Rome: Bulzoni Editore, 1988); Benedino Gemelli, *Isaac Beeckman. Atomista e lettore critico di Lucrezio* (Florence: Leo S. Olschki, 2002); H.H. Kubbinga, *L'Histoire du concept de 'molecule'*, 3 vols. (Paris: Springer, 2002), I, pp. 203-237; Richard Arthur, 'Beeckman, Descartes and the Force of Motion', *Journal of the History of Philosophy* 45:1 (2007), pp. 1-28.

the theory of music and natural philosophy in general. Similarly, Reyer Hooykaas studied the impact of Beeckman's religious beliefs on his natural philosophy.[7]

In the 1980s, however, the social history of science emerged as a serious new approach to the history of science of the early modern period. It resulted from a process, going back at least to the 1940s, in which an increasing number of social groups and their practices were identified as co-constitutive in the making of the new sciences of the early modern period. The introduction of hands-on or practical knowledge of matter (living and dead, natural and artificial) into the natural sciences has been attributed to 'superior artisans' (Edgar Zilsel and Paolo Rossi),[8] visual artists (Erwin Panofsky),[9] printers (Elizabeth Eisenstein),[10] merchants and explorers (Harold Cook)[11]; other groups that have been identified are medical practitioners and aristocrats or *virtuosi*.[12] Not only new groups of people, but certain previously ignored sites and practices also came under investigation as places for the making of knowledge about nature and the world, such as the cabinets of curiosities

7 R. Hooykaas, 'Science and Religion in the Seventeenth Century: Isaac Beeckman, 1588-1637', *Free University Quarterly* 1 (1951), pp. 169-183.

8 See the collection of Zilsel's essays in: E. Zilsel, *The Social Origins of Modern Science* (Dordrecht: Kluwer Academic Publishers, 2003); see also: Philip P. Wiener and Aaron Noland, eds., *Roots of Scientific Thought: A Cultural Perspective* (New York: Basic Books, 1957). For Rossi, see: Paolo Rossi, *Philosophy, Technology, and the Arts in the Early Modern Era*, trans. by Salvatore Attanasio, ed. by Benjamin Nelson (New York: Harper & Row, 1970); more recently: Pamela O. Long, *Openness, Secrecy, Authorship: Technical Arts and the Culture of Knowledge from Antiquity to the Renaissance* (Baltimore: Johns Hopkins University Press, 2001); Pamela H. Smith, *The Body of the Artisan: Art and Experience in the Scientific Revolution* (Chicago: University of Chicago Press, 2004).

9 See: Erwin Panofsky, 'Artist, Scientist, Genius: Notes on the "Renaissance-Dämmerung"', in: Wallace K. Fergusan et al., *The Renaissance* (New York: Harper Torchbooks, 1962), pp. 123-182; more recently Svetlana Alpers, *The Art of Describing: Dutch Art in the Seventeenth Century* (Chicago: University of Chicago Press, 1984); Brian W. Ogilvie, *The Science of Describing: Natural History in Renaissance Europe* (Chicago: University of Chicago Press, 2006); Tine L. Meganck, *Pieter Brueghel the Elder, Fall of the Rebel Angels: Art, Knowledge and Politics on the Eve of the Dutch Revolt* (Milan: Silvana Editoriale, 2014); Marisa A. Bass, *Insect Artifice: Nature and Art in the Dutch Revolt* (Princeton: Princeton University Press, 2019).

10 E. Eisenstein, *The Printing Press as an Agent of Change: Communication and Cultural Transformations in Early Modern Europe* (Cambridge: Cambridge University Press, 1980).

11 Harold J. Cook, *Matters of Exchange: Commerce, Medicine, and Science in the Dutch Golden Age* (New Haven: Yale University Press, 2007).

12 See, for example: E. Leong and A. Rankin, eds., *Secrets and Knowledge in Medicine and Science, 1500-1800* (Farnham/Burlington: Ashgate, 2011); Steven Shapin and Simon Schaffer, *Leviathan and the Air-Pump: Hobbes, Boyle, and the Experimental Life* (Princeton: Princeton University Press, 1985).

that became popular among princes and rich city dwellers in the sixteenth and seventeenth centuries.[13]

In general, these groups are associated with the making of a knowledge culture out of which the natural sciences developed, grounded in collecting, observation, and experiment. They contributed 'something' from their world of practice to the culture of 'experience' as it developed and gained status over the course of the seventeenth century. The argument is that in the new science the knowledge of 'things', *naturalia* and/or *artificialia* (how-to knowledge), met with theoretical philosophizing (knowledge of causes).[14] Put differently, practical manipulations of matter which had been the 'impure' (that is, irrational) domain of the mechanical arts were integrated with the theorizing on rules and causes, which had been the 'pure' (that is, rational) domain of the liberal arts and sciences. In particular, the tradition of 'books of secrets' and engineering expertise have been identified as important spheres of practical knowledge that contributed to this 'revolution' of both the practice of science and the practice of the arts.[15] Thus, hands-on dealing with nature, rather than just philosophical speculation, moved to the centre of attention in the history of science. Klaas van Berkel's 1983 dissertation *Isaac Beeckman (1588-1637) en de mechanisering van het wereldbeeld* (revised and translated as *Isaac Beeckman on Matter and Motion: Mechanical Philosophy in the Making*, 2013) marks the transition from the classical history of science to the social history of science. Commenting on the criteria Beeckman put forward for any decent explanation in natural philosophy and physics, Van Berkel claimed: 'He views the world like a craftsman inspecting a machine he is about to repair.'[16]

The success of the social history of science has been overwhelming, and in turn has evolved into the interdisciplinary history of knowledge of the

13 See especially: Oliver Impey and Arthur MacGregor, eds., *The Origins of Museums: The Cabinet of Curiosities in Sixteenth- and Seventeenth-Century Europe* (Oxford: Clarendon Press, 1985). For cabinets of curiosities in the Dutch Republic, see: Ellinoor Bergvelt et al., *De wereld binnen handbereik. Nederlandse kunst- en rariteitenverzamelingen, 1585-1735* (Zwolle: Waanders, 1992).

14 See: Anthony Grafton and Nancy Siraisi, eds., *Natural Particulars: Nature and the Disciplines in Renaissance Europe* (Cambridge, Mass.: MIT Press, 1999); Pamela H. Smith, Amy R. Meyers, and Harold J. Cook, eds., *Ways of Making and Knowing: The Material Culture of Empirical Knowledge* (Ann Arbor: University of Michigan Press, 2014); Lissa Roberts, Simon Schaffer, and Peter Dear, eds., *The Mindful Hand: Inquiry and Invention from the Late Renaissance to Early Industrialisation* (Amsterdam: KNAW, 2007).

15 See: William Eamon, *Science and the Secrets of Nature: Books of Secrets in Medieval and Early Modern Culture* (Princeton: Princeton University Press, 1994); Alison Kavey, *Books of Secrets: Natural Philosophy in England, 1550-1600* (Urbana: University of Illinois Press, 2007).

16 Van Berkel, *Isaac Beeckman on Matter and Motion*, p. 137.

early twenty-first century. In the history of knowledge approach, unlike the older social history of science, well-established disciplinary boundaries are being historicized and identified as the nineteenth-century outcome of the making of the new sciences. As the outcome clearly did not exist in the early modern world, these boundaries cannot be used to generate genealogies of the sciences, which means that the histories that were written of these sciences at the time of their making and the institutionalization of their boundaries can no longer be taken for granted. This is the fundamental belief of the new history of knowledge. It calls into question our current epistemic hierarchies and opens up ever new spaces, communities, and networks of knowledge-making. The new approach disregards and even dismantles these hierarchies and urges historians to trace the trajectories of their making and subsequent interaction.[17]

At the same time, social and cultural historians and sociologists uncovered 'non-scientific' types of knowledge, sometimes in relation to the then existing epistemic hierarchies called subjugated forms of knowledge. It is commonly understood that, in early modern Europe, previously subjugated or geographically distant forms of knowledge (from Asia, Africa, the Americas) were integrated into newly configured socially powerful and productive epistemic communities which eventually gave rise to the new sciences.[18] The history of knowledge aims to study the full range of knowledges and their relationships and hierarchies that humans have produced, with modern science being one of the forms among many, and the heir of many more. Recently, it has been argued (again) that (parts of) the humanities should also be recognized as the ancestors of modern science.[19] A potential problem with that claim, however, is that it seems to ignore older claims and reintroduces modern disciplinary hierarchies into the study of their making. It might also have the effect of re-enforcing the current status of the natural sciences at the top of the epistemic hierarchy. In order to write histories of knowledge, as a rule of method, one should

17 On the social history of knowledge in general, see the introductory texts: Peter Burke, *A Social History of Knowledge: From Gutenberg to Diderot* (Cambridge: Polity Press, 2004), and his *What Is the History of Knowledge?* (Cambridge: Polity Press, 2017).

18 The definition and demarcation of the history of knowledge (and what it is not) is discussed in: Lorraine Daston, 'The History of Science and the History of Knowledge', *KNOW: A Journal on the Formation of Knowledge* 1 (2017), pp. 131-154; Johan Östling, David Larsson Heidenblad, and Anna Nilsson Hammar, eds., *Forms of Knowledge: Developing the History of Knowledge* (Lund: Nordic Academic Press, 2020); see also the contributions in the *Journal for the History of Knowledge* 1:1 (2020).

19 Rens Bod, *A New History of the Humanities: The Search for Principles and Patterns from Antiquity to the Present* (Oxford: Oxford University Press, 2013).

maybe ignore or temporarily forget reigning disciplinary divisions and hierarchies.[20]

The shift from history of science to social history of science and then to history of knowledge has far-reaching (and sometimes controversial) consequences for the way in which historians now describe and analyse the rise of the mechanical philosophy in the seventeenth century, with which the name of Isaac Beeckman is inextricably linked. In the classical approach, the causes of the making of this philosophy of nature are sought in the world of theory, as a result of debates within an established field of natural philosophy. Since the history of knowledge questions this assumption and operates from the foundational belief that the modern sciences cannot simply be understood as the heirs of a pre-existing 'institution' of natural philosophy, the main concern is to establish exactly what types of knowing and what related practices made it into Beeckman's new mechanical philosophy of nature, and how, and what disciplinary boundaries and new epistemic hierarchies emerged from the merger of these previously disconnected worlds. This natural philosophy was an entirely new project, which certainly had not found any clear boundaries or any new epistemic hierarchy by the time of its constitution around 1600. One might even say that around 1600 it was less clear than ever before what natural philosophy was.

Bringing the study of Isaac Beeckman up to date with the current move toward the history of knowledge therefore requires a more broadly conceived contextual approach to his new ways of dealing with nature. Not primarily to explain the choices he made, but to understand what he was doing or what he thought he was doing. Beeckman was born in a part of Europe that underwent tremendous changes in its social, political, economic, religious, intellectual structure and orientation towards oceanic navigation, trade and warfare.[21]

20 As quoted in: Rens Bod, 'How to Open Pandora's Box: A Tractable Notion of the History of Knowledge', *Journal for the History of Knowledge* 1:1 (2020), art. 5, pp. 1-7, esp. p. 1, n. 3.

21 The literature on the Dutch Republic is enormous. Maarten Prak, *The Dutch Republic in the Seventeenth Century: The Golden Age* (Cambridge: Cambridge University Press, 2005), is an excellent introduction, with a strong focus on social and economic structures. More focused on cultural developments is: Willem Frijhoff et al., *Dutch Culture in a European Perspective, Vol. 1: 1650: Hard-Won Unity* (Assen: Van Gorcum/Basingstoke: Palgrave Macmillan, 2004). Slightly older, but no less useful is: Jonathan Israel, *The Dutch Republic: Its Rise, Greatness, and Fall, 1477-1806* (Oxford: Clarendon Press, 1998). On the immigrants from south to north, see: Oscar Gelderblom, *Zuid-Nederlandse kooplieden en de opkomst van de Amsterdamse stapelmarkt (1578-1630)* (Hilversum: Verloren, 2000). The role of immigrants in Zeeland, who dominated the province demographically and supported its rise to global power, has not (yet) been studied. The reverse movement of loyal Catholics leaving the Dutch Republic has been (largely) ignored, but see recently: Geert H. Janssen, *The Dutch Revolt and Catholic Exile in Reformation Europe* (Cambridge: Cambridge University Press, 2014).

Middelburg, the city in which Beeckman was born in 1588, was the capital of the province of Zeeland, one of the provinces that had managed to remove Philip II as their sovereign and to form a confederation of free provinces, also known as the Dutch Republic. During Beeckman's lifetime, the new Republic was engaged in a wide-ranging war against Philip II and his Habsburg successors both in the Low Countries and in the global East and West. Having become a centre for the global operations of the Dutch seaborne empire (through its chambers of the Dutch East India and West India Companies), Middelburg was a burgeoning and lively city with old contacts with Flanders, Brabant, Holland, England, France, and the Iberian peninsula. By the time of Beeckman's death, however, due to its role as a political and economic centre in the Dutch global trade and warfare, the city had also developed more extensive connections to trading posts and colonial settlements in Africa, Asia, and the Americas as well. This global centre, like the rest of the Republic, also absorbed an influx of thousands of immigrants from the southern provinces of the Netherlands after 1585. Immigrants poured in from the cities that were forced to surrender to the multinational 'Spanish' Army of Flanders (such as Brussels, Ghent, Bruges and Antwerp in the years 1584-1585).

These immigrants brought with them capital, knowledge and connections that further energized an already expanding Dutch economy. Since most of the newcomers adhered to the Reformed creed, they further strengthened the Protestant nature of the new state. From the beginning, the University of Leiden, founded in 1575 as a reward for the hardships endured during the siege of the city by the Spanish troops in the previous year, prided itself as a bulwark of freedom (*praesidium libertatis*). Because the nobility, with the exception of the most eminent noble family, the Oranges, to some extent lost power and the Catholic Church had gone underground, the young Dutch Republic was mainly a commonwealth of burghers, and especially in the early decades of the seventeenth century, social upward mobility was a real possibility for many of them; yet at the same time, as social and economic historians have pointed out, life was expensive and many inhabitants lived in dire poverty. The young state was also burdened with debt.

This globally connected part of Europe, the now transnational Scheldt region (divided in a Catholic Habsburg and a Protestant part) and the province of Holland, was the world in which Isaac Beeckman largely moved; first-born son of immigrants from the southern provinces who, after a stay as refugees in England, prospered in Middelburg. His father a staunchly Reformed artisan and practitioner with good connections to family and friends in England, a strong personality intent on giving his sons a good education and his daughters a good match. After the death of Isaac Beeckman the family, including Beeckman's daughter Catelyntje, quickly moved up in the ranks of Middelburg's elites. The Dutch Republic had taken over the leading role of the bustling cities of the southern Low Countries in the area. Many of the coastal cities in the Dutch Republic in particular now resembled sixteenth-century Antwerp, having become a multilingual country full of opportunities, open to the world, eager to incorporate new ideas, experimenting with new ways of life, and forced to be tolerant towards dissenters and heretics from elsewhere because of the minority status and strict membership policies of its public Reformed Church; yet rather intolerant towards Catholics and radically minded Protestants. From the perspective of the history of science and the study of nature in general the Dutch Republic has been compared to a laboratory.[22] Its western provinces, oriented towards the seas and the oceans, certainly can be compared with Deborah Harkness's London of the sixteenth century, a city she characterized as a 'proto-Baconian' sphere.

This volume brings together more classical studies of early modern scientific theory with contextual studies and exercises in the history of knowledge related to Beeckman. The chapters have been ordered in three major categories. The first part of the volume, entitled 'Assessing Beeckman', contains chapters that evaluate in general terms the place of Isaac Beeckman in the seventeenth-century world of nature study. John Schuster aims to establish what sort of philosophy Beeckman was doing, what prompted him to accept a mechanical and corpuscular philosophy of nature and what his exact place in the Scientific Revolution of the seventeenth century was. Responding to a suggestion by Schuster, Floris Cohen imagines how Beeckman would have looked back – had he lived until the 1660s – on his own philosophical career and especially his troubled relationship with Descartes. Finally, Klaas van Berkel deconstructs how Cornelis de Waard edited Beeckman's manuscript. He unravels the way in which De Waard framed Beeckman and thereby makes space for new ways of interpreting the philosopher.

The second part of the volume is devoted to chapters that analyse Beeckman's contribution to specific scientific disciplines or fields of interest. Tiemen Cocquyt discusses Beeckman's initial understanding of the telescope, which he characterizes as being technological, notwithstanding his acquaintance with Johannes Kepler's optics. Édouard Mehl details the

22 Klaas van Berkel, 'The Dutch Republic: Laboratory of the Scientific Revolution', *Bijdragen en Mededelingen betreffende de Geschiedenis der Nederlanden/The Low Countries Historical Review* 125 (2010), pp. 81-105.

reactions of Beeckman and his circle to Kepler's optics and astronomy, with consequences also for the interpretation of Descartes' philosophy. With Elisabeth Moreau we turn to medicine, a field also studied extensively by Beeckman. She shows how Beeckman's physiology is a clever combination of modern atomism with Galenic medicine. Dániel Moerman then discusses how Beeckman, who was trained as a medical doctor but never settled as such, treated himself and close friends in cases of illness and death. Samuel Le Gendre discusses the role of mechanics and more specifically the principle of the conservation of motion or inertia. The author makes the case that Descartes may very well have come across this basic principle of mechanics before he met Beeckman, thereby inviting us to add much more precision to the study of what exactly Descartes learned from Beeckman during their famous exchange in November-December 1618. In the last chapter of this section Fabrizio Baldassarri highlights some little known entries by Beeckman that show how he used a mechanical interpretation of the way plants grow and react to external circumstances.

The third part of the volume contains chapters that survey or discuss the intellectual, cultural, social and linguistic context in which Beeckman lived. Huib Zuidervaart delves into the networks of knowledge that existed in Middelburg during Beeckman's youth, including his family network. The appendix on the houses where Beeckman grew up also reveals some interesting things about the favourable financial and material conditions of his youth. Music has always been one of the recurring themes in the *Journal* and therefore Albert Clement provides an overview of the rich musical life in Middelburg before and during Beeckman's lifetime. Arjan van Dixhoorn discusses another aspect of cultural life in Middelburg and elsewhere in the Low Countries: the culture of the rhetoricians (*rethoryckers* in early modern Dutch, *rederijkers* in modern Dutch). He claims that Beeckman's special way of philosophizing is heavily indebted to the *consten*-culture (a vernacular culture of the arts and science) of which the rhetoricians claimed to be the core. In Fokko Jan Dijksterhuis's contribution the focus shifts from Middelburg to Rotterdam, where Beeckman lived from 1620 to 1627. In this booming port city, Beeckman devoted much time and energy to the discussion and manipulation of atmospheric machines, especially those inspired by the controversial inventor and projector Cornelis Drebbel. Dijksterhuis characterizes Beeckman's way of dealing with nature as 'thinking with machines'. Vera Keller then discusses Drebbel's often misunderstood habit of communicating his ideas and findings 'only to good friends and philosophers', which offers a clue to the problem many historians of science have raised in discussing Beeckman: why did he publish so little of his ideas and why did he even acquire a reputation of being 'incommunicative'? Communication is also central to the contribution by Semra Meray, who methodically studies the use of Stevinian terms and expressions in Beeckman's *Journal* (which was very limited) and Beeckman's switching between Latin and Dutch. The final contribution to this part is by Klaas van Berkel, who, in this second contribution to this collection, analyses Beeckman's use of illustrations in his *Journal* and concludes that in Beeckman's pictorial way of reasoning images sometimes acquired an argumentative force of their own.

In the concluding remarks two of the editors identify some directions for further research, conscious as they are that the richness of Beeckman's *Journal* has by no means been fully exhausted. Although the authors of this volume have contributed to a more precise, less anachronistic understanding of Beeckman and his unique position in the worlds of knowledge of the seventeenth century, there is so much more that remains to be explored. We hope that this volume invites the authors and others to do just that and join the quest.

Part I

Assessing Beeckman

2 Isaac Beeckman in the Context of the Scientific Revolution

John A. Schuster

Abstract
This chapter locates Beeckman in the most important, 'critical', phase of the Scientific Revolution, taken as a long process of several overlapping stages. His *de novo* invention of corpuscular-mechanical natural philosophy, a seminal event in that phase, offers a test case for analysing the contextual causes of this breakthrough. Beeckman's significance for the later stages of the process resided primarily in the little noticed Beeckmanian conceptual genes at the heart of Descartes' mechanism – in his vortex celestial mechanics, the 'engine room' of his system and key to his radical Copernican realism. Finally, to illustrate that the experimentally oriented corpuscular-mechanism of indirect Beeckmanian origin was central to the next phase of the Scientific Revolution, a counterfactual scenario is offered concerning the work of a 'Beeckman' still alive in the 1660s.

Keywords: Isaac Beeckman, Scientific Revolution, corpuscular-mechanical philosophy, René Descartes, contextual explanation

1 Riding Orders: 'What Was Beeckman Doing – in the Context of the Scientific Revolution?'

I have been concerned about Isaac Beeckman as a figure in the Scientific Revolution for just short of 50 years, having begun to read his Latin writings on mechanics in 1971, under the guidance of the late Michael S. Mahoney in the Princeton History of Science Program.[1] Mahoney and I

1 Michael S. Mahoney (1938-2008), a prolific scholar in the history of science, mathematics and technology, is also famed in early modern circles for his widely distributed typescript translations

Van Berkel, Klaas, Albert Clement, and Arjan van Dixhoorn (eds.), *Knowledge and Culture in the Early Dutch Republic: Isaac Beeckman in Context*. Taylor & Francis Group 2022
DOI: 10.5117/9789463722537_CH02

started terming the early mechanists such as Beeckman, Descartes, Hobbes and Gassendi, 'corpuscular-mechanists'. Stimulated by another faculty member, Theodore K. Rabb, we began to think of the generation in which corpuscular-mechanical natural philosophy was invented as *the* critical stage in a multiphase process of the Scientific Revolution.[2] Mahoney pointed out to me that in 1618 Beeckman had recalled the young Descartes to study, including the study of corpuscular-mechanical natural philosophy. Later, in my doctoral dissertation, the Beeckman/Descartes relationship covered 80 pages out of about 750.[3]

However, I am not a fully-fledged Beeckman scholar. Believing in the importance of Beeckman in the Scientific Revolution and having a particular interest in his relations with Descartes, I have had a watching brief on Beeckman scholarship. When the organizers of the Middelburg Beeckman conference invited me to deliver a plenary lecture on the topic that is now the title of this chapter, I requested more guidance about my assignment. Klaas van Berkel asked me: 'What do you think Beeckman was doing?' My answer in the plenary talk and this chapter is that Beeckman was practising natural philosophy, in a novel register. He was symptomatic of a crucial moment in the process of the Scientific Revolution and seminal in the weave of its later phases. I shall argue that he was a natural philosophical visionary: he knew what was then needed in natural philosophy, and later events bore him out. My argument will necessarily be painted in quick, broad strokes. It is suggestive and experimental, even to the point of ending with an exercise in counterfactual history. My tone, derived from the plenary talk, is declarative and necessarily terse. References to a number of prior works of mine may offer some guidance as to what resides behind the relatively few words that I can deploy here.

2 Sizing up the Scientific Revolution: Phases and Stages – Natural Philosophizing as the Central Field/Institution

Periodization is the indispensable armature of historical inquiry, but, only half the story. Historical understanding also requires conceptualization of

of sources in seventeenth-century mechanics, starting with excerpts from Beeckman's *Journal*, Descartes' *Principia philosophiae* and the mechanics treatises of Huygens.

2 Theodore K. Rabb, *The Struggle for Stability in Early Modern Europe* (New York: Oxford University Press, 1975).

3 John A. Schuster, *Descartes and the Scientific Revolution, 1618-1634: An Interpretation* (PhD diss., Princeton University, 1977), pp. 51-111, 565-579, 590-593.

the entities and structures in play. Only this allows serious narration *and* explanation to be slung across a periodization armature. My key, but not only category is 'natural philosophy'. By natural philosophy I mean not simply the new mechanical philosophy, nor the still institutionally entrenched neo-Scholastic Aristotelianism. Rather, I mean the wider institution and disciplinary field of natural philosophy in all its variety and contention – a field that included these and other variants, a field that had both a social structure (in flux in Beeckman's time) and certain rules of engagement (also changing).[4] When one 'natural philosophized' one was making claims regarding one or more of the dimensions of natural philosophical discourse: the nature of matter, the cosmological structuring of that matter, the principles of causation and the methodology for acquiring or justifying such natural knowledge.

There was a Europe-wide culture of natural philosophizing, because of a High Medieval development of world historical import – the establishment of a system of universities all teaching variants of a Christianized Aristotelian corpus in natural philosophy. For educated men, neo-Scholastic Aristotelianism provided the template for natural philosophizing. This template applied to all jostling species of the genus: to Scholastic Aristotelianism and to all its natural philosophical challengers. Additionally, Scholastic Aristotelianism entrained an entire geography of knowledge: the way in which other disciplines were conceived, and related to each other, and to natural philosophizing. Some disciplines were considered superior to natural philosophy (such as theology); others cognate with it (such as mathematics); or subordinate to it. One may think of the subordinate disciplines as an *entourage* of more narrow traditions of science-like practice. These included sciences inherited from classical antiquity: the subordinate mixed mathematical sciences, such as geometrical astronomy, geometrical optics, mechanics, statics, and music theory, as well as the bio-medical domains, such as anatomy, medical theorizing and physiology. From the early seventeenth century, these entourage members changed, were contested and related to natural philosophizing in new, non-Aristotelian ways, and some new ones were

4 Material in this and the following paragraph is treated more extensively in: John Schuster, *Descartes-Agonistes: Physico-mathematics, Method & Corpuscular-Mechanism 1618-33* (Dordrecht: Springer, 2013), Chapter 2; John Schuster, 'What Was the Relation of Baroque Culture to the Trajectory of Early Modern Natural Philosophy?', in: O. Gal and R. Chen-Morris, eds., *Science in the Age of Baroque* (Dordrecht: Springer, 2013), pp. 16-28; John Schuster, 'L'Aristotelismo e le sue Alternative', in: Daniel Garber, ed., *La Rivoluzione Scientifica* (Rome: Instituto della Enciclopedia Italiana, 2002), pp. 337-357; John Schuster, 'The Scientific Revolution', in: G. Cantor et al., eds., *The Companion to the History of Modern Science* (London: Croom Helm, 1990), pp. 217-242.

created. Finally, some non-Aristotelian natural philosophers embraced the values of the practical arts and tried to import them into the field.

Now a periodization of the Scientific Revolution. I have discussed this, modifying it over the years, in several works.[5]

I The Scientific Renaissance (1500-1590). Not only in the field of natural philosophy, but across the set of entourage sciences, and mathematics as well, we have recovery, translation, editing and publication of a wide range of classical sources – just as happened earlier in the Renaissance in the fields of classical languages, history and literature. These developments mark the first stage and essential precondition for the further process of the Scientific Revolution. These activities took place amid pedagogical and philosophical assaults on Scholastic natural philosophy and a revival of types of Platonism, which helped revalue mathematics as the key to knowledge. The ideal of knowledge as contemplative was challenged in terms of practice and utility. Much of this occurred beyond the universities, in princely courts, print houses and workshops of master artisans. Nevertheless 'orthodox' neo-Scholastic Aristotelianism, now in Catholic and Protestant variants, remained dominant. Nothing in this stage surpassed the scientific achievements of antiquity or the High Middle Ages, and even this high tide might have receded without the Scientific Revolution having further developed.

II The Critical Period (or phase of the civil war in natural philosophizing) (1590-1650). This second period of the Scientific Revolution saw a conjuncture unique in history, whether in classical antiquity, medieval Islam or Renaissance Europe. On the one hand Kepler, Galileo, and Descartes led an accelerated conceptual transformation in the subordinate entourage sciences – optics, mechanics and astronomy as well as the cognate, mathematics. Harvey did the same in the bio-medical domain. On the other hand, unprecedented challenges arose to Aristotelianism. There was heightened competition amongst types of natural philosophy (some tied to utopian programmes of religious and social reform). Many of these were broadly Platonic in nature. Just at that moment and responding to this situation, the corpuscular-mechanical natural philosophy was constructed.[6] Calls

5 Schuster, 'Scientific Revolution'; Schuster, *Descartes-Agonistes*, pp. 77-88; Schuster, 'Baroque Culture', pp. 19-21; and Schuster, 'L'Aristotelismo', pp. 337-357.

6 My characterization of the Critical Period predates, but nicely comports with, Floris Cohen's insights about the early seventeenth century in European natural philosophy and related sciences.

for the re-evaluation of practical knowledge and the domination of nature now sounded more urgently. Figures like Francis Bacon and Descartes assimilated them to high cultural natural philosophical programmes. The keynote was more heightened contestation than ever had existed inside the cosy if quibbling world of Scholastic Aristotelianism.

III The 'CMF' or 'Consensus, Muting and Fragmentation' period (1650-1720). This concluding phase is characterized by four main tendencies. First, early in this period the cultural dominance of Aristotelianism collapsed (although it continued supreme in most universities for another generation). Second, a consensus formed around corpuscular-mechanical natural philosophy with a strong emphasis on experimental evidence, which in turn was shaped by Baconian rhetoric of method, progress and utility. Third, public disputation about natural philosophy was muted by this consensus. The new experimental philosophy was held in non-dogmatic, non-systematic forms. Inside the new scientific institutions debates about natural philosophies and about specific theories were similarly muted, but not abolished as some erroneous historiographies hold. Finally, the entire field of natural philosophizing became more autonomous of other disciplines like theology, and other branches of philosophy, whilst it also began a long process of fragmentation into a number of more narrow domains, which start to look like sciences in our modern sense.

3 Beeckman: Symptomatic and Seminal in His Stage of the Scientific Revolution

Isaac Beeckman inhabited the Critical Period of the Scientific Revolution. He constructed a corpuscular-mechanical natural philosophy.[7] It

In Cohen's model, a series of early-seventeenth-century transformations (like the dramatic developments early in my 'Critical Phase') in the work of Kepler, Galileo, Descartes, and Bacon took European nature knowledge beyond anything ever achieved elsewhere (but still requiring several more phases of development before the late-seventeenth-century emergence of modern science). Floris Cohen, 'The Onset of the Scientific Revolution: Three Near-Simultaneous Transformations', in: Peter Anstey and John Schuster, eds., *The Science of Nature in the Seventeenth Century: Changing Patterns of Early Modern Natural Philosophy* (Dordrecht: Springer, 2005), pp. 1-33.

7 In focusing upon Beeckman as a natural philosopher, I am in no way denying that he was also many other things, for example, medical doctor, practical artisan, pedagogue and practical mathematician. As I have pointed out in many previous works, all of these domains need to be brought into relation to the dynamics and evolution of the field of natural philosophy in the period. For the case of Descartes' natural philosophizing and mixed and practical mathematics,

was not simply ancient atomism revived. Rather, it was a qualitatively new species of philosophy of nature. With the possible exception of Thomas Harriot, Beeckman was first on the scene with a corpuscular-mechanism. Being first makes Beeckman a prime object of historical study. Corpuscular-mechanism came to define much of the terrain of the succeeding CMF period. Hence Beeckman's work on corpuscular-mechanism is symptomatic of his stage and seminal within it. His *Journal* is a unique source of insight into the values and aspirations behind this initiative. By 1620 the visionary Beeckman had an unsystematized vision of the mechanical philosophy of nature that equates to what won out after 1650.

Beeckman's corpuscular-mechanism was certainly atomistic in matter theory, but that was not especially radical. Qualitative atomisms were then starting to spread. Matter theory was not the explosive core of corpuscular-mechanism. That resided elsewhere, in what I term the 'causal register' of the natural philosophy, where it was expected that a theory of mechanics would run the atomic show. Promoting any type of mechanics, a mixed mathematical science, into the nucleus of a natural philosophy was an epochal gambit. Beeckman's disciple, Descartes, embraced exactly this. Shortly after Descartes became acquainted with Beeckman's breakthrough, in November 1618, and worked with him on falling bodies and hydrostatics, Descartes twice mentioned to Beeckman that he was working on a 'mechanics', which obviously was meant to work the particles.[8] They differed on details. Beeckman's causal mechanics was premised on a dynamical interpretation of the theory of the simple machines.[9] Descartes would look for what Stephen Gaukroger and I have called a 'punctiform dynamics', its principles later extracted from what he thought he had found out about the behaviour of light, as an instantaneously transmitted mechanical impulse.[10]

I have done this in: John Schuster, 'Consuming and Appropriating Practical Mathematics and the Mixed Mathematical Fields, or Being "Influenced" by Them: The Case of the Young Descartes', in: Lesley Cormack, Stephen Walton, and John Schuster, eds., *Mathematical Practitioners and the Transformation of Natural Knowledge in Early Modern Europe* (Dordrecht: Springer, 2017), pp. 37-65.

8 Schuster, *Descartes-Agonistes*, pp. 124, 190-209.

9 *Journal tenu par Isaac Beeckman de 1604 à* 1634, publié avec une introduction et des notes par C. de Waard, 4 vols. (The Hague: Martinus Nijhoff, 1939-1953) [henceforth *JIB*], I, pp. 265-266; III, pp. 133-134; Schuster, *Descartes-Agonistes*, pp. 108-112.

10 Stephen Gaukroger and John Schuster, 'The Hydrostatic Paradox and the Origins of Cartesian Dynamics', *Studies in the History and Philosophy of Science* 33:3 (2002), pp. 535-572, esp. p. 569; Schuster, *Descartes-Agonistes*, pp. 11, 369-370, 409; John Schuster, 'Physico-mathematics and the Search for Causes in Descartes' Optics – 1619-37', *Synthèse* 185 (2012), pp. 467-499.

Elsewhere I have argued that putting a mixed mathematical science like mechanics inside a natural philosophy actually meant that the mixed science was 'physicalized', not that the natural philosophy was mathematicized. Beeckman and Descartes – the latter especially – called this kind of move 'physico-mathematics'.[11] Two examples: (1) in 1618 Descartes tried to show Beeckman how to reduce Simon Stevin's hydrostatic paradox to the mechanics-controlled motions of the particles making up the fluid; (2) in 1626/1627 Descartes read his discovery diagram of the law of refraction as a picture of the underlying dynamics of light corpuscles.[12] Beeckman did not exactly follow suit. He had a broader programme called *philosophia mathematico-physica*, defined by Klaas van Berkel as '[serving] to explain natural phenomena by integrating natural bodies' simple geometrical properties with their mechanical properties (primarily movement and inertia)'.[13] Beeckman could sympathize with Descartes' radical physico-mathematics, but it was not necessarily his preferred mode of radical natural philosophical innovation. Nevertheless, Beeckman produced two strong physico-mathematical initiatives: he brought a mixed mathematical science, mechanics, into the core of his natural philosophy, and as we shall see in Section 5, he later attempted to articulate a realist Copernican doctrine of celestial causation to the natural philosophy.

The Critical Stage of the Scientific Revolution was marked by a great deal of rule breaking and rule bending in the now rampant conflicts of the field of natural philosophizing. Many natural philosophically literate men understood that the narrow limits of contestation within neo-Scholastic Aristotelianism had been shattered. Many rules of the neo-Scholastic Aristotelian hegemony were explicitly or implicitly challenged, and Beeckman participated in the fun. For example, the hegemonic vision of natural philosophy said: *if you pursue the mixed mathematical sciences, do not do this with the aim of grafting anything from them into natural philosophical discourse of matter and cause.* Beeckman broke this. The hegemonic version said: *do not bring in inappropriate goals and values, particularly those of the crafts and craftsmen.* But, Beeckman's corpuscular-mechanism spoke for the value of practice, utility and active domination of nature. The hegemonic version said: *do not deny Aristotelian cosmology or bring in Copernicanism as anything other than a calculating device.* Beeckman was a realist Copernican, as we shall see.

11 On physico-mathematics, see: Schuster, *Descartes-Agonistes*, pp. 56-59, 108-128, 167-209.

12 Schuster, *Descartes-Agonistes*, pp. 190-193.

13 Klaas van Berkel, *Isaac Beeckman on Matter and Motion: Mechanical Philosophy in the Making* (Baltimore: Johns Hopkins University Press, 2013), p. 79.

4 Explaining Beeckman?

Obviously, we cannot explain Beeckman by emulation of (non-existent) earlier corpuscular-mechanists, nor due to 'influence' from a revived ancient atomism, which was not 'mechanistic'. Thus, as most historians of science would agree, he is a good specimen through which to study the wider, contextual shaping of this natural philosophy. But, we must carefully unpack the possibilities and pitfalls of contextual explanation. Two heuristic guidelines must be honoured when essaying a contextual explanation.[14] In Beeckman's case the first guideline runs as follows: Beeckman did not produce corpuscular-mechanism because he was 'influenced' or 'imprinted' or 'shaped' by contextual features. We have learned not to indulge in such vulgar, indeed impossible, historical notions. When an actor constructs something qualitatively new, he does it by active adoption and modification of resources – conceptual, material, normative – somehow available to and thinkable by him. Creators and innovators are actors, not cultural dopes.[15] This rule protects against arbitrarily selecting this or that bit of context as THE *explanans* which caused, shaped or influenced the *explanandum*. But it also points us to something else, more subtle, which amounts to a second heuristic rule: when attempting the contextual explanation of a significant discovery, claim or invention, always look first for the relevant 'proximate' context in which the actor was working; that is, the tradition, field or discipline in question. Contextual explanation must be executed 'from the inside out'. But this 'inside' must be conceptualized, indeed modelled by the historian – the structure and dynamics of the relevant field or discipline set out, grounded in evidence and open to revision as historical research and historiographical debate unfold. Such proximate contexts are never merely the old internalist universe of 'ideas only'.

In the case of Beeckman, the relevant discipline is what I have above termed the 'field of natural philosophizing'. Beeckman's invention of corpuscular-mechanism consists of a set of radical claims inside an already existing, highly structured, and, at that moment highly contested field. I hold

14 Schuster, *Descartes-Agonistes*, pp. 104-112. Also: John Schuster, 'Pitfalls and Opportunities of Contextual Explanation: The Case of Isaac Beeckman's Invention of the Mechanical Philosophy', in: *Isis* 110 (July 2019), pp. 308-311 [Focus Section on Explanation in the History of Science]. See also: H. Floris Cohen, *How Modern Science Came into the World: Four Civilizations, One 17th-Century Breakthrough* (Amsterdam: Amsterdam University Press, 2010), pp. 221-226, 238-242.

15 On the problems of asserting 'influence', see: Schuster, *Descartes-Agonistes*, p. 13, n. 25 (contributions of Quentin Skinner and post-Kuhnian sociology of scientific knowledge), and Schuster, 'Consuming and Appropriating Practical Mathematics', pp. 37-41.

that careful construction of proximate categories like natural philosophy paves the way for fruitful and defensible modes of wider contextual explanation, rather than arbitrary ones. Drawing upon the first heuristic rule, one can now say the following: quite macro entities – social structure, economic forces, political structures and forces – can be brought into the explanatory machinery, but not in the form of causing, imprinting or influencing actors' ideas and actions. Rather, natural philosophers responded to challenges and forces outside their proximate context and decided to bring them into play in the form of revised claims, skills, material practices and values in the field. To do that, the 'things' being brought in, or responded to, had to be represented to and by them (not us!) in suitable form. Appropriately thinkable/writable representations of matters concerning entities that we historians now call contextual structures were intentionally mobilized, shaped and deployed strategically by historical actors in natural philosophical claims.[16] A moment's reflection at this point reveals that everything in our explanatory strategy depends upon Beeckman's decision to inhabit and compete within the natural philosophical field. To construct his radical species of natural philosophy, he was importing resources – cognitive, normative and material – into this proximate context. We can now move forward while avoiding explanatory pitfalls.

First of all we can detect the error of simply invoking Beeckman's wider Dutch economic and cultural contexts. The blurb for the conference on Beeckman held at Middelburg in September 2018 – the occasion for the first version of the present chapter – read as follows: '[In Middelburg] he was deeply embedded in a cosmopolitan culture, a world in which sophisticated artisanal skills, riches from the overseas trading routes, humanistic culture and the study of nature were merging into a new culture of knowledge.' This appeals either to Beeckman's local context, or takes that context as an instance of even wider patterns of emerging commercial capitalism, imperial outreach and state formation – or both.[17] My point is that any

16 Schuster, *Descartes-Agonistes*, pp. 43-44, 65-70 (referencing what my approach owes to Marshall Sahlins). Paolo Rossi, *Philosophy, Technology, and the Arts in the Early Modern Era* (New York: Harper & Row, 1970), offered an early example of this type of argument. See: Schuster, *Descartes-Agonistes*, p. 80, n. 107. On the roles played by historians' models of the relevant macro structures and processes, see: Schuster, *Descartes-Agonistes*, p. 67, n. 83.

17 In using this passage from the organizers of the Middelburg Beeckman Conference in September 2018 in a discussion about avoiding the pitfalls of simplistic contextual explanation, I am in no way impugning their historiographical capabilities or their choice of words. The passage is one of the few available in the literature broadly asserting a relation between Beeckman and his social-economic-cultural context. I am arguing that contextual explanation is possible; but only on certain conditions, which are not met by headlines or blurbs. Thus, the cited passage

such a simple blurb or headline juxtaposing wider context to intellectual creation inside a particular tradition cannot in itself justify the claim that Beeckman's context(s) caused or brought into being his natural philosophical innovation. Nor can it suggest how to find a more fruitful and acceptable understanding of the role of context(s) in such a case.

It may readily be granted that if an individual of Beeckman's acumen and experience had thought about it, this lifeworld might have prompted sentiments against Aristotle and for material practice. Indeed, this undoubtedly occurred, and its plausibility tends to promote under-theorized contextual explanations. But, strictly speaking, none of this directly entails his becoming an active natural philosopher, advocating a new species of natural philosophy. Why not, like Stevin, be a maestro of the mathematical practical arts, yet avoid the game of natural philosophizing? Similarly, in the late sixteenth and early seventeenth century the values of practice and utility were widely embraced. But, why do this inside the field of natural philosophizing? Why not just assert them in the practical arts and practical mathematics and leave it at that? Moreover, it was one thing to practise mechanics and advocate its value and status, and even call for its closer relation to natural philosophy, but it was quite another to place one's mechanics at the causal heart of a new species of natural philosophy.

The foregoing points suggest that an absolutely necessary but far from sufficient part of the explanation has to be *why and how Beeckman elected to be an active player in natural philosophy*. Only his placement and strategies in that field allow us further to ask how he recruited resources from even wider contexts. Any explanation of how and why Beeckman entered the field requires a biographical reconstruction beyond our scope here.[18] Still, we can note that as a schoolmaster and rector Beeckman clearly accepted and coveted the high cultural status of natural philosophizing. The elevated place of natural philosophy in the geography of knowledge was a large contextual feature of Beeckman's world. But he had actively to decide to enter

is a useful example. Note, however, the use of the word 'embedded' rather than 'influenced' or 'shaped'. Klaas van Berkel subsequently informed me that 'embedded' was chosen after careful consideration. This suggests to me four conclusions: (1) 'embedded' does not literally speak of 'influence' or 'imprinting', so the organizers avoided asserting a simple causal relation; (2) 'embedded' has the virtue that it conveys for some listeners the promise that a yet to be found, improved manner of contextual explanation might eventuate; (3) 'embedded' can also convey the idea that there are multiple layers of contexts in which an actor is embedded, thus that the promised explanation is going to be nicely complex; (4) while 'embedded' may thus hint at the possibility of sophisticated contextual explanation, it remains the case that the term itself cannot indicate how such an explanation might in fact be articulated.

18 On scientific biography, see: Schuster, *Descartes-Agonistes*, pp. 13-19.

that arena and what his agenda would be within it. Beeckman's actions in natural philosophy are creative and rule-bending. They amount to cultural discoveries or inventions. It should be obvious that appealing to the rise of the commercial bourgeoisie, or the lifeworld of the artisan, engineer, or pedagogue in touch with the burgeoning networks of local and seaborne exchange does not explain individual creative moves inside the field of natural philosophy.

Next, after correctly placing Beeckman as an actor in the natural philosophy business, there emerges the question of why he did not insert his recruited values and aims into some existing species of natural philosophy. An ambitious natural philosopher could bring in values and goals from that Beeckmanian lifeworld of technics and commerce, and become Gilbert or Galileo (neither of them corpuscular-mechanists). Similarly, Francis Bacon was systematizing this galaxy of new norms, although his views on matter and cause remained a late Renaissance mash-up of atomistic, vitalistic and immaterialistic categories. Beeckman heavily criticized him for that. Corpuscular-mechanism, once created, spoke to the lifeworld which promoted these values and aims. But, again, why become the first corpuscular-mechanist? This suggests we need to refine the problem by putting it this way: granted (1) that Beeckman, for whatever biographical reasons had stepped into natural philosophizing, and (2) that he had relevant experience from that wider context from which he might borrow, the issue becomes, 'how exactly did he come to construct the kernel of his corpuscular-mechanism?'[19]

Following the pioneering studies of Klaas van Berkel, we must grant a large role to Beeckman's Ramism, or rather to his way of dealing with and exploiting the Ramism that he had imbibed from the elder Snel and Stevin.[20] In reviewing van Berkel's excellent book on Beeckman I wrote,

> The hallmark of Beeckman's mechanism, van Berkel plausibly argues, was the application of criteria of 'picturability' to the explanatory realm of particles and motions – where such picturability was explicated in terms of what would make imaginative sense to a master mechanician and craftsman, such as Beeckman, rather than, say, symbolic/metaphorical

19 To point out that an actor 'had relevant experience from a wider context from which he might borrow' is another way of saying the actor was 'embedded' in a certain context that can come into play in the explanation of his creation inside a given field or tradition. Beeckman was indeed 'embedded' in his Dutch techno-cultural lifeworld.

20 Van Berkel, *Isaac Beeckman on Matter and Motion*, pp. 148-162.

> sense to a neo-Platonist. But, why value such a version of picturability, and then transplant it to the heart of something so high cultural as natural philosophy? Here van Berkel brings into play Beeckman's deep commitment to Ramism as a philosophy, pedagogical policy and model of human cognition.[21]

Beeckman sharpened his version of Ramism through intercourse with machines, mechanics, and practical mathematics. Then, when doing natural philosophy he applied 'Ramism-as-a-view-of-those-enterprises'. Beeckman's lifeworld of commerce and technics was interpreted by him through his 'specified Ramism' and that bundle was then applied within the natural philosophical field. He wanted to avoid using metaphysics or theology to anchor this natural philosophy. His 'specified Ramism' could do that. It provided an epistemological thence pedagogical frame for this new kind of natural philosophy talk, corpuscular-mechanism. One can further suggest that Beeckman's leveraging of his specified Ramism within the field of natural philosophizing helps explain his radical move of making a theory of mechanics the causal register of his natural philosophy.

In sum, Beeckman's Ramism was the catalyst of, and an ingredient in, his corpuscular-mechanism. This helps to explain Beeckman, provided we view it not as an 'influence' or 'driver' but as a culturally available 'resource', picked up, adopted, adapted and deployed in constructing novel utterances in the field of natural philosophy.[22] In that delicate sense Ramism helped 'cause' Beeckman's corpuscular-mechanical natural philosophy. Similar points obviously hold for Beeckman's Dutch cultural and economic context. These can be taken as 'causes', but only along the nuanced trajectory of explanation outlined above. Elements in those larger contexts that Beeckman perceived as relevant to his moves in natural philosophizing were selected, adapted and deployed. Generalizing from the case of Beeckman, the historiographical advice becomes: examine actors in their proximate contexts first, and their importation and deployment of resources from wider contexts second. Contexts, viewed simply, should not be taken unreflectively as causes.

21 John Schuster, 'Review of Klaas van Berkel, *Isaac Beeckman on Matter and Motion*', *Isis* 105 (2014), p. 445.

22 Hence, to reiterate, there were many Ramists – such as Beeckman's mentor Rudolph Snellius – who did not operate in the field of natural philosophizing. Beeckman chose to do so, and to utilize his version of Ramism in designing the content of his natural philosophy. The socio-cognitive network of Ramism was another context in which Beeckman was fortuitously embedded.

5 Realist Copernicanism in the DNA of Corpuscular-mechanism: The Second Beeckman/Descartes Encounter – October 1628/Spring 1629

During the encounter of Beeckman and Descartes in 1628/1629, their first in ten years, Descartes revealed to Beeckman his remarkable discovery of the law of refraction of light. He explained it to Beeckman through a wonderful analogy of the behaviour of a bent-arm balance in air and water. Beeckman probably thought this discovery, and this illustration, fit nicely with his own *mathematico-physica* philosophy style. Descartes had also started to develop his lens theory. Beeckman helped him with a nice derivation of the functioning of a plano-hyperbolic lens. Descartes 'approved' this and used this exact construction in the *Dioptrique* (1637).[23]

But, Descartes also had problems at this stage. I have shown that his draft *Regulae ad directionem ingenii* were on the point of collapse. He had launched into composing the bulk of the *Regulae* in Paris shortly after the discovery of the law of refraction in 1626/1627. This did not involve a system of natural philosophy but rather a 'universal mathematics', a general technique for problem-solving in all fields of mathematics and physico-mathematics. To ground this universal mathematics, Descartes invoked his mechanistic optics and sketch of a mechanistic physiology; to show how our immaterial thinking power, the *vis cognoscens*, could grasp external reality in the form of 'dimensions' and how the mathematical operations of universal mathematics could be reliably performed by that knowing mind, operating upon corporeal brain loci. I have shown that by 1629 Descartes knew that this programme had failed on three fronts. When Descartes met Beeckman this second time, his career had reached what I call an 'inflection' point. Over the six months from mid-1629 Descartes slowly but surely evolved toward writing a system of corpuscular-mechanical natural philosophy, which would include realist Copernican cosmology and celestial mechanics. He had shown no inclinations this way at any time previously.[24]

As I have argued in several places, it is Descartes' discourse of celestial mechanics, his vortex theory, that is the engine room of his version of the mechanical philosophy. It is complex, it is serious and it makes more

23 Schuster, *Descartes-Agonistes*, pp. 199-203 (refraction of light and bent arm balance analogy), 186-188, 608-609 (lens theory).

24 John Schuster, 'Descartes' *Mathesis Universalis*: 1618-1628', in: Stephen Gaukroger, ed., *Descartes: Philosophy, Mathematics and Physics* (Brighton: Harvester, 1980), pp. 41-96, Schuster, *Descartes-Agonistes*, pp. 334-346.

sense than usually granted.[25] Descartes imagined whirlpools or vortices of spherical corpuscles of his second element, rotating around their respective central stars to sweep along their planets like boats in a strong current. However, this swishing along of the planets in the vortex was the least of his concerns. I have shown that what really interested him, and what shaped the conceptual architecture of the vortex celestial mechanics, were such non-trivial questions as: 'why does each planet maintain its own characteristic and yet relatively stable cosmic distance from its star?'; and, 'what physical role does a central star play in altering the fine structure of the vortex and thus allowing planets to attain stable orbits; and why do comets, unlike planets, oscillate between the higher regions of adjacent vortices?'

Descartes' vortex theory is in essence a science of equilibrium, but not in the classical sense of being a science of weights, volumes and densities. Instead, it deals with planetary orbital equilibrium through concepts of his own design, the 'massiveness' of a planet (the way its surface area relates to its quantity of matter), the centrifugal force the planet generates, the centripetal force exerted upon it by the nearby particles of the vortex, and the volume to surface ratios of those vortex particles, which ratio varies with distance from the central star according to a function determined by the presence of that star (it would differ if no star sat in the centre of a the vortex). To this base Descartes articulated his theories of the planetary satellites, comets, and multiple star-planet systems. Later, in the *Principles* (1644), it becomes even more elaborate, as theories of cosmic magnetism, sunspots, *novae* and variable stars fill out this Copernican vision – clearly the central point, aim and achievement of his corpuscular-mechanical philosophy.

What does all this have to do with Isaac Beeckman? Quite a lot, it turns out, as can be documented from an almost unnoticed aspect of their second meeting. The impetus for Descartes' realist Copernicanism came directly from Isaac Beeckman – no wonder Descartes soon became defensive and evasive about his intellectual debts to Beeckman.[26] He found Beeckman ploughing through the astronomical works of Kepler. Everywhere Kepler had invoked immaterial celestial forces or powers, Beeckman re-wrote these into corpuscular-mechanical terminology. As far as Beeckman was

25 John Schuster, '"Waterworld": Descartes' Vortical Celestial Mechanics – A Gambit in the Natural Philosophical Agon of the Early 17th Century', in: Peter Anstey and John Schuster, eds., *The Science of Nature in the 17th Century: Patterns of Change in Early Modern Natural Philosophy* (Dordrecht: Springer, 2005), pp. 35-79; Schuster, *Descartes-Agonistes*, pp. 453-477.

26 Klaas van Berkel, 'Descartes' Debt to Beeckman: Inspiration, Cooperation, Conflict', in: Stephen Gaukroger, John Schuster, and John Sutton, eds., *Descartes' Natural Philosophy* (London: Routledge, 2000), pp. 46-59.

concerned, the key issues in astronomy did not involve the traditional activities of observation or even Kepler's work on elliptical orbits. Rather, Beeckman saw in Copernican astronomy, especially as transformed by Kepler, a neglected field for natural philosophical explication, meaning a corpuscular-mechanical explanation. Beeckman wanted to translate into mechanistic terms the problematic of a discourse of celestial causation that Kepler had begun in neo-Platonic register – Beeckman was showing natural philosophical acumen and vision of the first order.[27]

Beeckman's review of Kepler starts with a mechanistic critique of Kepler's immaterialist theory of light: light is corporeal, consisting in a type of heat particle emitted by stars. Then Beeckman takes on Moon theory, followed by a couple of versions of planetary theory. In each case he wants to explain orbital stability, through a balance of various particles issuing from the Sun and the other stars. He postulates that different flows of particles weaken differently with distance, so that equilibrium loci, or orbits, are created, with slight perturbations in the causes explaining radial changes (for elliptical orbits etc.). Beeckman focuses on orbital equilibrium, and the physical roles of the Sun and stars in celestial causation – exactly as Descartes' vortex theory would soon do. By mid-1629, Beeckman had not achieved a settled view and in typical fashion, the matter was dropped. Beeckman's work just pre-dates and overlaps the period of renewed contact with Descartes from late 1628.

Arguably, after 1628 Descartes was interested in a corpuscular-mechanical Copernicanism because he had seen Beeckman pioneer the idea. No doubt Descartes thought his own corpuscular-mechanical celestial mechanics showed more comprehensiveness, coherence and theoretical rigour than Beeckman's jottings. *But, and this is the key, Beeckman had shown him that the royal road to any systematic articulation of corpuscular-mechanism ran through the challenge to explain Copernican theory in corpuscular-mechanical terms.* Such a gambit would marginalize Gilbert and Kepler, the older, towering natural philosophers intent on physicalizing astronomy. Beeckman had shown Descartes that their corpuscular-mechanism could trump Gilbert's magnetic philosophy and Kepler's neo-Platonist conceptualizations. Descartes saw his main natural philosophical chance right in this locale. He actualized it over the next few years in writing *Le Monde*, and he pursued the matter in his later systematic masterwork of corpuscular-mechanism, his *Principia philosophiae*. So, right here we have an even deeper connection of

27 For detailed treatment of material in this and the next paragraph, see: Schuster, *Descartes-Agonistes*, pp. 471-475; Schuster, *Descartes and the Scientific Revolution*, pp. 567-579.

Beeckman to Descartes' corpuscular-mechanism than in their 1618 interactions: *the idea that a complete corpuscular-mechanism is one containing and bespeaking a radically realist Copernicanism.*

6 Reprise: Beeckman's Unique Stature in His Time and Place

Everything we have explored thus far prompts some comparative reflections on Beeckman's stature as a natural philosopher in the early seventeenth century. Comparison with other figures is useful in historiography, especially for understanding a first-on-the-scene figure like Beeckman. Consider this set of important practical mathematicians and engineers who, with one exception, also had natural philosophical agendas: Giovanni Battista Benedetti, the young Galileo, and Simon Stevin.

Like Beeckman, they were adepts of practical mathematics, engineering, mechanics, and the study of machines. They were all pro-Copernican, seeking to extract anti-Aristotelian capital from their practical mathematics and mechanics. But none of them – not even the mature Galileo – put a mechanics, like a key into a lock, directly and essentially into a natural philosophy as its 'causal register'. Indeed, we know that from the late sixteenth century there had been scattered moves to put the mixed mathematical science of mechanics into closer relation with natural philosophy. But again, nobody, except Beeckman, had made some mechanics the causal register of his natural philosophy. The young René Descartes immediately got the point, even though, as I have said, he developed a different conception of the mechanics of corpuscles. Moreover, although Benedetti, Galileo and Stevin all advocated realist Copernicanism, none of them put a matter and cause account of Copernicanism centrally into a natural philosophy. Even the mature Galileo only did this in piecemeal and indirect ways by his telescopic claims and the theory of the tides. But Galileo had no enveloping natural philosophical account of matter and cause, nor a celestial mechanics discourse. Beeckman, in embryo, and Descartes, in full systematic bloom, supplied what Galileo always failed to deliver.

In sum there was something deeper, more dynamic and portentous in Beeckman compared to these earlier practical mathematicians, engineers, Copernicans (even Galileo). *Beeckman represented the deepest potentials of the contemporary turmoil over nature knowledge because he did his work in the field of natural philosophy and most likely because his creative moves there were shaped and informed by his manner of deploying Ramism.* Of course, he did not achieve wide notoriety; but he showed some others (Descartes,

Gassendi, Mersenne) what was at stake at that moment. We can conclude that compared to his contemporaries – excepting Descartes – Beeckman arguably was a 'visionary', albeit at the time a rather underground figure. Moreover, as we are about to see, he was also arguably a seminal figure in the next phase of the Scientific Revolution, again, albeit a somewhat occluded one.

7 Beeckmanian Threads in the Weave of the Later Process of the Scientific Revolution

Beeckman did not leave wildly popular publications. Nor did he leave a continuous school or tradition. So, can we say he was significant in the later weave of the Scientific Revolution? Yes we can, provided we are historiographically alert and imaginative. First of all, in a way, Beeckman stands behind each and every instance of the uptake of corpuscular-mechanism out of texts by Descartes or his followers. Of course Cartesian mechanism is Descartes' product, but, as we have seen, Descartes put more Beeckmanian resources into his corpuscular-mechanism than he granted – or we realize. Around 1675, if you were impossibly well informed, you could have said: 'Today, in a sense, we are all Beeckmanians.' To grasp his huge but occluded significance, try these aphorisms knitted together into a syllogism: (a) 'No Beeckman, no Descartes the systematic mechanist cum committed and public realist Copernican'; (b) 'No Descartes (of that type), a much different pattern to the next two generations of the Scientific Revolution'; (c) 'Hence, no Beeckman, a much different pattern to the next two generations of the Scientific Revolution.' Granted, these propositions are overly simple; but, they point toward fruitful insights and inquiries.

An amusing way to try to grasp Beeckman's significance in the later weaves of the Scientific Revolution is to try a counterfactual scenario.[28] Imagine Beeckman does not die in 1637, but like Thomas Hobbes, born also in 1588, he lives into the late 1670s. He writes and teaches in his own academy; he travels to England and to Paris (between relevant wars of each against the Dutch Republic). In 1645 he reads with some chagrin, and some satisfaction, the entire *Principia* of Descartes. He then carefully prepares his own highly systematic natural philosophical text, published

28 I have applied four cases of counterfactual scenarios to the biography of Descartes in: Schuster, *Descartes-Agonistes*, pp. 376-377, 413-415, 486, and 596-600 (where 'Descartes' himself lives into the CMF stage).

in 1650. It, too, is a thundering statement of corpuscular-mechanism, Copernicanism and the values of the practical arts. And no metaphysics or theology to be seen! The next year he publishes his own autobiographical pronunciamiento – entitled *The Road to Mechanism and the Domination of Nature* to rival René's *Discourse on Method.* Around his life story he articulates a Ramist rationale for linking corpuscular-mechanism and the world of the mechanical arts – like the one offered in Section 4. 'Beeckman's' system of natural philosophy makes use of clarified and systematized ideas in mechanics. Like Christiaan Huygens is about to do, he melds Galilean and Cartesian resources for the now crystallizing field of classical mechanics. He works on inertia, collision, circular motion and pendula. And, like Robert Hooke, he edges toward a mathematical approach to celestial causation by looking closely at circular motion and the likely celestial causal role of the Sun. Outliving Descartes, 'Beeckman' is able to explain his early relations with him and how Descartes was hampered by his attachment to a Jesuitical, neo-Scholastic mind set, and metaphysical speculation.

My 'Beeckman' has a wry sense of humour. Addressing the Royal Society in 1668 during their debates on the laws of collision, he jokes in front of Wallis, Boyle, Hooke, Oldenburg, Wren and others of having discovered the mechanical philosophy a mere 55 years earlier. He congratulates his English colleagues on having finally caught up with him, and closes by disingenuously asking about the whereabouts of Mr. Hobbes. 'Beeckman', of course, knows that Hobbes is not a Fellow, but he also knows that, like himself, Hobbes is one of the few individuals still alive who had met Descartes on something approaching equal terms.

A still-living Beeckman arguably could have done these things. Later historians would have recognized him as a giant in the rise of modern science. The batting order, Stevin-Beeckman-Huygens would constitute an all-star Dutch line-up, mirroring the 'Renaissance', 'Critical' and 'Consensus' phases of the Scientific Revolution. So, we have a fictional narrative, certainly; but, as Max Weber could have said, an 'objectively possible' one.[29] Viewing both the Scientific Revolution and Beeckman in the ways advocated here, one should be quite willing to promote him into the first rank of figures in that process – a mere 400 years since November 1618, when he first took the 22-year-old René Descartes under his natural philosophical wing.

29 Max Weber, 'Objective Possibility and Adequate Causation in Historical Explanation', in: Max Weber, *The Methodology of the Social Science,* trans. by Edward Shils and Henry Finch (New York: Free Press, 1949), pp. 164-188.

About the Author

JOHN A. SCHUSTER is Honorary Reader, School of History and Philosophy of Science, University of Sidney, and Honorary Research Fellow, Campion College, Sydney. He has published extensively on Descartes, the Scientific Revolution, the myth of the scientific method, and historiographical issues. He is a Fellow of the Australian Academy of the Humanities.

3 Isaac Beeckman at Gresham College in 1668

An Alternative 'As If' Scenario

H. Floris Cohen

Abstract

In this chapter Isaac Beeckman is imagined to have lived on to at least 1668 and to have been invited by the Royal Society to come to Gresham College in London and inform its fellows about his friendly yet at times troubled relationship with René Descartes since their first encounter 50 years earlier. The chapter consists of the speech Beeckman is imagined to have given there. The speech, though entirely fictional, is nonetheless based on solidly established, overall well-known historical facts about Beeckman and Descartes, and on the author's own interpretation thereof. The speech centres on the similarities and the differences between the two men's pioneering conceptions of the 'mechanical philosophy', and on the issue of priority that Descartes rather obsessively kept raising.

Keywords: Isaac Beeckman, René Descartes, 'as if' history, mechanical philosophy, priority dispute

On the final pages of the previous chapter, John Schuster undertook a delightful exercise in 'as if' history. In the present chapter I take up the idea by imagining Beeckman to accept an invitation by the Royal Society to detail his relationship with the late René Descartes.[1]

1 In this chapter I have availed myself in the first place of the following pages in my own work: *Quantifying Music: The Science of Music at the First Stage of the Scientific Revolution, 1580-1650* (Dordrecht: Reidel, 1984), pp. 116-179 and 187-204, and *How Modern Science Came into the World: Four Civilizations, One 17th-Century Breakthrough* (Amsterdam: Amsterdam University Press, 2010), Chapters 6 and 11. Everything I make Beeckman tell the Royal Society about his encounters

Van Berkel, Klaas, Albert Clement, and Arjan van Dixhoorn (eds.), *Knowledge and Culture in the Early Dutch Republic: Isaac Beeckman in Context.* Taylor & Francis Group 2022
DOI: 10.5117/9789463722537_CH03

Domini praeclari, Most learned Fellows of the Royal Society for the improvement of naturall knowledge by Experiment (it is, to say that right away, not clear to me how experiments could possibly help us improve our understanding of the natural world).

At the age of 80 back in your country for the third time, it is my honour and my great pleasure to stand in this festive hall at Gresham College and to consider from a variety of viewpoints the topic that you have invited me to address. What are the origins, as I see them, of what you here in Britain have of late accustomed yourself to calling 'the mechanical philosophy'? To be more specific, the wording of the very flattering letter that your most diligent secretary, Mr. Henry Oldenburg, Esq., has sent me leaves some room for the suspicion that your innocent-sounding question is really aimed at learning how, if asked, I would assess my own scholarly achievement in comparison to that of the man meanwhile known all over Europe as the great pioneer of the mechanical philosophy – my regretted friend, the late René Descartes.

In seeking to answer that underlying question, then, let me go back to when and where René and I first met, wholly unforeseen and by sheer coincidence. Scarcely could I have suspected, now exactly half a century ago, that I, the elder of the two of us by eight years, would survive him by no less than meanwhile eighteen years. Imagine, a professionally settled candle maker *cum* strictly Reformed theologian and an aristocratic military man fresh from the desks of a well-reputed Jesuit high school hundreds of miles away who in November 1618 make their acquaintance in the still heavily contested city of Breda! Do you, by any chance, wonder what men from so different backgrounds at so different stages in their lives could possibly have in common? Oh, how sweet it is to reminisce for a while about the mutual intellectual recognition that our conversations exuded from the very first! I have turned up the pages of the diary I had, by then, already been keeping for fourteen years to find what exactly I recorded in it during that fateful month-and-a-half. Not that I was really in any need to look up that diary passage for present purposes, as I still fondly remember it to the

and exchanges with Descartes is based on the available sources, though surely rendered in light of my own interpretation thereof (an interpretation that owes various insights to work by Klaas van Berkel and by John Schuster). Other things related here as factual go back to matters that historians uncontroversially acknowledge as facts, notably regarding British/French scholarly exchanges in Paris during the British Civil War and the Interregnum. I have, however, refrained from an effort to imitate Beeckman's unremarkable writing style.

letter, dry-as-dust as it may well sound to you but I beg you not to miss the underlying sense of exultation. At any rate, here is how it opens:

> This man from Poitou has associated with many Jesuits and other studious and learned men. Yet he says that he has never found anyone beside me who in this way as I like to do conducts his studies and accurately combines Physics with Mathematics. And I, too, have never talked with anyone beside him who studies in this way.[2]

Right away, then, René and I defined what we felt we shared as a combination of mathematics and natural philosophy (for you should by no means take the word 'physics' in the much narrowed-down sense it seems to have been acquiring of late). For brevity's sake I shall from here on speak of our physico-mathematics. Now what did we mean by so uncommon an expression? Uncommon indeed, since from the ancient Greeks onward natural philosophy and mathematics have been wide apart enterprises. Whereas the former was concerned with explaining the natural world from intuitively obvious first principles, the latter was occupied with bringing the geometrically reduced properties of a small, well-defined set of quite specific natural phenomena (just planetary trajectories, equilibrium states, light rays, consonant intervals, and geographical longitude and latitude) to highly abstract mathematical rule and order.

I shall address the question of what René and I meant by the expression first for myself alone, then for him, then for what we accomplished together during those six weeks in Breda that are still so dear to me. For me, then, to combine natural philosophy and mathematics stood for my personal determination to seek to explain all and sundry natural phenomena by means of nothing but quite specific, well-imagined mechanisms of atoms moving in the void in accordance with one basic principle of motion – the principle 'dat eens roert, roert altyt, soot niet belet en wort', that is, 'whatever moves, moves forever, unless it is prevented from doing so'.[3] I first posited the principle five years before I met René, and for the years c. 1613-1618 (as also after 1618) my diary is filled with notes in which I kept forever

2 *Journal tenu par Isaac Beeckman de 1604 à 1634*, publié avec une introduction et des notes par C. de Waard, 4 vols. (The Hague: Martinus Nijhoff, 1939-1953) [henceforth *JIB*], I, p. 244: 'Hic Picto cum multis Jesuitis aliisque studiosis virisque doctis versatus est. Dicit tamen se nunquam neminem reperisse, praeter me, qui hoc modo, quo ego gaudeo, studendi utatur accurateque cum Mathematica Physicam jungat. Neque etiam ego, praeter illum, nemini locutus sum hujusmodi studii.'

3 *JIB*, I, p. 44.

applying the principle to a large variety of natural phenomena of such a kind as presented themselves to me, more often than not, in my everyday craftsman's practice.

By 1618, then, I had been a full-fledged atomist for at least five years, though one with a decisive twist – that principle of motion of mine. Already as a student at Leyden University I had become dissatisfied with the standard Aristotelian doctrine served up there as at just about every other university at the time. After a brief intermediate stage, during which I introduced particles made of Aristotle's four elements, I switched for good to full-fledged atomism in the manner of the ancients. That is, I began to conceive of our world as made up of invisibly tiny, indivisible particles which differ in nothing but their sizes, their figures, and their position. Or rather, I switched to atomism in the ancient manner yet meanwhile enriched, from 1613 on, with that programme of mine in which I set out to show how, underneath their surface appearances, natural processes are brought about not just by minute particles of various shapes and sizes but in addition by their forever ongoing motion of one special kind – they retain it, unless constricted otherwise. Such, then, was, and still is, my own variety of physico-mathematics. For instance, I explained that, if on your keyboard you touch a string, another one that remains untouched may, seemingly spontaneously, begin to sound as well. The customary explanation used to be through sympathy – some hidden sense of the strings somehow belonging together, that has even given the phenomenon its name of 'sympathetic resonance'. To me such an almost magical account was, and is, as fanciful as it is vacuous. As a down-to-earth craftsman I accepted no explanatory mechanisms but such as we can readily visualize, which in almost all cases means: mechanisms of matter in motion. In the present case my explanation was, and is, that the vibrations set up by the finger that touches the former string are transmitted, through the regular cycles of condensation and rarefaction set up by its regular vibrations in the particles of the surrounding air, to the latter string. My full explanation, to be sure, accounts in addition for the circumstance that sympathetic resonance occurs only at certain intervals, not at certain others, but to explain that highly sophisticated mechanism of particles in motion would carry us much too far. My only point here is to give you an idea of how I used to reason in such matters throughout my diary. And I am sure that reasoning of this kind sounds thoroughly familiar to you, as this is exactly how by now, half a century later, just about every innovation-prone scholar is proceeding, whether you call it 'the mechanical philosophy' or not.

Now back to 1618. René arrived in Breda with rather a different view of the ultimate structure of the natural world. Naturally for a young man just fresh from the school desks, that view was the one he had just imbibed at La Flèche. The advanced Jesuits who had taught him there had adopted an approach pioneered (and this is a piece of history that is still fresh in my mind but that you from a later generation probably have no inkling of any more) by their late colleague, Christophorus Clavius SJ. In an effort to make Aristotelian doctrine, which had come under attack from a large variety of quarters, more acceptable to the mathematically inclined, Clavius had sought to amplify what little room Aristotelian doctrine left to its quite subaltern category of quantitative change. That room, customarily called 'mixed mathematics', used to be by and large confined to equilibrium problems. Over the second half of the previous century Clavius kept forever extending the topics that an Aristotelian natural philosopher could safely treat with a generous dash of quantities thrown in, such as perspective, sundials, the calendar, etc. Clavius, then, remained a person to whom the true state of things in the world is revealed by way of secure possession of solid, because intuitively obvious, first principles (which for any Aristotelian is change in four distinct varieties), with the mathematics added solely to fill in the quantitative details.

René quickly demonstrated his mastery in this kind of extended, really Clavian mixed-mathematics in a little treatise, entitled *Compendium musicae*, which he was kind enough to write for me personally by way of a New Year's gift. Then already I noticed with awe the facility with which he could write up a coherent, well-structured argument at booklet length – a capacity it still takes all my mental balance, and all my inclination to admire what deserves to be admired, not to envy him!

It is true, though, that my own meanwhile habitual way of philosophizing did creep through the pores of his treatise at one specific point – those numerous diary notes of mine in which I had been analysing musical intervals by means of the concept of *ictus* ('strokes') of the air. That particular approach to the basic issues of harmonic theory was wholly foreign to the 'mixed mathematics' context in which René inserted a brief remark to that effect, and there can be no doubt that he got the 'strokes' idea from me (nothing to be ashamed of, to be sure, least of all for a young man still in his formative years).

But the impact which my thought exerted on René went further. Together we undertook analyses of two further problems, and here he demonstrated two admirable features not readily apparent from *Compendium musicae*: his mathematical prowess, which by then already far exceeded mine, and his willingness, not to say his urgent drive, to seek new pathways rather than to settle for good for the Clavian approach of extended mixed-mathematics so familiar to him. Together we derived the squared proportion between time and distance in vertical descent. My own contribution was a proposal to conceive of the cause of free fall as a force that 'trect met cleyne hurtkens' ('which pulls by tiny jerks').[4] René contributed his acquired ability to handle geometric proportions that approximate the infinitesimally small. We further tackled Simon Stevin's strictly mathematical proof of the hydrostatic paradox, not because it would be wrong (it isn't) but because we wanted to understand the causal mechanism underlying the paradox. René devised a specific mechanism of particles in motion that could serve as the natural-philosophical counterpart to Stevin's purely mathematical handling of the issue. True, he worked this out in Breda on his own, when I had already returned to Middelburg, so the details are all his. Even so it is unmistakable that the inspiration for making the causal attempt and for how he undertook it, so foreign to both Stevin's strictly Archimedean approach and to René's own Clavius-style mixed-mathematics, came from his encounter with me, who had already for years been producing accounts of that very kind.

René kept sending me friendly, almost obsequious letters expressing his profound gratitude to me until he, too, left Breda. The next time I heard from him was when, ten years later, I actually heard him knock on my door in Dordrecht, where I had meanwhile come to serve as the principal of its Latin School. He reiterated that neither during his travels in Germany nor once resettled in France he had ever met anyone with whom he could talk so well about natural-philosophical matters as with me. I in my turn reiterated that in mathematics I esteemed René above anybody else. I also went so far as to allow him now to read in my diary. He was the very first to receive that permission, to be followed in later years only by my countryman Maarten van den Hove (Martinus Hortensius) and by his two countrymen Marin Mersenne and Pierre Gassendi when they descended from France to visit me. I still do not quite know whether or not I should regret that I ever did allow René access – so much is certain that it was to lead, in the next year, to a row that, though superficially healed a while later, could not fail to impair our former intimacy for good.

Should I go into any detail regarding the painful episode of the two deeply deceptive, deeply insulting letters that he wrote me in September and October of 1630 and of the one in between in which I plainly rebutted his insinuations? Should I set forth here the bare facts that expose him as an arrogant liar, using his outstanding gift for prevarication and a haughty

4 *JIB*, I, p. 264.

sense of his aristocratic and therefore far higher status in life to seek to nip in the bud the only attempt I have ever made in my entire career to write up a coherent account in terms of my own physico-mathematics? No, I shall let all that pass, though not without conceding that his attempt was quite successful. Yes, I stopped turning those diary notes of mine in which I provided Kepler's brilliant mathematical rules for our solar system with an underlying causal mechanism of particles in motion into a full-fledged treatise, and returned, but now for good, to the diary note as my one and only form of scholarly expression. Not until 1644, when I purchased a copy of René's treatise *Principia philosophiae* (he had not taken the trouble to send me one) and read it from beginning to end, did it occur to me that here might have rested the true ground underlying those insidious letters of his. There was more than sheer coincidence, so I now began to suspect, to the circumstance that in my diary he had encountered my notes on Kepler at the very time when he was considering to settle in the Netherlands for writing up his own meanwhile invented variety of an account of the solar system through physico-mathematics – a variety not at all so different from mine!

Let me be more specific. *Principia philosophiae* is much like the treatise I would have written in the absence of his intervention, or at least I would have done so if I had not been so timid and if I had also possessed the required talent in this regard – the talent, that is, to organize one's thoughts not only in one's head but also on the blank pages of a planned book. If I had possessed that talent (and apparently René did not want to gamble on its absence but rightly felt that some intimidation would do the trick in any case), I would around 1630 have published a book that would have preceded *Principia philosophiae* by close to fifteen years.

Not that that imaginary book of mine would have been near-identical with his, as *Principia philosophiae* counts certain features that are not really mine. First of all, Descartes makes much, in the book, of the vast chasm that distinguishes his natural philosophy from all preceding ones. He makes the reader believe in particular that his own variety and that of the ancient atomists are (almost literally) worlds apart. The reason he advances for this preposterous claim is that, unlike with the ancients' atomism (and also, I may add here, with mine), in his system of natural philosophy matter is endlessly divisible. As a consequence, he deals throughout his treatise in whirlpools of matter. Surely that is different from how I set up my particulate explanations, yet here in Gresham College I need hardly remind you that your own Lord Boyle has proclaimed what he calls 'the mechanical philosophy' as a way of coming to grips with the phenomena of nature by means of just the two 'Catholick Principles of Matter and Motion'. In other words, His

Lordship has quite rightly declared the question of the endless or limited divisibility of matter to be wholly subordinate to the overriding importance of explaining things through particles of matter in certain kinds of motion. I do not want to compare myself in any way with this high and very learned Lord, yet I cannot refrain from feeling that I could have said so myself, and in fact by and large did, although in my personal diary only.

There is a further difference between *Principia philosophiae* and my own approach. This is a difference that, unlike the previous one, I find of the highest possible significance, and that goes to show that I had not been mistaken when, in Breda in 1618, I at once saw incipient greatness in that young man. I am talking about the ways in which he and I dealt with motion. Surely we had in common that, whereas the ancient atomists had left the motion of their atoms wholly indeterminate, concentrating rather on their mutually different sizes and shapes only, both René and I were concerned with motion in far more specific ways. I posited my principle of rectilinear or circular motion retained in 1613, as I have related to you a few minutes ago. In *Principia philosophiae* Descartes now came forward with a splendid new idea, that of laws of motion. As you are aware quite as well as I am, that idea has since taken flight and has become part of the standard equipment of every innovative scholar out to gain knowledge of the natural world. It is true that I have my doubts about the other laws of motion that Descartes posited (I mean in particular his law of the relativity of motion, but also his laws of motion in percussion). But that is not the main point – the very idea of a conception of the natural world as one unified whole for which we can posit general rules valid over the entire universe was, in 1644, just as bold and stunningly novel and path-breaking as it has in the meantime become standard fare.

I am much inclined to attribute the ingenuity that Descartes displayed in conceiving of this masterful idea of natural law to what is perhaps the most admirable, yet also the trickiest aspect of his intellectual powers – his drive to find in the most varied domains universal rules, universal approaches, and, above all, universal methods. Take the various posthumous writings edited in (as far as I can tell) not too reliable a fashion by that wholly uncritical admirer of his, Mr. Claude Clerselier. These writings give us at least some idea of how, in the 1620s, René worked his way toward what future generations, in an effort to characterize his whirlpool mechanisms in a brief verbal formula, may well decide to call the 'water world' that marks not only his *Principia philosophiae* but also, years earlier, his *Le Monde* and the broadly phrased précis thereof in Part 5 of his *Discours de la méthode*. There, and in his *Meditationes*, his universalist inclination has led (to mention just two major achievements) to his brilliant yet risky 'malin génie' pathway towards defeating scepticism and to his metaphysics of a universe made up of no more than two substances, one material, one mental. It would surely be possible to demonstrate that, if you were to assemble my diary notes and bring them to regular order, you would encounter there, but solely implicitly and largely hidden to even my own thought, notions similar to these all-encompassing conceptions of Descartes.

However, so much is all I could possibly claim in this regard – I have never been a metaphysician, nor, ever, a mathematician in anything like the same class to which René belonged already in 1618. When his *La Géométrie* came out, peruse it superficially is all I could do with it – its highly advanced mathematics is quite over my head. Perhaps even more relevant in this regard, I have never held mathematics in so high esteem as René held it from the very start. For him mathematics has always signified the very model of solid reasoning, witness his 'clair & distinct' criterion, that he derives straight from how mathematical proofs proceed. For me, the position that mathematics ought to hold in our thinking is considerably lowlier. As I observed in the little address that I gave in September 1618 when I was in Caen to attain my doctorate,

> philosophy is divided into two parts: mathematics and physics. Physics is concerned with the corporeal things themselves, but mathematics with their quantity, and the dignity of physics is as much greater as the shade is more ignoble than the body itself. Even so, mathematics is so necessary for leading to physical knowledge that it can properly be called its hands, by which alone all that physics contains can be apprehended [...]. Physics is the science of natural bodies, and it investigates the nature of heaven and earth and everything contained therein.[5]

By now far past my creative years, I still think that this is the right way to see it. And yet I readily concede that René's so much more august idea of the power of mathematics has led him to some results that seem capable of holding their own for longer, perhaps, than we at present, with our 1668 outlook on things, can still confidently foresee.

5 *JIB*, IV, p. 41: 'Dividitur philosophia in duas partes: *mathematicam* et *physicam*. Physica circa ipsas res corporeas, mathematica vero circa earum quantitatem versatur, tantoque major est physicae dignitas quanto umbra ignobilior ipso corpore. Est tamen ad physicam cognitionem consequendam tanta mathematicae necessitas ut aptissime ejus manus vocari possit, qua sola quicquid physica continet, apprehenditur [...]. Physica est corporum naturalium scientia scrutaturque naturas caeli et terrae et omnium quae in ipsis continentur.'

With this little look forward I leave the history you have asked me to relate behind. Consequently, I have reached the end of my story. I have given you my assessment, frank and unadorned as you may expect from a Dutchman and most certainly from a 'well-rounded' man from Zeeland,[6] of my own scholarly achievement as compared to that of the famous René Descartes. I have also answered your implied question about how priority stands between the two of us. Not, to be sure, that I am much interested in matters of priority, let alone that I would ever have shared René's veritable obsession with it. In that nasty letter that he sent me on 17 October 1630 he went to remarkably far-fetched lengths to interpret my simple and quite obvious habit of placing the appropriate date in the margin of my consecutive diary notes as so many claims to priority – priority, in particular, over René himself. He also portrays himself in that letter as the humblest of men ('I, who have taken the habit to put myself after the least of men').[7] How interesting a ploy of this immensely vainglorious person, whose 'I' and 'me' fill the pages of his published work to an extent I have never encountered in the writings of any other author! I myself, in marked contrast to René, have never set store by issues of priority. It is great ideas that matter to me, not who first thought of them. I recall that, when I read Galileo's *Saggiatore* and encountered therein certain assertions identical content-wise with statements noted down in my diary years earlier, I only felt honoured that I, just a humble schoolmaster, had managed to arrive at some thoughts that so great a man had quite on his own conceived of as well.[8]

So much, then, for my answers to your two questions, the explicit one and the implied one. Now that I think of it, how in the world has it come into your learned heads to ask those questions? I have never published a word of my natural philosophizing, so how do you even know that I have dabbled in that domain, and that I have been of some significance in the making of the great Descartes' natural philosophy? Come to think of it, it must be by hearsay that you have become aware of all that.

Come to think of it further, I can conceive of no more than one possible source of that hearsay. A search for it brings you and me back to the Civil War and the Interregnum period you have, for eight years already, so mercifully behind you. During that period, more in particular in the aftermath of the battle of Marston Moor, several Royalist scholars fled to Paris, and mingled there with a bunch of innovative French scholars, René from time to time

6 After the Dutch expression 'goed Zeeuws, goed rond'.

7 *JIB*, IV, p. 201: 'ego qui me consuevi minimis quibusque postponere'.

8 *JIB*, III, p. 223.

being among them.[9] But he is, of course, the most unlikely person of all to have informed any Englishman about my existence, let alone about my work in natural philosophy. As I already told you, beside René there are two more Frenchmen who have read my diary, Pierre Gassendi and Marin Mersenne. And, now that I am reminiscing anyway, it was precisely Mersenne who, quite unwittingly so, brought about that nasty exchange that I had with René in 1630. For it was Marin who found, in coming across those marginal dates in my diary, that quite a few thoughts and conceptions in natural philosophy that René had written his old school-friend Marin about as if they were wholly new, had really been conceived by me at least a decade earlier. Marin went on to confront René with this discovery, which fuelled René's simmering concern over my book-like effort to provide Kepler's geometrical rules for planetary motion with a proper causal mechanism to the point where René decided to break up our friendship and successfully bark me back into the kennel of my diary.

That painful story apart, it must have been my good friend Marin, the most communicative of all those Parisian scholars in any case, who must have told one of your scholars-in-exile about my diary and what he had found there. Let me guess – who may have been the scholar-in-exile in question? Walter Charleton? An atomist for sure, a faithful disciple of Gassendi even, yet I believe that, although a Royalist, he kept quiet here in London during the Interregnum rather than fleeing to Paris. Or perhaps Thomas Hobbes? I greatly doubt it – he is not in your august ranks, and probably has no trust in your Society, either. But what about Hobbes' secretary at the time, William Petty? Years after his return from Paris, that much I do know, Sir William became one of the founders of your Society. Ah, he may even be sitting here in the audience! I have once seen his portrait, so let me look around – hey, there I believe I see him! Well, I am most honoured, Sir, to have been the recipient of Mr. Oldenburg's invitation through your apparent intermediary. Now that my talk, and therewith this session, has come to an end, let you and I by all means go to a coffee house and talk natural philosophy for the rest of the day, no longer fretting about my relationship with the late René Descartes but just talking business as scholars should always do rather

9 William Petty in 1674 in a dedicatory letter to Duke William Cavendish (as quoted from the manuscript source in: R.H. Kargon, *Atomism in England from Hariot to Newton* (Oxford: Clarendon Press, 1966), p. 69: 'Your Grace [...] did Encourage Me 30 years ago as to Enquiries of this kind. For about that time in Paris, Mersennus, Gassendy, Mr. Hobs, Monsieur Des Cartes, Monsieur Roberval, Monsieur Mydorge and other famous men, all frequenting and caressed by your Grace and your memorable brother Sir Charles Cavendish, did countenance and influence my studies as well by their Conversation as their Publick Lectures and Writings.'

than bringing up personal matters of the kind I have, at your invitation only, indulged in just for once!

About the Author

H. Floris Cohen is Emeritus Professor in Comparative History of Science at Utrecht University, and is affiliated with the Descartes Centre at that university. From mid-2014 to mid-2019 he served as the Society Editor of the History of Science Society. His opus magnum is *How Modern Science Came into the World: Four Civilizations, One 17th-Century Breakthrough* (Amsterdam: Amsterdam University Press, 2010).

4 Framing Beeckman

Cornelis de Waard as Editor of the Beeckman Papers

Klaas van Berkel

Abstract

By editing Isaac Beeckman's lost notebook as the *Journal*, Cornelis de Waard presented an interpretation of Beeckman that very much dominated further discussions of his contribution to the Scientific Revolution. De Waard modelled his edition of the notebook like Paul Tannery and he himself had done with their edition of the correspondence of Marin Mersenne. De Waard took the notebook to be a scientific diary, which documented the chronological development of his thought, and he acted as if the notebook was a collection of notes that should have resulted in a treatise on the mechanical philosophy (but did not). The result was the picture of Beeckman as a failed scientist, whereas renewed attention to the actual notebook may reveal other interests of Beeckman, his actual place in the networks of knowledge in the first half of the seventeenth century, and his self-image as a philosopher.

Keywords: Isaac Beeckman, Cornelis de Waard, Marin Mersenne, scientific diary, editorial decisions

No historian of science has contributed more to the study of Isaac Beeckman than Cornelis de Waard, the somewhat reclusive teacher of mathematics and physics from the city of Vlissingen in the Dutch province of Zeeland. Before 1905, the year in which De Waard rediscovered Beeckman's long-lost scientific notebook, the latter, a schoolmaster from Dordrecht, was just a shadowy figure in the margins of the study of Descartes, both in France and elsewhere. Through Adrien Baillet's biography of Descartes (1691), scholars knew that Beeckman was in some way involved in the maturing of Descartes' ideas, but little else was known. After 1905, however, Beeckman gradually

Van Berkel, Klaas, Albert Clement, and Arjan van Dixhoorn (eds.), *Knowledge and Culture in the Early Dutch Republic: Isaac Beeckman in Context.* Taylor & Francis Group 2022
DOI: 10.5117/9789463722537_CH04

became better known to historians of science and philosophy. Certainly after De Waard's masterful edition of the *Journal tenu par Isaac Beeckman de 1604 à 1634* in 1939-1953, it was possible to reconstruct Beeckman's own ideas about nature and its mechanisms, to study his dealings with Descartes and others, and to assess how he fits into what came to be called the Scientific Revolution of the seventeenth century.

De Waard's edition of the *Journal* was a major contribution to the history of science, but, as we know, every edited publication is in some ways a distortion, and the *Journal tenu par Isaac Beeckman de 1604 à 1634* is no exception. Every editor has to make choices about what to include and what not, how to arrange the material, and how to proceed with the annotation. Some of these choices are deliberately made, others are made on a subconscious level and are sometimes only detected by later generations. Now, more than a century since De Waard made his discovery and started transcribing the manuscript notes, the time is ripe for an assessment of his work as the editor of Beeckman's *Journal*. In what follows, I will first present an overview of De Waard's career as a historian of science. Special attention is devoted to the circumstances surrounding the discovery of Beeckman's notebook manuscript and the long and winding road that led to its publication in 1939-1953. Subsequently, I will compare De Waard's edition of Beeckman's notebook with his work on the *Correspondance du P. Marin Mersenne, religieux minime*, his other major editorial enterprise, the first volume of which was published in 1932. This will then lead to some conclusions about the way in which De Waard framed Beeckman as a 'failed scientist', that is, as a highly intelligent scientist who nevertheless was unable to do what every scientist – according to De Waard and his contemporaries – should do, that is, to publish his findings in a coherent way.

The Road to the *Journal*

In June 1905, as a student of mathematics and physics at the University of Amsterdam, Cornelis de Waard made a discovery that would fundamentally change his life and career. In the Provincial Library of Zeeland in his home town of Middelburg he stumbled upon a manuscript by a seventeenth-century natural philosopher from Zeeland that he recognized as the long-lost notebook of Isaac Beeckman, a correspondent of Descartes, Mersenne and Gassendi. The manuscript had been in the possession of the Provincial Library since October 1878, when it was acquired simply because its author had been born in Zeeland. On inspection of the notebook, De

Waard realized that the manuscript contained precious information not only on Beeckman, but also on major figures in the history of philosophy and science of the early modern period. The manuscript contained, for example, a number of letters to and from Descartes that were previously unknown. De Waard informed his professor at the University of Amsterdam, the mathematician D.J. Korteweg, who then communicated these findings to the surviving editor of the *Oeuvres de Descartes*, Charles Adam (the other editor, Paul Tannery, had died in the previous year). Beeckman now finally received the attention he so rightfully deserved.[1] However, we might ask, how did a mathematics student at the University of Amsterdam become interested in a seventeenth-century manuscript in Middelburg, and how did this lead to the publication of the full manuscript in 1939-1953?

Cornelis de Waard was born on 19 August 1879, the son of Cornelis de Waard Sr., who at the time was an officer in the Dutch Army at Bergen op Zoom (in the province of Brabant). Two years later, De Waard Sr. was relocated to the city of Middelburg, the capital of the neighbouring province of Zeeland. Soon afterwards, he left the army and became an assistant archivist at the Provincial Archives at Middelburg. In that capacity he completed several inventories that were deemed to be exemplary, which were thus recommended to fellow archivists. De Waard Sr. also published some of his findings in local journals and in small monographs.

Cornelis de Waard Jr. received his primary and secondary education in Middelburg and in 1898 matriculated at the University of Amsterdam to study mathematics and physics. In 1903, he passed his *kandidaatsexamen* (BSc) and continued his studies in mathematics under the guidance of Diederik Johannes Korteweg, one of the leading mathematicians in the Netherlands. Apart from his contributions to mathematics, Korteweg was also well known for his contributions to the history of mathematics and physics. He was a member of the committee that oversaw the publication of the *Oeuvres complètes* of Christiaan Huygens, which had been underway since the late 1880s. Korteweg's interest in the history of mathematics presumably inspired his student Cornelis de Waard, whose father was also a historian of sorts, to pursue historical studies while preparing for his

1 For more details see: Klaas van Berkel, 'La doctrine de la société zélandaise – De wetenschapshistoricus Cornelis de Waard jr. en het Zeeuwsch Genootschap der Wetenschappen', in: Arjan van Dixhoorn, Henk Nellen, and Francien Petiet, eds., *Een hoger streven. Bouwstenen voor een geschiedenis van het Zeeuws Genootschap, 1769-2019* (Middelburg: Koninklijk Zeeuwsch Genootschap der Wetenschappen, 2019), pp. 410-439.

concluding exams, the *doctoraalexamen* or MSc, which would qualify him to teach mathematics at secondary school.

The first topic De Waard Jr. explored was the history of the invention of the telescope. This was of special significance for someone from Zeeland as the invention of the telescope in 1608 was usually ascribed to a lens grinder from Middelburg, either Sacharias Jansen or Johannes Lipperhey (while Jacob Metius from Alkmaar was also sometimes mentioned). De Waard wanted to clear up the confusion once and for all and embarked on the study of the literature and the relevant archival sources. In June 1905, while consulting some papers at the Provincial Library in Middelburg, he was approached by the librarian, the historian Jacobus Broekema, who showed him a seventeenth-century leather-bound manuscript with brass locks that contained a passage that he thought might be of interest to De Waard.[2] This passage contained a remark from 1634, made by the son of one of the alleged inventors, Sacharias Jansen, that his father had been the first to construct a telescope in the Netherlands after a model of an Italian, with the inscription '1[5]90'. To De Waard this was a clear indication – at the time he had no reason to question the truth of the remark – that the invention of the telescope dated from before 1600 and was probably by an unknown Italian. This became the central thesis in the book that he published two years later: *De uitvinding der verrekijkers. Een bijdrage tot de beschavingsgeschiedenis* (The invention of the telescope: A contribution to the history of civilization) (The Hague 1906). In 1907, a synopsis of the book was published in the journal *Ciel et terre*, which ensured that others outside the Netherlands were also informed about De Waard's findings. De Waard had also corresponded with Antonio Favaro, the editor of Galileo's *Opere*, who spread the word in Italy about the history of the telescope. Thanks to Favaro, the discovery by the Amsterdam student of mathematics and physics was even recorded in Roman newspapers.

2 In the 'Note sur le manuscrit' in the first volume of the *Journal*, De Waard offers a slightly different, but less accurate account of the discovery of the notebook. There he states: 'Ce fut en juin 1905 que nous le trouvions dans cette bibliothèque [the Provincial Library in Middelburg] au cours des autres recherches'. *Journal tenu par Isaac Beeckman de 1604 à 1634, publié avec une introduction et des notes par C. de Waard*, 4 vols. (The Hague: Martinus Nijhoff, 1939-1953) [henceforth *JIB*], I, p. xxxii. Jacobus Broekema (1846-1928) published regularly on the history of Zeeland and must have been an acquaintance of the father of Cornelis de Waard. He was provincial librarian in Middelburg from 1878 and may very well have been responsible for buying Beeckman's manuscript in late 1878. For the acquisition of the manuscript see: *JIB*, I, xxxii. The manuscript is still preserved in the Zeeuwse Bibliotheek, Middelburg, ms. nr. 6471.

As suggested above, De Waard was in fact not the first to 'discover' Beeckman's notebook. It was Broekema, the proverbial 'invisible librarian', whose role is so often overlooked and disregarded, who must be credited with discovering the manuscript. De Waard was not even the first to whom Broekema pointed out its potential importance. H.A. (Henri) Naber, a teacher of physics from Hoorn (in the province of North Holland), who devoted all his spare time to the study of the life and works of Cornelis Drebbel, had visited Middelburg shortly before De Waard and was shown Beeckman's manuscript. Broekema had drawn Naber's attention to certain remarks by Beeckman about Drebbel, which then were published by Naber in an extensive article on Drebbel at the beginning of 1905.[3]

However, while Naber was only interested in the details about Drebbel, De Waard became interested in the manuscript as a whole and soon realized that it contained material that was even more important than the few remarks on the invention of the telescope. For example, the manuscript included several letters to and from Descartes dating from 1618 and 1619, which as far as De Waard knew, were not known to Descartes scholars. The manuscript also reported extensively on the meeting between Descartes and Beeckman in autumn 1618, as well as later meetings between 1628-1634. De Waard checked with his mentor in Amsterdam, Korteweg, just to be sure that he had not discovered something that was already known to the specialists, and when he received confirmation that the early letters to and from Descartes were indeed not yet known, he published two articles in the Dutch mathematics journal *Nieuw Archief voor Wiskunde*. The first dealt with Descartes and the law of refraction, the second with the correspondence of Descartes with Beeckman between 1618-1619.[4]

As the manuscript turned out to be a treasure chest for students of early modern science, the idea developed to transcribe the entire manuscript and publish it in an annotated edition. The Haarlem-based Hollandsche Maatschappij der Wetenschappen (Holland Society of Sciences), which was already engaged in the publication of the *Oeuvres complètes* of Christiaan Huygens, took the initiative to explore this idea. The secretary of the

3 H.A. Naber, 'Cornelis Jacobsz. Drebbel, 1572-1634', *Oud-Holland* 22 (1904 [published in 1905]), pp. 195-237. At he end of the article, Naber states that he finished it in 'Zierikzee Dec. 1904'. Since the Beeckman manuscript is only mentioned in the footnotes, it is likely that he only learned about the manuscript in autumn 1904, but that is still well before De Waard was introduced to the manuscript.

4 For the correspondence of De Waard with Korteweg, see the documents in the archive of Johannes Bosscha, the secretary of the Holland Society of Science (Hollandsche Maatschappij der Wetenschappen) in: Noord-Hollands Archief, Haarlem.

Hollandsche Maatschappij, Johannes Bosscha, editor-in-chief of the *Oeuvres complètes* of Huygens and known as an ardent advocate for the recognition of Simon Stevin (in his view, a mathematician and physicist more important than Galileo), was very much in favour of the idea, since the manuscript also appeared to contain quite a few references to the work of Stevin. In 1624, Beeckman had also visited Stevin's widow and had made an inventory of the manuscripts by Stevin that were in her possession.

Korteweg also strongly supported the idea of publishing Beeckman's manuscript, although he was more than a little concerned about De Waard's involvement. Korteweg knew that De Waard was not, to put it mildly, his most gifted student and feared that the effort De Waard would put into transcribing and studying Beeckman's manuscript might postpone the completion of his studies in mathematics. Nevertheless, in May 1906, the general assembly of the Hollandsche Maatschappij made the considerable sum of 500 guilders available for De Waard to transcribe the complete manuscript. At this meeting, it was also decided that a committee should be established to investigate the possibilities of publishing a complete edition.

De Waard immediately began to transcribe Beeckman's manuscript, the last part of which he handed over to the Hollandsche Maatschappij in 1909. However, the idea of publishing a complete edition of the manuscript was soon given up. Bosscha discovered that it did not contain new material about or by Stevin (there were no unknown manuscripts, just an inventory of unpublished material) and therefore withdrew his support. Furthermore, the financial state of the Hollandsche Maatschappij simply did not allow the society to take on any additional obligations, with the edition of Huygens's *Oeuvres complètes* having drained the finances of the society. In 1905, Bosscha had seen the tenth volume of the series sent to press and there were still quite a few to follow. In addition, potential sources of government support were also unlikely to provide the necessary funds, so the Hollandsche Maatschappij dropped the idea of a complete edition of Beeckman's manuscript. It was only through constant pressure from external parties, among whom was the well-known historian of science Eduard Jan Dijksterhuis, that in the 1930s the publishing firm Martinus Nijhoff in The Hague took upon itself to publish the edition De Waard had prepared.[5]

5 Little is known about the role of the publisher Martinus Nijhoff in the process of editing and publishing the *Journal*. The latest history of the publishing firm by Bob Jongschaap, *Martinus Nijhoff N.V. 1853-2002. Opkomst, bloei en ondergang van een boekenimperium* (Nijmegen: Vantilt, 2019), does not provide any information about the publication of the *Journal*. It seems that Nijhoff was financially compensated for a publication of which only 200 copies were printed and thus must have been very expensive. This can at least be inferred from a remark in a review of the final

Thus, in 1939, the first volume of the *Journal tenu par Isaac Beeckman de 1604 á 1634* was published, soon to be followed by three more volumes, in 1942, 1945 and 1953, respectively.

De Waard's Further Career as an Editor

As Korteweg had feared, the work on the transcription of the Beeckman manuscript caused serious delay to De Waard's studies. He only passed his MSc exam (*doctoraal*) in 1913, which by all standards is rather late, even if we take into account that he had taken on a temporary job as a teacher of mathematics in The Hague. After obtaining his MSc diploma, he moved to the city of Winschoten in the province of Groningen to become a full-time teacher of mathematics. Two years later he married Helena Christina Schwarz, a former nurse in Amsterdam, with whom he had two children, the twin brothers Maarten and Cornelis (b. 1917). In 1919, De Waard moved back to Zeeland, where he became an unobtrusive, even reclusive teacher of mathematics and physics at the Rijks-HBS (State Higher Burger School) in Vlissingen. He would remain in Vlissingen, not far from Middelburg, until his retirement in 1944 and his death in 1963. De Waard never obtained a PhD, which effectively precluded him from an academic position.

While a teacher in Winschoten, De Waard continued to spend many hours in local archives and libraries, and in 1914 he made another major discovery. In the University Library of Groningen he discovered unknown papers by the seventeenth-century mathematician Pierre de Fermat (1607-1665). He reported this to historians of mathematics with whom he had been in contact since the discovery of the early letters of Descartes, and this resulted in an additional fifth volume of the *Oeuvres de Fermat*, published in 1922.[6] The first four volumes had been published between 1891 and 1912 and had

volume by George Sarton, who said that the four volumes 'are a credit not only to the learned editor, but also to his sponsors' (*Isis* 46 (1955), pp. 71-72, esp. p. 72). In a review of the first three volumes, Sarton mentioned a letter written to him by De Waard, dated Vlissingen, 25 July 1947, stating that the manuscript of the fourth volume had already been sent to the publisher in 1945, immediately after the war. It is not known why it took Nijhoff so long to actually print and publish the volume (*Isis* 38 (1948), pp. 249-250, esp. p. 250). The archives of the Martinus Nijhoff firm at the Municipal Archives in The Hague do not contain information about this particular project of the firm.

6 *Oeuvres de Fermat. Supplément aux tomes I-IV. Documents inédits publiés avec notices sur les nouveau manuscrits*, ed. by C. de Waard (Paris: Gauthier-Villars, 1922).

mainly been the work of the historian of mathematics (and former tobacco manufacturer) Paul Tannery. While Tannery had died in 1904, his widow, Maria Alexandrine Prisset (Marie) Tannery (1856-1945), tried to continue the good work of her deceased husband by urging other scholars to carry out the many projects that he had started but had been unable to finish or execute himself.

Among these projects was the edition of the letters of the French friar, philosopher and mathematician Marin Mersenne (1588-1648), whose extensive correspondence with scholars across Europe had earned him the title of 'secretary of learned Europe'. Tannery had already collected much of the material for this edition and had published some of it before his death, but the main part of the work was still to be done. Madame Tannery had been so pleased with De Waard's work on Fermat that she paid him a visit in Vlissingen in 1922 – perhaps to personally present to him a copy of the supplement to the *Oeuvres de Fermat* – and offered him either the editing task of minor works of several sixteenth- and seventeenth-century French mathematicians, among whom was François Viète, or the correspondence of Mersenne. De Waard, 'le savant professeur hollandais', as Marie Tannery called him, chose the latter option, although he was well aware of the specific problems involved with gathering the huge volume of correspondence by Mersenne.[7]

For many years, the edition of Mersenne's correspondence took up almost all the time De Waard could spare from his teaching obligations. Most of the work was done in Vlissingen, but every year De Waard spent the long summer holidays in Paris, where he enjoyed the hospitality of Madame Tannery in her house at 16 Rue Bouchut (on the border of the 7th and the 15th arrondissements). After her death in 1945, De Waard fondly remembered his stays in Paris:

> where we worked from morning to evening in the various public libraries or even visited private collections. Generally, the evenings were reserved for the revision of our notes, but it was also a great joy for Madame Tannery to receive old friends of her husband.[8]

7 Cornelis de Waard, 'A la memoire de Mme Tannery', *Revue d'histoire des sciences* 2 (1948), pp. 90-94, esp. p. 92.

8 'Où nous travaillions du matin jusqu'au soir dans les diverses bibliothèques publiques ou visitions même des collections particulières. En général les soirées étaient réservées à la révision de nos papiers, mais c'était aussi une grande joie pour Mme Tannery de recevoir les anciens amis de son mari.' De Waard, 'A la mémoire de Mme Tannery', p. 93. Madame Tannery was very fond of De Waard and called him 'mon cher grand Ami'.

In 1932, this collective work, in which De Waard was assisted by René Pintard, resulted in the publication of the first volume of the *Correspondance du P. Marin Mersenne*. The appreciation of De Waard's labours had by then been expressed in his nomination as a 'membre correspondent' of the Académie Internationale d'Historie des Sciences in 1931.

During the later 1930s, De Waard worked on both the correspondence of Mersenne and the edition of Beeckman's *Journal*. The latter profited enormously from the former, as is clearly visible in the fourth volume of the *Journal* (1953), which includes Mersenne's correspondence with Beeckman. In 1939, De Waard went to Paris for the last time, but he kept in contact with Madame Tannery through correspondence. In 1943, she moved to the countryside because of her deteriorating health, where she died in January 1945, a couple of months after the liberation of France.[9]

Subsequently, De Waard continued his work on Mersenne's correspondence, as always with the help of other scholars, such as Robert Lenoble and Bernard Rochot. Shortly after De Waard's death in 1963, the eighth volume was published. In even later volumes, the annotations compiled by De Waard were still used extensively and his contribution was duly mentioned on the title page. In 1988, the final volume of the series was published, compiled and edited by Armand Beaulieu. The complete seventeen volumes of Mersenne's correspondence are generally seen as one of the monuments of twentieth-century scholarship in early modern intellectual history.[10]

The *Journal* and the *Correspondance* Compared

It is worthwhile to compare the way De Waard edited the Beeckman manuscript and the correspondence of Mersenne. Both editions have been praised for their accuracy and the efforts that De Waard took to identify the correspondents, the names mentioned in the text and places and events referred to. It appears that the volumes of the *Correspondance du P. Marin*

9 One of her last letters to De Waard dates from 8 March 1944, in which she is still hopeful to return to Paris and to resume her work on the correspondence of Mersenne: 'Moi-même, si je vais mieux, j'espère bien encore vous recevoir dans le petit appartement de la rue Bouchut quand vous pourrez vous-même y venir. Faut-il espérer que vous viendrez à Pâques [...]. Je me remets très doucement et j'espère malgré tout, sur le beau temps qui approche; je voudrais tant et tant vous revoir encore à Paris [...]. Encore à vous et de tout mon coeur.' De Waard, 'A la mémoire de Mme Tannery', p. 93.

10 A. Rupert Hall, 'Concluding a Correspondence' [Essay review of *Correspondance du P. Marin Mersenne, religieux minime*, vol. XV], *Isis* 76 (1985), pp. 75-80, is a first evaluation of the series.

Mersenne contain more typos or erroneous attributions, but in general the level of scholarship is equally high. More important than the differences are the remarkable resemblances between the two editions. It seems that De Waard took the *Correspondance* as his model and edited the *Journal* according to the rules established for the Mersenne edition. The fact that the first volume of the *Correspondance* was published in 1932, seven years before the first volume of the *Journal*, also makes it reasonable to assume that the rules set down for the edition of the *Correspondence* also guided the edition of Beeckman's *Journal*.

There are also major differences. One major characteristic of the *Correspondance* is the inclusion of extensive in-text notes by De Waard on certain scientific or technological issues referred to in the source. These *éclaircissements* had been a feature of the edition from the very start, not because De Waard chose to insert them, but because this had been the explicit wish of Paul Tannery, or at least of his widow.[11] Not all reviewers, however, thought these *éclaircissements* were particularly helpful. For example, in his review of Volume 15 of the *Correspondance du P. Marin Mersenne*, A. Rupert Hall, editor of the unpublished works of Isaac Newton, was quite outspoken:

> De Waard, in my view, overdid it – introducing long monographs on topics that took his fancy, which obstructed both his progress and the reader's. It is *not*, in my view, upon the appearance of the word *telescope* (say) in a seventeenth-century text, the editor's duty to introduce the definitive essay on the invention, introduction, nomenclature, use and development of this instrument.[12]

Later editors of the *Correspondance* showed more restraint and their *éclaircissements* were no more than long notes, although even Beaulieu inserted a four-page essay on 'Recherches sur les tirs du canon au XVIe and XVIIe siècle' ('Investigations on cannon fire in the sixteenth and seventeenth centuries').[13]

In the *Journal*, such *éclaircissements* are absent. Each volume contains a few appendices, but usually these are notes drawn from the notebook itself or from other contemporary sources. Volume 1 contains an appendix entitled,

11 De Waard, 'A la memoire de Mme Tannery', p. 92: 'Il fallut [...] faire des recherches pour les notes et éclaircissements que Mme Tannery, selon le désir de son mari, voulait y ajouter.'

12 Hall, 'Concluding a Correspondence', p. 79.

13 As noted by Hall, 'Concluding a Correspondence', p. 79.

'Études sur la chainette', a topic discussed by Beeckman and Descartes in 1618. In this appendix, De Waard simply puts together diverse remarks by Beeckman on the topic, without commenting on the content. The second appendix in the first volume includes some notes by Descartes on his meeting with Beeckman in 1618, drawn from the tenth volume of the *Oeuvres de Descartes* (1908). In the second volume of the *Journal*, we find 'Extraits des manuscrit de Stevin' and 'Notes du "Collegium Mechanicum"', while at the end of the third volume, De Waard included a letter from Robbert Robbertsz le Canu to Willem Jansz. Blaeu, referred to by Beeckman, and a letter from Drebbel to King James I of England, copied by Beeckman into his notebook. Hall would have had no objection to these appendices, since they mainly contain material from the notebook manuscript itself. In editing the *Journal*, De Waard did not have to take into account the wishes of Mme Tannery and her husband.

Of course, in itself, the fourth volume of the *Journal* is one great collection of *éclaircissements* to the edited notebooks. It presents a wide variety of texts, from letters and genealogical extracts to archival documents relating to Beeckman and his family. It also contains the text of Beeckman's dissertation (1618), a partial copy of which was found in the British Library, as well as his inaugural lecture as headmaster of the Latin School in Dordrecht (1627), extracted from Beeckman's notebooks. In general, De Waard put all the texts that he could not fit into the chronological format of the notebooks, or that were not found in the bound notebook, in this fourth volume. Nevertheless, De Waard did not include short essays commenting on the material found in Beeckman's notebooks or the archival material, as he did in the *Correspondance du P. Marin Mersenne*. On some topics, such as the concept of air pressure, he chose to write a separate monograph, *L'Expérience barométrique, ses antécédents et ses explications* (1936).[14] De Waard published this unpretentious, but richly documented booklet three years before he finally managed to have the first volume of the complete *Journal* printed. It was partly intended to establish Beeckman's priority in introducing the concept of air pressure. In order to make his case, De Waard included a facsimile of the first and the last pages of Beeckman's dissertation and numerous carefully annotated extracts from Beeckman's as yet unpublished notebook.[15]

14 C. de Waard, *L'Expérience barométrique. Ses antécédents et ses explications. Étude historique* (Thouars: Impremerie nouvelle, 1936). See the critical review by George Sarton in: *Isis* 26 (1936), pp. 212-215.

15 De Waard, *L'Expérience barométrique*, pp. 78-79, 145-168.

Chronology

Chronology is an issue for both editions, but paradoxically less so in the *Correspondance* than in the *Journal*. This relates to the fact that while the *Journal* is an edition of an existing text, the *Correspondance* is put together from different sources that were not even found in one place but were scattered across Europe. Beeckman started his notebooks on paper he intended to use as a collection of *Loci communes* (this is also what the title page of the manuscript actually says), and in 1628 had the loose cahiers bound together in leather as one huge book. This is how De Waard found the notes when the librarian in Middelburg pointed out the manuscript's potential value for someone like De Waard who was interested in the history of the invention of the telescope. The *Correspondance*, in contrast, had never existed as a physical entity before Tannery and De Waard started to collect the letters to and from Mersenne (whether the original letters or copies of them).[16] The *Correspondance* is not even a reconstruction of something that once existed, it is an assemblage of original sources that resulted in a physical unity (the seventeen volumes) that Mersenne never saw or imagined.

The fact that the *Correspondance* never existed as a physical entity made it easy to structure a letter-by-letter edition of the actual correspondence, with the chronological order as the main guideline. This chronology was therefore never an issue for Tannery and De Waard. They did not have to consider the integrity of the original source since there was none, and it was therefore not a problem to have a letter written by Descartes followed by a letter written by someone unknown to Descartes (or even to Mersenne). However, chronology was a much more complicated issue for De Waard in editing Beeckman's notebook, given, moreover, that Beeckman had not respected chronology very much. Although the manuscript was roughly chronological in its organization, Beeckman had also filled in blank spaces with annotations at a later date. He used these blank spaces, for example, to insert genealogical notes, relating the story of his family, of which he had become the head, the *pater familias*, in the late 1620s. De Waard, therefore, had to decide what to do with these insertions. Should he keep them in place or move them to an appendix? The general guidelines in matters such as these was to respect chronology as much as possible.

16 The complications involved in gathering all the letters to and from Mersenne are enumerated in: C. de Waard, 'A la recherche de la correspondance de Mersenne', *Revue d'histoire des sciences et ses applications* 2 (1948), pp. 13-28.

From the very start, De Waard considered the manuscript to be more or less a scientific diary; a compilation of notebooks in which Beeckman had recorded, in chronological order, what he had seen, thought and experienced. In the very first publication on Beeckman's notebook he calls it (in Dutch) 'het journaal van Isaac Beeckman'.[17] Also, in reviews of the *Journal*, the word 'diary' is often used in translation, for example, by George Sarton, who wrote about 'a kind of scientific diary, *Loci communes*, wherein he noted scientific reflections, the results of his experiments, occasionally biographical data etc.'[18] Beeckman himself, however, never used the word 'journal' or 'diary'. In his days, the word 'journael' was commonly used to designate a travelogue. Instead, he referred to the manuscript as 'meditata mea' ('my reflections').

It should also be noted that he was not very systematic in dating his reflections. Every now and then, he noted down a time and a place, but there are large portions of the manuscript that remain undated, especially in the first years of his writing. Between 12 April 1614 and the end of January 1615 (from fol. 17 verso to fol. 29 verso), no dates are given and the latter date was only established incidentally because Beeckman inserted some meteorological observations.[19] Between March 1615 and February 1616 (fol. 30 verso to fol. 39 verso), Beeckman once again did not date his reflections.[20] Thus, during some of his most creative years, he did not take the trouble of dating the entries, which makes it a little odd to call the manuscript a 'journal' or diary.

That De Waard still called his edition of Beeckman's notebook the *Journal* may perhaps be explained by the need he felt to establish Beeckman's priority in several scientific issues. De Waard projected the modern obsession with priority back onto Beeckman, who lived at a time when the desire to be the first was only beginning to become important. It was only in the seventeenth century that priority disputes became a common ingredient of the world of science and learning. By the second half of this century, priority disputes

17 C. de Waard, 'Descartes en de brekingswet', *Nieuw Archief voor Wiskunde*, tweede reeks, 7 (1905), pp. 64-68, esp. p. 64.

18 George Sarton, [Review of *Journal tenu par Isaac Beeckman de 1604 à 1634*, vols. I-III], *Isis* 38 (1948), pp. 249-250, esp. p. 250. See also: George Sarton, [Review of *Journal tenu par Isaac Beeckman de 1604 à 1634*, vol. IV], *Isis* 46 (1955), pp. 71-72. In his review of the first three volumes of the *Journal*, Bernard Rochot, editor of *Les Travaux de Gassendi sur Épicure et sur l'atomisme, 1619-1658* (1944), avoids calling Beeckman's manuscript a diary, but instead simply refers to 'ces notes', 'le manuscrit' or 'un gros registre'. Bernard Rochot, [Review of *Journal tenu par Isaac Beeckman de 1604 à 1634*, vols. I-III], *Revue philosophique de la France et de l'Étranger* 137 (1947), pp. 233-239, esp. p. 234.

19 *JIB*, I, pp. 34-59.

20 *JIB*, I, pp. 62-87.

had become quite common, such as the dispute over who had discovered the true nature of the female reproductive organs, Reinier de Graaf, Jan Swammerdam or Niels Stensen. Even more famous is the bitter controversy between Newton and Leibniz over the invention of calculus.[21]

In the first half of the century, Descartes, in particular, made a great fuss about priority issues. On hearing (from Mersenne) that Beeckman had claimed that he had been Descartes' master in certain respects, Descartes vehemently denied having learned anything of value from his Dutch friend. To him this was sufficient reason to write some very nasty letters to Beeckman and end his friendship with him.[22] It is comprehensible that De Waard, by finding Beeckman's report on his conversations with Descartes, attempted to rectify the belief current at the time that Descartes had devised the mechanical philosophy all by himself. He could now demonstrate that Beeckman had actually been the first to introduce Descartes to this new philosophy (and that Descartes had in fact acknowledged this in his 1619 letters to Beeckman). Much of the literature on Beeckman and Descartes since then turns on the question of whether or not Descartes misrepresented (intentionally or not) his early relationship with Beeckman. In order to establish priority, dates become very important, which explains why De Waard heavily emphasized the dating of entries in the notebook.

Beeckman himself, however, was not so much interested in establishing priority, at least not until the disruption to his friendship with Descartes. When he started scribbling down his reflections, he did not always see himself as a natural philosopher with a completely new notion of the material world. Beeckman had studied at Leiden University and was thus familiar with modern currents in the academic world. However, after moving back to Middelburg and Zierikzee, as far as we know he had no intention of contributing to scholarship. He was flattered, of course, on hearing Descartes saying, in 1618, that he had never met someone who could so fruitfully combine mathematical and physical reasoning. However, in the following years, Beeckman's real interests lay elsewhere, especially technology. It was only when he moved to Dordrecht and mingled in scholarly circles that he fully realized that his notebook contained insights that were very new and interesting and worth sharing with others. Even when he encountered a line of reasoning in the works of Johannes Kepler that was quite like his

21 A. Rupert Hall, *Philosophers at War: The Quarrel between Newton and Leibniz* (Cambridge: Cambridge University Press, 1980).

22 See: Klaas van Berkel, *Isaac Beeckman on Matter and Motion: Mechanical Philosophy in the Making* (Baltimore: Johns Hopkins University Press, 2013), pp. 60-65.

own, he expressed no regrets in not publishing his findings before Kepler, but rather joy that a great mind like Kepler had actually confirmed what he had been thinking himself. He toyed with the idea of publishing some of his findings as a supplement to Kepler's works, but then the brawl with Descartes began and Beeckman, disillusioned as he was, abandoned all thought of publishing his ideas.

Therefore, it is fair to say that De Waard, simply by calling his edition the *Journal tenu par Isaac Beeckman de 1604 à 1634*, framed the manuscript in a way that would have been foreign to Beeckman. He also relegated certain documents that Beeckman had pasted into his notebook to the appendices or the fourth volume, or completely omitted certain parts of the manuscript that he found uninteresting from a scientific point of view – mainly meteorological observations and genealogical entries. He also omitted complete documents that Beeckman had inserted, such as the copy he had made of the nautical observations of Jan Jacobs, who had travelled to the East Indies in 1606-1610 and had become an inhabitant of the city of Zierikzee at the time that Beeckman lived there; and Beeckman's copy of the *Compendium musicae*, the small treatise on music written by Descartes at the end of 1618, and presented to Beeckman as a token of their friendship.

Of course, we can fully understand why De Waard omitted these documents. The observations of Jan Jacobs do not include any information on Beeckman. The fact that Beeckman found them interesting enough to insert the document in his notebook was worth mentioning – and this was duly done in a footnote (as well as in the 'Avertissement au premier volume').[23] The same applies to the *Compendium musicae*, which had already been published in several places. However, by leaving out these documents and restricting the edition as much as possible to the chronological flow of Beeckman's reflections, De Waard framed the notebook as a scientific diary comparable to the laboratory journals of modern scientists more than is justified.

Conclusion

By framing the notebook as a scientific diary, De Waard also, unknowingly, framed Beeckman as a modern scientist more than is warranted – which he certainly was not. To frame him as such is not only historically misleading, it also leads to further misunderstandings. From the very beginning, De Waard's edition of Beeckman's notebook suggested that Beeckman in one

23 *JIB*, I, p. xxxvii; p. 59, n. 2.

way or another failed to do what every decent scientist should have done, that is to publish what he had discovered. In 1948, George Sarton already depicted Beeckman in this way:

> He [Beeckman] was a scientifically-minded person, equally familiar with the experimental and the mathematical methods, open-minded and truth-loving, but he lacked the ambition and the tenacity needed for writing treatises, and remained all his life an amateur, albeit a very distinguished one.[24]

Dijksterhuis was even more critical when discussing Beeckman's ideas in his famous *The Mechanization of the World Picture*:

> Beeckman showed the same defects in the matter of science as Leonardo da Vinci. Both were deficient in the tenacity of purpose and powers of concentration required to systematize, finish, record, and publish their enquiries, even if only in one field. Of Faraday's motto: 'Work, Finish, Publish', they only took to heart the first injunction.[25]

As a consequence, Dijksterhuis added, Leonardo did not advance science at all, while Beeckman did so to a much smaller extent than he might have done.

However, this is only one way of looking at Beeckman's work. If the main question is who discovered and published what first, then timing and dating are important, and people who discover things or new ideas but do not publish them become unimportant. As soon as the main question concerns the circumstances in which people in the past discovered things, what the road to discovery actually was and, most importantly of all, what it meant *to them*, then those who did not publish become as equally important as those who did. In the latter case, context becomes as important as results, and the self-understanding of the people involved becomes even more important than what we think of them. In this respect, it also becomes highly important to respect the integrity of the sources, which, as will be clear by now, is not exactly what De Waard did as editor of Beeckman's manuscript. Of course, De Waard painstakingly justified all his editorial decisions but, nevertheless, the result of his edition of Beeckman's notebook, from a twenty-first-century perspective, gives a misleading picture of who

24 Sarton, [Review of the *Journal*, vols. I-III], p. 249.

25 E.J. Dijksterhuis, *The Mechanization of the World Picture* (Oxford: Oxford University Press, 1961), p. 330.

the man was. In the future, we should therefore refrain from writing about Beeckman's *Journal* if we actually mean his notebook.

In addition, a digitalized copy of the manuscript in its current state would be very helpful to redress the slightly misleading impression one might obtain from consulting the *Journal* as edited by De Waard. Perhaps we should refer to the manuscript as 'the Beeckman Papers', in analogy to the 'Hartlib Papers'. Then we would no longer frame what is essentially a miscellaneous collection of notes, a hybrid dossier of sometimes unrelated documents and observations, as a scientific diary that one day might have provided material for a scientific paper or a book. From this perspective, we will no longer see Beeckman as a failed scientist but as a highly intelligent philosopher whose thinking reveals certain traits of the early-seventeenth-century culture of knowledge that will remain unacknowledged as long as we concentrate on the published works of that period.

About the Author

KLAAS VAN BERKEL is Emeritus Rudolph Agricola Professor of History at the University of Groningen. He has published widely on cultural history and history of science and has a special interest in the history of early modern science and the history of scientific institutions. He is the author of *Isaac Beeckman on Matter and Motion: Mechanical Philosophy in the Making* (Baltimore: Johns Hopkins University Press, 2013).

Part II

Understanding Beeckman

5 'Like Water, That Is Forced to Flow through a Narrow Opening'

Isaac Beeckman's Early Conceptualization of the Telescope

Tiemen Cocquyt

Abstract

This chapter sets out how Isaac Beeckman developed an understanding of the telescope and its observation-enhancing power. In order to make sense of the instrument, Beeckman borrowed from a heterogeneous palette of optical principles. By merging these with his developing mechanistic ideas, Beeckman arrived at a coherent and adequate understanding of the telescope. Increased experience with practical optics, and an exposure to the writings of Kepler, subsequently brought about an evolution in the way Beeckman thought about optics. Yet, these conceptual innovations never jumped over completely to the domain of telescopes. Real and virtual lens imaging only found a common ground within the context of camera obscura usage. The latter also served as a means for lens quality innovation in the 1630s.

Keywords: Isaac Beeckman, telescope, geometrical optics, practical optics, camera obscura

The telescope was crucial in bringing about a methodological shift in seventeenth-century natural philosophy. It legitimized the use of observational aids in the study of natural phenomena, and thus furthered the very role of observation in seventeenth-century inquiry.[1] Nonetheless,

1 Albert Van Helden, 'The Telescope in the Seventeenth Century', *Isis* 65 (1974), pp. 38-58; Albert Van Helden, 'The Birth of the Modern Scientific Instrument, 1550-1700', in: John Burke, ed., *The Uses of Science in the Age of Newton* (Berkeley: University of California Press, 1983), pp. 49-84, esp. pp. 50-53.

Van Berkel, Klaas, Albert Clement, and Arjan van Dixhoorn (eds.), *Knowledge and Culture in the Early Dutch Republic: Isaac Beeckman in Context*. Taylor & Francis Group 2022
DOI: 10.5117/9789463722537_CH05

many questions about its emergence, and its interaction with the evolving science of optics, remain unanswered.[2] Isaac Beeckman's notes carry all the potential to shed light on a uniquely early and local reception of the telescope. In this chapter I will analyse how Beeckman gradually developed an understanding of the working principle of the telescope.

My analysis partly picks up on the recent historiography of the telescope. I will argue how Beeckman's familiarity with late-sixteenth-century optical innovations is larger than his modern editor, Cornelis de Waard, assumed.[3] Still, if these innovations in optics were characterized by an increased attention for the mechanism of refraction of light in lenses, Beeckman's understanding of the telescope was only partially embedded in these. From his first notes onwards, Beeckman's reasoning based on refraction was complemented with a working principle resulting from a physical interpretation of the agents of vision, *species*, borrowed from other branches within the multifaceted sixteenth-century optical tradition. In Beeckman's early notes, these distinct contributions blended into a coherent whole. Interestingly, it was chiefly his physical interpretation of species that Beeckman subsequently merged in his maturing mechanistic ideas, arriving at a conception of the telescope that was original, and that was surprisingly adequate, too. It allowed Beeckman to understand the telescope as a device that brings remote things nearby, rather than an instrument that produces larger images. Remarkably, this conception inevitably led him to a preference for large lenses. As a side effect, we are offered a fresh view on an intervention that has recently been identified as an essential step in making the first telescopes viable: the diaphragm. It turns out that Beeckman had good reasons to deploy a diaphragm, yet covering grinding errors was not his motivation, nor did this cross his mind for many years.

Subsequently, I set out how Beeckman's conception of the telescope in the 1620s gradually grew to include more dioptrical innovations in optics. The seventeenth-century science of dioptrics established the study of visual perception as a process of refraction in lenses, and arrived at novel concepts such as geometrical images having spatial extension. Beeckman's deployment of such notions obviously bears relation to a growing familiarity

2 Albert Van Helden, *The Invention of the Telescope*, Transactions of the American Philosophical Society 67, nr. 4 (Philadelphia: American Philosophical Society, 1977; repr. in 2008 with a new foreword), pp. 1-67, remains authoritative for the main historiographical issues. An updated overview, answering some questions but also raising new ones, can be found in: Albert Van Helden et al., eds., *The Origins of the Telescope* (Amsterdam: KNAW Press, 2010).

3 De Waard looked in vain for references to Johannes Kepler, but recent studies have shown that we have to look for other elements in early-seventeenth-century optical literature.

with the foundational works of dioptrics, those by Johannes Kepler. Yet, Beeckman's notes are interesting because in a unique way they testify to a piecemeal appropriation of Kepler's innovations, hand in hand with other concurrent factors such as the phenomenology of lenses, and a growing awareness of material aspects of optics. Thus, I argue how an earlier focus by Beeckman on large lens diameter keeps lingering on in his thoughts, at a time when notions such as 'intensity' and 'size' semantically start to crystallize in optical literature. Importantly, I set out how it is the *status aparte* of virtual images, as recently suggested by Antoni Malet, that makes such a two-track approach by Beeckman understandable.[4] Finally, I discuss the role of lens phenomenology in the shaping of Beeckman's optical thoughts. Beeckman's witnessing of, and experimentation with, projection through lenses allowed him to bridge the domains of virtual and real lens imaging. Eventually this led to a 'technological' grip on lenses. That is: projection made Beeckman aware of material imperfections in lenses, and it gave him access to procedures through which these imperfections could be mitigated, thereby effectively improving the quality of lenses. The latter is important as advances in observational astronomy have earlier been attributed primarily to innovations in lens grinding, rather than to theoretical advances in optics.[5] Yet, a mechanism for this technological process has not yet been described, and it is in Beeckman's notes that we find such an innovation practice clearly documented.

The Science of Optics and the Telescope at the Onset of the Seventeenth Century

Before embarking on an analysis of Beeckman's notes, it is worthwhile to sketch the status of the field of optics at the onset of the seventeenth century.[6] When Beeckman wrote down his first notes on the telescope, the science of optics was going through a tremendous revision. By this time,

4 Antoni Malet, 'Early Conceptualizations of the Telescope as an Optical Instrument', *Early Science and Medicine* 10 (2005), pp. 237-262.

5 Van Helden, 'The Telescope in the Seventeenth Century', p. 45; Rolf Willach, 'The Development of Telescope Optics in the Middle of the Seventeenth Century', *Annals of Science* 58 (2001), pp. 381-398.

6 This sketch follows the lines set out in two major reference works on the history of optics: David C. Lindberg, *Theories of Vision from Al-Kindi to Kepler* (Chicago: University of Chicago Press, 1976); A. Mark Smith, *From Sight to Light: The Passage from Ancient to Modern Optics* (Chicago: University of Chicago Press, 2015).

optics had evolved into a rich synthesis of many past traditions. The field was rich, vibrant, and productive, but – exactly because of its multifaceted character – it faced underlying contrasts.[7] Anomalies manifested themselves most clearly over the approach optics was to follow: a physical versus a purely mathematical one, but such discussion was intimately intertwined with issues over intro- versus extramission, and over the range of phenomena the science of optics should cover. In the long run, the only rigid solution to these tensions proved to be a radical revision of the science of optics' fundamental assumptions.

By and large, sixteenth-century optics was the result of a perpetual reinterpretation of two major traditions. On the one hand there were the classical sources on optics.[8] The prime aim of classical optics was to explain vision. Consequently, authors such as Euclid and Ptolemy posited an extramissionist model that takes vision to originate with the observer, and to propagate as rectilinear rays and visual cones towards the perceived object. While this tradition was originally characterized by its geometrical approach, in developing, these models gradually gained additional layers of physical interpretation. For instance, Ptolemy took Euclid's rectilinear rays of vision to constitute a visual 'flux', thus adding subtle physical meaning to an otherwise purely geometrical model. Not unimportantly, earlier Greek authors had taken an entirely different stance and had posited an atomist theory, which considered an efflux of corpuscles emanating from the object, and was therefore intromissionist. These atomist theories, however, remained outside the slipstream of geometrical optics, and only saw a revived interest much later in history.

A second major tradition developed in Islamic culture.[9] Starting from the ninth century, the reworking of classical sources soon evolved into a highly creative intellectual dynamic which saw alternative, intromissionist models being spearheaded. These models take vision to originate from a luminous object, and proceed towards the eye. While these models explicitly remained models of vision, the increased attention for light rays spurred the development of advanced analytical tools, such as point-ray analysis, and punctiform models of perception (an approach which analyses the

7 In addition to the works mentioned above, some in-depth discussions of specific issues at stake in the field of optics at the turn of the seventeenth century are given in: Isabelle Pantin, '*Simulachrum, species, forma, imago*: What Was Transported by Light into the Camera Obscura?', *Early Science and Medicine* 13 (2008), pp. 245-269; Sven Dupré, 'Optics without Hypotheses: Kepler's Optics between Natural Philosophy and Mathematics', *Synthese* 185 (2012), pp. 501-525.

8 Lindberg, *Theories of Vision*, pp. 1-17; Smith, *From Sight to Light*, pp. 23-129.

9 Lindberg, *Theories of Vision*, pp. 18-86; Smith, *From Sight to Light*, pp. 155-227.

perception of an object as the sum of the perceptions of its constituent points). While many of these innovations originated in distinct, not always mutually compatible contexts, a 'great synthesis' of this Islamic branch of optics was brought by the tenth-century polymath Alhazen. His works propagated a coherent model of vision based on intromission and point-ray analysis.

The legacy of Alhazen was of great influence to the Western, late medieval revival of optics, but not before it was harmonized with reigning medieval-Aristotelian philosophy.[10] This task was taken up by the thirteenth-century 'perspectivist'[11] school of optics, which included scholars like Robert Grosseteste, Witelo, and, most notably, Roger Bacon. Exemplifying for the perspectivists' integration of traditions, is Bacon's model of 'multiplication of species'.[12] Bacon took the intromissionist orientation of Alhazen, but – rather than vigorously assimilating visual information with geometrical lines – posited this visual information to be perceived by the observer as non-corporeal 'species', a concept Bacon borrowed from Aristotle. These species are entities that embody the perceived qualities of an object and, by successively multiplying a likeness of themselves, spread through the transparent medium that separates object and observer. With perspectivist authors, we see the advanced ray-tracing tools of Islamic science maintaining a strong role, but being juxtaposed to a physical interpretation that essentially belonged to a different visual model. A fundamental tension this created was that the perspectivist authors put the geometrical ray representation of vision from the Arabic school in the spotlight, while denying these geometrical rays any physical existence. The intromissionist/species model initiated by Bacon held sway as the dominant optical model of vision up to the early modern period.

It is worth stressing that for the perspectivists, the programme of redistributing Alhazen's optics in the West was driven by an incentive to create a harmonious compromise of the many past contributions to the field of optics.

10 Lindberg, *Theories of Vision*, pp. 87-121; Smith, *From Sight to Light*, pp. 228-321.

11 The term 'perspectivist optics' derives from the three almost identical titles of the thirteenth-century foundational works in this tradition: *Perspectiva* by Roger Bacon (1260s), *Perspectiva* by Witelo (c. 1275), and *Perspectiva communis* by John Pecham (c. 1278). Optical works in the perspectivist tradition thrived together with sixteenth-century printing culture, until the seventeenth-century science of dioptrics came to challenge some of the perspectivist literature's core axioms. For a discussion, see: Lindberg, *Theories of Vision*, pp. 104-120.

12 Lindberg discusses Bacon's 'multiplication of species' in detail in: David C. Lindberg, 'Roger Bacon on Light, Vision, and the Universal Emanation of Force', in: Jeremiah Hackett, ed., *Roger Bacon and the Sciences: Commemorative Essays* (Leiden: Brill, 1997), pp. 243-275. This article also discusses the usage of 'species' more extensively.

They felt they could not pass over any of their predecessors' authorities. Many obstacles were overcome by realigning the hierarchy of arguments. On the other hand, anomalies and ambiguities in definition remained hidden under the surface of perspectivist optics. For this reason, perspectivist optics has notably been described as residing in relative harmony, but being, in the end, only 'outwardly compatible'.[13]

Moreover, during the Renaissance, while maintaining its overall stronghold, the perspectivist's model increasingly came under tension through external developments which realigned the scope of optics, and made its playing field even more complex. The rising to prominence of linear perspective[14] and other branches of 'practical mathematics', because of their strong focuses on a geometrical modelling of sight, fostered renewed interest in the classical extramissionist models. In addition, hybrid or 'mixed' models were being postulated.[15] Although these geometry-oriented models belonging to linear perspective were not always inherently concerned with physical interpretation, to some extent they adopted the terminology of Euclid's geometry of vision. Yet another professional group that held a distinct view on perception were the physicians. Their accounts leaned more towards holistic and, increasingly, anatomy-informed expositions on vision. The authority of Galen's (intromissionist) account held sway for long, although alternative views were postulated – some of them extramissionist which, for instance, linked vision with bodily fluids or *spirites*.[16]

It is important to realize that the study of visual perception that fell within the boundaries of the canonical science of 'optics', the mathematical descriptions that made up the study of (linear) perspective, and the physicians' accounts, largely remained confined to distinct domains and distinct professions, and therefore co-existed in relative harmony. Underlying tensions were only amplified when domains were being crossed. What was available at the turn of the seventeenth century, to sum up, was a science of optics that had evolved into a rich, accessible, and elegantly but arduously

13 Lindberg, *Theories of Vision*, p. 120; also see: Pantin, '*Simulachrum, species, forma, imago*', p. 247.

14 Not to be confused with the term 'perspectivist optics' mentioned before, which denotes the thirteenth-century appropriation of Alhazen's optics into Bacon's 'multiplication of species' model.

15 On the importance of these mixed models and their implications for the status of optics: Dupré, 'Optics without Hypotheses'.

16 Lindberg, *Theories of Vision*, pp. 168-177. Some sixteenth-century extramissionist accounts are set out in: Katrien Vanagt, 'Suspicious Spectacles: Medical Perspectives on Eyeglasses: The Case of Hieronymus Mercurialis', in: Van Helden et al., eds., *The Origins of the Telescope*, pp. 115-127.

harmonized synthesis of contributions, but retaining underlying discrepancies nonetheless. These discrepancies did not only concern the direction of vision, but also the scope of optics, and the domains it was considered applicable to. Optics was divided between several fields, each of them with different aims and definitions. Underlying, too, were contrasting visions and latent uncertainties about the nature of light.

Likely, for the typical student of optics at the turn of the seventeenth century, these tensions were not immediately felt. Only when one attempted to approach the science of optics as a natural philosophical system did the anomalies turn into real problems.[17] This is what happened in the early seventeenth century when Johannes Kepler entered the scene. His revision of optics, propounded in his works *Paralipomena* (1604) and *Dioptrice* (1611), was a fundamental departure from perspectivist optics on many levels.[18] Perspectivists had gone to great lengths to numb any incompatibility between the physics of species and the geometry of light, by assigning them different rules. Kepler, instead, showed that they were actually the same thing.[19] He abandoned the perspectivists' preferential treatment of primary rays of vision, and, through his concept of a 'retinal image', carried through the treatment of vision as a science of *light* rays all the way up to the observer's retina. Kepler's 'dioptrical' approach centred upon a mathematical treatment of image formation through the mechanism of refraction in lenses. Any psychological aspect of vision was explicitly delegated away from Kepler's model of the eye as an optical instrument. While Kepler did acknowledge a role for cognition in the overall process of perception, the mechanisms taking place behind the observer's retina could not be made compatible with his assimilation of light and geometrical rays, and were therefore excluded from 'optics'. In the hands of Kepler, optics was redefined as a science of light.

While Kepler's optics was eagerly taken up by contemporary scholars, its revision was so fundamental in character that not everyone was confident to

17 Dupré, 'Optics without Hypotheses'.

18 Johannes Kepler, *Ad Vitellionem Paralipomena quibus Astronomiae Pars Optica Traditur* (Frankfurt: Claudius Marnius, 1604); English translation in: William H. Donahue, *Optics: Paralipomena to Witelo & Optical Part of Astronomy* (Santa Fe: Green Lion Press, 2000); Johannes Kepler, *Dioptrice, seu Demonstratio eorum quæ visui & visibilibus propter conspicilla non ita pridem inventa accidunt* (Frankfurt: David Frank, 1611).

19 This is the main argument from: Smith, *From Sight to Light*. Note that Lindberg and Smith slightly differ in their appreciation of Kepler's accomplishments. Lindberg presents Kepler as being much indebted to the perspectivist tradition. Smith presents Kepler's optical work as a far more radical switch.

understand, or to accept, its full implications.[20] One of these hesitations dealt with the nature of light, and the need to adhere to the perspectivists' concept of 'species'. Species *had* acquired a particularly ambiguous definitional status in late-sixteenth-century optics. Depending on the context, they could signify a quality, aspect, form, image, likeness, but also a force or a power. Kepler wittingly played with this ambiguity in his works, referencing the term when commenting on earlier authors, but abandoned species altogether when advancing his own theory.[21] They had no meaning in his understanding of vision. To some of his contemporaries, this seems to have been a bridge too far. Christoph Scheiner, for instance, as pointed out by Isabelle Pantin, in his appreciation of the camera obscura and the images it could cast, could not possibly conceive how an 'objective' representation could be produced by the instrument without the intervention of species. Irrespective of his enthusiasm for Kepler's theory of light, Scheiner considered species a prerequisite for the transfer of visual information.[22]

The latter brings us to the role of optical instruments in the upheavals in optics at the turn of the seventeenth century. Increasingly, starting from the sixteenth century, the emergence of optical experiments and devices, and the curiosity these aroused, challenged specific aspects of optical theory. One instrument that burdened the science of optics with new demands at the turn of the seventeenth century was the telescope. About the emergence of the telescope much background has been uncovered in the past decades, although its exact 'invention' will likely remain shrouded in mystery forever.[23] For all we know, the telescope's invention was a chance discovery. Although the telescope materialized at a time when optical science witnessed one of its most fundamental revisions, it is not known whether the science of optics had anything to do with the instrument's emergence at all.

What we do know is that the telescope, in turn, greatly stimulated the development of the science of dioptrics in the seventeenth century. An early successful application of Kepler's optics was his treatment of the telescope, presented in *Dioptrice*, a mere three years after the instrument's invention.[24] The dioptrical approach, through the mechanism of refraction, firmly associated lenses with their focal parameters, and eventually substituted

20 See in particular: Pantin, '*Simulachrum, species, forma, imago*'.
21 Pantin, '*Simulachrum, species, forma, imago*', pp. 255-256.
22 Pantin, '*Simulachrum, species, forma, imago*', pp. 263-267.
23 Van Helden, *The Invention of the Telescope*; Van Helden et al., eds., *The Origins of the Telescope*.
24 Antoni Malet, 'Kepler and the Telescope', *Annals of Science* 60 (2003), pp. 107-136.

some earlier, more preliminary accounts of the telescope's operation that explained the instrument in terms of changing visual angles.[25] The dioptrical mode of explanation, through a mathematical description of the mechanism of refraction, was to stay dominant throughout the seventeenth century. It is therefore correct to state that, while the telescope's sudden emergence makes the instrument's initial embedding within the study of optics difficult to assess, there is ample proof that it soon aligned successfully with the science of dioptrics. This success, in turn, has had its repercussions for the historiography of the telescope.

The key issue in the historiography of the early telescope is understanding how the telescope was conceived as a magnifying device, and how it was subsequently improved in that capacity. Often, it is tacitly assumed that understanding the working principle of the telescope (in a sufficient manner to improve upon it) required an understanding of the instrument's magnifying power as defined by the focal ratio of the convex and concave lens. Hence, it is reasoned, understanding the telescope required access to the focal properties of both constituent lenses. Knowledge of these focal properties, in turn, required adherence to a mathematical model of refraction. This focus on refraction ultimately leads us to the innovations of Kepler. Such expectations explain, for instance, why De Waard, in discussing the reading list that the Leiden professor of mathematics Rudolph Snellius gave to Beeckman in 1609, showed himself surprised that 'novel' works such as Kepler's *Paralipomena* were not included.[26] De Waard considered Kepler's optical theory to have immediately become inevitable once it appeared.

Recent literature, instead, has come to question both the impact of Kepler's writings on the early conceptualization of the telescope, and the requirement that both constituent lenses, the convex and the concave one, needed to be

25 On the development of focal parameters for lenses in *Dioptrice*, see: Antoni Malet, 'Kepler's Legacy: Telescopes and Geometrical Optics, 1611-1669', in: Van Helden et al., eds., *The Origins of the Telescope*, pp. 281-300, esp. p. 285. On the initial understanding of magnification in terms of visual angles, before a fully geometrical account based on image size (eventually) was adopted: Malet, 'Kepler and the Telescope', pp. 112-116. On Kepler's indebtedness to Galileo's account of angular magnification, see: Sven Dupré, *Galileo, the Telescope, and the Science of Optics: A Case Study of Instrumental Practice in Art and Science* (PhD diss., Ghent University, 2002), pp. 281-289; Sven Dupré, 'Ausonio's Mirrors and Galileo's Lenses: The Telescope and Sixteenth-Century Practical Optical Knowledge', *Galilaeana* 2 (2005), pp. 145-180, esp. pp. 174-177. The long-term development of a dioptrical theory of lenses in the seventeenth century is set out in detail in: Fokko Jan Dijksterhuis, *Lenses and Waves: Christiaan Huygens and the Mathematical Science of Optics in the Seventeenth Century* (Dordrecht: Kluwer Academic Publishers, 2004).

26 *Journal tenu par Isaac Beeckman de 1604 à 1634*, publié avec une introduction et des notes par C. de Waard, 4 vols. (The Hague: Martinus Nijhoff, 1939-1953) [henceforth *JIB*], IV, 18, n. 10.

associated with magnification. In particular, Malet has shown how a 'true' geometrical treatment of the telescope – i.e., founded on spatially extended images, fully incorporating the innovations of Kepler – only emerged in print in the 1660s. Instead, earlier discussions did lean heavily on the process of refraction in lenses, but kept explaining the telescope's magnification in terms of visual angles.[27] Moreover, Malet argues that the *virtual* character of the early telescope's images gave rise to a two-track approach, in which the telescope was not awarded the same treatment as other optical phenomena were.[28]

Sven Dupré has shown how, to an early telescope practitioner, the instrument's magnifying effect would have been thought to reside in the convex lens, and not in the combination of lenses.[29] In contrast to the convex lens, the concave lens was assigned a merely sharpening role. From these observations, Dupré traces a sixteenth-century prehistory of 'telescopic experiments' that emerged from the mix of perspectivist optics and practical mathematics. Handles to the magnifying properties of convex lenses, indispensable for the telescope to emerge, he identifies in the late-sixteenth-century assimilation of the perceptive effects of lenses with their effects on ray propagation.[30] This move opened up quantitative control of the magnifying properties of lenses and, eventually, shaped the experimental conditions in which the telescope could emerge.

Within the framework of the present chapter, Dupré's and Malet's accounts are in a sense complementary. If Dupré shows that key elements required for understanding the telescope's operation had a long sixteenth-century preamble, Malet points out that the one conceptualization of the instrument that eventually proved to be the most successful – the dioptrical one – only

27 Malet, 'Kepler's Legacy'; Also see: Malet, 'Kepler and the Telescope', pp. 125-127; Malet, 'Early Conceptualizations of the Telescope as an Optical Instrument', pp. 250-259.
28 Malet, 'Early Conceptualizations of the Telescope as an Optical Instrument'. Designating the telescope's images as 'virtual' acknowledges that when looking through a telescope, no 'real' images are formed that exist independently from the observer, or that can be cast on a screen, in contrast to e.g. the camera obscura.
29 Dupré, 'Ausonio's Mirrors', p. 174.
30 Dupré points out that up to the sixteenth century the description of (solar) ray propagation, in lenses and burning mirrors, remained distinct from, and subordinate to, the study of vision. An innovative feature of sixteenth-century optics was the identification of the *punctum inversionis* (the location in front of a concave mirror where the perceived image 'blows up', before it becomes inverted) with the *punctum concursus* (the location where solar rays unite). The 'jumping over' from mirrors to lenses of this identification first appeared in manuscript form in c. 1560, and in print in Giambattista della Porta's *De refractione* (1593). See: Dupré, 'Ausonio's Mirrors', pp. 160-170.

saw a delayed and piecemeal adoption. Together, they suggest that the evolution of the telescope's conceptualization was a more complex, but also more continuous and sustained process, that basically lasted from the mid-sixteenth to the mid-seventeenth century. It is in the analysis of this process that Beeckman's notes on the telescope offer unique opportunities. They invite us to consider how his understanding of the working principle of the telescope may have been embedded in a late-sixteenth-century conceptual framework. Moreover, they may further clarify our understanding of the broader scholarly attitude towards the telescope in its early years.

Beeckman's First Note on the Telescope (1612)

At first sight, Beeckman seems to be the perfect candidate to inform us about the telescope's early history. Born in Middelburg, the cradle of the telescope, Beeckman was 20 years old when the instrument made its first appearance. Moreover, he was a student with an active interest in mathematics, as is illustrated by the lessons he took from Rudolph Snellius. If we then look at Beeckman's first entry in his notebook about the telescope, dating from 1612, one ends up somehow disappointed. It is striking that Beeckman refers to the instrument by its French name 'lunette'.[31] It seems to indicate that the local events surrounding the emergence of the instrument went largely unnoticed to Beeckman, and he only got to know of it during his studies at Saumur in the summer of 1612.[32] Furthermore, the list of book recommendations by Rudolph Snellius, jotted down by Beeckman during his studies in Leiden, also suggests that optical innovations were hardly taken up by the scholarly community. For the fields of optics and catoptrics, Snellius recommends his student the works of Euclid, Ptolemy and Witelo – all spokesmen of a long past optical tradition.[33] Of course, one may ask: why would we expect Snellius to show an interest in optical innovations? For sure, 1609 was a very early point in time – both in terms of the shaping and consolidation

31 *JIB*, I, p. 12: 'Instrumentum quo è longinquo res parvae videntur et gallicè *lunette* vocatur.'
32 Klaas van Berkel, *Isaac Beeckman on Matter and Motion: Mechanical Philosophy in the Making* (Baltimore: Johns Hopkins University Press, 2013), p. 16. Note that Van Berkel does advance the possibility that Beeckman learned about the telescope in Middelburg. I feel that the indications that Beeckman *did not* know of the instrument outweigh those that suggest he did. In 1608, when the telescope made its first appearance, Beeckman was not living in Middelburg anymore, but studied at Leiden University.
33 *JIB*, IV, pp. 17-19. On the popularity of these works in the second half of the sixteenth century, see: Smith, *From Sight to Light*, pp. 328-329.

of Kepler's optical theory, and in the context of the telescope's diffusion. On the other hand, the interests revealed by Snellius's course material stand in sharp contrast with how the telescope was demonstrated almost immediately in 1609 in one of his optical courses.[34] The question we need to ask, therefore, is not so much why Snellius didn't embrace the optical innovations taking shape at that moment. Rather, we need to ask how we can reconcile his documented enthusiasm for the telescope with an adherence to sources in which lenses do not even appear. Was Snellius's demonstration of the telescope embedded in a perspectivist teaching context? If so, how exactly? And was Beeckman receptive for this context provided by Snellius?

Beeckman's 1612 note teaches us that for a scholar initiated in optics through the works of Euclid, Ptolemy and Witelo, his description of the telescope is surprisingly modern.[35] What dominates in Beeckman's account is an exposition in terms of refraction, in line with the direction optics was taking at that time. On the other hand, there are references to visual angles, which, it must be noted, remain largely cursory. Thus, if his 1612 note confirms an awareness of recent developments, his description also testifies to a heterogeneous borrowing from the science of optics. Beeckman sets off by stating how he has observed that the telescope operates by virtue of a burning glass, that enlarges the visual angles, and a second glass that reduces these angles.[36] Subsequently, he discusses the contributions of both lenses, one after the other. The burning (convex) glass he attributes the focusing of species through the mechanism of refraction. He states that, if the eye was placed at the combination point where the species unite, only 'the angle of one single point', and therefore nothing, would be seen.[37] This suggests an

34 Obviously, for Beeckman's understanding of the telescope it would be particularly relevant to know whether he attended this lesson or not, but it seems this cannot be ascertained. We only know of Snellius's demonstration from a later reference by Théodore Deschamps, see: *JIB*, I, p. 12, n. 3. Beeckman never makes a reference to such a demonstration, nor to Snellius in relation with the telescope. Most likely he missed this event because of a stay in Middelburg: he resumed his Leiden studies on 29 September 1609, while Deschamps enrolled as early as July. See: *Album Studiosorum Academiae Lugduno Batavae, 1575-1875* (The Hague: Martinus Nijhoff, 1875), pp. 92, 96; Van Berkel, *Isaac Beeckman on Matter and Motion*, pp. 15-16. For the attribution of this telescope to the Delft lens grinder Evert Harmansz. Steenwijck, see: Huib J. Zuidervaart and Marlise Rijks, '"Most Rare Workmen": Optical Practitioners in Early Seventeenth-Century Delft', *British Journal for the History of Science* 48 (2014), pp. 53-85, esp. pp. 58-59.

35 This note is in *JIB*, I, pp. 12-14: 'Telescopij ratio.'

36 *JIB*, I, p. 12: 'Instrumentum [...] reperi compositum esse ex vitro comburente, hoc est angulos visuales majorante, et ex vitro angulos visuales minorante.'

37 *JIB*, I, p. 13: 'Incidentibus enim speciebus visibilibus in vitrum majorans, refringuntur introrsum et tandem concurrunt. In concursu autem si oculus locatus sit, tantùm angulum unius puncti recipit, ut nihil nisi hoc punctum videat, ergo nihil videt.'

acquaintance with punctiform ray analysis (the approach where each point of an object is analysed as a separate source of rays). That is, Beeckman follows this approach at least up to the point where the species enter the observer's eye, but not further. Hereafter, Beeckman deduces, in a rather discursive manner, that species will also combine in one point, though a different one, if the object is situated closer to the lens.[38] The focusing to one point appears to be important to Beeckman. Next follows the reducing glass, about which Beeckman is much more concise: it is placed a little before the burning glass's combining point, in which position it gathers the species and unites them, and ensures that they are more clearly seen (more on this in the next section of this chapter).[39] Finally, Beeckman elaborates on what the combination of these glasses yields: a large quantity of species enter the burning glass, and the reducing glass, through its hollowness, subsequently receives and unites these species.[40] Beeckman concludes the entry by contrasting these glasses' refraction behaviour with how they would perform in the case of reflection. Beeckman associates these mechanisms with the production of either remote, or nearby, 'images' (representations) by these glasses. For the case of the burning glass, reflection yields a remote representation, while for the concave glass it works the other way around.[41]

A few things stand out in Beeckman's 1612 account of the telescope that indicate an awareness of novel developments. First is how he, in a

38 *JIB*, I, p. 13: 'Quòd si res visa sit nimis propinqua vitro, ita ut species in vitrum incidentes angulum obliquum faciunt cum vitro, fit ut illae species aliquando parallelae sint et nunquam concurrant, aut in alio puncto coeant praeter id, quod vitro proprium est. Cùm enim anguli refractionis in illo vitro positi sint et semper ejusdem magnitudinis, cùmque quantitates istorum angulorum aptatae sint ad species secundum angulos rectos incidentes conjugendas, patet angulos obliquos non posse refringi ad positum punctum coitûs. Quod fit in rebus longè dissitis, ubi omnes species rectè in vitrum incidentes esse videntur. In unum ergo facilè refringuntur.'

39 *JIB*, I, p. 13: 'Alterum vitrum voco vitrum minorans, quod locatur paulò ante concursum specierum incidentium in vitrum majorans. Hoc igitur excipit omnes species illasque conjugit, facitque ut visui magis pateant, per illud: *virtus coacta fortior etc.*'

40 *JIB*, I, p. 13: 'Tales species multò plures unius rei incidunt in vitrum majorans quàm in vitrum communè, quia unum punctum in vitrum majorans species immittens, quamquam reverâ non inmittat rectè, tamen omnes habentur pro rectè immissis propter refractionem; in communi verò vitro absque refractione transeunt. Vitrum ergo minorans, cùm sit subcavum, recipit omnes illas species, illas conjungit, angulos minores faciens.'

41 *JIB*, I, pp. 13-14: 'Quod ad reflectionem istorum vitrorum attinet, invenies duplices res representare, unam prope vitrum, alteram magis elongatam. Videtur fieri secundum rationem speculorum communium, speciebus saltem leviter superficiem vitri tangentibus. Quae verò propinquior apparet species, fit ratione refractionis, speciebus interiora vitri magis penetrantibus, et ita secundum angulos refractionis reflectentibus. Quod ita naturâ videtur comparatum. In minorante vitro verò propinquior species causatur a concavitate quam habet, secundum rationem speculorum concavorum.'

somehow probing manner, yet with confidence, applies the principle of refraction to lenses – or, to be precise, how he applies this to the convex, burning lens. The references to visual angles, as stated, remain cursory, and almost feel like obligatory remarks.[42] Beeckman's confidence in refraction is further illustrated by his remark that 'it [refraction] is arranged by nature like that'. He considers it a fundamental natural principle.[43] Second is Beeckman's discussion of shifting 'combining' points for reduced object distances – though, again, he only applies this to the convex lens. The relating of object distance with focus behaviour was quite novel within the optical theory of that time. Third is that Beeckman's 'species', though superficially reflecting older terminology,[44] are effectively used interchangeably with rays. They perform exactly the same function. Finally, Beeckman uses ray pencils (he has rays enter the lenses in quantities), rather than prioritizing any 'primary' or preferred ray. The latter was common in sixteenth-century optics because of the 'cathetus rule' that was intimately associated with any account of refraction, but was completely overturned in the dioptrical tradition.[45] Beeckman's usage of ray pencils strongly suggests a familiarity with dioptrical developments.

On the other hand, in Beeckman's account, what exactly takes place in the eye with the species, remains hidden. Beeckman's species do not enter the eye in any spatial capacity; they are simply 'perceived'. Thus, modern notions such as the 'retinal image', in which the eye operates as a camera obscura, are absent. The latter model was the fundament of Kepler's *Paralipomena* (1604), but had a prehistory that can be traced back to the anatomical writings of the late sixteenth century.[46] In addition, another aspect that can be considered 'archaic' is Beeckman's qualitatively unequal

42 Arguably, Beeckman carries his reference to 'visual angles' no further than the first sentence of the 1612 entry. All subsequent references to 'angles' in this entry concern *refraction* angles, related to ray theory, rather than visual angles, related to perception.

43 'Quod ita naturâ videtur comparatum.' The elevation of refraction to a universal law of nature can only be discerned in the later sixteenth century. See, in particular: Arianna Borrelli, 'Thinking with Optical Objects: Glass Spheres, Lenses, and Refraction in Giovan Battista Della Porta's Optical Writings', *Journal of Early Modern Studies* 3 (2014), pp. 39-61, esp. pp. 43-45: 'The fact that the law of refraction quantitatively depends on materials would have already disqualified it as a general principle of nature.'

44 I will address Beeckman's interpretation of species further on in this chapter.

45 For an accessible account of the cathetus rule, see, for instance: Dupré, *Galileo, the Telescope, and the Science of Optics*, pp. 128-129. In fact, the cathetus rule was overturned by Kepler precisely because it could not be made to work with refraction in lenses.

46 Lindberg, *Theories of Vision from Al-Kindi to Kepler*, pp. 168-177.

treatment of the convex and the concave lens. In Beeckman's account, both lenses operate on different grounds.

Of relevance is that none of the conceptual innovations that Beeckman's account testifies to, can be traced back to the works recommended to him by Snellius. Euclid, Ptolemy and Witelo are sources that, essentially, do not mention lenses at all. It suggests that Beeckman was at least informed by optical works quite novel at that time. Given his focus on refraction, likely candidates are Giambattista della Porta's *De refractione* (1593)[47] or even Kepler's *Dioptrice* (rather than, as De Waard suggested, his *Paralipomena*).[48] Nevertheless, Beeckman's access to these works cannot be validated by any source, and his indebtedness to them remains puzzling.[49]

Compressed Force Is Stronger

As we just saw, Beeckman treats the convex and the concave lens on different grounds. While he associates the convex lens's operation with refraction, he assigns the concave lens a perception-enhancing role through its compression of species. This qualitative distinction mimics Dupré's discussion of late-sixteenth-century proto-telescopic experiments, in which the concave lens was given a mere sharpening role. Beeckman's note indeed gives ample

47 Giambattista della Porta, *De refractione, optices parte, libri IX* (Naples: Johannes Jacobus Carlinus, 1593). In many respects, Della Porta's *De refractione* can be considered a forerunner of Kepler's work on optics. Della Porta, however, holds on to the fundamental rules of perspectivist optics, e.g. in the analytical tools he chooses to tackle perception. On *De refractione*: Smith, *From Sight to Light*, pp. 344-350; Borrelli, 'Thinking with Optical Objects'; Arianna Borrelli, Giora Hon, and Yaakov Zik, eds., *The Optics of Giambattista Della Porta (ca. 1535-1615): A Reassessment* (Cham: Springer, 2017).

48 Beeckman's discussion of combination points varying with object distance offers the strongest grounds for considering Kepler's *Dioptrice* as an influence, see: Malet, 'Kepler's Legacy', p. 283. Note that Kepler in fact omitted references to his earlier developed 'retinal image' when discussing the telescope in his *Dioptrice* in 1611. Therefore, a focus on Kepler's *Dioptrice*, rather than on his *Paralipomena*, conforms well with Beeckman showing awareness of varying combination points, while *not* showing any familiarity with the concept of retinal images.

49 Identifying the sources available to Beeckman in these early years remains difficult. We do have the list, compiled by De Waard, of works referred to by Beeckman throughout his notebook, but – comprehensive as it is – this does not exclude the possibility that Beeckman took notice of other works without referring to them. See: *JIB*, IV, pp. 293-304. The *Catalogus librorum* of Beeckman's library after his death, published by Canone in 1991, offers many new insights but does not inform us about the starting dates from which moment onwards Beeckman had access to these works. Furthermore, as Canone notes, a large number of natural philosophical works seem to be missing from the catalogue. See: Eugenio Canone, 'Il *Catalogus librorum* di Isaac Beeckman', *Nouvelles de la République des Lettres* (1991), pp. 131-159.

justification to question a dioptrical understanding by Beeckman of the concave lens, or knowledge of its associated focal parameters. On the other hand, Beeckman's note also indicates that a satisfactory alternative explanation for his reasoning about concave lenses should not be confined to the mathematical realm.

Associating Beeckman's understanding of the concave lens with contemporary sources is difficult. Although concave spectacles did exist since at least the mid-fifteenth century,[50] optical accounts of concave lenses remained extremely scarce for a long time. Bluntly put, before the seventeenth century, opticians ignored concave lenses completely in their works,[51] writers on practical mathematics or telescope builders hardly understood them because no focal point could be assigned to them,[52] and physicians distrusted them.[53]

Nevertheless, there are parallels between Beeckman's description and a medical account that appeared shortly before, in the medical writings of the Delft physician Pieter van Foreest (Forestus). As stated, in the medical tradition, up to that point, spectacles were barred from the literature on eye therapy because vision problems needed to be dealt with at their cause: the eye. Only when spectacles were assigned a correcting action, rather than a distorting, magnifying one, did spectacles turn up in medical-optical expositions. Katrien Vanagt has identified such an early account of concave spectacles in Forestus's 1591 *Observationum et curationum medicinalium libri quinque*.[54] Forestus explains how a concave spectacle lens, not unlike the eye pupil, through its hollow shape pushes the *spirites*[55] towards the centre (where the glass is thinner), and

50 Vincent Ilardi, *Renaissance Vision from Spectacles to Telescopes*, Memoirs of the American Philosophical Society (Philadelphia: American Philosophical Society, 2007), pp. 82-95.

51 Ilardi, *Renaissance Vision*, pp. 235-239; Smith, *From Sight to Light*, pp. 338-341.

52 For instance, in both his *Magia naturalis* (1589) and his *De refractione* (1593), Giambattista della Porta described how concave lenses do not kindle fire, because they do not bundle light rays, see: Giambattista della Porta, *De refractione*, bk VIII, prop. 18. This point is further addressed in: Dupré, 'Ausonio's Mirrors', p. 172. Dupré points to this fragment to argue that concave lenses were not attributed focal properties, and thus could in that capacity not end up in the telescope.

53 Ilardi, *Renaissance Vision*, pp. 216, 247-252; Smith, *From Sight to Light*, pp. 338-341.

54 Petrus Forestus, *Observationum et curationum medicinalium libri quinque* (Leiden: Franciscus Raphelengius, 1591), pp. 162-163; Vanagt, 'Suspicious Spectacles', pp. 123-124. Forestus's treatises were present in Beeckman's library – though not verifiable in 1612, as he referenced them only between 1616 and 1618, see: *JIB*, I, p. 146. Beeckman refers to Books XIX and XX of Forestus. The fragment on myopia is in Book XI, 'De morbis oculorum'.

55 In (some parts of) the medical tradition, 'spirites' (or visual spirits) were bodily entities that are sent from the body to the eyes, and subsequently are ejected towards the observed object. See: Vanagt, 'Suspicious Spectacles', p. 120.

as such squeezes them together, as a result of which their power grows. The explanation is in line with the observation, voiced since longer in ophthalmology, that hollow lenses cause a painful sensation for (young) people who are not suffering from myopia. In Forestus's model the thicker rim of the lens literally is a wall through which the spirites cannot pass, and consequently they subside to the thinner centre of the lens. This compression results in a gathering of the spirites that propagate from the eye, making them proceed further, thus enabling a more remote observation. Forestus's account is original in that it attributes the spectacle's function to its physical shape, and this mechanical characterization would most likely have appealed to Beeckman. A medical indebtedness to Beeckman, moreover, would also be far from surprising. Still, taking everything into consideration, I believe that Forestus's description of concave spectacles only superficially, or at most partly, covers the mechanism of the concave lens that Beeckman sets out in 1612. First of all, Forestus's account deals with extramissionist vision, while Beeckman's deals with intromissionist species. Second, the 'spirites' in Forestus's account really are bodily agents specific to the medical tradition, and don't translate all that easily to the optician's 'species'.

This also prompts the question what species really are to Beeckman. In fact, this question is highly relevant for our analysis of Beeckman's ponderings over the telescope. Establishing what the agents are that are being processed in Beeckman's telescope allows us to determine what the applicability of the telescope was to Beeckman, and what mechanisms he associated with its workings. This gives us insight in what Beeckman considered to be the scope of optics. The latter is required to judge whether Beeckman sees the telescope as operating on optical principles, and thus whether he sees the telescope as an optical instrument.

We have observed that in his discussion of the convex lens, Beeckman uses the term 'species' throughout, but in an interpretation that is virtually interchangeable with rays of light. First of all, Beeckman's usage of the term 'species', by itself, does not imply any adherence to a strict perspectivist interpretation of this term, that is, an adherence to Bacon's 'multiplication of species'. Throughout his career Beeckman keeps using the term 'species', even after (as we shall see) his reasoning had turned entirely mechanistic in the 1620s. This implies that to Beeckman, 'species' have no fixed meaning that is stable throughout his career. On the other hand, Beeckman does often consider light and species in relation to a more general form of celestial radiation. This conclusion emerges when we look at Beeckman's maturing natural philosophy on a broader level.

The role of celestial radiation in Beeckman's cosmology – his 'cosmic economy of power' as Vermij aptly styles it – has hitherto been analysed particularly in the context of Beeckman's heliocentrism for the period from about 1616 onwards,[56] and for the period in the 1630s, when Beeckman mainly comments on the cosmological views expressed in Kepler's *Mysterium cosmographicum* (1596) and *Astronomia nova* (1609), which Beeckman then juxtaposes to his own ideas.[57] These later notes also reveal that since early on in his career, Beeckman had been developing his own, original, mindset about a universe that is governed by celestial radiation. Notebook entries from the 1610s show that also in these early days, to Beeckman, both light and species were intertwined with celestial radiation. For instance, in 1613 Beeckman introduced his concept of 'igniculi', entities of stellar radiation, of which light is a manifestation.[58] The igniculi radiate from the stars, and are reflected by the Sun, the pressure they exert on the planets dictating the order of the solar system. Later on, Beeckman saw stellar radiation as a continuous flux that affects mechanical processes on earth and, if reaching a certain threshold, manifests itself as visible light.[59] In a note dating from 1613, Beeckman judges vision to take place by celestial radiation that, being reflected on the perceived object, is bounced off to our eyes as species. Yet he doubts if this also holds for luminous objects, and, wonders why this doesn't allow us to see at night.[60] Therefore, given that light and species were always intertwined with celestial radiation in his natural philosophy, there is little reason to assume that species would strictly denote light and images in Beeckman's early optical notes. Moreover, these examples show that Beeckman is supportive of a causal influence that is being brought about by species.

In his 1612 entry as well, there are indications that suggest a more extended applicability of species than one would initially assume. Firstly, Beeckman gives a surprising amount of attention to the number of species that enter the telescope by virtue of its relatively large-sized convex lens. Beeckman subsequently attributes the power of the telescope to this amount of species. In other words, he considers species to enter the telescope in bundled rays, and compressing these bundles leads to a stronger perceptual effect. It is

56 Rienk H. Vermij, *The Calvinist Copernicans: The Reception of the New Astronomy in the Dutch Republic, 1575-1750* (Amsterdam: KNAW, 2002), pp. 113-115.

57 Vermij, *Calvinist Copernicans*, pp. 115-117; John Schuster, *Descartes-Agonistes: Physico-mathematics, Method & Corpuscular-Mechanism 1618-33* (Dordrecht: Springer, 2013), pp. 471-475.

58 Van Berkel, *Isaac Beeckman on Matter and Motion*, pp. 97-103.

59 See in particular: Van Berkel, *Isaac Beeckman on Matter and Motion*, p. 98.

60 *JIB*, I, 28: 'Lux reflexa a rebus est visûs materia, imo et colorum.'

Figure 5.1 The harmful effect of concentrating sunlight by means of a burning mirror

Engraving by Adriaen van de Venne in: Johan de Brune, *Emblemata of Zinne-werck* (Amsterdam: Jan Evertsen Kloppenburch, 1624). Library Vrije Universiteit Amsterdam, XH.00046. Although misinterpreted by the artist, the accompanying text makes it clear that the mirror is meant to be concave. This ought to have resulted in converging, rather than in diverging the solar rays.

essential to point out that none of the traditional optical theories that dealt with *perception* that were en vogue in the early seventeenth century would have suggested an interest in ray quantity or, more accurately, ray density, to Beeckman.[61] On the other hand, treatises dealing with the astrological or metaphysical aspects of radiation were much more inclined to do so.[62] Secondly, Beeckman writes that the concave lens 'gathers all the species [from the convex lens], unites them, and makes that they are more clearly seen'. Thus, he again relates the compression of species with an enhanced visual sensation. The concave lens, writes Beeckman, gathers the species

61 Within the confines of *perception*, perspectivist optics did not consider 'bundles' or 'pencils' consisting of a large number of rays; it prioritized one preferential ray. Also, neither the 'visual angle' accounts of the telescope that appeared shortly after the telescope's emergence, nor the dioptrical accounts of a somewhat later date, did relate telescopic power with the *amount* of rays that is processed through the instrument. Rather, such theories judged magnification by the ray *angles* that are affected by the telescope.

62 That is, because they could fully engage in *ray* theory (cfr. burning mirrors), without having to deal with the subtleties of perception. This distinction was fundamental in pre-1600 optical theory.

'through the hollowness it has'. Beeckman expressly summarizes the mechanism that takes place in this concave lens with the saying 'compressed force is stronger' ('virtus coacta fortior etc.').

Beeckman's usage of this phrase is as fascinating as it is enigmatic. While the brevity with which he references it appears to indicate that the phrase must have been rather well-known, its exact source is difficult to establish. Beeckman *may* have borrowed this phrase from a mid-sixteenth-century medical work by the Dutch physician Johannes Wier, and there are in fact good grounds for such attribution,[63] but since Wier's discussion deals with the disease progression of malaria, we are left in the dark about any action in concave lenses.

On the other hand, *virtus* (power) was a notion that had gained a more detailed meaning in the context of celestial radiation in the sixteenth century. It was used to signify the actualization of *vis* (force). In his early notes, we do occasionally see Beeckman referencing *virtus* it in the context of stellar or planetary influence. In particular, the term turns up denoting the causal influence that Beeckman assigns to such celestial radiation.[64] Given the extended scope we should confer on Beeckman's 'species', this leads to a more refined interpretation of what Beeckman considers to take place in the concave lens of the telescope. Primarily, the concave lens increases ray density, because of the gathering action that results from its hollow shape. This concentration increases the causal effect of the gathered species. This, in turn, leads the observer with an enhanced perceptual sensation. Moreover, we can extend this mechanism to the entire telescope, because the convex lens also takes part in the collection of many rays, through its large aperture. Many rays are required in order to sufficiently 'feed' the concave lens for its compressing action, therefore the convex lens needs to be large. It enables us to categorize Beeckman's telescope as a 'hybrid', yet fully optical instrument.[65] Its operation is based on two complementary

63 Johannes Wier, *Medicarum observationem rararum* (Basel: Johannes Oporinus, 1567), p. 43. The exact phrase 'virtus coacta fortior quàm dispersa' appears in the index at the end of this work which, moreover, was present in Beeckman's library.

64 For instance: *JIB*, I, p. 110: 'Terra est in medio virtutum planetarum' (1616); *JIB*, I, p. 101: 'Planetarum lux qualis' (1616).

65 This interpretation is in agreement with the observation made recently by Dijksterhuis, that in situating the telescope in its early years, one should resist a classification of the instrument as a refractive or even an optical device, as contemporary references testify to a similar multidisciplinary interpretation of 'optical' as 'a general sense of instruments to create and manipulate light and images'. See: Fokko Jan Dijksterhuis, 'Magi from the North: Instruments of Fire and Light in the Early Seventeenth Century', in: Borrelli, Hon, and Zik, eds., *The Optics of Giambattista della Porta*, pp. 125-143, esp. pp. 129-130.

principles: the mechanism of refraction in the convex lens, and the act of increasing the causal influences that are embedded in the species, an act that takes place in the concave lens.

It is left to explain how Beeckman would have developed his occupation with a cosmology built on celestial radiation in the first place. Pinpointing an exact source is, once more, difficult, because attributions of effects to celestial radiation were not uncommon in sixteenth-century (astronomical and, obviously, astrological) literature. Ultimately, many of these accounts derive from the concept of a 'universal radiation of force' presented in Al-Kindi's *De radiis stellarum* (ninth century), a cosmology that deeply influenced the physical aspects of Grosseteste's and Bacon's optics, and that saw a further distribution as a (somehow disconnected) part of the perspectivist tradition.[66] If Beeckman associated celestial radiation with optics however, and lenses in particular, this has far fewer precedents. In this respect, though more conjectural in nature, it is worth pointing out similarities between Beeckman's celestial causality and the concept of 'astrological causality' that the English polymath John Dee had set out some 50 years earlier, in his *Propaedeumata aphoristica* (1558).[67] In this work, in 120 aphorisms Dee sets out the astrological behaviour of 'pre-eminent virtues of nature', signifying a.o. the imprinting power of species on matter.[68] Explicitly moving away from prognostication,[69] Dee considers astral virtues to be completely predictable, regular, natural, and fit for mathematization. Moreover, Dee advances the idea that mirrors can manipulate the effects of astrological causality, very much like they can bundle light, by focusing and combining their rays of influence. And while a link between Beeckman and Dee's *Propaedeumata* unfortunately cannot be established, these similarities become all the more striking in the light of Zuidervaart's identification of Dee's mirrors in Middelburg around 1600, described elsewhere in this volume.[70]

66 Lindberg, *Theories of Vision from Al-Kindi to Kepler*, p. 19; Nicholas H. Clulee, 'Astrology, Magic, and Optics: Facets of John Dee's Early Natural Philosophy', *Renaissance Quarterly* 30 (1977), pp. 632-680, esp. pp. 665-671; Lindberg, 'Roger Bacon on Light, Vision, and the Universal Emanation of Force', pp. 244-246.

67 John Dee, *Propaedeumata aphoristica. De praestantioribus quibusdam naturae virtutibus* (London: H. Suttonus, 1558).

68 Clulee, 'Facets of John Dee's Early Natural Philosophy', pp. 652-664.

69 This remark is warranted especially in the case of Beeckman who, at a later date, repeatedly expressed his contempt for prognostication. See: Rienk H. Vermij, 'The Marginalization of Astrology among Dutch Astronomers in the First Half of the 17th Century', *History of Science* 52 (2014), pp. 153-177.

70 Dee's *Propaedeumata aphoristica* was not present in Johan Radermacher's library. On the other hand, Dee's *Monas hieroglyphica* (1564) was. In the *Monas* Dee elaborated on the

Finally, the redefinition of Beeckman's telescope as an instrument that complies with celestial radiation as well, offers a different perspective on an issue that was raised earlier: how can we reconcile Rudolph Snellius's enthusiasm for the telescope with an adherence to 'archaic', perspectivist sources? Both Beeckman and Snellius were very supportive for Ramist ideas. Although Ramism essentially is a pedagogical approach, rather than a doctrine that implies adherence to any ontology of light, Ramist (optical) followers specifically aimed at broadening the scope of optics, making it encompass more phenomena that the 'rules of optics' were applicable to. By elevating the science of optics to a more general mathematical science, it was hoped to give optics a firmer footing in the universities' curricula. A close look at the works recommended by Snellius to Beeckman shows that, notwithstanding the age of their *original* authorship, a Ramist connotation was indeed present in all these works.[71] In particular, the sixteenth-century, staunchly Ramist editors of these works had added introductions on the utility (i.e. universal applicability) of the optical method, and treatises on burning mirrors.[72] 'Ptolemy', on Snellius's list, was *Tetrabiblos*, Ptolemy's quintessential work on astrology, which was also supplemented with two treatises on burning mirrors.[73] This shows that Snellius saw astrology to be an integral part of optics. Moreover, the note about Snellius's demonstration of the telescope in 1609 clearly states that this took place during his

Propaedeumata's content matter by giving it a more Hermetic and alchemical embedding.

71 This identification was already made by De Waard, see the notes to *JIB*, IV, 17-19. For discussion, also see: Van Berkel, *Isaac Beeckman on Matter and Motion*, p. 15.

72 That is, Euclid's classic work on optics and catoptrics was likely the 1557 edition to which Ramus's pupil Jean Pena added his *De usu optices*, a treatise which argued for the utility of optics for mathematical descriptions of many other types of radiation exhibiting physical effects, and highlighted the study of concave mirrors. See: Dupré, *Galileo, the Telescope, and the Science of Optics*, pp. 39-42. 'Witelo' referred to the popular *Opticae thesaurus* by Ramus and Friedrich Risner, a volume that combined the optical works by Alhazen and Witelo and, through its selections and cross-references, presented these in a sixteenth-century jacket. See: Dupré, *Galileo, the Telescope, and the Science of Optics*, pp. 50-53; Smith, *From Sight to Light*, pp. 328-329. 'Ptolemy', if we follow De Waard's interpretation, was the 1548 edition of the *Tetrabiblos* or *Quadripartitum*, the quintessential textbook on astrology. This very edition had acquired an optical connotation through Antonio Gogava's addition of two treatises on catoptrics. See: Steven Vanden Broecke, *The Limits of Influence: Pico, Louvain, and the Crisis of Renaissance Astrology* (Leiden: Brill, 2003), pp. 177-178.

73 In fact, it has been suggested that John Dee's visit to Louvain in 1548, when Antonio Gogava was editing this very edition of Ptolemy, was a direct inspiration for Dee to develop the astrological-catoptrical interest that resulted in his *Propaedeumata*. See: Clulee, 'Facets of John Dee's Early Natural Philosophy', pp. 638-639; Vanden Broecke, *The Limits of Influence*, pp. 174-178.

lessons on Ramist optics.[74] Given that lenses are conspicuously absent in the Ramist-inclined optical works that Snellius recommended, we may need to reconsider how the utilitarian aspect of Ramism, for followers like Snellius, was associated with the telescope. Earlier discussions of Snellius's courses that mentioned his demonstration of the telescope in the context of Ramist education, assumed Snellius's telescope to serve as an example of practical utility[75] or, alternatively, as a means to reach the Ramist goal of regaining the wisdom of the ancients through observation.[76] Our association of Beeckman's optics with celestial radiation then suggest another possibility: that the telescope *did* in fact serve as an illustration of Ramist optical theory – be it an illustration not pertaining to refraction in lenses, but to the metaphysics of (celestial) light, and how this is affected by optics. Quite possibly, it was not the *telescope*, in being the product of practical application, which was useful per se. Rather, it was the *optical approach* that was useful – useful for allowing an explanation of how the telescope affected celestial radiation.

Both Van Berkel and Schuster, in the present volume, have pointed to Ramism as a resource for Beeckman to engage in corpuscular-mechanistic thinking. They lift out the picturability and the programmatic approach of

74 It is reasonable to assume that these courses were based on the newly published *Optics* by Ramus and Risner: Friedrich Risner, *Opticae libri quatuor ex voto Petri Rami* (Kassel: Wilhelm Wesselius, 1606). Rudolph Snellius was actively involved in the (delayed) publication of this book, see: Liesbeth de Wreede, *Willebrord Snellius (1580-1626), a Humanist Reshaping the Mathematical Sciences* (PhD diss., Utrecht University, 2007), pp. 141-142. The background of its publication is further discussed in: Dupré, 'Optics without Hypotheses', p. 508. To this work, too, an introduction on the utility of optics was added. The first of the four books which constitute Ramus's *Optics*, including manuscript annotations by Willebrord Snellius, was later published as: Johan Adriaan Vollgraff, *Risneri Opticam cum annotationes Willebrordi Snellii* (Ghent: Plantijn, 1918). Vollgraff was particularly interested to find any mathematical or experimental origin of Snellius's law of refraction. As for Rudolph Snellius's teaching of Ramus, we also have to take into account that in 1592 Snellius was summoned by the university curators to soften the Ramist character of his lessons, see: De Wreede, *Willebrord Snellius*, p. 38. This, in fact, makes it surprising, and in a sense revealing, that Deschamps refers to Snellius's lessons of Ramist optics. A reference in Beeckman's notebook no later than 1612 shows that he was familiar with Ramus and Risner's *Optics* at the time when he started pondering over the telescope's working principle, see: *JIB*, I, p. 6: 'Specierum in obscuram cameram erectio.' Surprisingly, in this entry Beeckman contemplates the reversal of a (pinhole) camera obscura image by means of an additional convex lens. This reversal method is not described in Ramus and Risner's *Opticae libri quatuor* but rather traces back to Della Porta's *Magia naturalis* (1589), references to which only turn up much later in Beeckman's notebook.

75 Tabitta van Nouhuys, *The Age of Two-Faced Janus: The Comets of 1577 and 1618 and the Decline of the Aristotelian World View in the Netherlands* (Leiden: Brill, 1998), p. 331.

76 De Wreede, *Willebrord Snellius*, pp. 162-170, esp. p. 165.

Ramism as key elements that moved Beeckman into the natural philosophical arena. Our survey of Beeckman's sources reveals that, in addition, the Ramist-optical ideas that Beeckman encountered in his Leiden years were characterized by being favourable of causal explanations, and by their tendency to address not only light, but other fields, such as cosmology, as well. These characteristics may have hinted at natural philosophy but, admittedly, did not offer any clear-cut suggestion or motivation to engage in mechanistic thinking. Nevertheless, the contribution of optical sources to Beeckman's early mindset becomes all the more relevant when we take into account the position that optical matters had, later, in Beeckman's physico-mathematical exercises, and conversations with Descartes, around 1618 and especially in 1628.[77]

Telescopium, the Diaphragm, and Corpuscular Optics

In the next entry that Beeckman made on the telescope, in 1618, it appears that a mathematical framework based on the theory of refraction moved even further to the background. Admittedly, Beeckman's vision on nature had evolved significantly in the intervening years. As Klaas van Berkel notes, Beeckman's corpuscular ideas came to maturation during the years leading up to his promotion in Caen.[78] Indeed, if Beeckman's interpretation of celestial influence through *species* had carried a strong suggestion of mechanical causality since the beginning, by 1618, it had become fully corpuscular. 'What the opticians call *species visibiles* are bodies', he wrote as one of the corollaries to his thesis, plainly and clearly.[79] What I then want to argue, is how Beeckman's corpuscular ideas enabled an integration of his earlier understanding of the telescope with an essential technological characteristic of early telescopes that was recently identified by the historian of optics Rolf Willach: the diaphragm. On the other hand, what it also shows is that in this period, Beeckman's understanding of this diaphragm was *not* based on technological grounds.

In recent years, the discussion about the telescope's invention has shifted from lens combinations to lens quality. Willach has identified the addition of a diaphragm – a cardboard ring that covers the edges of a lens – as an

77 Schuster, *Descartes-Agonistes*, pp. 153-163, 471-475.
78 Van Berkel, *Isaac Beeckman on Matter and Motion*, pp. 85-86.
79 *JIB*, IV, p. 44: 'Quas vocant optici species visibiles sunt corpora.'

essential step for the emergence of the telescope. This intervention softened the influence of material defects (grinding errors) in early-seventeenth-century spectacles by covering all but a small central portion of the lens, and made their application as telescope lenses viable.[80] Willach's interpretation of this evolution is primarily a technological account. That is, supported by a large number of measurements on early modern (spectacle) lenses, Willach convincingly shows that the problem of lens quality was indeed solved at the time when the telescope emerged. Yet he leaves open the question whether the problem was actually *understood*, and, by not addressing this, perhaps implicitly suggests that it was. Whether the application of a diaphragm originally was *motivated* by an awareness of material lens defects therefore remains an open issue, and it should not come as a surprise that alternative motives have been proposed.[81]

Beeckman's 1618 notebook entry sheds new light on this matter. In this note, Beeckman refers to a book he has seen in Caen in which 'the telescope of Galileo' is described.[82] De Waard identified this book as the Milanese author Girolamo Sirtori's *Telescopium*, published in 1618 but written around 1612, and Beeckman's discussion in the note puts further weight to this attribution. Sirtori's *Telescopium* is a remarkable work. It undoubtedly is one of the earliest printed works that deals in such length with the telescope. The many details Sirtori gives on production techniques make it an essentially technological

80 Rolf Willach, 'Der lange Weg zur Erfindung des Fernrohres', in: Jürgen Hamel and Inge Keil, eds., *Die Meister und die Fernrohre. Das Wechselspiel zwischen Astronomie und Optik in der Geschichte. Festschrift zum 85. Geburtstag von Rolf Riekher* (Frankfurt am Main: Harri Deutsch, 2007), pp. 34-126; translated into English as: Rolf Willach, *The Long Route to the Invention of the Telescope*, Transactions of the American Philosophical Society 98, nr. 5 (Philadelphia: American Philosophical Society, 2008). Also see the slightly condensed version in: Rolf Willach, 'The Long Road to the Invention of the Telescope', in: Van Helden et al., eds., *The Origins of the Telescope*, pp. 93-114.

81 Empirical knowledge about diaphragms seems to be suggested by a remark that Daniel Barbaro made about the camera obscura in his *La pratica della perspettiva* (1567): 'if you want to cover the lens as much as to leave a bit of circumference in the middle, and that part which is clear is not covered, you will see an even more vivid effect'. See: Ilardi, *Renaissance Vision*, pp. 220-221. An alternative, non-technological motivation for a diaphragm in the specific case of Galileo's telescope is set out in: Sven Dupré, 'Galileo's Telescope and Celestial Light', *Journal for the History of Astronomy* 34 (2003), pp. 369-399. Dupré argues how the diaphragm was added to counter the effect of irradiation, the phenomenon in which shiny objects are perceived larger than they really are. The application of the diaphragm then pertains to the discussion whether the Moon reflects light, or was a source of light itself. Both the debate about moonlight and the concerns about irradiation predate the emergence of the telescope, and therefore fit in well with Galileo's early adoption of the instrument.

82 *JIB*, I, pp. 208-209: 'Telescopium Galilaei pictum a me visum et examinatum.' In the entry itself, Beeckman refers to the instrument not as a telescope but as 'tubus ocularem'. However, the title of the entry was added later to the manuscript, probably in 1628.

Figure 5.2 Propagation of rays of light in a telescope according to Sirtori's *Telescopium*

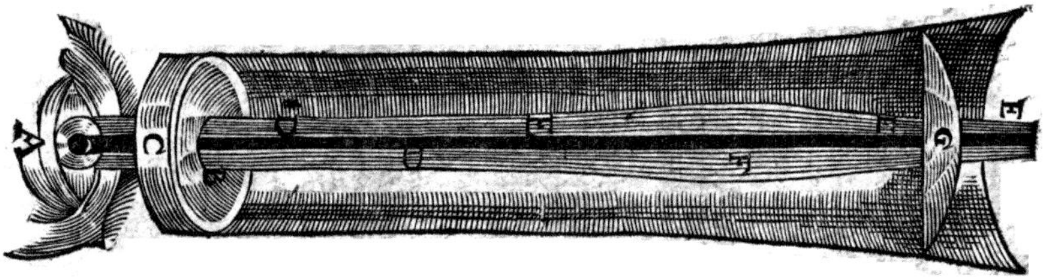

From: Girolamo Sirtori, *Telescopium* (Frankfurt, 1618). Bayerische Staatsbibliothek, München. Beeckman consulted this book during his stay in Caen in 1618.

book. Furthermore, some core elements in the historiography of the telescope derive from Sirtori's account.[83] Yet, what has remained relatively unnoticed is that Sirtori also describes the working principle of the telescope, and in all likeliness this is the fragment Beeckman is referring to in 1618, as it is one of the few chapters that are adorned with an illustration (fig. 5.2).[84]

Although Sirtori dutifully refers to recent optical innovations by Kepler in the introduction of his *Telescopium*, the working principle Sirtori sets out in Chapter 19 is entirely pre-Keplerian. Within an extramissionist framework, Sirtori describes how from the eye pupil a most vehement mass of rays proceeds, unrefracted, through the telescope tube, where it normally would lose power as it gets further from the eye. Yet, the concave lens near the eye supports the rays with a painful force ('intensa et dolorosa vi') by which means they eventually do reach the convex lens. Inside the tube, the rays spread out nonetheless, but as they approach the convex lens the latter furnishes them with extra force, changing their nature, such that they proceed much further than they normally would. Next, Sirtori deals with the peripheral rays (which in the perspectivist tradition are fundamentally different). Although these do undergo refraction – among others the convex lens bends them back to the axis – above all they are dredged along with the central ray, which dictates their propagation. Thus, the combined action of the telescope makes all the rays proceed further, and this is what makes remote vision possible.

Interestingly, Sirtori's account shares certain details with another, earlier treatise, recently uncovered by Huib Zuidervaart, in which the

83 These are reproduced and discussed in detail in: Van Helden, *The Invention of the Telescope*.
84 Girolamo Sirtori, *Telescopium. Sive Ars Perficiendi novum illud Galilaei Visiorum Instrumentum ad Sidera* (Frankfurt: Paulus Jacobus, 1618), pp. 67-68: 'Tubus et obiter de refractione.' In his 1618 note Beeckman refers to an illustration. See: *JIB*, I, pp. 208-209: 'Telescopium Galilaei pictum a me visum et examinatum.'

telescope is mentioned in relation to the Twelve Years' Truce.[85] Here the author, Johannes Walchius, describes how 'the visual rays of a man [...] are gathered and aided by an appropriate instrument made for this purpose, such that their force [*vis*] gathers, as the pupil otherwise emits these rays in a spread out manner'. The compression that this ingenious 'spectacle instrument' then yields, makes the power, which is otherwise distributed over a wide area, be concentrated on only one thing.[86] Walchius makes further references to the 'lynx eyes' of those who use the telescope, and, while acknowledging the role of the Dutch, credits Della Porta with the instrument's very idea. As Christoph Lüthy has shown, such references should prompt us to read Walchius's presentation of the instrument as the fulfilment of a long-anticipated 'telescopic dream', rather than as a neutral description of its workings.[87] On the other hand, Walchius's account shows such rhetoric undergoing integration of early thoughts on the instrument's working principle, and gives us rare insight in the role bestowed upon both individual lenses in this working principle. Sirtori's and Walchius's accounts, both extramissionist in character, have in common that they associate the telescope with extending the *reach* of vision. Also, for having the instrument achieve this, they assign a significant role to either the concave lens, or both lenses, in, through compression, aiding the rays of the eye by conferring them an extra force. More important, and unique to Sirtori's 1618 account, is the latter's subsequent empirical observation on the diaphragm, the operation of which Sirtori also attributes to a compressing effect: 'experience clearly proves it, that if you accommodate the tube in the location F with a narrow aperture, by which the rays are more compressed, and gathered in a more parallel manner, they extend vision more vividly and remotely'.[88]

85 Johannes Walchius, *Decas fabularum humani generis* (Strasbourg: Zetner, 1609). There is no indication that Beeckman would have known this work. I am grateful to Huib Zuidervaart for bringing it to my attention.

86 Walchius, *Decas fabularum*, p. 249: 'Radii hi si per aliquod conveniens atque arte ad hanc rem paratum instrumentum colliguntur ac juvantur, illorum vis ut coeat, quos antea sparsim pupilla ejaculabatur, totoque in latum se ubique hemisphaerio explicabant'; resp. 'si nimirum omnis illa latè ac circumquaque dispersa vis, docti hujus specilli instrumento contrahitur, atque latitudinis amplitudini quod dabatur, longitudini nunc conceditur soli, ac seclusis plurimis adeoq; infinitis, uni rei adplicatur tantum.'

87 Furthermore, the references to 'lynx eyes' testify to an appropriation of the instrument's early history in the institutional setting of the Accademia dei Lincei. See: Christoph Lüthy, 'Atomism, Lynceus, and the Fate of Seventeenth-Century Microscopy', *Early Science and Medicine* 1 (1996), pp. 1-27, esp. pp. 6-11.

88 Sirtori, *Telescopium*, pp. 67-68: 'Experientia id clare probat, quia in situ F si tubo angustum foramen accomodes per quod radii strictius, et magis paralleli cogantur transire, vividius, et remotius protrahunt visum.'

Beeckman's reading of the 'picture' he saw in Caen, in addition to his discussion in a context of diaphragms, virtually rules out any other possible source than Sirtori's *Telescopium*. In his 1618 note, Beeckman describes how the diaphragms are there

> because the light is gradually dispersed in the tube: because hitting the air, it bounces somewhat and deviates from a straight path, making it approach the other glass somewhat scattered, and insufficiently compressed. But when in the middle a diaphragm is placed, perforated in the centre with a round hole, nearly all the rays, that initially were dispersed, are compressed again, just like water, that has proceeded little in a wide river bed, but is then forced to flow through a narrow opening.[89]

In a following entry, 'on the diaphragms of a telescope, what they are capable of', Beeckman formalizes his idea.

> After these considerations, I believe that one can gain in transparency [*perspicuitas*], by means of interposed diaphragms in the above described manner: since then, the dispersed light is gathered more and from this, more species enter the eye, and from their quantity more clarity [*claritas*] ensues.[90]

89 *JIB*, I, pp. 208-209: 'diaphragmata ideo interposita credo, quia lumen longitudine viae dispergitur in tubo: impactus enim aeri, impingitur paululum et aberrat a rectitudine viae, unde fit id ad alterum vitrum dispersius, nec satis coacte, pervenire. At cum in medio diaphragma positum est, in medio rotundo foramine perforatum, omnes fere radij, qui antea paululum dispersi erant, iterum coguntur in illo foramine, non aliter quam aqua, quae amplo alveo aliquantum processit, at tum per angustiorem locum tota cogitur transire.' It should be noted that Beeckman considers the diaphragm(s) to be positioned in the middle of the telescope tube. This differs from contemporary sources and from preserved instruments, where the diaphragm is always placed on top of, or close to, the objective lens, where its masking of grinding errors is optimal. That being said, Beeckman's diaphragms should not be confused with *field* stops, which were placed in the focal plane of (later) telescopes, and delimited the field of view, rather than masked lens errors. Telescopes of the Galilean type, and therefore any telescope before c. 1630, through their optical configuration did not possess a focal plane and therefore were not compatible with field stops. Any diaphragm placed in the tube of such an early telescope would retain its function as an aperture stop, and would to some degree have masked grinding errors. In short: if Beeckman had put his idea of intermediate diaphragms into practice, he would effectively have observed the effects of a lens mask, mitigating grinding errors, in agreement with Willach's analysis of the diaphragm in the earliest telescopes.

90 *JIB*, I, p. 209: 'His vero animadversis, credo perspicuitatem posse conciliare, interjectis diaphragmatibus in morem praedictum: tum enim lux dispersa magis colligitur ob idque plures species oculum incidunt, a quarum multitudine claritas oritur.'

Beeckman reinterprets Sirtori's description in an original way. While refraction arguably still had a role in Sirtori's account, this mechanism is now lacking in Beeckman's reading of the fragment. Furthermore, Beeckman dismisses the empirical character of Sirtori's observation about the diaphragm, and extends Sirtori's understanding of the *concave lens* into a governing principle for the entire telescope, including any diaphragm. Beeckman's thoughts are converging towards a single working mechanism. Indicative of this, but remarkable as well, is Beeckman's comparison of the flow of light rays with water in a river bed. What is evidently at stake here is that Beeckman borrows from his experience as a craftsman. Between 1612 and 1618 he was heavily involved with water supply systems, and this background, as Van Berkel has shown, left clear traces on his developing natural philosophy.[91] Yet there is more to it than the inspiration of practical utility, or an interpretational embedding in the context of engineering, a field so familiar to Beeckman. What Beeckman is doing here is creating natural philosophy. This is characterized by the fact that Beeckman takes the gist of the argument from Sirtori, but seamlessly transposes this from an extramissionist framework to an intromissionist one.[92] To Sirtori, but also to Forestus and to Walchius, the visual rays originated in the eye, and propagated towards the object under observation. Beeckman instead has the light rays proceed from the object to the eye. The reason he does so is that to him, these light rays are real and corpuscular, just as water is. In other words, the direction of light propagation is dictated by his natural philosophy, for this requires a physical interpretation of light.

Telescopic Configurations

The previous fragment offers an intriguing insight into Beeckman's thoughts on the action of the diaphragm. Yet, care must be taken not to interpret the notions Beeckman uses, such as 'transparency' or 'clarity', from too modern a point of view. As Fokko Jan Dijksterhuis has pointed out, in the seventeenth-century optical discourse, a notion such as 'distinct' or 'sharp'

91 Van Berkel, *Isaac Beeckman on Matter and Motion*, pp. 136-140.

92 In fact, it was not uncommon for adherents to the sixteenth-century optical tradition to show preference for an intromissionist model, while in their *mathematical descriptions* of light propagation, they did not see too much of an issue in shifting to an extramissionist model, borrowed from the mixed mathematics of 'perspectiva'. An engagement in natural philosophy, then, required this ambiguity to be resolved. See: Dupré, 'Optics without Hypotheses'. This obligation to settle for one preferred model is also what we see here with Beeckman.

lent itself rather well for mathematical treatment, leading to a relatively continuous shift of meaning, while 'clear' was a much more problematic concept, causing discussion about its definition to surface only haphazardly, and generally within mutually incompatible contexts.[93] Possibly, Beeckman simply *meant* to say that images seen through a telescope are perceived with more intensity. It is essential, nevertheless, to realize that for Beeckman, telescopic action relates unequivocally to the quantity, or density, of species that enter the eye, and that any notion of 'magnification' for Beeckman in the 1610s necessarily falls back on this basis. A notion of *image* is unavailable to Beeckman at this early point. While gradually an idea of spatially extended images starts to surface in his journal entries in the 1620s, this idea does not take part in Beeckman's thoughts on the magnifying power of a telescope. Nor does Beeckman employ Galileo's notion of magnification in terms of altered visual angles, which, as Malet sets out, saw an influential adoption in the writings of Kepler – even though Beeckman serves himself of this *terminology* as early as 1612.[94]

What I want to point out is how his core concept of telescopic magnification, as hinted at in the previous pages, and building on the density of observed species, in Beeckman's hands successfully lent itself to quantification, and consequently was adequate and coherent. A peculiarity of Beeckman's magnification model, nonetheless, was how it logically led to a requirement for large lenses. Furthermore, it calls for a reduction of the concept of 'magnification', in which the magnifying power of a telescope is not to be seen as intrinsic to the instrument alone, but should be considered in relation to the surrounding space, as dictated by a telescopic configuration.

In both optics and astronomy, the notion that species radiate spherically from a luminous source had been a well-established axiom for a long time. Consequently, intensity was considered to be uniform at any sphere with a constant radius from this source. In Beeckman's notes on perception through a lens or telescope from around 1620, we see him generally maintaining a constant relationship between the quantity of rays or species that enter the eye, and the 'strength' with which an object is perceived. Beeckman implicitly assumes that the perception of a higher quantity of rays, as they emanate from the object, is equivalent to an observation of the object from nearer, where the ray density has not yet decreased that much along

93 Fokko Jan Dijksterhuis, 'Clair & Distinct: Seventeenth-Century Conceptualizations of the Quality of Images', in: Wolfgang Lefèvre, ed., *Inside the Camera Obscura: Optics and Art under the Spell of the Projected Image* (Berlin: Max-Planck-Institut für Wissenschaftsgeschichte, 2007), pp. 105-113.
94 See note 42.

its propagation. An essential observation is that Beeckman carries this through in a consistent manner, and thus considers this equivalence to be unconditional. This allows him to model telescopic action in a quantitative manner, without requiring a concept of spatially extended, 'retinal' images, nor to focal lengths. Beeckman's telescope, in short, is an instrument that does not primarily offer an enlarged view, but one that yields an observation as if the observer is standing *closer* to the object, because more rays are being captured. It is an instrument that brings nearer – the larger or enhanced observation being only a logical consequence. This, then, is in striking agreement with the earliest references to, and naming of the telescope in the Dutch language area.[95] That is, references as early as the 1608 patent application to the 'gecommiteerde raden' in Middelburg (the first reference to a telescope in history), up to far into the 1630s, describe the telescope as an instrument that shows things that are far away 'as if seen from nearby', but remain silent on any *enlarging* power. The same is true for designated names such as 'verrekijker' ('far-seer') or 'ver-siende bril' ('far-seeing spectacles'), names that emerge in 1620 and 1633, respectively.[96]

A similar situation holds, of course, for the Latin term 'telescopium' (also signifying 'far-seer') and its derivations. This name had been coined in 1611, replacing earlier names that generally reflected the tubular shape of the instrument. In this respect, Lüthy has shown how the name 'telescopium' had – in a manner of speaking – been in the making long before the instrument was invented, exactly because once the telescope emerged, its magnifying qualities fulfilled long-held anticipations about extended vision.[97] Arguably, the Latin name 'telescopium' therefore only partly reflects the properties of the actual instrument, and tells a great deal more about its anticipation. Still, backed by the earliest Dutch names for the instrument that I mentioned, the point I want to make is rather a semantic one. None of these names or descriptions refers to magnification or size, but rather to distance, and this holds for 'telescopium', too.[98] It illustrates that also once the instrument

95 For the earliest Dutch references to the telescope, see: Van Helden, *The Invention of the Telescope*, pp. 20, 35-44; Huib J. Zuidervaart, 'The "True Inventor" of the Telescope: A Survey of 400 Years of Debate', in: Van Helden et al., eds., *The Origins of the Telescope*, pp. 9-44, esp. pp. 11-15.

96 Huib J. Zuidervaart, personal communication.

97 This stands in contrast to the microscope, the magnifying capacity of which came unannounced. See: Lüthy, 'Atomism, Lynceus, and the Fate of Seventeenth-Century Microscopy', pp. 1-13.

98 Moreover, nothing forbids that different conceptions of magnification could have co-existed. For instance, as we saw, Kepler hinged upon Galileo's idea of magnification that, being rooted in visual angles, was more compatible with (angular) size. That Galileo's concept of angular magnification did indeed find a fluent adoption, and thus that Beeckman's ideas should be

had emerged, early practitioners did not associate it with a magnification of the *size* of images. Beeckman's notes corroborate this and, in addition, show that an alternative model based on the spatial diffusion of ray density could successfully be quantified.

If there is one thing that worries Beeckman in applying this model, it is a reservation that stems from his corpuscular view of light. In Beeckman's optical notes, often an uneasiness comes up that light particles bounce against the air during their propagation, and lose even more of their ray density. If they would indeed bounce, Beeckman no longer could rely on a geometrical expression of diminishing ray density with distance, and this would threaten his conception of telescopic action. Photometric notes and experiments by Beeckman, also later, in 1622 and in 1633, can probably be traced back to this concern.[99]

In Beeckman's desire to enhance observation by guiding as many rays as possible to the eye, the convex (object) lens of a telescope plays a central role. In part because of the qualitative role that the concave (ocular) lens retains in Beeckman's thoughts, the key to telescopic power is the diameter of the convex lens. For a more powerful observation, a large lens diameter is not only a desideratum, it becomes a requirement. This is illustrated by notes in Beeckman's notebook dating from 1622. Under the heading 'a large convex lens is *required* to see far away with a telescope', Beeckman describes how

> those that want to make telescopes, with which the most remote things can be seen, necessarily have to position a very large lens as far away from the eye as possible; since an object point emits its rays in all directions and fills all the air that surrounds it, and thus the air closer [to the object] will be much more packed than the air further away from it.[100]

considered as a contemporary alternative to this, is also attested by the explanation of the Galilean telescope in: Marco Antonio de Dominis, *Tractatus de radiis visus et lucis in vitris perspectivis et iride* (Venice: Thomas Baglionus, 1611), pp. 34-43.

99 See, for instance: *JIB*, II, p. 210: 'Radij ad aerem et atomos ejus reflexi, pereunt.' Of particular interest are Beeckman's 1633 suggestions to either evacuate the air out of a telescope to avoid scattering of light against air particles, or to open up part of the tube along its circumference to 'release' light that has become superfluous through collision. See: *JIB*, III, pp. 318-320: 'Telescopium promovere' resp. 'Telescopij diaphragma quale esse debeat.' For the photometric experiment that the Dordrecht surveyor Jacob Spoors carried out in 1633 at the request of Beeckman, see: *JIB*, III, p. 321; Zuidervaart and Rijks, '"Most Rare Workmen"', pp. 66-67.

100 *JIB*, II, p. 209: 'Lens magna convexa requiritur ad remota per telescopium videnda', resp. 'Qui tubos oculares cupiunt facere, quibus longissima possunt conspici, debent necessario lentem ab oculo remotissimo valde magnam constituere, nam cum punctum visibile in omnem partem

The requirement is stressed as Beeckman continues: 'who thus, at a location 100 times as far from the object point as the eye pupil, wants to see it as bright [*tam clare*] as when observed with his own eyes, needs to capture the rays with a lens 100 times as large as the eye pupil'. Beeckman subsequently expresses doubt whether such a lens can be manufactured at all, and therefore questions earlier claims about lenses and telescopes that allow the reading of letters from as far as a mile away. The excerpt is important because here Beeckman for the first time expresses an awareness of material restrictions in telescope production. On the other hand, this remark also refers to the 'telescopic dream', the sixteenth-century rationale in which imaginary instruments are assigned startling magnifying qualities. Here, such claims are soberly refuted by Beeckman.[101]

This being said, Beeckman's preference for large-diameter lenses is remarkably reminiscent of Dupré's characterization of Leonard Digges' and William Bourne's proto-telescopic experiments in the 1570s. In Elizabethan England, argues Dupré, the polymaths Digges and Bourne chased the telescopic dream by innovatively approaching convex lenses from a phenomenological angle. In particular, they experimented with the 'blown-up' images that these lenses produce when being observed from their *punctum inversionis*.[102] Unfortunately, in hindsight, they associated magnification with lens *diameter*, and, in demanding extremely large lenses, Digges and Bourne posed criteria that became a 'dead end' on the technological level (for making such large-diameter lenses was destined to end in failure at this time). Beeckman's notes seem to indicate that on the conceptual level this attention for large-diameter lenses could linger on for a longer time, and in fact allowed one to approach the power of a telescope from an alternative angle, at a time when a model fully based on refraction was yet crystallizing out. Furthermore, Beeckman's approach to the problem, in contrast to Digges' and Bourne's, stands out in being rooted in a undeniably clear theoretical, geometrical

radios suos emittat totumque circum stantem aerem ijs impleat, erit proximus aer multo plenior quam remotior.'

101 On the 'telescopic dream', see: Van Helden, *The Invention of the Telescope*, p. 19; Lüthy, 'Atomism, Lynceus, and the Fate of Seventeenth-Century Microscopy', pp. 6-11; Dupré, *Galileo, the Telescope, and the Science of Optics*, pp. 234-236. There is a possibility that Beeckman is referring to Cornelis Drebbel's 1613 claim that he could make a telescope by means of which letters could be read up to 8 or 10 miles distance. Letter from Drebbel to King James, 1613, reproduced by Beeckman in his notebook in 1631 (this would imply that Beeckman possessed the letter in 1618, which cannot be ascertained). See: *JIB*, III, p. 440.

102 Sven Dupré, 'William Bourne's Invention: Projecting a Telescope and Optical Speculation in Elizabethan England', in: Van Helden et al., eds., *The Origins of the Telescope*, pp. 126-145.

Figure 5.3 Telescopic configuration according to Beeckman

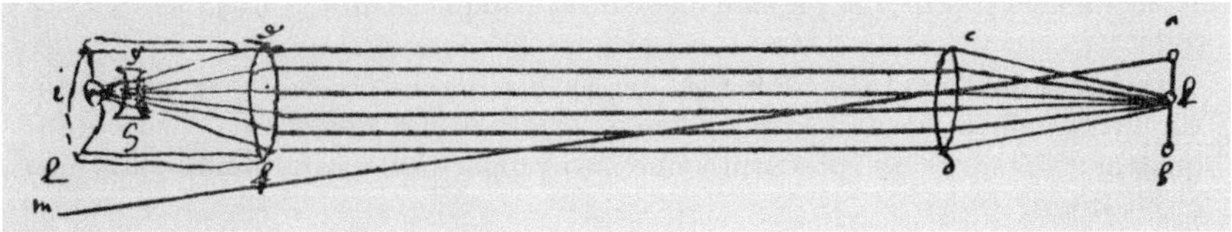

From: *JIB*, II, p. 209 [July-October 1622]. A convex lens is placed close to the object (to the right), relaying its image to a remote observer (to the left), who observes it through a common telescope.

footing, and – importantly – Beeckman knows where to stop dreaming as soon as he realizes that too large lenses can fall short on material grounds.

Large-diameter lenses were fundamental to Beeckman's idea of the telescope. A similar tendency to collect as many rays as possible can be found in another note dating from 1622.[103] Based on the accompanying drawing (fig. 5.3), De Waard interpreted the entry to describe a three-lens Keplerian telescope, but this interpretation is not correct.[104] What Beeckman argues is that a high quantity of rays ideally should be collected as close to the source as possible. He proposes to place a convex lens close to the observed object (in the figure: on the right-hand side). Aware that the rays, which leave the first lens in a parallel manner, can essentially be guided infinitely far away without dispersal, Beeckman lets the observer pick up these rays in a remote location by means of an ordinary telescope. What is remarkable is the oblique line *ma* in the drawing. In an unvoiced manner, here Beeckman shows a sudden attention for the spatial extension of the observed object. A consequence of this is that not all the rays radiate from the same point. Given this extra condition, the observer can no longer be placed arbitrarily far away, an outcome that Beeckman appears to realize. In an outburst of creativity, Beeckman concludes that the set-up can work nevertheless, if a 'helper' moves the first lens as instructed. When even this leads to nothing, the configuration condenses to a communication set-up in which messages are sent as coloured blotches.

A third application of his optical conception that Beeckman describes is the simple microscope.[105] Here, too, we can detect an equivalence with

103 *JIB*, II, pp. 209-210: 'Telescopio mediocri et vitro prope visibile remotissima videre.'

104 *JIB*, II, p. 210, n. 1.

105 *JIB*, II, p. 210. Of interest for the historiography of the microscope is that in 1620 (!) Beeckman refers to this instrument as 'such a spectacle through which one can see that a flea has a tail', suggesting a medical application in the same note: *JIB*, II, p. 33: 'Microscopii usus in medicina', where we read 'Twelck sonder schade van de sieckten soude konnen geschieden met sulck eenen bril daerdoor men sien kan dat een vloo eenen steert heeft.' The name 'microscope' was only coined in 1625, indicating that this title was added later. There is no reason to assume that

Figure 5.4 Diagram sketched by Beeckman for investigating ray density

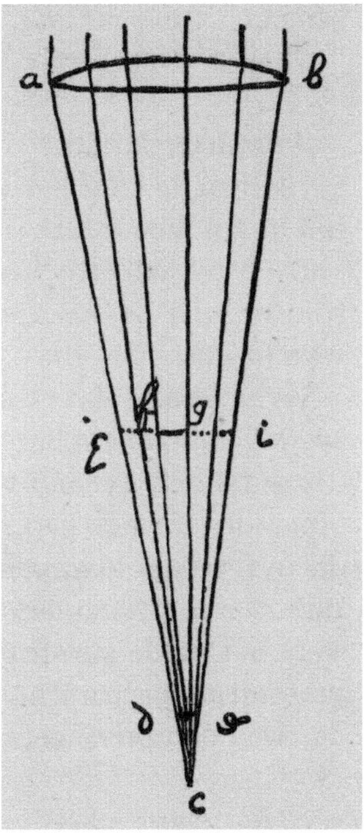

From: *JIB*, II, p. 211 [July-October 1622]. The diagram also functions as a representation of the simple telescope.

an observation from 'nearer'. The rationale is that a flea might be observed very close to the eye, if not for the fact that the eye cannot handle the high divergence of the rays the flea emits. A convex lens, placed close to the flea, collects these rays in sufficient quantity before they spread out too much, and sends them in a parallel fashion to the observer's eye. There, they cause a sensation which is equivalent to an observation from nearby.

The Phenomenology of Lens Projection

In the previous entry by Beeckman on the simple microscope, some hesitation can be sensed. This is because his description serves a higher cause, that

in this context Beeckman refers to anything but a single magnifying glass. The interpretation as 'microscope' stems from the application that Beeckman suggests. This takes into account Lüthy's observation that the emergence of the microscope should not be correlated with any technological event, but rather with the unfolding of a philosophical framework in which its usage could be embedded. See: Lüthy, 'Atomism, Lynceus, and the Fate of Seventeenth-Century Microscopy', esp. p. 2.

is, to find out 'how large the quantity of rays is that enter the eye through a telescope'. Beeckman sets off: 'For a long time I have doubted whether long telescopes do not show objects larger than the number of rays that enter the eye. If this is the case, things will be observed larger, but also darker, as the rays don't relate to the proportion of size.'[106] Geometrically, Beeckman deduces that rays bundled by a lens increase in density quadratically with distance (fig. 5.4). This, he subsequently compares using a measure of magnification he borrows from Kepler's *Dioptrice* (1611), that also diminishes 'with doubled proportion', i.e. quadratically, with the distance from the eye to the lens.[107] In this manner, Beeckman seems to keep intact the connection he demands between 'strength' of observation, or magnification, and ray density. And while notions such as 'size' and 'clarity' semantically start to take shape here, as yet they do not seem to undermine Beeckman's conceptualization of telescopic power.

What *does* contradict this conceptualization, Beeckman continues, is an empirical observation that he has obtained from projection through lenses: 'And nevertheless, shallower lenses make bigger figures on a white paper held nearby, with the same amount of rays, that is, when the lenses have the same size.'[108] It is important to realize that up to this point, Beeckman has no access to the Keplerian notion of a projected image, 'pictura', and therefore in his model of magnification he does not take the geometrical extension of images into account.[109] Beeckman assumed that two lenses with a different curvature, but with equal diameters, collect the same quantity of rays, and thus would yield an identical magnification. What is lacking is the inference that unequal curvatures lead to different focal lengths, at which distance the images are projected in focus but with *different* sizes. The phenomenology of lens projection gives Beeckman access to this phenomenon, which contradicts his reasoning, and, arguably, serves as an 'experimentum crucis' for his understanding of lens imaging and telescopic observation.

106 *JIB*, II, p. 210: 'Dubitavi diu an in tubis ocularibus longioribus res visibiles non magis augeantur quam multitudo radiorum in oculum incurrentium. Quod si fiat, videbuntur quidem res majores, sed obscurius, cum pro proportione magnitudinis radij non respondeant.'
107 It has been suggested by De Waard that proposition 83 of Kepler's *Dioptrice*, which Beeckman refers to, would not be applicable to the problem. The context makes it clear, however, that Beeckman uses Kepler's notion of perceived size, which is set out in propositions 80 and 82.
108 *JIB*, II, 211: 'Et nihilominus tamen lentes minoris convexitatis, in subjecto papyro albo, majores figuras faciunt cum non pluribus radijs, videlicet si lentes sint aequalis quantitates.'
109 Note that Kepler's notion of 'pictura' takes on a much more important role in his *Paralipomena* (1604) than is the case in his *Dioptrice* (1611). See: Malet, 'Kepler and the Telescope'.

Beeckman's optical notes from the 1620s are characterized by an increasing appropriation of the optics of Kepler.[110] In particular, in Beeckman's writings of this period, we can detect a Keplerian influence in the way he integrates spatial extension in his notion of imaging. Thus, in 1624 he explains how the imaging (projection) of the Sun through a lens has 'a nearby point [...], through its rays, create a different gathering point behind the glass, close to the first one, and so forth, until everything is found on the wall like it is in reality, but inverted' and also that 'where the Sun is a hundred times as large, so its appearance should also be hundred times as large'.[111] Both statements underwrite an understanding of Kepler's concept of a spatially extended 'retinal image'. Yet, as for the *domain* onto which Kepler's notion of image was considered to be applicable, we have to take into account an essential observation made by Malet. As already touched upon in the introduction, Malet points out that, particularly in the analysis of telescopic observation as written down in Kepler's *Dioptrice*, the latter's innovative concept of 'pictura' remained absent from modes of explanation. Rather, until well into the seventeenth century, a twin-track approach existed for descriptions of the formation of *real* images caused by projection, and the observation of virtual images, as produced by optical instruments such as the telescope. Thus, in case direct observation through optical instruments was approached through geometry, the associated magnifications were expressed using 'visual angles' as a quantitative handle, rather than through the geometrical size of the perceived images.[112]

Beeckman's notes from the 1620s, which by this time grow in frequency, endorse Malet's point. The distinction explains how Beeckman on the one hand manages to describe imaging through a single lens in strictly geometrical terms, soon also incorporating the quantification of the law of refraction which is developing in circles close to him,[113] but on the other hand in 1629 still struggles to coherently link the observation of sunspots through short versus long telescopes, with varying lens apertures, with

110 In his notebook, Beeckman starts to make references to Kepler's *Paralipomena* from 1616 onwards, and to the *Dioptrice* from 1619 onwards. The references grow in frequency in the 1620s, indicating that Beeckman studied these works more thoroughly around that time. See also Édouard Mehl's chapter in this volume.

111 *JIB*, II, pp. 295-296: 'ende het naeste punt maeckt door syn stralen een ander vergaerpuntken achter het glas, dicht by het eerste, ende so voorts naer advenant, dat alles op een muer staet gelyck in der waerheyt is, doch verkeert [etc.]', resp. 'ende waer de Sonne hondertmael grooter, so soude dien schyn oock hondertmael grooter moeten syn'.

112 Malet, 'Kepler and the Telescope', pp. 112-116; Malet, 'Kepler's Legacy'.

113 Beeckman had access to (Descartes') law of refraction in 1628, see: Schuster, *Descartes-Agonistes*, pp. 184-203.

Figure 5.5 Beeckman's drawing of a 'facet' lens

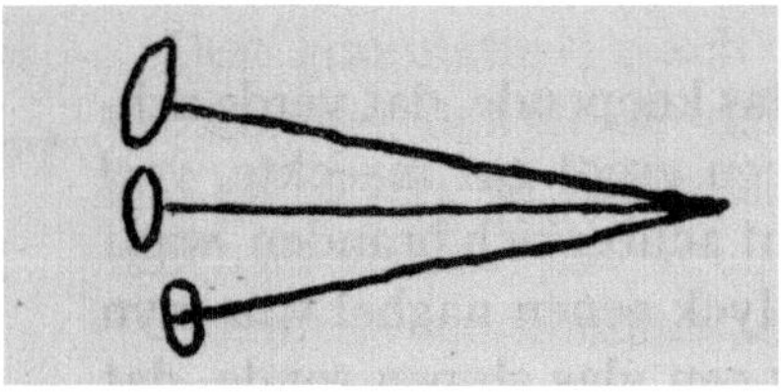

From: *JIB*, II, p. 296 [June-July 1624]

the projection properties of these devices.[114] Also, a belief in enhanced vision through large-diameter lenses keeps lingering on in Beeckman's mind. When material challenges put limits on this diameter, around 1624-1626 Beeckman initiates a trajectory to make 'facet' lenses, made of many common spectacle lenses bound together with copper wire (fig. 5.5). 'Thus joining many of these spectacles in one point,', Beeckman explains his motivation, 'the rays of one point will multiply towards the eye.'[115] In 1631, for the case of observation of virtual images through a telescope, we still see such considerations surfacing in his notes.[116]

After Philippus Lansbergen urged Beeckman in 1622 to procure a telescope 'as it appears, that Galileus à Galilæo had in his *Nuntio sidereo*', and the commercial lens supply does not seem to meet his demands, in the late 1620s Beeckman takes up lens grinding himself. The number of optical entries in his journal increases dramatically. By analysing these notes, Dijksterhuis has shown how Beeckman's grip on lens grinding developed between simple homemade observations and a growing involvement in a network of craftsmen. Yet, Dijksterhuis also points out a remarkable discrepancy between the time Beeckman invests in lens grinding, and the absence of a real purpose for his lenses.[117] Evidently, in Beeckman's notes a growing familiarity with the influence of material defects on lens performance can be discerned. Expressions of this kind generally originate in Beeckman's frustration to obtain a good quality telescope lens. For instance, in 1624 Beeckman concludes that a lens he commissioned in Middelburg does not satisfy, after

114 *JIB*, II, p. 121: 'Telescopium quo Solis maculae conspiciantur.'

115 *JIB*, II, p. 296: 'Ergo veel sulke glasen op een punt aeneen voechende, sullen de stralen eens punts aen het oogh vermenichvuldighen.' The project eventually fails on theoretical grounds. Observation teaches Beeckman that placing the individual glasses under an angle will not alter their 'linea refractionis', and will not make their gathering points coincide. Beeckman subsequently tries to reach his goal with concave mirrors. See: *JIB*, II, p. 357.

116 *JIB*, III, pp. 228-229: 'Telescopia longoria meliora ob duas rationes.'

117 Fokko Jan Dijksterhuis, 'Labour on Lenses: Isaac Beeckman's Notes on Lens Making', in: Van Helden et al., eds., *The Origins of the Telescope*, pp. 257-270.

he judged the gathering point to be 'so big, that it could not be brought to perfection'. To blame, according to Beeckman, is 'the boy who had ground it, not having a perfect and sufficiently large grinding dish for our lens'.[118]

We thus learn that Beeckman evaluates lens quality by projecting light and judging the size of the focal point. The appreciation for projection with lenses in the early seventeenth century was all but reserved to Beeckman. Descriptions of a camera obscura that make use of a lens (rather than a pinhole) trace back to 1550.[119] In the seventeenth-century Low Countries, camera obscura projection became part of a broader cultural scheme, in which the pursuit of a naturalistic representation of landscape and surroundings was paramount.[120] Well-known is Constantijn Huygens's testimony about the camera obscura he bought from Cornelis Drebbel, that 'all painting is dead by comparison, for here is life itself or something more elevated if one could articulate it'.[121] Closer to Beeckman, in his Dordrecht circle, we encounter this practice as well. In Johan van Beverwijck's *Schat der Ongesontheyt* (1644), the author describes the projection of harbour scenes, in a small tower on his Dordrecht house (colour ill. 1).[122] The aspects that make Beeckman's references to projection stand out, however, are twofold. First, they show that from an early date onwards, projection was applied in lens-grinding circles for the quality evaluation of lenses. Second, this practice shows that, although the role of the concave lens and the eye in telescope observation escaped a geometrical analysis for a long time, in the practical setting of projection, the boundary between real images and direct, virtual observations started to get crossed, quite some time before this rapprochement took place in the theoretical models of (telescope) imaging. Essential for this domain hopping is that the suitability of a lens for telescope use was read from its behaviour in a projection context.

118 *JIB*, II, p. 295: 'Ick bevondt dat dit vergaerpunt so groot was, dat ment tot gheen perfectie en konde brenghen', resp. 'So docht ick doen dat dit quam by foute van de jonghen, die het geslepen hadde, niet hebbende een perfect ende so groot cirkelstick daer hy ons glas in slypen konde.'
119 Girolamo Cardano, *De subtilitate* (Nuremberg: Johannes Petreius, 1550).
120 See, for instance: Svetlana Alpers, *The Art of Describing: Dutch Art in the Seventeenth Century* (Chicago: University of Chicago Press, 1983). For a critical evaluation of the role of the camera obscura, see: Sven Dupré, ed., *Optics, Instruments and Painting, 1420-1720: Reflections on the Hockney-Falco Thesis.* Special issue of *Early Science and Medicine* 10:2 (2005).
121 'Toute peinture est morte aux prix, car c'est icy la vie mesme, ou quelque chose de plus relevé, si la parole n'y manquoit.' On this quote, see: Arthur K. Wheelock, 'Constantijn Huygens and Early Attitudes towards the Camera Obscura', in: *History of Photography* 1 (1977), pp. 93-103; Alpers, *The Art of Describing*, p. 12.
122 Johan van Beverwijck, *Schat der Ongesontheyt* (Dordrecht: Jasper Gorissz., 1642), II, pp. 9-11. On Beeckman and Van Beverwijck, see: Van Berkel, *Isaac Beeckman on Matter and Motion*, pp. 54-55.

The practice of lens evaluation through projection is documented by Beeckman from circa 1633 onwards. Thus, he describes how 'a glass, as large as another one that was ground on a shallower dish, will not make the *species rerum exterarum* in a darkened chamber as perfect', concluding that 'therefore it is not surprising that long telescopes are the best'.[123] References to material (grinding) imperfections that require evaluation soon turn up as well.[124]

Noteworthy is that Beeckman seems to share this evaluation practice with peers. For instance, in 1635 Beeckman wins a lens-grinding competition against Johannes Sachariassen after both their lenses are tested in the 'dark chamber of Dr. [Jacob] Lansbergen', medical doctor and son of Philippus Lansbergen.[125] About Sachariassen, one of the earliest actors in the Dutch telescope making tradition, another note by Beeckman indicates that he made use of projection in 1634.[126] In Deventer, too, Beeckman's friend Henricus Reneri informs us by letter that he puts the same lenses to use in both a camera obscura and in a telescope.[127] The practices of Beeckman and his peers show that they transformed camera obscura experiments into a setting for quality innovation.

Finally, one note by Beeckman written down in 1635 can elegantly be linked with recently identified grinding properties that troubled the quality of the earliest telescopes. Beeckman describes how he intentionally introduces grinding defects in a lens, by applying extra pressure on the left and the

123 *JIB*, III, p. 258: 'Een glas, even groot met een ander, dat op een vlacker becken geslepen is, en sal de species rerum exterarum in een doncker kamer so perfect niet doen konnen als het ander voorss.', resp. 'Daerom en ist niet vrempt dat de langhe verrekyckers de beste syn.' Beeckman argues that the shallower lens shows the image points mutually more separated, noting that in this manner the eye can discern them more easily.
124 *JIB*, III, p. 387.
125 *JIB*, III, p. 430: 'Ick sleep teghen Joh. Sacharias om best, van gelyck glas, maert myne was veel beter, so hyselve seyde, beproeft synde in een doncker kamer tot Dr. Lansbergens.'
126 *JIB*, III, p. 397: 'Johannes Sacharias maeckt de schilderye in een doncker kamer recht met een platte spieghel, die hy in het backxken boven teghen het verhemelte horisontael vast maeckt' ('Johannes Sacharias makes the paintings in the camera obscura erect by means of a flat mirror that he attaches horizontally against the ceiling of the case'). On the position of Johannes Sachariassen, son of Sacharias Jansen, see: Huib J. Zuidervaart, 'The "Invisible Technician" Made Visible: Telescope Making in the Seventeenth and Early Eighteenth-Century Dutch Republic', in: Alison D. Morrison-Low et al., eds., *From Earth-Bound to Satellite* (Leiden: Brill, 2012), pp. 41-102, esp. pp. 67-71. While the reference to Jacob Lansbergen shows Sachariassen was involved in lens evaluation, the 1634 reference mentions Sachariassen reversing camera obscura images by means of a mirror – a method that traces back to Della Porta's *Magia naturalis*.
127 Robin Onno Buning, *Henricus Reneri (1593-1639): Descartes' Quatermaster in Aristotelian Territory* (PhD diss., Utrecht University, 2013), pp. 108-116.

right of the lens surface with his fingers, such that 'those areas are abraded the most, and therefore are thinner than above and below'. Subsequently deploying the lens for projection, Beeckman observes that 'holding the paper closer, one only saw the *erecta*, as if there were no *transversa*; and holding it further, one only saw the *transversa*'. Next, Beeckman relates the local 'saddle-shaped' deformation of the lens surface with the variation in focal distance that it induces on the affected rays.[128] What Beeckman is experimenting with, is a grinding error known as 'astigmatism'. Recently, this type of lens error has been identified as a prime characteristic of spectacle lenses from the late sixteenth century, and this is what impeded the early innovation of the telescope.[129] Beeckman's grip on this material phenomenon can be described as 'closed loop'. The causality between lens shape and resulting image provides Beeckman with all the necessary ingredients to mitigate the defect.

This becomes evident when we contrast Beeckman's awareness of lens-grinding errors, at this stage, with how such errors were approached around 1608, when the telescope first emerged. As we have seen, the application of a diaphragm on the objective lens of the first telescopes masked the worst grinding errors, and made an instrument composed of common spectacle lenses viable for telescope use. Nevertheless, even if this happened by trial and error, it is doubtful whether the link with grinding errors was actually made at that point. To the contrary, we have seen that references to diaphragms in the 1610s, also by Beeckman, rather attributed its effect to the enhancement of observation through compression. Consequently, while the recipe for good telescopes was certainly available around 1608, in the form of a diaphragm, what was missing at that time was an identification of

128 *JIB*, III, p. 422: '15^{en} Junij 1635 hebbe ick een glasken geslepen ende gepolyst op myn metalen becken in deser voeghe: Aen d'een syde gelyck ick gewoon was, maer aen d'ander syde stelde ick geduerich myn vynghers op de slyncker ende rechterkant; in de midden noch boven noch onder niet [...] waerdeur gebeurt is dat die plaetskens meest geschuert syn geweest ende daerom dunder dan onder ende boven [...] Dit glasken dan stellende int gat van den veynster van myn doncker kantoor [...] so schenen de erecta (dat is t'gene dat recht overeynde stondt) buyten in de locht door het glasken opt pampier kommende, klaerder alsmen het pampier wat naerder het glasken brocht ende de transversa schenen klaerder als men het pampier verder afhielt.' On this experiment, also see: Dijksterhuis, 'Labour on Lenses', pp. 257-259. On the effect on the focal point, Beeckman writes: 'Waeruyt blyckt dat de concursus radiorum verder of naderby kommen naerdat het glas hier of daer meer of min gedouwt is' ('from which it follows that the *concursus radiorum* comes further or nearer depending on whether more or less pressure is exerted on the glass here and there').

129 Willach, 'The Long Road', pp. 103-104. The dominance of astigmatism in early modern spectacles and telescope lenses from the early seventeenth century has been confirmed by measurements by the present author.

the cause of the problem, grinding errors, and thus an incentive to resolve these and make even better instruments. This situation differs from how Beeckman approached lens quality in the 1630s. The projection setting offered Beeckman and his peers almost tangible access to grinding errors in lenses, in which the impact on image quality could readily be evaluated. Moreover, Beeckman integrates a reasoning based on refraction in his hands-on experiment. He associates the variation in gathering points with local defects on the lens's surface. This allows him not only to explain, but also predict, how an image produced by a badly shaped lens is affected, and how this may be avoided. In short, it leads us to the mechanism by means of which lens innovation in the early seventeenth century could take place. Eventually, this mechanism allowed telescope lenses to achieve a good quality over their entire surface, making a diaphragm superfluous.[130] It implies that, through the access to an innovation trajectory it offered, we can categorize Beeckman's understanding of lens imaging in his later years as 'technological'.

Conclusion

Beeckman's notes offer us a unique view on the early conceptualization of the telescope in general, and, more particularly, on how Beeckman incorporated an understanding of the instrument in his developing natural philosophy. For one thing, they show how a single, preferred framework for understanding the telescope did not exist in the 1610s. Rather, different modes of explanation circulated, borrowing from the metaphysics of light or from medical literature, and adherence to an understanding based on the principle of refraction initially was but one of the available options. Such alternative modes of explanation did not necessarily compete – rather, aspects borrowed from seemingly incompatible sources could be brought to converge by early commentators in their often highly original conceptualizations of the telescope.

This heterogeneity of sources and explanatory frameworks was essential for Beeckman to construct his proper understanding of the telescope in 1618. Beeckman's piecing together of a working principle for the telescope was original, too, and, furthermore, it was driven by his desire to explain phenomena in a natural philosophical system. This move allowed Beeckman

130 On the improvement of lens quality: Willach, 'The Development of Telescope Optics in the Middle of the Seventeenth Century'.

to explain the telescope from a single underlying principle, that of corpuscular optics. It also enabled him to merge knowledge about the diaphragm, which came to him as an empirical observation, into his conceptualization of the telescope – as his analogy of water flowing through a river bed so beautifully illustrates. On the other hand, Beeckman's interpretation also shows that the masking of grinding errors initially was not a motivation for him to apply a diaphragm. At stake was the compression of species, not the material imperfections of lenses. More generally, it confirms that the application of a diaphragm by early telescope practitioners, while having solved real quality issues, may have been motivated by completely different reasons, not necessarily grounded in the quality of lenses.

The framework that Beeckman subsequently arrived at shows that telescopic vision could to a surprising degree be modelled quantitatively without having access to focal parameters of a telescope's lenses. This finding contradicts earlier historiography, which assumes that access to the power of a telescope required knowledge of the focal ratio of its lenses. Surprisingly, for Beeckman, the convex lens's *diameter* is key to a telescope's power. The diameter determines the amount of rays, emanating from an object, that is captured by the telescope, and is subsequently guided into the eye. It logically leads to a preference for large-diameter lenses. Interestingly, while such an inclination has been described as a (technological) 'dead end' in c. 1570 proto-telescopic experiments, it seems that on the conceptual level this attention for large-diameter lenses could linger on for a longer time. Importantly, what Beeckman's notes show is that the power of a telescope did not need to be associated with magnification as a measure for the *size* with which an object is perceived. This is key to understanding how Beeckman manages to model telescopic vision in a quantitative manner. He unconditionally relates the number of rays that enter the eye with the *distance* at which an object is observed. The telescope captures more rays, and therefore provides a view equivalent to an observation from nearby. It virtually translates the observer towards the object. Such an interpretation is in line with the majority of (Dutch) references to the telescope in the first decades of the seventeenth century, which refer to it as an instrument 'for seeing far', *not* an instrument for seeing 'large'.

The relation with distance, then, implies that we cannot dissociate the telescope from the spatial context in which it was deployed and understood. This is evidenced by the many optical 'configurations', which are clearly embedded in a spatial, observational setting, that Beeckman thinks up and draws up in his notebook in order to study the telescope. The spatial setting is significant, as Beeckman's conception of telescopic vision is founded on the mathematical-optical proposition that the intensity of radiation diminishes quadratically with the distance from the source. Thus, the core axiom of Beeckman's understanding of the telescope was conceptually rooted in a geometrical treatment of light that also came to prominence in astronomical and astrological literature of the sixteenth century, fields that, moreover, were increasingly being appropriated by the science of optics. On the other hand, to Beeckman it intertwined with his corpuscular reasoning – witness of which are his concerns that light corpuscles may collide with air, in which case the quantitative character of his modelling would come under threat.

Beeckman's notes on projection reveal an inquisitive mind that operates in the social setting of the seventeenth-century Netherlands. Much has been written on the use of camera obscura projection in this time frame, yet Beeckman's notes offer us, often in relation to his experience with lens grinding, some surprisingly detailed insights into this practice. Of concern here is the question how projection affected Beeckman's understanding of the telescope and of lenses. I have argued that optical projection gave Beeckman access to the notion of spatially extended images. Next, I advanced an experiment in 1622 as a crucial step for Beeckman. Up to that point, he had shown himself unaware of the influence of lens curvature on the size of a projected image. The projection phenomenon that Beeckman observed required him to reconsider the primacy of lens diameter for his optical conceptualization and opened the door for novel optical notions, such as curvatures and images, that would become the core vocabulary of seventeenth-century dioptrics. Still, Beeckman did not apply images and curvatures to the telescope immediately, nor did he do so unequivocally. This supports Malet's observation that Kepler's innovations were integrated in the field of optics along a two-track approach. Indeed, we see that Beeckman kept associating large lens diameters with better observation *in telescopes* throughout much of his career, while he had left this path behind him for the case of real images produced by single lenses.

Subsequently, I showed how it was in the context of projection that this distinction started to get lifted. The practice of projection created a setting in which real and virtual images reached a common ground. The generation of both virtual and real images by the same lens made it possible to correlate the effects of lens properties for both camera obscura and telescope use. In Beeckman's notes, we encounter this crossing over predominantly in the context of lens quality evaluation. The optical quality of a lens, and hence its suitability for telescope use, was read from the lens's performance in a camera obscura setting. This reveals that quality criteria were tacitly believed to be comparable for both contexts. Moreover, the experimental setting of projection gave tangible access to lens quality and lens defects, as it allowed for easy manipulation with direct visibility of the resulting effects. An experiment described by Beeckman in 1635 shows that in this context, all the required ingredients were present to improve the quality of lens grinding in a controlled manner. In particular, Beeckman's deliberate 'misgrinding' of an astigmatic lens, and subsequent association of the surface defects with focal variations and with the image imperfections he observed, testifies to a technological understanding of lens performance. Considered in its broader social context, it suggests that the warm reception and deployment of lens projection, using camera obscuras, stands in relation to an increase in lens quality that can be observed around c. 1630. This quality increase yielded better lenses, made the application of a diaphragm to cover lens defects eventually redundant, and brought about a coming of age of the intriguing instrument that got Beeckman pondering over optics early on in his career, in the 1610s.

About the Author

Tiemen Cocquyt is curator in early modern science at the Rijksmuseum Boerhaave, Leiden. He specializes in the history of optical instrumentation, combining historical research with modern measurement techniques. The research for the article in this volume was made possible through the support of NWO-Humanities, Museumbeurzen, grant no. 333.54.004.

6 Optics, Astronomy, and Natural Philosophy

Beeckman, Descartes, Kepler, and the Dutch Connection

Édouard Mehl

Abstract

The place of Isaac Beeckman in the history of philosophy and science is paradoxical. On the one hand, through recent works (Klaas van Berkel) there is no longer any doubt that he was one of main promoters of the mechanical philosophy; on the other hand, his name still remains in the shadows of the great names (Descartes, Kepler, Galileo, Gassendi). This chapter attempts to relate Beeckman's way of thinking as closely as possible to the reality of his intellectual network and his scientific activity. Starting from a strictly local analysis, and from questions that have been partly forgotten by historians of the long term, this chapter tries to recover a part of an original train of thought that cannot be reduced to the thought patterns (e.g. abstraction and idealization) that are traditionally presented as the drivers of the Scientific Revolution.

Keywords: Isaac Beeckman, René Descartes, Johannes Kepler, Martinus Hortensius, *potentia dei*, reflecting telescope

The discovery and publication, in the first half of the twentieth century, of Isaac Beeckman's *Journal* made it possible to understand the role played by Beeckman in the development of the new 'mechanistic' physics, of which the principal actors were already well-known: Marin Mersenne, René Descartes, Pierre Gassendi, and Thomas Hobbes.[1] In this context, Beeckman was first

1 Even though the great narrative of modern science that we owe to Edmund Husserl (*Die Krisis der europäischen Wissenschaften*, 1936) does not mention Beeckman, two of Husserl's close

Van Berkel, Klaas, Albert Clement, and Arjan van Dixhoorn (eds.), *Knowledge and Culture in the Early Dutch Republic: Isaac Beeckman in Context*. Taylor & Francis Group 2022
DOI: 10.5117/9789463722537_CH06

perceived as the missing piece in an already fully-formed puzzle, one in which he simply had to be fitted, with perhaps some rounding off at the edges where the new piece failed to fit perfectly into the space allocated for it. Since then, studies more specifically centred on Beeckman – including Klaas van Berkel's eminent and pioneering work – have made it possible to better understand the particularities of Beeckman's 'mathematico-physical' philosophy without measuring him against the standards of this 'greater picture' in which his philosophy had haphazardly been integrated, as was done previously.

A third step, which I aim to sketch here, might involve trying to understand how Beeckman's original philosophy was created by situating itself into a fabric of local relationships, through intensive and ongoing exchanges on specific well-defined questions that bore not so much on the great scientific discoveries of Galileo Galilei or Johannes Kepler, but on what made these discoveries possible in the first place. My purpose, therefore, is to shed new light on the Dutch-German exchanges of the late 1620s and early 1630s concerning telescopic observations of sunspots, the survey of solar motion and magnitude, and their cosmological consequences, which are not always explicit in the technical debates surrounding Beeckman, Martinus Hortensius (Maarten van den Hove), Philippus Lansbergen, Kepler, Wilhelm Schickard, Gassendi, Mersenne, and, of course, Descartes. This intense activity of observation, experience, and exchange focuses less on the objects of experience and the phenomena themselves than on theorizing the means and conditions that make this experience possible. This applies, for example, to the lively exchange between Hortensius and Kepler concerning the respective advantages of the 'optical tube' and the lens telescope that is mentioned, among other things, in this chapter. There is nothing spectacular and no sensational discovery in this exchange, which is probably one of the reasons as to why it has so far attracted little interest from historians of science, despite the fact that it constitutes one of the essential clues explaining the shift from Kepler's *Dioptrice* (1611) to that of Descartes (1637). It is this scientific network that constitutes, beyond any doctrinal divergences, the community of 'good minds' to whom Descartes gladly recognizes his debt, even as he denies being indebted to any of the

disciples highlight his importance: Alexandre Koyre, *Études Galiléennes* (Paris: Hermann, 1939), and Jacob Klein, 'Phenomenology and the History of Science', first published in *Philosophical Essays in Memory of Edmund Husserl* (Cambridge, Mass.: Harvard University Press, 1940), and later in: Robert B. Williamson and Elliott Zuckerman, eds., *Jacob Klein: Lectures and Essays* (Annapolis: St. John's College Press, 1985), pp. 65-84, esp. p. 83.

individuals in it.[2] It is therefore less Beeckman, or his relationship with Descartes, than the lattice of this scientific network that I propose to explore a little further here.

1 Beeckman and Descartes, Readers of Kepler: The 1620s

Beeckman's reading of some of Kepler's great texts in the summer of 1628 and his influence on the creation of Cartesian physics has been the object of a number of studies.[3] Throughout this essay, there will be opportunity to see that Descartes' and Beeckman's critical stance on the imposing life work of the imperial mathematician is comparable to that of the Dutch mathematician, and student of Beeckman and Lansbergen, Martinus Hortensius (1605-1639),[4] one of the main protagonists of this 'Kepler affair', whose attitude – quite independently of the 'Galileo affair' – is representative of the very attentive reception, both admiring and highly critical, given to

2 Descartes, *Dioptrice*, in: *Oeuvres de Descartes*, publiées par Charles Adam et Paul Tannery, 12 vols. (Paris: L. Cerf, 1897-1910; new ed. in 13 vols., Paris: Vrin, 1974-1986) [henceforth *AT*], VI, p. 82. See Philippe Hamou's commentary in: *La mutation du visible. Essai sur la portée épistémologique des instruments d'optique au XVIIe siècle. Vol. 1: Du* Sidereus Nuncius *à la* Dioptrique *cartésienne* (Villeneuve d'Asq: Presses Univiversitaires de Septentrion, 1999), pp. 244 sq. On Descartes' 'debt' to Beeckman, see Klaas van Berkel's seminal article, 'Descartes' Debt to Beeckman: Inspiration, Cooperation, Conflict', in: S. Gaukroger, J. Schuster, and J. Sutton, eds., *Descartes' Natural Philosophy* (London: Routledge, 2000), pp. 46-59.

3 This reading mainly concerns Kepler's *Astronomia nova* (1609) and the fourth section of his *Epitome astronomiae Copernicanae* (1621). For the *Astronomia nova* (referred to in Beeckman's *Journal* by its subtitle: *De motu Martis*), see: E.J. Aiton, *The Vortex Theory of Planetary Motion* (London: Macdonald/New York: American Elsevier, 1972), p. 13; R.S. Westman, *The Copernican Question: Prognostication, Skepticism and Celestial Order* (Berkeley and Los Angeles: University of California Press, 2011), p. 497; Édouard Mehl, 'Descartes a-t-il critiqué les lois de Kepler?', in: É. Mehl, ed., *Kepler: La Physique céleste. Autour de l'*Astronomia nova *(1609)* (Paris: Les Belles Lettres, 2011), pp. 231-259; S. Gaukroger, *Descartes' System of Natural Philosophy* (Cambridge: Cambridge University Press, 2002), p. 145; John Schuster, 'Kepler', in: Lawrence Nolan, ed., *The Cambridge Descartes Lexicon* (Cambridge: Cambridge University Press, 2015), p. 421; Klaas van Berkel, *Isaac Beeckman on Matter and Motion: Mechanical Philosophy in the Making* (Baltimore: Johns Hopkins University Press, 2013), p. 64 et passim – quoting John Schuster: *Descartes and the Scientific Revolution, 1618-1634: An Interpretation* (Ann Arbor: University Microfilms International, 1977), a work I have not had the opportunity to read. And finally, my own *Descartes et la fabrique du monde. Le Problème cosmologique de Copernic à Descartes* (Paris: Presses Universitaires de France, 2019).

4 On Hortensius, see the biographical notes in: Klaas van Berkel, 'De illusies van Martinus Hortensius. Natuurwetenschap en patronage in de Republiek', in: Klaas van Berkel, *Citaten uit het boek der natuur. Opstellen over Nederlandse wetenschapsgeschiedenis* (Amsterdam: Bert Bakker, 1998), pp. 85-110.

the work of the imperial mathematician by the members of the scientific network surrounding Beeckman.

Hortensius claims that there is a standard of mathematical evidence, and his main criticism of Kepler is that the imperial mathematician fails to reach this goal or general condition of geometrical evidence. This may seem, at first sight, highly paradoxical since Kepler is the first known astronomer to introduce the 'a priori' in astronomy and to claim to demonstrate, on purely mathematical grounds, the necessity of the Copernican hypotheses.[5] Kepler seeks the causes of planetary motions in a theory of harmonic proportions, a kind of speculative mathematics that both Beeckman and Hortensius believe to have no foundation in astronomical phenomena, nor any correspondence with them. Up to this point, Hortensius's criticism of Kepler is almost identical to Aristotle's attitude toward the Pythagoreans, who preferred to accommodate phenomena to their so-called 'reasons', rather than the other way around.[6]

Beeckman would not have disagreed with Aristotle: first look at the phenomena, and then find the causes or reasons that explain them.[7] Instead of studying the book of nature, as he should do and as he pretends to do, Kepler bases himself on the innateness of mathematical reasons, which the human mind finds first within itself before recognizing them in nature. In fact, he believes that he can only recognize these reasons in nature because they dwell first and foremost inside his mind, just as the acquisition of grammar precedes reading and makes it possible.[8] Hence Kepler's great faith in pure geometry, which he sees as the *modus operandi* of

5 See, for example, Michael Mästlin in the preface to Kepler's *Mysterium cosmographicum* (1596). Mästlin insists that the Keplerian *inventum* is not a geometric invention, but the invention of a norm and rule for restoring astronomy's certainty (Johannes Kepler, *Gesammelte Werke*, ed. Walther van Dyck and Max Caspar, 22 vols. (Munich: Beck, 1937-2017) [henceforth *GW*] 1, $82_{39\text{-}46}$). He therefore insists on the methodological value of Kepler's discovery: 'Ab hoc igitur tempore, qui coelorum motus plenius inquirere, et quae in Astronomia adhuc mancae sunt, reficere et redintegrare volet, habet iam a priori patentem ianuam, qua ingrediatur, habet rectissimam normam, ad quam, ceu ad Lydium Lapidem, omnes suas observationes, totumque calculum examinet.'

6 Aristotle, *De caelo*, II, 13, 293a25.

7 On the distinction between *causa* and *ratio*, see: *Journal tenu par Isaac Beeckman de 1604 à 1634*, publié avec une introduction et des notes par C. de Waard, 4 vols. (The Hague: Martinus Nijhoff, 1937-1953) [henceforth *JIB*], I, p. 215 (which opposes a unique cause to a multiplicity of 'reasons' deduced not from things, but 'from axioms and all arguments').

8 See *Harmonice mundi* (1619), IV, 7, in: *GW* 6, $277_{4\text{-}14}$: all mathematical genera are universals that are, neither more nor less than other genera, abstracted from sensible things. However, among the mathematical species, the circle has a special status: it is not only in the soul as the Idea of things external (*non tantum ut Idea rerum externarum*), but also as a certain form of the

the world's creation. Beeckman, on the other hand, believes that the creative act precedes the reasons that explain it, and that Kepler's main failure is therefore to try and make these reasons (the *explanans*) independent of the creative act (the *explanandum*), and submit the latter to the former. God does not do what he *is capable of*, but what he *wants* to do, and the things we conceive of as 'possible' are only so because God wished them to be, following the divine decree that institutes the order of possibility, i.e. the order of creation. In a very early text (1617-1618), Beeckman returns to a classical theme in theology relating to the debate with the Ancients, primarily Galen, whose position, which aims to occupy the middle ground between 'Moses' and Epicurus, does not attribute absolute power to God, but only recognizes God's power to do His utmost within the limits of possibility itself as defined by the various dispositions of matter. This leads Galen to conclude that God cannot 'turn a stone into a man', nor 'give a man the strength of a lion'. 'Why was man not born with the strength of a lion? Because all bodies are created in number and in order [*numerus et ordo creata sunt corporum*], an order and number outside of which nothing can be done [*extra quae nihil fit*].' This order defines the limits within which bodies act and produce effects that are not infinite, but 'indeterminately finite' (*non infinita, sed finita non determinata producant*). This order, which proceeds from the Divine, defines nature and at the same time circumscribes the domain of things that we can understand. The domain of intelligibility is in effect circumscribed and instituted by divine decree: 'Indeed, we only understand as possible the things that God wants to be possible.'[9]

Beeckman revisits this theme for the second time in June 1620, and insists once again that possibilities can be reduced to well-defined combinations such as number sequences and series, or syllabic combinations.[10] Number (in which bodies are created) is therefore what limits or even defines natural or ordered

soul (*ut forma quaedam ipsius Animae*), and as the unique breeding ground of all mathematical sciences (*ut promptuarium unicum omnis Geometricae et Arithmeticae scientiae*).

9 *JIB*, I, pp. 163-164 [23 December 1616-16 March 1618]. I do not retain the correction of *fieri* and *intelligi* by C. de Waard. Galen's locus is referred to and commented on by Pierre Hadot, *Le Voile d'Isis. Essai sur l'histoire de l'idée de nature* (Paris: Gallimard, 2004), p. 182. Hadot's only failing is that he never mentions Beeckman, and orchestrates a debate between Descartes and Galen that would never have taken place had Beeckman not inspired the young Poitiers resident to it.

10 *JIB*, II, pp. 56-57 ('Numerus etiam certus formarum certa composita et finita praescribit'). This entire paragraph can be compared to what Aristotle says about the Abdera School in *Metaphysics*, A, 4, 985b13, where he relates Democritus's triad ῥυσμός, διαθιγή, τροπὴ to σχῆμα, τάξις, θέσις. See: Aristotle, *Metaphysics, Alpha Volume*, introduction, translation and notes by J.-F. Pradeau (Paris: Presses Universitaires de France, 2019), p. 157 (*Met.* H, 2, 1042b10 sq.).

possibility. This is the only possibility that we can conceive of.[11] Beeckman therefore believes that Galen is right to deny, contrary to the atomists, that all things are born of the accidental interaction of atoms, since this interaction is not accidental but pre-determined. The arrangement of possibilities is determined *a priori* by the parameters that govern and determine quantity: numbers and shapes. It is therefore impossible to have a mountain of infinite height with a fixed shape.[12] However, Beeckman continues, Galen is wrong in affirming that God always does His utmost, because possibility derives from matter, which allows certain transformations and forbids others (whence the logic of species, which results from a range of combinations equally governed by the inviolable law of numbers).[13] Matter's creative act cannot be submitted to the laws of matter since matter precedes the creative act, and not the other way around. And since *possibility itself presupposes matter*, of which it is the

11 *JIB*, II, pp. 56-57, commenting on Galen, *De usu partium corporis humanis*, XVII, 1. In this text, Galen argues that we should not blame nature for having configured animals in one way rather than another: 'One must consider from the outside the entire body of the animal, and contemplate the action of each one of its parts [...] and not wrongly accuse nature of ill will. And for all that some have affirmed and maintained that the first elements of substance are such that by nature's artifice they cannot be conjoined and assembled, this has given them occasion to make war on, blame and correct nature.'

12 Everything that is represented as a thing must therefore have a well-defined shape that delimits it, thereby making it thinkable. Descartes mentions the example of a mountain without valley (i.e. without boundaries to delimit it and give it its well-defined shape) as an example of an absolute impossibility: *Meditationes, V, AT*, VII, 66, 2-15. Descartes and Beeckman had a direct exchange, no later than in 1630, on the question of what 'absolute impossibility' could mean, and it was Descartes himself who apparently first experienced some difficulties in the matter, with Beeckman criticizing him for prescribing impossibilities to angels, an accusation Descartes defended himself against by objecting that his use of the *ab Angelo fieri non posse* formula was intended precisely to avoid falling into the trap of theologians who attributed impossibility to God (Descartes to Beeckman, 17 October 1630, *AT*, I, p. 165, 12-29). Descartes and Beeckman apparently kept throwing the ball back and forth, and we can see a posthumous follow-up to this discussion in the *Letter to Gibieuf* dated 19 January 1642 (*AT*, III, p. 476, 8; p. 477, 15), according to which one could easily have an (incomplete) idea of a mountain by abstracting away from its shape, but not the idea of a material surface that is defined and at the same time indivisible, since this implies a contradiction: 'We can therefore say that the existence of atoms implies a contradiction.'

13 *JIB*, II, p. 57 [24-29 June 1620]: Rather than saying (as does Galen) that God is a demiurge who creates each thing by submitting to the laws of nature, 'it is more appropriate to say that all things are born of nature and the constitution of a place [*a natura et constitutione loci esse nata*], but that in the beginning God created the principles that in joining cannot fail to produce this result. When particular primary elements [*primordia*] co-occur, they produce birds, when others co-occur, dogs, and when still others co-occur, fish. This concurrence [*concursus*] does not allow for more variation than that required to make three or ten syllable words from the 24 letters of the alphabet.'

effect or consequence, we cannot say that God only does what He is capable of, or that He can do the impossible. Rather, in creating matter *ex nihilo*, God creates the very possibility of the possible and the impossible. We can therefore not say that 'even God' is incapable of turning a man into a stone, but rather that He has made it materially impossible to turn a man into a stone. Let us remember this essential point: that possibility derives from matter, as number combinations derive from the existence of quantity. In other words, and more fundamentally, there is only possibility where there is first space, which is the universal form of all possibility. Independently of Descartes, and even before their first meeting, Beeckman therefore formally articulates in his *Journal* the key aspects of the famous Cartesian doctrine of the creation of eternal truths set out in a letter from Descartes to Father Mersenne in the spring of 1630[14]: the divine decree does not depend on possibility; it is possibility that depends on the divine decree. God is therefore essentially incomprehensible: possibility presupposes matter, i.e. space, which is not *a* body, but the essence of all bodies or, if you prefer to use the language of metaphysics, the possibility of possibility.[15]

The fact that Beeckman's early reflections on Galen (1616) and the creation and layout of the *primordia rerum* (June 1620; *Journal*, II, 57) partly preceded his first meeting with Descartes is of an importance that cannot be fully apprehended but that has at the very minimum the following three major consequences: (i) Beeckman was fully aware of the ontology (i.e. the doctrine of possibility) underlying his conception of nature as an object of 'physico-mathematical' science, (ii) Descartes' 1630 position on the 'creation of eternal truths' – which cannot be isolated from his *Discourse* on nature as the norm of possibility[16] – itself constitutes an extension of Beeckman's

14 Descartes to Mersenne, 15 April 1630, *AT*, I, p. 145, 7-10; 6 May 1630, *AT*, I, p. 149, 21-30; 27 May 1630, *AT*, I, p. 152, 2-9. On the notion of possibility/impossibility, there was indeed a confrontation between Descartes and Beeckman in 1630 (see Descartes to Beeckman, 17 October 1630, *AT*, I, p. 165, 12-29).

15 See, as a complement to previously quoted texts, the text in: *JIB*, I, p. 131 [23 December 1616-16 March 1618]: 'Igitur nihil fuit nisi Deus incomprehensibilis absque tempore et spacio.' See also my comments in: Mehl, *Descartes et la fabrique du monde*, pp. 267-269.

16 I therefore not only maintain that Beeckman's 1617-1618 writings 'anticipate many aspects of the 1630 doctrine' (Mehl, *Descartes et la fabrique du monde*, p. 373), I also believe that we must recognize the presence of this ontology of possibility in Descartes' considerations on the 'order of the world' in the *Discourse*, which ensures that 'we no more desire to be healthy when ill, or free when in prison, than that we now desire to have bodies of a matter as incorruptible as diamonds, or wings to fly like birds' (*AT*, VI, p. 26, 10-14, my translation). Indeed, if we do not desire this 'now', it is because we do not consider it possible. Unsurprisingly, Descartes draws the example of bodies made of diamond from the same lecture by Galen as Beeckman: 'Is it not possible for all body parts to be constructed without any inconveniences? If it were possible,

reflections on the divine decree as a principle for defining this effectiveness of possibility that we call *nature*, and (iii) Beeckman's concise reflections therefore envelop the de facto shift from the Greek notion of *phusis* as pure coming into being to the modern logico-mathematical concept of nature, understood from the perspective of a concept of *order* and similar to what Husserl identifies as a 'defined multiplicity' (a closed system of axiomatically regulated relations), which in turn defines a purely combinatorial and ordinal concept of possibility as we conceive of it. It is Beeckman who takes this giant leap, but it is Descartes who, having taking it alongside him, then thinks through its ultimate metaphysical consequences.[17]

These philosophical beliefs, which predate Beeckman's first encounter with Descartes (10 November 1618), also guide Beeckman's critical reading of Kepler's *Epitome astronomiae Copernicanae* in the spring of 1629 – maybe this time following a suggestion of Descartes, who read *Epitome IV* when it was first published, as is evident from his notes in *Cogitationes privatae*, written between 1619 and 1621.[18] Thus, towards the end of the dozen folios

they would all be: but since there is no artifice that can circumvent the depravation, vice and imperfection of matter, in creating a masterpiece as hard and as impervious to everything as a diamond, one is left to arrange, accommodate, and appropriate it to the extent that it allows itself to be manipulated [...]. We and the stars are not made of the same substance' (Galen, *De l'usage des parties du corps humain. Traduit du grec et du latin*, V, 4 (Paris, 1559), p. 209). Since Descartes developed his 'moral by provision' when he was in Germany, sometime in 1620, we cannot exclude that the two authors may have exchanged letters in June 1620.

17 On the Greek concept of nature, see: M. Heidegger, 'Die Physis bei Aristoteles', French translation in: *Questions II* (Paris, 1968), p. 227. In translation: 'The other essential moment of οὐσια is lost: *the entrance into presence*. Yet it is this, which in the Greek notion of being constitutes the decisive point.' I take the term 'multiplicity' to have the meaning assigned to it by Husserl in his *Ideen I* (§ 72). See: Husserl, *Idées directrices pour une phénoménologie pure et une philosophie phénoménologique*, trans. by Jean-François Lavigne (Paris: Gallimard, 2018), pp. [135], 211-212. His thesis, developed earlier in his *Krisis der europäischen Wissenschaften*, is that the modern era proceeds to the 'substitution' (*Unterschiebung*) of a 'multiplicity' for original nature. According to Husserl, this defines the very object of Galileo-Cartesian science, but, to be accurate, one should note that it was Beeckman who led Descartes to discover this object.

18 We can therefore not exclude that it was Descartes who led Beeckman to closely read Kepler's *Astronomia nova* and *Epitome IV*, rather than the opposite. For this early reading of the *Epitome IV* by Descartes, at a time when he himself was writing his *Cogitationes privatae*, compare *AT*, X, p. 218, 6 ('Una est in rebus activa vis') and Kepler's remark: 'Una est actio seu ἐνέργεια naturalis movendi corpus planetae' (*GW* 7, $333_{24\text{-}31}$). In 1619-1620, Beeckman only quoted Kepler's *Dioptrice* (1611), and in particular, in the margin of a reflection by Galen on the usefulness of mathematics for medicine, Kepler's example demonstrating the hyperbolic shape of a lens (*JIB*, II, p. 56; cf. Kepler, *Dioptrice*, prop. LX, in: *GW* 4, 372). However, we cannot exclude that at the time of their meeting in 1618, Beeckman and Descartes were already familiar with the first part of the *Epitome* or Kepler's 'spherical doctrine', announced since 1614, and published by Planck in 1618 with a dedication dating from August 1617 (*GW* 7, 574, 'Nachbericht'). In this volume, Kepler outlines

Beeckman devotes to Kepler, between his reunion with Descartes [*JIB*, III, p. 94] in the autumn of 1628 and his meeting with Gassendi in the summer of 1629 [*JIB*, III, p. 123], he writes:

> Kepler, in book 4 of his *Epitome astronomiae Copernicanae*, p. 532 [*GW* 7, 307_{25} sq.] and before [p. 530/*GW* 7, 306], looks accurately for the cause [*causa cur*] according to which the periodic times of the planets are in the proportion *sesquialtera eversa* of the intervals [i.e. distances from the Sun], and why there is for the same planet a proportion between velocity and distances, to the perihelion and to the aphelion. But he introduces here the variable density of the planetary body [*GW* 7, 283_{10}], although he could have done the same by considering only the variable surface area of the largest and smallest planets, which are somehow hindered by the matter floating around them, so that their speed does not increase infinitely. Indeed, he would have seen that the bodies themselves have a triple reason in the continuation of their movement, and the surface area as a double reason to prevent it. He could have thus, without appealing to the variable of density, obtained the same results, satisfying appearances – even though I do not deny in any way that all the planets can have different degrees of material density. But how admirable these harmonies are! Harmonies, I say, everywhere so well-proportioned and adjusted to his meditations. We admire it, for we do not consent to such harmonies: these bodies indeed have been constituted by a divine decree much more casually [*fortuito jussu*] than he thinks, like our mountains, the oceans and all the things that can happen on Earth.[19]

his purely geometric concept of motion (*Epitome* I-V, in: *GW* 7, 82_{13-15}): 'Nam cum motus sit res geometrica, non minus quam magnitudo; magnis igitur tarditas, parvis celeritas respondet, non vicissim.' We should also note that Descartes' definition of motion explicitly echoes *Epitome* (*GW* 7, 81_{26-27}: 'Motus est separatio mobilis, quatenus mobile, de loco suo, et translatio in locum alium'), but also radically modifies it: motion is still a reciprocal separation from the parts of a body, but *from those of a contiguous body*, 'non autem ex uno loco in alium' (*Principia philosophiae*, II, art. 28, in: *AT*, VIII-1, 55_{12-21}).

19 *JIB*, III, pp. 120-121: 'Keplerus Lib. 4 Epit. astron., pag. 532 [*GW* 7, 307_{25} sq.] et ante [p. 530/*GW* 7, 306], sollicite quaerit causam cur motus planetarum periodici sint in proportione intervallorum sesquialtera eversa; et cur idem planeta in perihelio habeat motum in proportione intervallorum ad motum in perihelio [sic, read *aphelio*]. At introducit variam corporum raritatem [*GW* 7, 283_{10}-284_{47}] cum idem praestare potuisset collimando ad varias superficies majorum et minorum planetarum, quae varie in motu suo propter aliquam ibi volitantem materiam impediuntur, ne infinite eorum celeritas crescat. Vidisset enim corpora ipsa triplicatam habere rationem in motu continuando, superficies vero duplicatam duntaxat in impediendo. Facillime igitur poterat, densitate non mutata, tales magnitudines exhibere, quae apparentijs satisfecissent, cum tamen interim non negaverim omnes planetas posse habere diversas in raritate et densitate

Curiously, Beeckman does not contest the material validity of the law of periods, which he seems to accept without examining it, but only its justification or 'cause',[20] which he refuses to seek, as does Kepler, in the theory of 'harmonic proportions' (*ut causa formalis*) and which he transfers to the current of celestial matter (*ut causa efficiens*) that slows down bodies in proportion to their surface area, given that *the smaller a body, the more surface area it has in proportion to its size*. This is an idea that Beeckman defended for a long time and it gained the status of a mathematical principle of physics,[21] as did his 'theorem of motion' (*quod semel movetur, semper movetur*). If Kepler had paid attention, he could have used it to deduce his law of periods (*vidisset enim*). This principle, which Beeckman argued since at least 1614 to be crucial in explaining the motion of projectiles and gravity, therefore constitutes the only 'cause', if cause there is, of periodic times. This means first of all that contrary to Aristotle, Beeckman only recognizes a single unique principle that explains both the motion of heavy bodies and that of celestial bodies. Note that the explanatory scheme Beeckman proposes strictly inverts that of Kepler: whereas Kepler believes that an external motive power overcomes the resistance to motion (or a tendency to inertia) characteristic of all bodies as material bodies (the *vis inertiae* of the

consistentias. At miramur tantas et ubique tam proportionatas, et ejus harmonicis meditationibus adeo accommodatas; miramur, inquam, quia isti harmoniae non assentimur et magis fortuito Dei jussu haec corpora esse constituta quam ille existimat, exemplis montium apud nos et marium omniumque omnino, quae in Terra proveniunt.' For the proportional ratio of volume, mass and density of planetary bodies (abstracting away from the physical cause that makes them such), see: Max Caspar, in: *GW* 7, 587. If V_1 and V_2, M_1 and M_2, D_1 and D_2, r_1 and r_2 designate respectively the volume, mass, density and radius of two planetary bodies, Kepler states that since $V_1 : V_2 = r_1$ and r_2, and since $V_1D_1 = M_1$ and $V_2D_2 = M_2$, we obtain the ratio $\frac{d1}{d2}=\frac{\sqrt{r2}}{\sqrt{r1}}$. What Beeckman criticizes here is as much the introduction of a density variable, which cannot be determined a priori, as the assumption of a *concinnitas geometrica* as the formal cause.

20 Kepler, *Epitome*, IV-2, 4 [De causis proportionis periodicorum temporum], in: *GW* 7, 306_{33}.

21 *JIB*, I, p. 25 [July 1613-April 1614]: On the difference between the motion of projectiles: if a stone thrown with the same force as a feather rises higher, it is because the feather, being more 'rare', has more surface area, and therefore more resistance to air. Beeckman therefore uses this theorem as the mathematical explanation of the fall of heavy bodies (see later *JIB*, I, p. 267 [26 December 1618]). See also *JIB*, I, p. 31 (on the movement of the spinning top); I, p. 171 ('notum est mathematicis ut proportio superficierum a soliditatis pro magnitudine corporum variet'). One of the crucial uses of this theorem is its application to celestial mechanics. See: *JIB*, I, p. 196 (27 June-7 July 1618): the larger the bodies, the less surface area they have in proportion to their size, and therefore the less resistance to the continuation of their motion. It should therefore not be surprising that the motion initially imparted to celestial bodies proceeds without fail for thousands of years. Compare on this point Descartes, *Principia philosophiae*, III, art. 144 (*AT*, VIII-1, p. 194), the reasoning of which seems to come straight from Kepler and Beeckman's *Journal*.

Earth, or of any other celestial body, proportionate to its material density),[22] Beeckman sees inertia as the absence of this tendency: that which resists the motion of a moving object does not come from the object itself, but always from outside – in this case from a *species motrix* emanating from the Sun.

However, Kepler only applies the scheme of solar motive power to the planets' annual motion, or orbital revolution, while seeking another explanatory principle for their daily axial rotation, which cannot, in his opinion, emanate from the Sun, and must therefore be found in the bodies of the planets in question – in this case, the Earth. On this issue of the cause of axial rotation, the first part of his *Epitome astronomiae Copernicanae* (1618) attempts to bypass, by means of a daring deduction, the Aristotelian opposition between violent motion (produced by an extrinsic cause) and natural motion (*ab intrinseco*). In 1618, Kepler uses the notion of *forma corporea* to explain that the Earth's motion is perennial because the motion originally imparted to the Earth was incorporated in it, and it has so much arranged and conformed its matter to fit this motion that the motion has replaced and, in a way, eliminated the Earth's material inertia.[23] The Earth's motion is, as it were, its life and soul.[24] The comparison with the movement of a spinning top, which Kepler comments on at length,[25] is mostly used to mark the essential difference between the two: the spinning top is a

22 Kepler, *Epitome astronomiae Copernicanae* IV-2, 1 (De causis motus planetarum), in: *GW* 7, $296_{30\text{-}37}$: 'Even if any celestial globe is not as grave a body as is, on Earth, rock, nor as light, as fire is with us, it nevertheless has of itself, because of its matter, a certain impotence to transit from one place to another, it has a natural inertia or rest, by which it rests in any place where it is placed in an isolated state.' Cf. A. Koyre, *La révolution astronomique. Copernic, Kepler, Borelli* (Paris: Les Belles Lettres, 2016), p. 290.

23 Kepler, *Epitome* I-V [1618], in: *GW* 7, 89_{42}-90_{2}: 'Verisimile est, hanc primae rotationis continuatam speciem in terra, transformatam esse, seu coaluisse in talem facultatem corpoream, et sic in fibras terrae, dispositas secundum ductum motus ipsius, inolevisse.'

24 This alone suffices to situate Descartes' early thoughts, in his *Cogitationes privatae*, on Kepler's horizon and more specifically in his *Epitome*, since in the *Cogitationes* Descartes refers to *forma corporea*, a term that belongs exclusively to the language of *Epitome*. Descartes, *Oeuvres complètes*, under the direction of Jean-Marie Beyssade and Denis Kambouchner (Paris: Gallimard, 2016), I, p. 273: 'Any bodily form acts according to harmony.'

25 Beeckman returns to this problem in 1629, based on Kepler's text (*JIB*, III, pp. 118-119), in order to elaborate on his comparative study of the motion of the spinning top and the Earth: 'Aer enim, vel aqua turbinem circumstans, movetur a turbine in eam plagam, in quam turbo movetur. Turbine igitur vel inaequali existente vel ejus motu deficiente, ita ut non ampliùs turbine exactè erectum servare possit, turbo inclinatur; aer verò jam antè in gyrum motus, extremitatibus ejus extra gyrum exstantibus impingit et occursans secum rapit. Nunc hic ultimus motus motui praecessionis accommodandus est ac ostendendum praecessionem aequinoctiorum causari a motu substantiae circa Terram volitantis, eodem modo quo extremitates axis Terrae circa polum eclipticum moventur.'

heavy body and its parts tend towards the centre of the Earth, gravity being nothing more than the union of all parts to the whole. But the parts of the Earth do not tend towards union with the Earth since they are already in union with it. They therefore have no gravity of their own, except 'in force' (i.e. the Earth's parts only become heavy bodies and develop resistance to motion when they are separated from the Earth). The initial impulse towards motion has progressively insinuated itself into the body of the Earth and has taken root in its fibres, which are no longer content communicating a received motion, but now impart this motion as much as they receive it, and therefore receive it from themselves. Thus the Earth has progressively made the motion that it received its own. This motion has become the Earth's own action; it is no longer a guest on the terrestrial body but its 'entelechy'. The original motive power no longer has to conquer the inertial resistance of the material subject since, in animating it, it becomes this subject. Kepler is therefore very ambiguous, both leaning on a strictly mechanistic paradigm to explain celestial motion (that of a clock, scales, or spinning top), and at the same time maintaining that motion always depends on a motive power, which forces him to retain a form of soul or 'entelechy' in celestial bodies. It is this ambiguity that Beeckman attempts to outlaw and eradicate from pure natural philosophy.

In a way, Beeckman's theorem of motion (*quod semel movetur, semper movetur*) very simply resolves all the difficulties Kepler encounters in his attempt to explain the possibility and perennial nature of axial rotation: he even eliminates the problem itself. As a result, Beeckman has no difficulty concluding in 1628 that this entirely confused metaphysical theory of terrestrial animation is, like many of Kepler's other assumptions, 'not worthy of a philosopher',[26] thus agreeing – although for different reasons – with the Roman censors who suspected Kepler of repeating the errors of Origen and relegated the *Epitome* to the Index as early as in the first half of 1619.[27]

26 *JIB*, III, p. 29.

27 For the Roman censure of Kepler's *Epitome*, examined in February 1619, see: Pierre-Noël Mayaud S.J., *La condamnation des livres coperniciens et sa révocation à la lumière des documents inédits des Congrégations de l'Index et de l'Inquisition* (Rome: Editrice Pontificia Università Gregoriana, 1997), pp. 59, 65ff. Beeckman's general criticism of Kepler's postulation of a ruling intelligence in the Sun that allows it to govern the planets appears notably in his reading of Kepler's *Astronomia nova* (chap. 33), where solar radiation is understood as the emission of directing rays towards celestial bodies with the intention of moving them: *JIB*, III, p. 100: 'What would be this kind of Wisdom, which does not move anything out [sc. light] except to a certain preconceived goal?' ('Quae enim haec foret in Sole sapientia quâ nihil nisi ad praevisum usum eijceret?').

2 Did Descartes and Beeckman Read Kepler in 1618?

The fact that Beeckman's reading of Kepler was decisive for Descartes and played a crucial role in the genesis of Descartes' *The World* (*Le Monde*) is today a well-documented and little contested fact. But the question we would like to raise here is whether Beeckman could have had access to a copy of Kepler's *Epitome* as early as 1618, or at the time of his first meeting with Descartes. Is it not strange that around the time of their meeting, Beeckman and Descartes both undertook once again the study of the motion of the spinning top (following a first attempt in April 1614)[28] in order to call into question the role ascribed by others to the idea of a 'centre of gravity' that varies depending on the speed of rotation?[29] When Beeckman and Descartes imagine that a spherical container turning at great speed can be lifted into the air by the slightest force, including a simple draught, they are not trying to make physics amusing but to understand how planets find in the vortex a point of equilibrium at which their orbital trajectory stabilizes. It is therefore no coincidence that in the problems mentioned in their first collaboration in November-December 1618, Beeckman returns to the Earth's third motion or *motus trepidationis*, a very light and slow vacillation of the terrestrial rotation axis, which Kepler explains in his *Epitome* using precisely the example of the spinning top.[30] The approach in folio 103 of Beeckman's notebook is complex.[31] Beeckman begins with the theoretical case of motion in a vacuum, where this *motus trepidationis* cannot, according to him, occur. If it does occur, therefore, it is because of the presence of air, or the environment in which planetary revolutions take

28 *JIB*, I, pp. 30-32 [see fig. 17.1 – editor's note].

29 *JIB*, I, pp. 242-243. For the translation of this text, see: Descartes, *Oeuvres complètes* (2016) (*Notes et opuscules du Journal de Beeckman*, presentation, translation and notes by Frédéric de Buzon, Dutch translation by Theo Verbeek), I, p. 86: 'When a spinning top is spinning, the reason it remains vertical is not immediately the fact that it rotates around its own *centrum gravitatis*, but the result of the spinning motion I long attributed to the pin traversing the spinning top to the extent that it rests on the ground.'

30 *JIB*, I, p. 253: 'Terrae motus annuus, bene intellectus, tertium motum omnino abolet.' Beeckman may have obtained this information from Rudolph Snellius (*JIB*, I, p. 21), whose lectures he followed in Leiden in 1609. On Snell, see: Liesbeth Cornelia de Wreede, *Willebrord Snellius (1580-1626): A Humanist Reshaping the Mathematical Sciences* (PhD diss., Utrecht University, 2007), http://igitur-archive.library.uu.nl. Note also that Rudolph's son Willebrord Snellius had met Kepler in Prague a few years earlier, at the time they were both working with Tycho, just before Tycho died (October 1601). Later, Kepler suspected that Snell had taken all of Brahe's last observations from 1600 to 1601 to Holland, and thus he advised Brahe's heirs to claim them from him. See: Kepler to Georg Brahe, 17 August 1628 (letter n. 1088), in: *GW* 18, 361.

31 *JIB*, I, pp. 253-225 [23 November-26 December 1618].

place. Yet this page concludes with a paragraph that anticipates very precisely Beeckman's response to reading *Epitome IV* ten years later. It is not the laws of collision that are presented here, but the laws of the relationship between the revolution times of bodies that follow unequal circles in equal time, a question Beeckman indicates he has not yet had time to explore further:

> But if two bodies are so arranged that they travel in air of the same [density] in unequal circles [in size] in equal time, it is because they are unequal either in weight or in surface. But the one with the greater weight, if it is moved faster, will be much less impeded by air. If it is moved more slowly, then it will seem that the two bodies will approach the proportionality of their movement: so that the heavier and slower one can travel in parts through the small circle at the same time and in the same way [through its parts] as the lighter and faster one travels through the large circle. But since the areas are equal, it seems to me that the proportion cannot remain the same. Yet I have not yet had time to explore this question further. If, of two bodies of equal weight, the one with the greater surface is moved more with greater or lesser speed, we will see how the other will behave in relation to it.[32]

Let us also note that Descartes' first essay on the falling of heavy bodies rests solely on the assumption that the cause of motion is not to be found in the body itself but in the Earth's 'power of attraction' (*vis attractiva*), which is once again a way of hypothetically accepting the validity of celestial physics, of which this new definition of gravity by power of attraction constitutes the pivot and, one might argue, the Archimedean point.[33]

32 *JIB*, I, p. 254: 'Si enim duo corpora hoc momento disposita sint ut inaequales circulos aequali tempore percurrerent in eodem aere, id fit, quia sunt inaequalia pondere vel superficie. At quod est majus pondere, si celerius moveatur, multo minus ab aere impedietur; si tardius, magis quidem videntur dicta duo corpora ad motûs proportionalitatem accedere, ita ut gravius et tardius eodem tempore et eodem modo per partes parvum circulum percurrere possit, quo levius et celerius majorem percurrit. At cum superficies sunt aequales, mihi tamen videtur proportionem non perpetuo eandem permanere; non tamen jam vacat mihi diligentius inquirere. Si vero duorum corporum, pondere aequalium, id quod majus est superficie, tardius aut celerius moveatur, videat alius ut se res sit habitura.'

33 Descartes, *Oeuvres complètes*, I, p. 98 (*AT*, X, p. 75; *JIB*, IV, pp. 49-52). On the doctrine of gravity and the *vis tractoria telluris*, see the foundational text of the introduction to Kepler's *Astronomia nova*, in: *GW* 3, 25, and his letter addressed to David Fabricius, 11 October 1605, in: *GW* 15, 241. In his *Études Galiléennes* (fasc. II, p. 26), Koyré maintains that Descartes and Beeckman follow here in the steps of William Gilbert, which is rather unlikely from a doxographic perspective: anything they seem to borrow from Gilbert in fact came to them from its use and adaptation by Kepler.

It therefore seems that Beeckman's reflections in 1628 on the causes of periodic times are limited to revisiting the issue where he left off in November 1618, and adapting it to Kepler's data. It goes without saying, however, that Beeckman would not have been able to discover the sesquialteral ratio of distances and times and that the fact that Kepler was able to do so, thus achieving one of the greatest victories of the human mind, is not due to his superior observation skills but precisely to the fact that his investigations were led by a belief in a 'harmonic a priori', in accordance with the Copernican axiom of *symmetria mundi*,[34] an axiom for which Kepler never ceased to seek a specific mathematical meaning – something Copernicus never attempted – and which he always used as a point of focus or regulatory idea. In this sense, Kepler's harmonic causes do have a heuristic function and, even if we have to agree with Beeckman, Hortensius and Descartes that they do not sufficiently account for planetary eccentricities, we cannot deny Kepler the merit of having effectively discovered, by means of these 'causes', one of the most fundamental laws governing the universe.

3 From Planetary Eccentricities to Determining the Sun's Diameter: The Dissemination of an Originally Tychonian Criticism

It is hard to know whether, in rejecting the theory or harmonic causes, Beeckman truly understands to what metaphysical use Kepler wishes to put them, or whether it is precisely because of this use that Beeckman criticizes Kepler's theory and rejects what he considers to be arbitrary mathematics and pseudo-physics. What we do know is that Kepler greatly evolved on this issue and his discovery, in the spring of 1618, of the true reason or ratio of distances and velocities (i.e. Kepler's third or periodic law) opened his thinking to an entirely new field of possibilities.

Since 1596, Kepler had been keenly aware of the fact that phenomena did not always correspond precisely to the mathematical reasons assumed

34 Nicolas Copernicus, *De revolutionibus orbium coelestium/Des révolutions des orbes célestes*, I, 10, ed. M.-P. Lerner et al. (Paris: Les Belles Lettres, 2015), pp. 391-393: 'Invenimus igitur sub hac ordinatione admirandam mundi symmetriam, ac certum harmoniae nexum motus et magnitudinis orbium: qualis alio modo reperiri non potest.' This wording echoes G.J. Rheticus's *Narratio prima* [*and other texts*], edited and translated by H. Hugonnard-Roche with J.-P. Verdet, in collaboration with M.-P. Lerner and A. Segonds. Studia Copernicana, 20 (Wroclaw: Polish National Academy of Science, 1982), p. 59/112 ('admiranda […] motuum et orbium symmetria ac nexus').

to define them. For example, celestial distances did not coincide with the neo-Pythagorean theory of regular solids, the interlocking of which was supposed to explain intervals between planetary orbits. In 1596, Kepler, who had not yet broken away from the principles of circularist astronomy, accounted for this inexactness by postulating the need to manage a playing field between solids and inscribed orbits, a field in which he believed planetary eccentricities took place. It is easy to see how his great confidence in the vitality of a model that is in reality purely imaginary leads him to perceive as a cause something that is in fact an intellectually pleasing but perfectly false similarity, as he himself later recognizes:

> [*Mysterium cosmographicum* (1596), chap. XVIII] But either if these thicknesses of the spheres were investigated with complete certainty, or at least if the probable reasons why such large thicknesses were allocated by the Creator to each of them were revealed, then I pledge that I should produce from the solids angles [*arcus*] which would agree with the motions in all cases. For it is my opinion that after this discovery of this proportion in the heavens everything which still prevents us from attaining exact knowledge of the motions is to be attributed to errors in the eccentricities; and if those were removed [9]*, I think that the five solids would be of great assistance [...] for the corrections of the motions.
> [9]* [nota authoris, ed. altera, 1621] [*The five solids would be of great assistance for the correction of the motions.*] Of no assistance, in fact, not even the smallest, because they do not regulate the spheres, nor prescribe the limits of the eccentricities. But now that the eccentricities have already been found, as knowledge 'that', from the observations of Brahe, at last there is room for a search for causes, or knowledge 'why', from these five figures and the linked [*iunctis*] harmonic proportions.[35]

Kepler is and remains less interested in facts than in the explanation of their causes, even if over the course of 20 years this leads him to completely overturn his understanding of eccentricities (following his *Astronomia nova*, 1609) and their causes (starting from his *Harmonice mundi*, 1619). But Beeckman and his circle apparently fail to see why they should value 'harmonic reasons' more than the geometric reasons derived from Platonic

35 Johannes Kepler, *Mysterium Cosmographicum: The Secret of the Universe*, trans. by A.M. Duncan, introduction and commentary by E.J. Aiton, with a preface by I.B. Cohen (New York: Abaris, 1981), pp. 181, 189.

solids and applied in Kepler's *Mysterium cosmographicum* (1596) to distance ratios in a way that Kepler himself recognizes as impertinent.

Indeed, in the early 1600s, in the face of violent criticism from Tycho Brahe for his pseudo-deductive theory of eccentricities in the *Mysterium cosmographicum*[36] and with the rise of the new 'celestial physics', Kepler seems after a number of fruitless attempts to abandon the search for a precise and geometrically determined ratio between the *distances* of celestial bodies. With the elliptic orbits and the law of areas in his *Astronomia nova*, Kepler apparently renounces his youthful fantasies and concedes that planetary motion obeys purely mechanical reasons sufficiently accounted for by balance and weight (according to the analogy of the radius vector of a planet and the arms of scales). Kepler insists that there is therefore no intelligence at work in determining distances and their mutual relations but a necessity, the workings of which are adequately captured by 'celestial physics'. Instead, with the discovery of periodic times, Kepler convinces himself that he has once again found the perfect commensurability of distances and times, which can only be the work of the human mind (*opus mentis*).

Kepler's *Epitome IV* goes further still than his *Harmonice mundi* and the third law in applying the principle of proportion to all the world's bodies, allowing Kepler to attempt to deduce once again the distance of the sphere of fixed stars,[37] an undertaking Beeckman and Descartes rightly consider foolhardy and unfounded, for the simple reason that the sphere of stars ... does not exist! But the fact that Kepler uses the *same* reason – the general theory of proportions – to prove the existence of God or an intelligent design and the existence of this purely imaginary sphere makes it even more difficult to take these reasons seriously. It is therefore understandable that both Beeckman and Descartes refuse to see in these supposedly eternal truths anything but an effect of matter and its reflection in the mind co-created with the body.

While Beeckman studied the entire body of Kepler's work (autumn 1628-summer 1629), his former student Martinus Hortensius embarked on a Latin translation of Lansbergen's *Bedenckingen op den dagelijkschen*

36 See: Tycho Brahe to Kepler, 1 April 1598 (o.s.), letter nr. 92, in: *GW* 13, pp. 197-201.

37 See: *Epitome*, IV-I, 285, which, immediately following on its speculation on the density of planets, extends the principle of proportionality to three areas of the world, corresponding to the source of the motion (the Sun), bodies in motion (planetary bodies) and the locus of the motion (fixed bodies). This attempt to provide a specific mathematical meaning for cosmic proportions is also an attempt to provide a definitive answer to the objection raised from the onset by Tycho against Kepler concerning the disproportion of the sphere of fixed stars implied by the Copernican hypotheses (see: Tycho to Kepler, no. 92, in: *GW* 13, $199_{73\text{-}81}$).

en jaarlijkschen loop van den aerdtcloot (*Commentationes in motum terrae diurnum et annuum*), which he subsequently published with a preface that is highly critical of Kepler.[38] This preface echoed most of Beeckman's points of criticism. On the question of determining celestial distances, as well as their causes, Hortensius notes that Kepler customarily uses the kind of pseudo-physical explanations exemplified above in the *Mysterium*'s theory of the 'thickness' of spheres and of which he finds another example in the determination of the solar diameter. Kepler explains the discrepancy between his own theory and observations regarding solar magnitude by means of physical causes, such as the material density of planets or the nature of the aether surrounding the Sun.[39] This is the same kind of reasoning as in his 1596 treatise on measuring eccentricities. However, it is clear that these physical reasons are only assumed in order to explain the discrepancy between theory and observations, an approach that boils down, *mutatis mutandis*, to that of pre-Copernican astronomers who assumed new spheres to justify movements they could not explain in any other way. Kepler undoubtedly thinks that he is *deducing* the physical causes, which he believes can be mathematically determined. However, Beeckman and Hortensius argue that this pseudo-deduction is rather poor reasoning and that Kepler uses these physical causes, against all rules of common sense and evidence, simply as an adjustment variable: where his observations fail to meet his expectations, Kepler claims some physical cause is responsible for the anomaly. However, Hortensius argues that this is nothing but conjecture and a 'convenient excuse' (*frivolam excusationem a causis physicis*).[40] And yet, this criticism of Kepler, which was first expressed by Tycho Brahe, does not seem to spontaneously have

38 *Philippi Lansbergii commentationes in motum terrae diurnum et annuum et in verum adspectabilis caeli typum, ex Belgico sermonem in latinum versae a Martino Hortensio* [...] *una cum ipsius prȩfatione* (Middelburg: Zacharias Romanus, 1630).

39 Martinus Hortensius, *Responsio ad additiunculam D. Ioannis Kepleri* [...] *praefixam ephemeridi eius in annum 1624* (Leiden: J. Maire, 1631), pp. 52-53. 'Nam ut hoc fiat, requiritur ut in profunditate aurae aetheriae circa Solem, corpus aliquod adsit, & densius aethere, & certâ ac distincta superficie à reliquâ aura aetheriâ discretum [...]. Sed quis talia unquam in coelo observavit? [...] Tibi quidem facile erit, Keplere, sphaeram aethere densiorem, puta aërem, Soli circumdare; cui non tantum montes, maria, aër, & ne haec frustra sint, viventes etiam creaturae, in globo Lunae finguntur, sed & animalia somniantur in globo Solis [...]. Nam nisi haec firmissimis rationibus demonstres, nunquam a me impetrabis, ut relictis observationibus & demonstrationibus Geometricis, ad vanas illas conjecturas descendam.'

40 Hortensius, *Responsio ad additiunculam D. Ioannis Kepleri*, p. 47. See also p. 50: 'Ego, quo diutius haec verba tua considero, eo magis miror, tam frivolas conjecturas posse animo tuo satisfacere.'

occurred to Beeckman and Hortensius: rather, they inherited it from their teacher and elder, Lansbergen.

Behind Hortensius's virulent attack on Kepler, first published in Middelburg in 1630, one can see the shadow of Lansbergen. Lansbergen holds very original views in the field of Copernican astronomy and cosmo-theological ideas.[41] First of all, he belongs to the generation of Tycho Brahe and the young Kepler: Lansbergen was 35 years old when Kepler published his *Mysterium cosmographicum* (1596). As far as we can ascertain from a letter of Georg Herwart von Hohenburg to Kepler 1598, Lansbergen read the *Mysterium cosmographicum* early on and did not approve of it. At the time, Lansbergen was not following Copernicus and he flatly refused, as did Tycho Brahe, to place the Earth on the *orbis magnus* and to grant it an annual revolution around the Sun. As Kepler notes a few weeks later in his response to Herwart, Lansbergen belongs among geo-heliocentric system makers, together with Tycho, Helisaeus Roeslin, Nicolaus Reimarus Ursus and likely others.[42]

It is not known precisely when Lansbergen became a Copernican,[43] but the point is that although he converted quite late, at any rate after the 1600s, he remained an opponent of Kepler and clearly refused to accept anything that Kepler claimed as the foundation (either physical or metaphysical) of Copernican astronomy. He clearly rejected the first law of the *Astronomia nova* on the elliptical orbits of planets, as apparent from his diagram of the world system.

In this diagram, planets still complete circular orbs (or orbits) but the planetary orbits are not centred on the Sun: the centre of Saturn's orbit, as the planet furthest away from the Sun, is located precisely on the orbit of Venus and the centres of the revolutions of Mars and Jupiter, the two other highest planets, are located near Mercury's orb. This is rather unusual

41 On Philippus Lansbergen, see the essential chapter by Rienk Vermij, 'Philipp Lansbergen's Christian Cosmology', in: R. Vermij, *The Calvinist Copernicans: The Reception of New Astronomy in the Dutch Republic, 1575-1750* (Amsterdam: KNAW, 2002), pp. 75-99.

42 Kepler to Georg Herwart von Hohenburg, 26 March 1598, in: *GW* 13, 193: 'Quod de Philippo Landsbergio scribis: idem ante ipsum Reinmarus [sic] Ursum fecit.' On this episode, see: N. Jardine and A.-P. Segonds, *La Guerre des astronomes. La Querelle au sujet de l'origine du système géo-héliocentrique à la fin du XVIe siècle* (Paris: Les Belles Lettres, 2008), I, pp. 208-209.

43 It is important to mention here that Lansbergen's name appears for the first time in Beeckman's *Journal* in December 1616: *JIB*, I, p. 106. Beeckman notes a measurement of the meridian altitude of the Sun during the winter equinox he made using Lansbergen's quadrant and it is not impossible that the two men made this measurement together. As shown by Vermij, Beeckman's heliocentric turn (which rests on physical rather than astronomical causes and is, therefore, not strictly speaking Copernican in its approach) may have occurred during his meeting with Lansbergen, around 1616. See: Vermij, *The Calvinist Copernicans*, pp. 113-118.

since it means that the further a planet is from the Sun, the more its centre of revolution is eccentric. Moreover, Lansbergen is one of the very few Copernican astronomers to deny the slow motion of the Earth's axis and he still believes, as did Ptolemy and Al-Battani, that the slow motion of the equinoctial points really belongs to the sphere of the fixed stars.[44] Lansbergen refers to *De revolutionibus*'s third book, and shows that his own tables correspond more accurately with the ancient observations than either the *Prutenic tables* of Erasmus Reinhold, or Kepler's 1627 *Tabulae Rudolphinae*. Both Lansbergen and his former disciple Hortensius are very insulting to Kepler, claiming that *Tabulae Rudolphinae* lacks any predictive value – which is at least partially false.

It is not possible here to elaborate on all of the implications of these quite exotic and somewhat strange astronomical positions. At the very least, they imply that the Sun cannot be the centre of the magnetic force holding and moving the planets in their orbits, precisely as Kepler claimed since 1609. In other words, Lansbergen's position makes it impossible to maintain the physical foundations of the helio-dynamic theory of motion formulated by Kepler between 1609 and 1621 and which he referred to as *physica coelestis*. As far as we know, Lansbergen's critical position relies on astronomical observations of the motion of the Sun, the determination of its magnitude and its distance from the Earth. This must have required very precise observations of solar eclipses of the kind undertaken by Hortensius and Beeckman in mid-June 1630.[45]

44 M.-P. Lerner, *Le Monde des sphères* (Paris: Les Belles Lettres, 1997), II, pp. 129-134. Here is a translation of the justification offered by Philippus Lansbergen, *Tabulae motuum celestium pepertuae* (Middelburg: Zacharias Roman, 1634), pp. 17-18 [trans.]: 'Nicolas Copernicus makes the sphere of fixed stars immobile: proposing that the ecliptic and equinoctial sections are moved by some pre-existing kind of slow motion. From this he rightly concludes that the locations of stars fixed in appearance are as much held as a consequence as the motion of the ecliptic and equinoctial sections is pre-existing. But Copernicus's position in this is not very plausible, for two reasons. First of all the immobility of the sphere of fixed stars is not appropriate to nature. Just as the brain that God placed at the summit of the small world is not immobile but moves by a slow and almost imperceptible motion, which it communicates throughout the little world, similarly it is suitable that the supreme sphere of the larger world should not be immobile, but moved by a delayed and slow motion, and that it should communicate this motion to all other spheres. Secondly, it is not very likely that the motion of ecliptic and equinoctial sections is pre-existing, since if this were the case each and every place on Earth would have a different climate and be in a different part of the world than they were when the world began. Which is absurd.'

45 *JIB*, III, pp. 153-155 [10 June 1630]. Ten days later, during summer solstice (20 June 1630), Beeckman once again recorded a measurement of the meridian altitude of the Sun as seen from Dordrecht (*JIB*, III, p. 156). For Hortensius's observation, see his *Responsio ad additiunculam D.*

It is well-known that Kepler was obsessed with 'harmonic reasons/ratios' and, as shown above, both Hortensius and Beeckman find the Pythagorean background of Kepler's cosmological thinking highly suspicious. So far, so good, but the problem is far more complex and, we believe, rooted in the hypotheses underlying the *Astronomia nova*, in particular those regarding planetary eccentricities, which represent Kepler's reintroduction (in opposition to Tycho) of the *punctum aequans* and the bisection of eccentricity.[46] At least this is suggested both in Beeckman's *Journal* and in young Hortensius's aggressive response to Kepler (1631).[47] Clearly, the reintroduction of a Ptolemaic *punctum aequans* is a decisive point leading to the law of elliptical orbits and the law of areas.

The observational grounds for this conclusion are the measure of solar size and distance as observed in solar eclipses. These two measures formed, as we know, the basis for Lansbergen's highly unorthodox views on the dimensions of the heavens, which he believed to be far larger than assumed by Kepler. In his *Ouranometria*, which Beeckman read shortly upon its publication in 1631, Lansbergen offers some incredible distances for fixed stars, with diameters much larger than the Sun and with volumes 20,000 times larger than the *sphaera terrae*. According to Lansbergen, *sphaera terrae* refers to 'sphaericum corpus mente conceptum, cujus circulus maximus est magnus orbis Terrae'. It is unclear whether Hortensius himself acknowledged these incredible dimensions, but Descartes and Beeckman clearly disapproved of them; and they were right since Lansbergen had made some miscalculation.[48]

4 Hortensius as a Supporter of the Accuracy of the Telescope

As mentioned already, Hortensius wrote an important and very famous preface to his Latin translation of Lansbergen's *Commentationes in motum*

Ioannis Kepler, pp. 27-28: 'Mihi certe Deliquium Solis anni 1630 Iunii 10. S.N. [*stylo novo*] Dordraci Batavorum accurate observanti, tale nihil contingit animadvertere.'

46 See: Koyré, *La Révolution astronomique*, pp. 170-180.

47 This is one of the arguments put forward by Beeckman in his criticism of Kepler to Pierre Gassendi, in July 1629: *JIB*, III, p. 123: 'Ostendi quoque illi [sc. P. Gassendi] Keplerum frustra laborare, ut inveniat punctum ad quod planetae respicientes semper eundem situm retinent, ac demonstravi id per se necessarium esse.' It should be noted that on this visit, Gassendi gave Beeckman a copy of his observation of Roman parhelias (*JIB*, IV, pp. 149-151), the very observation Descartes did not wish to include in his *Meteors*.

48 See: A. Van Helden, *Measuring the Universe: Cosmic Dimensions from Aristarchus to Halley* (Chicago: University of Chicago Press, 1985).

Figure 6.1 Hortensius explains why Kepler claims that a telescope misrepresents the Sun's diameter

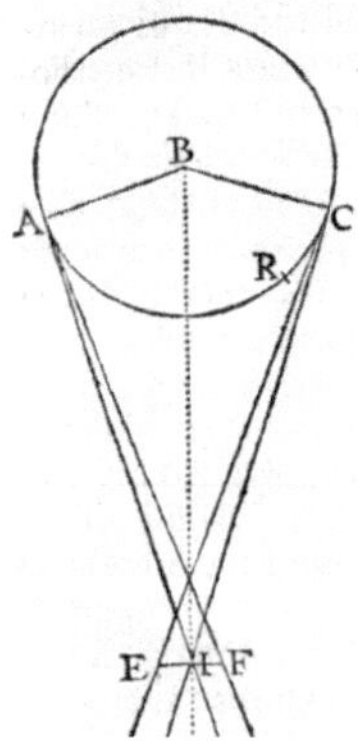

From: Martinus Hortensius, *Responsio ad Additiunculam Kepleri* (1631), p. 33

Terrae diurnum et annuum.[49] In this preface, Hortensius very precisely targets Kepler's observations of the Sun, the determination of its diameter and, consequently, its size and distance from the Earth. If these are false, he says, it is due to the tools used by Kepler, who claims that the lens-telescope cannot be more precise than a simple *tubum opticum* with a simple aperture (*foramen*).[50]

Following Kepler's brief reply in an *Additiunculam*, Hortensius repeats his attack in his *Responsio* (published in 1631) (unfortunately, Kepler died in October 1630 and never read the *Responsio*, nor did he receive the book, edited by Mersenne in 1630, with Gassendi's *Exercitatio* against Robert Fludd and some important observations of Gassendi on snow, sunspots and other matters).[51] In his *Responsio*, Hortensius explains quite well why Kepler believes that the telescope cannot be as accurate as a 'tubum' (which works on the same way as a camera obscura) in making solar observations (fig. 6.1).

49 See note 38.

50 Kepler, *Additiuncula*, in: *GW* 11, 1, $204_{40\text{-}42}$: 'Certissimis enim demonstrationibus dioptricis evinco [...] diametros Luminarium vitiari tubo dioptrico.' On the variable quality of lenses, see: *JIB*, III, p. 121. Beeckman evokes two causes: the length of the tube, and the density of the light rays after the passage of the concave lens. This note comes just before the report of the meeting with Gassendi (see above, n. 47).

51 See my essay: 'Gassendi in the Philosophical Debate: Stakes of the *Essay Concerning the Principles of Robert Fludd's Philosophy* (1630)', in: Delphine Bellis, Daniel Garber, and Carla Rita Palmerino, eds., *Pierre Gassendi: Humanism, Science and the Birth of Modern Philosophy* (London: Taylor & Francis, in press).

In this image, E-I-F represents the diameter of a lens or foramen. Kepler claims that the portions GH (the rays coming from the opposite side, C) and MN (coming from A) must be removed; therefore, Kepler claims that the diameter of the Sun's image must be corrected: the corrected value should be HM rather than GN. Moreover, Kepler claims that the Sun's rays are weakened by their inter-crossing in the foramen and the telescope can therefore not offer a precise image of these lateral sides, called the 'limbo'. Therefore, he claims, the Sun, like the Moon, is surrounded by an atmosphere with optical properties that make it appear larger than it is. Finally, Kepler claims that another cause lies in the convex surface of the observer's eye.

Hortensius's response is very clear: the telescopic images of the Sun are *perfectly clear* and do not require any correction. There is no dilatation at the periphery of the Sun's image. Kepler offers no geometrical (or optical) demonstration and the physical causes he appeals to (such as the inflammation of the aether around the Sun) are nothing but *ineptiae* or *vanas conjecturas*. It is significant that Hortensius and Beeckman take different, if not directly opposed, positions here. Both criticize Kepler's speculation on archetypal causes but Beeckman looks for these physical causes and reasons, for example, to explain how planets gradually grow to their full size thanks to solar exhalations, whereas Hortensius steers away from any physico-cosmological conjectures. It is also clear that some of Descartes' physical ideas and speculations about the solar atmosphere are directly linked to the problem discussed here.[52]

5 Conclusion

There is no doubt that Beeckman's and Hortensius's critical attitude towards Kepler profoundly influenced Descartes, including his metaphysical positions in the spring of 1630, which seem, as shown by Jean-Luc Marion, to primarily target Kepler.[53] From a doxographic perspective, this seems perfectly true and irrefutable. The fact that Descartes criticized and denounced anyone who attempted to submit the power of the Creator to logical possibility (or to possibility in general) is something no commentator

52 See: Descartes, *Principia philosophiae*, III, art. 100, in: *AT*, VIII-1, pp. 150-151.

53 See: Jean-Luc Marion, *Sur la théologie blanche de Descartes* (Paris: Presses Universitaires de France, 1981, 1991^2), pp. 178-203. The only missing element in this masterly work is a chapter specifically devoted to Isaac Beeckman, a shortcoming we hope to have remedied here.

has ever questioned. But the question is whether this is the deepest meaning and ultimate purpose of the doctrine known as the 'creation of eternal truths'. Another way of interpreting these texts in the light of Beeckman's precedent as discussed here is to show that 'possibility' as such presupposes matter, which precedes it in the same way that actuality precedes potentiality. 'Possibility' therefore designates nothing else than the virtual possibilities produced by possible variations in *quantity*, variations that are themselves defined by number, weight and size. 'Possibility' is therefore wrapped in the idea of matter, in other words size and quantity. As we know, Descartes places the proof for the world's singularity on the singularity of this idea of matter, i.e. the unequivocal *or rather unique* concept of quantity, the only concept of matter available to us.[54] Descartes' 'new' world is therefore unique and necessary for us as humans and there can be no question for us of a different world than the one in which we have life, being and motion.

About the Author

ÉDOUARD MEHL is Full Professor of Philosophy and Science in the Early Modern Period at the University of Strasbourg (France). He recently published a work on Descartes' cosmology: *Descartes et la fabrique du Monde. Le Problème cosmologique de Copernic à Descartes* (Paris: Presses universitaires de France, 2019). The book was distinguished with the Moron Prize of the Académie Française (2020).

54 Descartes, *Principia philosophiae*, II, art. 22, in: *AT*, VIII-1, p. 52: 'Nec ullius alterius materiae ideam in nobis reperimus.'

Appendix: Bibliography and Chronology

	Kepler	Beeckman	Hortensius	Lansbergen	Others
1596	*Mysterium cosmographicum* (MC), [Pref. Michaël Mästlin]				
1609	*Astronomia nova (De motu Martis)*				
1610	*Dissertatio cum nuncio sidereo*				Galileo Galilei, *Sidereus nuncius*
1611	*Dioptrice*				
1613		*JIB*, I, p. 25 [July 1613-April 1614]			Galileo Galilei, *Tres epistolae de maculis solaribus*
1618	*Epitome astronomiae Copernicanae I*	*JIB*, I, pp. 163-164, fol. 66 v [23 December 1616-16 March 1618]: on Galen			10 November: First meeting between Descartes and Beeckman *JIB*, I, pp. 242-243; *AT*, X, p. 75/ *JIB* IV, pp. 49-52
1619	*Harmonices mundi libri V*			*Philippi Lansbergii progymnasmatum astronomiae restitutae. Liber I. De motu solis* (Middelburgi: ex officina Richardi Schilders, 1619)	
1620	*Epitome* IV	*JIB*, II, pp. 56-57 [June 1620]: on Galen			Descartes, DM: 'We would no more wish […] to have bodies made of a matter incorruptible like diamonds than wings to fly like birds.' *Cogitationes privatae*: 'Una est in rebus activa vis', *AT*, X, 218

	Kepler	Beeckman	Hortensius	Lansbergen	Others
1621	*Mysterium cosmographicum*, ed. altera				
1627	*Tabulae Rudolphinae*				
1628		*JIB*, III, pp. 120-121 (*Epitome* IV)			
1629					*JIB*, III, p. 123: 'Ostendi quoque illi [sc. P. Gassendi] Keplerum frustra laborare, ut inveniat punctum ad quod planetae respicientes semper eundem situm retinent, ac demonstravi id per se necessarium esse.'
1630	*De raris mirisque anni 1631 phenomenis [...] admonitio ad astronomos* *Additiuncula ad ephemiridi [...] ad [...] annum [...] 1624* (*GW* 11, 1, 204-205): 'Ego hoc anno 1630 reversus sum ad foramen nudum tubi bene longi [...] plane ut P. Scheinerus, Cysatus, Galilaeus et alii.'	Beeckman to Mersenne, 30 April 1630 (*JIB*, IV, pp. 179-189) *JIB*, III, p. 153: *Eclipsis solis 10 Junij 1630 a me observata* *JIB*, III, p. 172: *Telescopij maculas solis manifestantis ratio*	*Philippi Lansbergii commentationes in motum terrae diurnum et annuum et in verum adspectabilis caeli typum. ex Belgico sermonem in latinum versae a Martino Hortensio, [...] una cum ipsius prętatione* (Middelburg: Zacharias Romanus, 1630)		Descartes to Mersenne, April-May June 1630 Descartes to Beeckman, 17 October 1630

	Kepler	Beeckman	Hortensius	Lansbergen	Others
1631			*Responsio ad additiunculam D. Ioannis Kepleri […] praefixam ephemeridi eius in annum 1624:* 'Mihi certe deliquium solis anni 1630 Iunii 10 […] Dordraci Batavorum accurate observanti, tale nihil contingit animadvertere.'	*Philippi Lansbergii uranometriae libri tres in quibus lunae, solis et reliquorum planetarum et inerrantium stellarum distantiae a terra et magnitudines hactenus ignoratae perspicue demonstrantur* (Middelburg: Zacharias Romanus, 1631)	
1632				*Tabulae motuum caelestium perpetuae* (Middelburg: Zacharias Romanus, 1632)	P. Gassendi, *Mercurius in sole visus* (Paris) W. Schickard, *Pars responsi ad epistolas P. Gassendi de Mercurio sub sole viso* (Tübingen)
1633			*Dissertatio cum Gassendo de Mercurio in sole viso et venere invisa*		
1637					Descartes, *Discourse on Method*

7 Combining Atomism with Galenic Medicine

The Physiological Theory of Isaac Beeckman (1616-1627)

Elisabeth Moreau

Abstract

Although he obtained a medical degree at the University of Caen in 1618, Isaac Beeckman never practised medicine. Instead, he developed an atomistic conception of Galenic physiology by discussing, throughout his notebook, the constituents and functioning of the living body. Interestingly, Beeckman applied his atomistic interpretation to the notion of temperament as the balanced proportion of elemental qualities, which defined the state of health. In this chapter, it is shown how his atomistic views on health and temperament amalgamated the Galenic theory of elements, mixture, and digestion. In appraising related interpretations of the body by late Renaissance *novatores*, Beeckman proposed an original theory of the organism, which put forward a mechanistic conception of metabolism as characterized by the rarefaction and condensation of atomic matter.

Keywords: Isaac Beeckman, matter theories, Renaissance, digestion, corpuscular philosophy, mechanism

In the last 30 years, historians of science have shown an increased interest in Isaac Beeckman's physical-mathematical approach to mechanism in the context of the 'Scientific Revolution'.[1] What we know about Beeckman

1 See: Klaas van Berkel, *Isaac Beeckman on Matter and Motion: Mechanical Philosophy in the Making* (Baltimore: Johns Hopkins University Press, 2013); Eio Honma, 'Beeckman's Natural Philosophy', *Historia Scientiarum* 5 (1996), pp. 225-247; Frédéric de Buzon, 'Beeckman, Descartes and Physico-mathematics', in: Daniel Garber and Sophie Roux, eds., *The Mechanization of*

Van Berkel, Klaas, Albert Clement, and Arjan van Dixhoorn (eds.), *Knowledge and Culture in the Early Dutch Republic: Isaac Beeckman in Context*. Taylor & Francis Group 2022
DOI: 10.5117/9789463722537_CH07

comes from his notebook that reports his thoughts about experiments, tools, and scientific theories, which he discussed with his circle of friends including Descartes, Gassendi, and Mersenne, among others.[2] However, Beeckman was not only a learned engineer, he was also trained in medicine, and obtained a medical degree from the University of Caen in Normandy in September 1618.[3] He began to study medicine in 1616 in preparation for the dissertation defence at Caen, mostly by reading medical treatises at his hometown, Middelburg.[4] At that time, Beeckman likely had access to these books through the library of his friend Philippus Lansbergen (1561-1632), a Dutch astronomer and Calvinist minister who lived in Middelburg from 1613.[5] Among Beeckman's early medical sources, one can find – aside from Galen – a significant number of Dutch and French authors. In fact, Beeckman's first mention of a medical treatise points to the *Universa medicina* of the French physician Jean Fernel in 1613-1614, which he continued to study until at least 1621.

Beeckman's first inclination for French medical literature – as evidenced by the references to Fernel, Joseph du Chesne, Jean Tagault, Jean Riolan the Elder, and Guy de Chauliac – might have prompted his choice to obtain a medical degree in France. For geographical, confessional, and financial reasons, he was likely compelled to choose the University of Caen as an accessible, religiously tolerant, and affordable institution. Besides, Beeckman was disposed to travel to north-west France, where he previously studied theology, mathematics, and philosophy at the Huguenot Academy of Saumur in the spring of 1612.[6] Beyond these assumptions about Beeckman's motive for graduating from Caen, the references in his notebook and the auction catalogue of his library show his undeniable interest in ancient, medieval,

Natural Philosophy (Dordrecht: Springer, 2013), pp. 143-158; Floris Cohen, *Quantifying Music: The Science of Music at the First Stage of the Scientific Revolution, 1580-1650* (Dordrecht: Reidel, 1984), pp. 116-161; Fokko J. Dijksterhuis, 'Understandings of Colors: Varieties of Theories in the Color Worlds of the Early Seventeenth Century', *Early Science and Medicine* 20 (2015), pp. 515-535.

2 On the nature of Beeckman's notebook, see the chapter 'Framing Beeckman: Cornelis de Waard as Editor of the Beeckman Papers' by Klaas van Berkel in this volume.

3 On Beeckman's medical theory, see: Elisabeth Moreau, 'Le Substrat galénique des idées médicales d'Isaac Beeckman (1616-1627)', *Studium* 3 (2011), pp. 137-151; Mart J. van Lieburg, 'Isaac Beeckman and His Diary-Notes on William Harvey's Theory on Blood Circulation (1633-1634)', *Janus* 69 (1982), pp. 161-163.

4 I am grateful to Klaas van Berkel for making me aware of the essentially bookish nature of Beeckman's medical training, and the need for further investigating the provenance of his medical sources.

5 Van Berkel, *Isaac Beeckman on Matter and Motion*, pp. 85, 130-131.

6 Van Berkel, *Isaac Beeckman on Matter and Motion*, p. 16.

and Renaissance medicine from the formative period of 1616-1618 until his last years.[7] Even as late as in the 1630s, Beeckman proved his up-to-date knowledge of medical debates among his contemporaries such as Santorio Santori, William Harvey, and Daniel Sennert.

Although Beeckman could not be called a 'doctor' to the extent that he never practised medicine, he was certainly a 'physician' in the sense that he applied natural philosophy to the study of medicine. The area of theoretical medicine Beeckman was interested in, corresponded to the field of physiology. As part of the university training in medicine, physiology was centred on the healthy body's structure and functioning, in particular its mere components (elements, humours) and their role in vital functions (generation, growth and nutrition).[8] For his medical education, Beeckman sought to explain how basic constituents such as elements could mingle and result in an organic living body. In this regard, digestion, because it implied the decomposition of food matter and its rearrangement to replenish the body, took an important part of the medical reflections in his notebook.

Interestingly, it was in the physiological moments of his notebook that Beeckman developed most of his atomistic theory of matter.[9] While his interpretation of atomism could be seen as an extension of the corpuscular theory he applied to physics and mathematics, it was also nourished by Lucretius's poem *On the Nature of Things* and by medical works related to the composition of bodies. Between 1616 and 1627, Beeckman read Galen

7 *Journal tenu par Isaac Beeckman de 1604 à 1634*, publié avec une introduction et des notes par C. de Waard, 4 vols. (The Hague: Martinus Nijhoff, 1939-1953) [henceforth *JIB*], IV, pp. 293-304; *Catalogus* [...] *librorum* [...] *Isaaci Beeckmanni* (Dordrecht: Isaac Andreas, 1637). See: Eugenio Canone, 'Il *Catalogus librorum* di Isaac Beeckman', *Nouvelles de la République des Lettres* (1991), pp. 131-159, esp. pp. 131-138; Van Berkel, *Isaac Beeckman on Matter and Motion*, pp. 73-74.

8 On Beeckman's corpuscular and atomistic theory of matter, see: Henk Kubbinga, 'Les Premières théories "moléculaires": Isaac Beeckman (1620) et Sébastien Basson (1621). Le Concept d'individu substantiel et d'espèce substantielle', *Revue d'histoire des sciences* 37 (1984), pp. 215-233; Henk Kubbinga, *L'Histoire du concept de 'molécule'* (Paris: Springer, 2002), I, pp. 203-225; Benedino Gemelli, *Isaac Beeckman. Atomista e lettore critico di Lucrezio* (Florence: Olschki, 2002); Norma E. Emerton, *The Scientific Reinterpretation of Form* (Ithaca: Cornell University Press, 1984), pp. 109-116.

9 On early modern physiology, see: Vivian Nutton, '*Physiologia* from Galen to Jacob Bording', in: Manfred Horstmanshoff, Helen King, and Claus Zittel, eds., *Blood, Sweat and Tears: The Changing Concepts of Physiology from Antiquity into Early Modern Europe* (Leiden: Brill, 2012), pp. 27-41; Andrew Cunningham, 'The Pen and the Sword: Recovering the Disciplinary Identity of Physiology and Anatomy before 1800, I: Old Physiology – the Pen', *Studies in History and Philosophy of Biological and Biomedical Sciences* 33 (2002), pp. 631-665; Thomas S. Hall, *Ideas of Life and Matter, Vol. 1: Studies in the History of General Physiology, 600 B.C.-1900 A.D.: From Pre-Socratic Times to the Enlightenment* (Chicago: University of Chicago Press, 1969).

as well as late Renaissance physicians such as Jean Fernel (1497-1558), Giovanni Argenterio (1513-1572), and Andreas Libavius (c. 1555-1616). The latter offered original Galenic theories that promoted Platonic philosophy, the revision of traditional pathology, and the promotion of medieval alchemy, respectively. However, the importance of these medical *novatores* has been unexplored in historical studies dedicated to Beeckman, whereas the Galenic tradition is key to contextualizing his atomistic theory. For this reason, this chapter explores Beeckman's medical theory of matter from a broader perspective, including the late Renaissance. Such an approach offers the advantage of clarifying how Beeckman could effortlessly reconcile Galen and the atomistic philosophy despite their presumed incompatibility. Galen, indeed, rejected atoms and corpuscles in *On the Elements According to Hippocrates*, and extensively criticized the corpuscular philosophy of the Greek physician Asclepiades of Bithynia throughout his works.[10] Nevertheless, this did not prevent Beeckman from reinterpreting Galenic physiology from an atomistic and mechanistic viewpoint.

The first part of this chapter examines Beeckman's medical theory of matter and how it integrates the traditional theory of elements and qualities into an atomistic framework. The next part discusses how his atomistic theory of elements applies to fundamental physiological notions, such as temperament and digestion.

1 Atoms, Elements, and Homogeneous Parts

From the beginning of his notebook, Beeckman's matter theory is remarkable for its conciliation of the traditional theory of elements with Lucretian atomism. Beeckman had the opportunity to study the physics of elements during his university training when he read up on the Aristotelian theory of matter-form or 'hylomorphism'.[11] According to Aristotle, all beings of the natural world were composed of four elements (air, water, earth, fire) which

10 Galen, *On the Elements According to Hippocrates*, trans. by Philip de Lacy (Berlin: De Gruyter, 1996). For the Latin edition, see: Galen, *De elementis ex Hippocrate*, in: *Claudii Galeni opera omnia*, ed. Karl Gottlob Kühn (Hildesheim: Georg Olms, 1821-1833), I, pp. 413-508. In this chapter, I will use Kühn's edition of Galen's works.

11 See: Craig Martin, 'Elements and Qualities', in: Thomas Glick, Steven J. Livesey, and Faith Wallis, eds., *Medieval Science, Technology and Medicine: An Encyclopedia* (Abingdon-New York: Routledge, 2005), pp. 157-158; Graig Martin, 'Hylomorphism', in: Glick, Livesey, and Wallis, eds., *Medieval Science*, pp. 234-236.

each were characterized by two of the four primary qualities (hot, cold, dry, moist). The elements were made of two principles: 'matter' which played the role of material substrate, and 'form' which determined their essence. During the generation of natural beings, the elements united through a process of 'mixture' during which they mingled their qualities and obtained a new 'substantial' form.

To this Aristotelian framework, Beeckman applied an atomistic conception of matter based on Lucretius's *On the Nature of Things*. As one of the first sources of the 'atomist revival' in the Renaissance, the poem was first printed in 1473 with numerous re-editions in the sixteenth and seventeenth centuries.[12] Beeckman possessed several copies of *On the Nature of Things*, including Denys Lambin's edition and commentaries, and started discussing Lucretius in his notebook in 1614-1615.[13] In this section I consider how Beeckman developed an atomistic conception of Aristotelian physics, which applied to elements and compounds bodies.

Four Atomic Elements

In his notebook, Beeckman followed the Aristotelian and Galenic tradition by reporting the four elements as basic components of bodies, to which were related the four primary qualities. At the same time, he merged this framework with his atomistic and corpuscular views. This interpretation came early in his notebook, in between 1616 and 1618, when he explained that bodies were composed of atoms surrounded by interstitial vacuum. The latter was described as 'intermediate empty spaces' forming pores of diverse size.[14] The same interstitial void was discussed in Beeckman's correspondence as early as in 1613.[15] In the corollaries of his medical thesis defended in 1618 at Caen, he also stated the existence of *vacuum intermixtum* and even identified to air pressure the concept of *fuga vacui* caused by pump suction.[16] Although the idea of void was contrary to the tradition, Beeckman was familiar with it through his professional experience in hydraulic

12 See: Christoph Lüthy, 'Atomism in the Renaissance', in: Marco Sgarbi, ed., *Encyclopedia of Renaissance Philosophy* (Cham: Springer, 2018); Ada Palmer, *Reading Lucretius in the Renaissance* (Cambridge: Harvard University Press, 2014), pp. 192-232.

13 Gemelli, *Isaac Beeckman*, p. xi.

14 *JIB*, I, p. 132: 'Deus corpora atoma primò movit non minus quàm creavit; motis semel nunquam quiescebant, nisi ab invicem impeditis. Ergo congredientes et cum vacuo misto, convenienter materia et forma extiterunt omnium compositorum coeli et Terrae.'

15 *JIB*, IV, p. 27.

16 *JIB*, IV, p. 44.

engineering. From his apprenticeship in candle making in 1610-1611, he had worked on water pipes and pumps for the construction of breweries and fountains.[17] In addition, Beeckman read of Hero of Alexandria's *Pneumatica* (*Liber spiritalium*) around 1616. In this treatise, Hero supported the particulate structure of matter and the existence of dispersed vacuum following the example of pumps, which prompted Beeckman's commentaries in his notebook.

For Beeckman, the four elements were atoms endowed with four types of shapes which determined their primary qualities.[18] It was their motion, shape (*figura*) and number that caused the 'forces' of bodies, that is their properties.[19] Heat and cold were due to the motion, speed, and size of atoms.[20] Moistness and dryness were related to the round or sharp shape of atoms. The same reasoning applied to sensory qualities such as taste. Following a Lucretian topos, Beeckman explained that pleasant and unpleasant flavours were due to round or sharp atoms, and their resulting accordance with the pores.[21]

Moreover, Beeckman revisited the Aristotelian notion of form related to the elements. In the Aristotelian tradition, form was the active principle that determined the essence of the elements, particularly during their mixture, for which the elemental compound acquired a 'substantial form'. However, for Beeckman, form was nothing but the arrangement of atoms, more precisely their 'situation' (*situs*), an equivalent of the Lucretian notion of 'position' (*positura*), which designated the spatial position of atoms with respect to each other.[22] The form of a compound varied according to the diverse arrangements of atoms at geometrical and spatial levels, for example, in a square or in a cube. As explained in Beeckman's notebook, two compounds

17 Van Berkel, *Isaac Beeckman on Matter and Motion*, pp. 16-19.

18 *JIB*, I, pp. 152-153: 'At calor, frigus, humiditas, siccitas tactu apprehenduntur absque specie figurarum, tametsi intellectu solâ figurarum ratione videantur. Unde verisimiliter concluditur omnes omninò rerum differentias ex figurâ atomorum petendas esse; et quia dictae qualitates solae tactui sunt subjectae, omninòque quatuor tantùm corpora simplicia, in totâ rerum naturâ, terram, aquam, aerem, ignem animadvertimus.'

19 On the Lucretian notion of shape (*figura*), see: Gemelli, *Isaac Beeckman*, pp. 90-96; Lucretius, *De rerum natura*, 1, 685, and 2, 1021.

20 *JIB*, I, p. 216.

21 *JIB*, I, pp. 149-150: 'Praeterea multa sunt insipida, calida, frigida, humida, sicca; sapida verò sunt onmia quae aliquam corpori nostro compositionem similem adepta sunt, id est cujus cavitates et asperitates cavitatibus et asperitatibus ita respondeant, ut ea suaviter nos afficiant.' See: Gemelli, *Isaac Beeckman*, pp. 59-61; Lucretius, *De rerum natura*, 2, 402-407.

22 On Lucretius's notion of *positura*, see: Gemelli, *Isaac Beeckman*, p. 79 et passim; Lucretius, *De rerum natura*, 1, 685; 2, 1017-1022. For Beeckman, *situs* relates to the position of particles, while *dispositio* refers to the proportionate arrangement of the compound.

of different nature might have the same 'portions' and 'particles' of fire, air, water and earth, but differed in their disposition.[23] Thus, the distance between the pores also defined their specificity ('essential difference'), that is their form.

Minimal Particles and *Homogenea*

In 1620, Beeckman further developed his atomistic theory of matter by positing different structural layers such as *minima*, particles, and homogeneous parts. His explanation drew on the Aristotelian and Galenic philosophy, which defined the body as structured into different levels of 'parts', comprising organic, homeomerous, and elemental parts. Body parts included organic parts (limbs and organs) – also called 'anhomeomerous' – which were made of 'homeomerous' parts such as nerves, flesh, muscles, and other tissues.[24] The 'homeomerous' parts were homogeneous compounds resulting from the achieved union ('mixture') of elements. This sophisticated framework allowed Beeckman to enrich his atomistic theory in order to investigate the specific properties of bodies and living beings.

As Beeckman explained in his notebook, atoms agglomerated into different levels of composition, first, *minima*, and then, 'minimal particles'.[25] The latter operated the actions of an organic body part; when destroyed, they were decomposed into their own *minima*.[26] Beeckman's terminology was common in the medical tradition. A similar definition of elements as 'minimal particles' – in the sense of minute portions – was endorsed in Renaissance medicine according to Galen's definition of the element in *On the Elements According to Hippocrates*. Following this treatise, physicians

23 *JIB*, I, p. 153: 'Fieri enim potest ut duae res aequalibus constent portionibus corporum ignis, aeris, aquae et terrae, suntque tamen dissimilis naturae. Nam hisce sita est ignis particula inter terram et aerem, et etiam inter aerem et aquam, omninòque multae sunt quatuor simplicium figurarum in unâ lineâ dispositarum aut in formâ cubi redactarum, positurae diversitates.'

24 See: Martin, 'Hylomorphism', pp. 234-236.

25 *JIB*, II, p. 96: 'Impraesentiarum autem sciendum est ignem purum non esse atomum (non enim atomus in aere ascenderet, quia ubique corpore plenus est ideòque gravis), sed ignis minima particula composita est ex multis atomis, ita junctis ut multum inter eas sit vacui.'

26 *JIB*, II, p. 117: 'Particula minima dupliciter dicta est sumi. Primò pro eâ minimâ quae primò possit perficere membri actiones, quamquam secundò ea constet ex multis absolutè minimis secundùm membri substantiam, ita ut hac divisâ substantia propria membri pereat, illâ verò divisâ actiones vel omnes vel hae, ita ut intelligantur quaedam actiones majoribus, quaedam minoribus particulis perfici.'

tended to consider the elements as the smallest parts of bodies subject to a union in order to constitute the body parts.[27]

To his conception of bodies as aggregates of particles, *minima* and atoms, Beeckman added that compounds were structured in primary and secondary *minima* as 'homogeneous' parts (*homogenea*).[28] By calling these *minima* 'homogeneous', Beeckman implicitly identified them with the homeomerous parts of bodies. In the same way, the 'minimal particles', which were composed of *minima*, corresponded to the traditional organic parts. Following this reasoning, the *minimum* designated the finite number of atoms that a body part needed to function. It was only to this extent that Beeckman adopted the Aristotelian terminology of 'natural *minimum*' as a body part that could not be indefinitely small in the same way as it could not be indefinitely large.[29] But Beeckman's *homogenea* also referred to other sources which were mentioned in his notebook. Among them, one can find several treatises on alchemy, physics, and logic.

Beeckman's penchant for alchemy overall reflected his deep interest in matter theories such as the one expounded in the *Alchymia* (1606) of the German physician Andreas Libavius (c. 1550-1616).[30] From Libavius's treatise, Beeckman retained the definition of the alchemical art as the 'separation' – the alchemical process of extraction – of homogeneous bodies from a substance.[31] With this definition of *homogenea*, Libavius aimed to continue the medieval alchemical tradition by merging the

27 See: Gweltaz Guyomarc'h and Stéphane Marchand, eds., 'Studies on Galen's *De elementis*', *Aitia* 7:2 (2017).

28 *JIB*, II, pp. 117-118: 'Sic quoque interdum in unâ re diversorum homogeneorum minima conjuncta minimum sunt alicujus virtutis. Hoc verò minimum, conjunctum cum ejusdem generis minimo, aliam virtutem exerit. [...] Sit igitur medicis id minimum, quod non minori quàm haec est particulâ opus, et vim optatam exerit. Liceat verò hoc minimum secare in alia, et haec in alia usque ad humores, elementa et atomos.'

29 Aristotle, *Physics*, 1.4, 187b35-188a13. See: John E. Murdoch, 'The Medieval and Renaissance Tradition of *Minima Naturalia*', in: Christoph Lüthy, John E. Murdoch, and William R. Newman, eds., *Late Medieval and Early Modern Corpuscular Matter Theories* (Leiden: Brill, 2001), pp. 91-132.

30 On Libavius, see: Bruce T. Moran, *Andreas Libavius and the Transformation of Alchemy: Separating Chemical Cultures with Polemical Fire* (Sagamore Beach: Watson, 2007); William R. Newman, *Atoms and Alchemy: Chymistry and the Experimental Origins of the Scientific Revolution* (Chicago: University of Chicago Press, 2006), pp. 66-84; Owen Hannaway, *The Chemists and the Word: The Didactic Origins of Chemistry* (Baltimore: Johns Hopkins University Press, 1975).

31 *JIB*, II, p. 127. See: Andreas Libavius, *Syntagmatis selectorum* [...] *Alchymiae arcanorum tomus primus*, 5.18 (Frankfurt: Nicolaus Hoffmann/Peter Kopff, 1615), pp. 193-194; Andreas Libavius, *Examen sententiae Parisiensis scholae contra alchymiae latae*, in: Libavius, *Alchymia recognita, emendata, et aucta* (Frankfurt: Johann Saur/Peter Kopff, 1606), pp. 8-9.

Aristotelian physics with the explanation of material change.[32] In turn, Beeckman used Libavius's reasoning with a somewhat different objective: to describe the body parts as homogeneous compounds according to the regular union of their atomic components. Interestingly, Libavius's late works, too, presented these homogeneous bodies as composed of minimal particles and atoms, but this aspect was not commented on in Beeckman's notebook.[33]

Rooted in early medicine and alchemy, Beeckman's notion of homogeneous part was also inspired from natural philosophy and logic, which he studied for his philosophical training at the University of Leiden between 1607 and 1610. Beeckman likely borrowed the term *homogenea* from the works of the German theologian Bartholomew Keckermann (c. 1571-1609), which he commented on in 1618.[34] Two treatises of Keckermann, in particular, showed the physical-logical counterpart of Beeckman's account of *homogenea*. In his *Systema physicum*, Keckermann defined elements as simple homogeneous bodies. By taking the example of heat, which assembled homogeneous parts and disintegrated heterogeneous parts, Keckermann specified that the term 'homogeneous', which was familiar to all logicians, designated things that shared the same nature and denomination. In his *Systema logicae*, Keckermann also presented as *homogenea* the bodies whose parts had the same name as the whole.[35] He anchored the term to the Greek notion of 'homeomerous' body such as water, wine, blood, gold or wood, whose *minima* and particles had the same name as the whole.

With his explanation of homogeneous parts made of particles and *minima*, Beeckman aimed to propose a theory of matter that worked at logical and physical levels. In his view, the *homogenea* were structured in 'primary' and 'secondary' levels, corresponding to minimal particles and *minima*,

32 See: Moran, *Andreas Libavius*, pp. 40-43; Elisabeth Moreau, 'Reforming the *Prisca Medicina*: Libavius' Axioms of Elements and Mixture', in: Pietro D. Omodeo and Volkhard Wels, eds., *Natural Knowledge and Aristotelianism at Early Modern Protestant Universities* (Wiesbaden: Harrassowitz Verlag, 2019), pp. 255-270.

33 See: Newman, *Atoms and Alchemy*, pp. 66-84.

34 On Keckermann, see: Joseph S. Freedman, 'The Career and Writings of Bartholomew Keckermann (d. 1609)', *Proceedings of the American Philosophical Society* 141 (1997), pp. 305-364; Cees H. Leijenhorst, 'Place, Space and Matter in Calvinist Physics: Petrus Ramus, Clemens Timpler, Bartholomæus Keckermann and Johann Heinrich Alsted', *The Monist* 84 (2001), pp. 520-541.

35 See: Bartholomew Keckermann, *Systema physicum septem libris adornatum* (Hanover: Wilhelm Antonius, 1612), pp. 128 and 133; Keckermann, *Systema logicae tribus libri adornatum* (Hanover: Wilhelm Antonius, 1611), p. 190.

respectively.[36] In other words, the first 'union' or 'conjunction' of elements resulted in a *minimum*. Then, the first level of *minima* formed the first or 'primary' homogeneous part. In turn, the primary *homogenea* constituted the *minima* of the 'secondary' homogeneous parts. In case of division, the secondary *homogenea* lost their particular force and fell back to the level of the primary *homogenea*. If further divided, the primary *homogenea* regressed to the elemental level of which they consisted. Conversely, atoms and minimal particles mingled to form complex and various *homogenea*. If Beeckman could not determine the exact number of *minima* which composed primary *homogenea*, he assumed that they existed in a finite number sufficient to produce a great diversity of things. They did so in the same way as the letters of the alphabet were able to produce an infinite number of words.[37]

So far, it has been shown that, whereas Beeckman adopted the traditional terminology of elements, qualities, and form, his theory of matter had little to do with the Aristotelian notion of matter-form or 'hylomorphism'. In fact, Beeckman eliminated the traditional distinction between substance and accident, and between primary and secondary qualities, since all these notions derived from the arrangement and motion of atoms. This reflected the corpuscular and mechanistic framework that Beeckman early proposed in his notebook, which has been much explored by historians of science. On the other hand, his conception of elements as particles and portions paid tribute to the Galenic definition of elements as the 'smallest' or 'minimal' particles of bodies. Beeckman, indeed, conceptualized a structural layering of atomic elements, which was strongly indebted to the Galenic approach to the body's composition in elements, homeomerous, and organic parts. Having explored Beeckman's account of atomic elements, *minima* and *homogenea*, I shall now turn to the application of this theory to physiology.

36 *JIB*, II, pp. 118-119: 'Nam prima elementorum conjunctio efficit hujus compositi aliquod minimum, quae multa simul sumpta, statuunt unum et primum homogeneum. Hujus primi homogenei minimum, conjunctum cum alterius primi homogenei minimo (quod ex aliâ mixtione elementorum existit) efficit minimum secundi homogenei, quo primò omnium et propriè continet suam vim; tum si tenuiùs secetur, etiam vim primi homogenei; ac tertiò, adhuc tenuiùs sectum, profert vim elementi.'

37 *JIB*, II, p. 122: 'Non autem existimandum est multa esse homogenea tam exiguorum minimorum. Cùm enim ea proximè constent ex elementis, necesse est pauca duntaxat esse homogenea, aptè mixta, à se invicem differentia. Haec verò homogenea pauca, inter se mixta ita ut res magnae inde fiant, constituunt multas res à se mutuò differentes. [...] Quot autem primum homogeneum minimis elementorum constet, nobis est ignotum.'

2 Temperament as a Geometric Proportion of Particles

Throughout the *Journal*, Beeckman's medical questioning is centred on the structure and functioning of the living body. In Galenic medicine, this theme corresponded to physiology as a theoretical branch of medicine rooted in natural philosophy. Developed in late medieval Latin-Arabic medicine, physiology was based on the Galenic and Aristotelian account of 'natural things'.[38] It examined the body's first components – elements, humours – and their balance, which determined the body's state of health or 'temperament'. These notions were fundamental to understanding vital functions such as generation, growth, and nutrition. Beeckman followed this physiological framework by studying 'temperament' as a way to investigate the living body at the level of its smallest constituents.

In the Galenic tradition, temperament – also named 'complexion' – resulted from the balance of the elements and their primary qualities into a moderate state. This notion also relied on the Aristotelian concept of 'mixture', that is the homogeneous union of elements into a new compound. The compound or 'mixt' was considered as acquiring a new 'substantial' form, while its constitutive elements remained in potentiality. This definition of mixture raised many debates on the status of elements, particularly their form (essence) and qualities, during and after mixture. In late Renaissance medicine, a successful interpretation of mixture and temperament was that of the French physician Jean Fernel, whose *Universa medicina* (1567) was several times re-edited in the early modern period.[39] Fernel stated that elements equalled to minute particles which juxtaposed during mixture and acquired a form of divine origin.[40] While the prominent place given to the celestial nature of the form pointed to a Platonic inclination, Fernel's account also drew on a longer tradition rooted in Avicennian medicine.[41]

38 See: Nutton, '*Physiologia*'; Nancy Siraisi, *Medieval and Early Renaissance Medicine* (Chicago: University of Chicago Press, 1990), pp. 78-80, 101-109.

39 On Fernel, see: José Kany-Turpin, ed., 'Jean Fernel', *Corpus. Revue de Philosophie* 41 (2002), pp. 5-197; John Henry and John M. Forrester, 'Introduction: Tradition and Reform: Jean Fernel's *Physiologia* (1567)', in: *The* Physiologia *of Jean Fernel*, trans. by John Forrester (Philadelphia: American Philosophical Society, 2003), pp. 1-13; John Henry and John M. Forrester, 'Jean Fernel and the Importance of His *De abditis rerum causis*', in: *Jean Fernel's* On the Hidden Causes of Things*: Forms, Souls, and Occult Diseases in Renaissance Medicine* (Leiden-Boston: Brill, 2005), pp. 3-65.

40 'The *Physiologia* of Jean Fernel (1567), ed. and trans. by John M. Forrester', *Transactions of the American Philosophical Society* 93 (2003), pp. 210-212.

41 Hiro Hirai, *Medical Humanism and Natural Philosophy: Renaissance Debates on Matter, Life and the Soul* (Leiden-Boston: Brill, 2011), pp. 46-79; Elisabeth Moreau, 'Elements, Mixture and

This medieval and Renaissance Galenic tradition formed the broader context of Beeckman's medical theory of matter and the conceptual foundation for his atomistic interpretation of the living body. From the beginning of his study on Galenic physiology, Beeckman had been aware of the Renaissance debates on mixture and temperament. The first physiological work he started to read for his medical studies in 1616 was Fernel's *Physiologia*, which was included in the *Universa medicina*. As Beeckman later noted in 1618, Fernel stated that the elements remained intact after their mixture in the compound, which brought about the body's temperament.[42] From this account, Beeckman took the idea of a juxtaposition of elemental particles, which he combined with his own atomistic matter theory. As a result, his account of temperament kept the traditional terminology of matter, form, and elements, though in a different sense. His reflections on this theme were developed between 1616 and 1627 and edited in the first and second volumes of the *Journal*. Afterwards, his physiological investigation became centred on digestion, which will be examined in the last section of this chapter.

Well-Connected and Arranged Particles

In addition to reinterpreting the Aristotelian account of elements and matter-form from an atomistic viewpoint, Beeckman revisited the Galenic concept of temperament. While the medical tradition defined it as the healthy constitution resulting from the balanced mixture of elements, Beeckman considered it as a correct arrangement of particles. In his view, the body's form (essence) was nothing but the 'disposition' (*dispositio*) and 'binding' (*connectio*) of its material parts.[43] As discussed in the previous section, these material parts were described as atomic elements arranged in different levels of *minima* and particles. For Beeckman, this implied that the form of the healthy body was the correct 'disposition' or 'binding' of its parts.[44] Such a disposition referred to the proper union of the elements,

Temperament: The Body's Composition in Renaissance Physiology', in: Chiara Beneduce and Denise Vincenti, eds., *Oeconomia Corporis: The Body's Normal and Pathological Constitution at the Intersection of Philosophy and Medicine* (Pisa: Edizioni ETS, 2018), pp. 51-58.

42 *JIB*, I, pp. 168-169. See: 'The *Physiologia* of Jean Fernel', 2.6 and 2.8, 200-204, and 210-211.

43 *JIB*, I, p. 203: 'Sanitatis quaedam dispositiones sunt visus, quaedam auditus. Materiaque visûs nihil est aliud quàm dispositiones quaedam ejusmodi quae visum constituunt, forma verò visûs istarum dispositionum apta compositio. [...] Sic sanus habet pro materiâ proximâ corpus animalis, pro formâ ejus partium aptam dispositionem; sic morbi materia sunt dispositiones quaedam corporis, forma verò mala earum connectio.'

44 *JIB*, I, p. 203: 'Materiaque visûs nihil est aliud quàm dispositiones quaedam ejusmodi quae visum constituunt, forma verò visûs istarum dispositionum apta compositio. [...] Sic sanus habet

and no longer to their mixture in the Aristotelian sense of the term. In consequence, the notions of matter and form took a different meaning adapted to Beeckman's atomistic theory. The form of the compound equalled to the ad hoc arrangement of its atoms, while its matter consisted of the elemental matter of which it was made.

For this interpretation of health as the correct arrangement of parts, Beeckman referred to the Italian physician Giovanni Argenterio, who was famous for his criticism of Galen regarding the notions of disease, cause, and symptom.[45] From Argenterio's treatise *De morbis* (1548), Beeckman borrowed the definition of health and illness as a correct or incorrect binding and disposition of the main body parts (limbs).[46] However, he applied this approach to his matter theory in such a way that the 'binding' and 'disposition' were related to the elemental particles that constituted body parts. Consequently, health was determined by the correct arrangement of the body from an atomic – rather than anatomic – viewpoint.

Interestingly, Beeckman's insistence on the *connectio* and *dispositio* of material units also pointed to his training in logic and dialectics. Among his sources, the German theologian Philip Melanchthon (1497-1560) gave a logical definition of form as the order, disposition, and binding (*connectio*) of the parts of an argumentation in the *Erotemata dialectices* (1547).[47] Keckermann, from whom Beeckman's partly derived the notion of *homogenea*, also used this formulation in his treatises on logic. In the same way, Beeckman considered the healthy constitution as a correct binding and disposition of its minimal parts.

Regular Polyhedra

In 1618, when discussing the nature of a healthy constitution, Beeckman provided his own interpretation of a classical question in Galenic medicine: how to define the most appropriate temperament (*temperatura*)?[48] For his

pro materiâ proximâ corpus animalis, pro formâ ejus partium aptam dispositionem; sic morbi materia sunt dispositiones quaedam corporis, forma verò mala earum connectio.'

45 Nancy Siraisi, 'Giovanni Argenterio and Sixteenth-Century Medical Innovation: Between Princely Patronage and Academic Controversy', *Osiris* 6 (1990), pp. 161-180.

46 Giovanni Argenterio, *De morbis libri XIIII* (Lyon: Sébastien Honoré, 1558 [1548]), pp. 652-654.

47 Philipp Melanchthon, *Erotemata dialectices* (Wittenberg: Johan Crato, 1556 [1547]), p. 142; Bartholomew Keckermann, *Systema logicae minus* (Hanover: Wilhelm Antonius, 1606), pp. 247-248.

48 *JIB*, I, p. 347: 'Dico igitur id in unoquoque genere eucraton esse, cujus omnes actiones etc. omnium individuorum optimae sunt. Sic homo aliquis est temperatissimus; qui verò homines ab hujus temperaturâ deficiunt, contrarijs juvantur. Leo quis est temperatissimus, multò quidem

medical training, he was familiar with the traditional account of temperament as a proportionate state resulting from the balance of primary qualities. Such a proportion was defined according to an ideal model, the Galenic notion of *eukratos*, from the Greek 'well-mixed'. Beeckman adopted this framework by stating that the balanced constitution varied from one species to another according to the traditional notion of 'latitude of temperament'.[49] For instance, the ideal temperament was different for a fish, a lion or a man. Each species had a particular moderate status (*medium*) achieved by the mixture of elements.

As he continued his discussion on the ideal temperament, Beeckman added that it was structured *ad pondus*, that is, as an arithmetically equal distribution of primary qualities.[50] This statement tended to go against the Galenic tradition, which established that the constitution *ad pondus* was purely theoretical and could not be found in the physical world. Instead, it was held that the ideal temperament of each species was a proportionate qualitative state (*in justitiam*). By contrast, Beeckman believed that temperament precisely consisted in a quantitative balance of elements and qualities related to a medium point. Thus, the notion of ideal constitution designated a proportionate union and a quantitative disposition of elemental particles and *minima*. In 1620, Beeckman refined this definition of temperament *ad pondus* by specifying that it was also determined by the situation and shape of its *minima*. The particles of the compound had a geometrical proportion and a particular situation (*situs*) resulting in the formation of regular polyhedra.[51] For instance, the human being was formed of polyhedral *minima*, whose shape was ordered in 20 triangles which formed 'suitably connected' icosahedra. If Beeckman also posited that dogs were formed of octahedral *minima*, he did not clarify how the five types

calidior homine [...]. Sic piscis aliquod genus temperatissimum est multòque homine frigidius, ideòque et multò frigidioribus quàm homo recreatur.'

49 On the latitude of temperament, see: Per-Gunnar Ottosson, *Scholastic Medicine and Philosophy: A Study of Commentaries on Galen's Tegni (ca. 1300-1450)* (Naples: Bibliopolis, 1984), pp. 167 et passim.

50 *JIB*, I, pp. 296-297: 'Sic uniuscujusque speciei est aliquis status temperatissimus: is in leone est calidior, in piscibus frigidior, in homine temperatus ad pondus. [...] Sic uniuscujusque hominis est temperamentum aliquod medium peculiare, ad quod ubi perveniat, optimè habet.'

51 *JIB*, II, pp. 124-125: 'Si igitur primordia nostra forent tales pyramides ordinatae, et ad constitutionem speciei virtutes activas exerentes, requiretur compositum ordinatum, circulo inscribendum. [...] Constituant igitur icosahedra, aptè sibi invicem conjuncta, hominem vel hominis semen; octahedra verò canem. [...]. Videmus enim canum diversa genera esse infinita et indies inter se commutari, quod indicat canum omnium minimum naturale idem quidem esse, sed positionis diversitatem esse variam.'

of regular polyhedra were distributed among the different animal species. Nonetheless, he considered that the connection of triangular units produced a geometrically ordered shape, whose diverse arrangements defined the particular features of each individual.

Among the possible sources for Beeckman's notion of polyhedral units of matter, one can find a range of ancient and early modern treatises. At first, Plato, in his *Timaeus* 55a-56c, described the four elements as polyhedra made up of triangular units, whose proportion in number, motion, and qualities had been harmoniously arranged by God. In a mathematical context, Euclid developed a demonstration of the Platonic solids in his *Elementa*. Moreover, the concept of a polyhedral configuration of the natural world was tackled by the *Six-Cornered Snowflake* (*Strena seu de nive sexangula*) (1611) of the German astronomer Johannes Kepler.[52] In this inquiry on the hexagonal structure of snow crystals, Kepler envisaged, among other possible explanations, that living beings might be composed of regular solid figures in a pentagonal proportion (dodecahedron or icosahedron).[53] According to this theory, the geometrical figure was related to an internal organizing principle responsible for the propagation of living beings: a 'seminary' or 'formative' faculty emanating from the earth's 'vapour'.

Although Beeckman shared Kepler's geometric and corpuscular reasoning, his primary objective was to show the mathematical possibility of defining temperament with a finite number of constituents.[54] For his strictly mathematical concern, he thus deviated from Kepler's supposition of a 'formative nature'. In a commentary on the *Six-Cornered Snowflake* around 1628, Beeckman even noted that this concept was 'ridiculous and unworthy of a philosopher'.[55] In the same way, Beeckman broke with the Renaissance Platonic tradition by explaining that the particular virtues of compounds were not due to an incorporeal entity of celestial origin, which was related to the seed or to the substantial form. In his view, all these notions pertained to the atomic composition of bodies and to the shape of their homogeneous parts. Consequently all living beings were provided with a particular atomic composition, a proportionate shape, and a correct disposition, although the exact configuration of each species was

52 Johannes Kepler, *The Six-Cornered Snowflake*, trans. by Colin Hardie (Oxford: Oxford University Press, 2014 [1966]); Johannes Kepler, *L'Étrenne ou la neige sexangulaire*, trans. by Robert Halleux (Paris: Vrin, 1975).

53 Johannes Kepler, *Strena seu de nive sexangula* (Frankfurt: Gottfried Tampach, 1611), p. 12.

54 On Beeckman and Kepler, see the chapter 'Optics, Astronomy, and Natural Philosophy: Beeckman, Descartes, Kepler, and the Dutch Connection' by Édouard Mehl in this volume.

55 *JIB*, III, pp. 33-34.

unknown to Beeckman. Following his previous statements about health and temperament, this entailed that the polyhedral *minima* corresponded to homogeneous parts. On the other hand, their form – in the Aristotelian sense of the term – equalled the regular shape of their *minima*, which gave specific characteristics to each species.

While it would be tempting to consider Beeckman's theory as materialistic, his explanations of the body's composition did take the intervention of divine providence into account. At first, Beeckman followed Galen's teleology expounded in *On the Usefulness of the Parts of the Body*. According to this treatise, the physiological processes were not associated to any divine intervention but presupposed a demiurge having created matter.[56] Nonetheless, the divine providence was evidenced by the determined functioning and usefulness of each body part, and, more broadly, by the body's organism whose structure was perfectly adequate to its function. As Klaas van Berkel has pointed out, Galenic teleology was particularly suited to the Calvinist dogma of predestination in which Beeckman believed.[57] As Beeckman explained in his notebook, God 'skilfully' created atoms so that their concourse was not accidental.[58] The divine creation ordained particular atoms, *minima*, and corpuscles which determined the organization and functioning of nature.[59] The achieved atoms combined in a favourable situation according to specific conditions inscribed in all constituents of nature. Consequently, a limited number of principles was able to produce the whole diversity of nature, just like an infinite number of words could be created from the letters of the alphabet.[60] This teleological reasoning formed the background of Beeckman's approach to physiology.

56 Galen, *De usu partium*, 11.14, ed. Kühn III, pp. 899-911; *JIB*, I, pp. 163-164: 'Cùm Gal. [...] probet hominem certâ prudentiâ, non fortuitò constructum esse, ita ut nulla pars magis illi conveniret quàm quas habet. [...] Sic numerus et ordo creata sunt corporum, extra quae nihil fit; apta tamen facta sunt ut concursu suo non infinita, sed finita non determinata producant. [...] Cùm enim opifex sit omnipotens, quidni posset, quod nos non intelligimus? id enim tantummodo intelligimus fieri posse, quod Deus intelligi posse voluit.'

57 Van Berkel, *Isaac Beeckman on Matter and Motion*, pp. 140-147.

58 On the necessity of an ordained universe in Lucretius, see: Gemelli, *Isaac Beeckman*, pp. 53-59; Lucretius, *De rerum natura*, 1, 159-204, and 2, 700-710, 720-729, 1067-1069.

59 *JIB*, II, p. 43: 'Adhaec mirari potiùs convenit Dei sapientiam qui naturae primordia, minimaque corpuscula ex nihilo creata, talem figuram dederit ut ex ijs non quidvis possit nasci, sed ea duntaxat quae convenientia toti universitati futura erant. Atomorum igitur, ut ait Lucretius, figurae sunt finitae idque ex finitis formis et speciebus rerum rectè probat; at nos harum figurarum in atomis causam Dei providentiae attribuimus.'

60 *JIB*, II, p. 57: 'Quantò igitur satiùs est dicens omnia haec a naturâ et constitutione loci esse nata, Deum verò ejusmodi principia creasse in principio, quae sibi mutuò juncta, non possint non hoc facere. Si enim conveniant haec primordia fit avis, si illa, canis, si alia, piscis. Non verò

3 Digestion as Rarefaction and Condensation of Food Matter

Besides supporting a general explanation of health, Beeckman's atomistic conception of elements and body parts shaped his understanding of physiological functions. Among them, digestion had a prominent place as an organic process relying on the transformation of matter.

In the latest phase of his notebook, Beeckman's medical account mostly explored digestion, which was a central question in Galenic physiology. As a vital function, digestion was considered as responsible for maintaining life by assimilating the nutritive properties of food during their conversion into humours. What raised Beeckman's attention was the decomposition of food into its smallest ingredients, its circulation through the digestive organs, and the very process of digestion as a transformation of matter by the body heat. Most notably, Beeckman reinterpreted in an atomistic way some major concepts introduced by Galen in *On the Natural Faculties*: heat, natural faculties, and food 'concoction' in the digestive organs.[61]

By developing the role of natural faculties during digestion, Galen centred the discussion on the 'attracting' faculty. Attraction was considered as ensuring the passage of ingested food into the digestive organs where food was subject to concoction. Galen considered two causes of attraction: a magnetic force related to the specific properties of the body's substance or 'total substance', and *fuga vacui*, that is the natural motion of beings in order to avoid vacuum. In addition to expounding his own interpretation of physiological attraction, Galen debunked Erasistratus's mechanistic interpretation of digestion as a process of contraction and dilation. In the same way, he rejected Asclepiades' corpuscular theory of digestion as a phenomenon of rarefaction and condensation.[62] Nonetheless, Galen conceded that the attraction of humours by *fuga vacui* during digestion was caused by the contraction and dilation of the organs. Following this reasoning, the body's vessels were comparable to water pipes of different

concursus hic in infinitum magis variat quàm ex 24 litteris infinita vocabula possunt fieri trisillaba aut decem syllabarum etc.' The analogy between the arrangements of principles and that of letters is borrowed from Lucretius, *De rerum natura*, 2, 688-689, 1013-1022.

61 Galen, *De naturalibus facultatibus*, ed. Kühn II, pp. 1-214.

62 On Asclepiades of Bithynia, see: J.T. Vallance, *The Lost Theory of Asclepiades of Bithynia* (Oxford-New York: Clarendon Press-Oxford University Press, 1990); David Leith, 'The Qualitative Status of the *Onkoi* in Asclepiades' Theory of Matter', *Oxford Studies in Ancient Philosophy* 36 (2009), pp. 283-320; David Leith, 'Pores and Void in Asclepiades' Physical Theory', *Phronesis* 57 (2012), pp. 164-191.

size in a garden.[63] Such an explanation undoubtedly struck Beeckman's attention and buttressed his hydraulic understanding of the digestive system.

What Beeckman proposed, in turn, was a synthesis of Galen's account of digestion which, interestingly, included the views attributed to Erasistratus and Asclepiades.[64] Whereas his conception of digestion relied on a Galenic framework and terminology, it supported the body's atomic structure and porosity. A major theme in his interpretation was the role of heat and *fuga vacui* in the phenomenon of attraction. In reconsidering Galen's account in *On the Natural Faculties*, Beeckman stated that heat 'compressed' the body, hence causing the attraction of humours by dilating the pores.[65] This process also caused the rarefaction of humours, which were transformed into vapours while penetrating the pores.[66] Thus, for Beeckman, it was the body heat that operated physiological functions by creating a movement of attraction by *fuga vacui*.[67]

Beeckman further applied this reasoning to the transformation of food in the digestive system. He understood the Galenic notion of cooking or 'concoction' of food stuff by the body heat as a phenomenon of *fuga vacui* caused by the dilation and rarefaction of heat. According to Beeckman, the liver was subject to a process of dilation due to the formation of vapours. By way of suction movement, it attracted the food 'concocted' in the mesenteric

63 Galen, *De naturalibus facultatibus*, 3.15, ed. Kühn II, pp. 206-214.

64 *JIB*, I, pp. 159-160: 'Cùm Gal., Lib. 1 Περι φυσικῶν <δυνάμεων>, multa contra Epicurum et Asclepiadem disputat. Concludit in fine Libri, et in principio secundi, viscera et partes omnes trahere sibi familiaria. Sed τὸ συστέλλειν hoc pacto exornari poterit: omnia terrestria undique premuntur ab incumbente aere, [...] ergo multò magis ea, quae sunt in corpore, accedente coincidentiâ circumjacentium corporum.'

65 *JIB*, I, p. 145: 'Calor attrahit etiam hac ratione. Dilatantur calore pori alicujus partis. Cùm autem totum corpus perpetuò contenta premat continuendo, non est absonum humorem, pressum in locum patentiorem, vijs amplioribus factis, detrudi, etiamsi concederemus partem calefactam non minus solito premere.'

66 *JIB*, I, p. 149: 'Moderatus calor in corpore humores, ut decet, attenuat perque poros transmittit. Major verò calor plus attenuat quàm transmittit, ideòque partem distendit. Minor autem calor non sufficienter attenuat, ita ut vapor spiracula non possit penetrare, atque idcirco etiam distendit. At minimus calor non magis distendit quàm lagenam vitream, aquâ plenam, ignis paucus disrumpit.'

67 *JIB*, II, p. 123: 'Non mirum est putrescentibus humoribus ad cor rapi. Trahit enim cor fluvidam materiam ratione caloris. Est etiam viscus omnium calidissimum in corpore nostro, ideòque calor trahit per fugam vacui, ut ignis magnus minorem ignem et aerem ad se trahit.'

veins.[68] In the corollary of his medical thesis, Beeckman considered that suction movement was due to *fuga vacui*, which he assimilated to air pressure. As he integrated this statement into his medical theory, he broke with the Galenic interpretation of attraction – either as a magnetic force or as *fuga vacui* – since it was pressure, in his view, which was responsible for the attraction of food in the digestive organs.[69] He further explained that the digested food was filtered through the wall of the digestive organs, which were pierced with pores of various shapes, just as if it passed through a sieve.[70]

Following his reinterpretation of heat, attraction, and concoction, Beeckman formulated in atomistic terms the Galenic natural faculties. While maintaining the Galenic terminology and accent on the primary qualities, he emphasized the dilation and rarefaction of matter, as well as the shape of its particles.[71] He began by reporting the four natural faculties as 'attracting', 'concocting', 'retaining', and 'expelling', respectively.[72] The attracting faculty caused the dilation of pores by the body heat so that food matter could be received in the digestive organs. The concocting faculty, which was stimulated by moistness, achieved food transformation. The retaining faculty, which was stimulated by cold and dryness, prevented digestive matter from slipping by grasping it with the 'hooks' of digestive wall particles. By tightening the digestive wall, the retaining faculty facilitated the expulsion of the remaining nutriment by the expelling faculty. Nonetheless, Beeckman expressed his doubts on the notion of natural faculty later in his notebook. To explain the passage of food from the stomach to the liver

68 *JIB*, I, p. 102: 'Iecur sugit alimentum per venas meseraicas, quia ab alimenti spiritibus corpus jecoris dilatatur calore; sedato calore id, quod suxerat, in venas dimittit coincidendo, per quas illud simili coincidentiâ partium undique aequaliter distribuitur.'

69 Even when Beeckman maintained the Galenic view of attraction as a magnetic force during digestion, he explained it as a phenomenon of pressure on food particles, see: *JIB*, I, p. 309.

70 *JIB*, I, pp. 159-160: 'Cùm ullum viscus, praeter proprium, possit unumquodque penetrare, quia pori visceris uniuscujusque respondent corporis uniuscujusque formis; non aliter quàm si cribrum diversis foraminibus perforatum sit rotundis, triangulatis, lunaribus etc. [...] Immisso cibo ventriculus et intestina etiam supra generalem dictam pressionem se contrahunt, quodque hepar potest penetrare, expellitur; quod verò hepatis poris non respondet, alio vergit. Idem etiam fit in venis post hepar.'

71 On Beeckman's corpuscular explanation of natural faculties between 1616 and 1618, see: *JIB*, I, p. 165.

72 *JIB*, II, p. 116: 'Hîc sitae sunt quatuor facultates universales. Attractrix non est aliud quàm pororum in particulâ conveniens apertio, ut materia legitima possit, aliunde in eam commodè expressa, a particulis recipi. [...] Coctrix facultas fit particulâ humidiore existente. [...] Retentrix facultas requirit siccitatem, ne ob fluxibilitatem humidam contenta excidant, sed firmiter velut duris uncis comprehendatur. [...] Expultrix verò amat frigus.'

during digestion, he eventually substituted the compression of the stomach to the attracting faculty, which he judged 'incomprehensible'. In his view, it should rather be understood as a phenomenon of rarefaction as the stomach evacuated digested food after discharging the liver.[73]

For his atomistic account of digestion, Beeckman might have found a source of inspiration in the medical philosophy of Asclepiades – through the lenses of Galen in *On the Natural Faculties* – especially to describe the formation of the humours. Asclepiades founded his physiological theory on the movement of corpuscles through invisible pores of varying size and shape, which played a key role in the rarefaction of the humours. According to Galen, his conception of corpuscles and pores was comparable to Epicurus's notions of atoms and empty spaces.[74] Similarly to Asclepiades, Beeckman explained that particles united to form natural *minima* and homogeneous bodies as a result of the dilation of the pores under the action of heat.[75] These homogeneous bodies constituted the four humours that were contained in the blood mass, namely blood, bile, melancholy, and phlegm.[76] To maintain the functioning of the body parts, these humours were constantly produced from the minimal particles of ingested food.[77]

Following this reasoning, Beeckman described digestion as a process of rarefaction that consisted in a 'separation' of food matter by the body heat.[78] The concoction of chyle was ensured by the liver through the rarefaction of food matter, whose useless residue was transformed into vapours evacuated

73 *JIB*, II, p. 133: 'Sed procul dubio voluit significare ventrem potiùs sese comprimendo alimentum visceribus tradere, quàm id ab ijs trahi vi attractrice, tam incomprehensibili. Exonerato igitur hepate et rarefacto, occasio datur ventri sese in hepar recepturum et vacuum exonerandi.'

74 Leith, 'Pores and Void', pp. 164-191. Leith has noted that Lucretius counted nutrition as a proof of the existence of vacuum: Lucretius, *De rerum natura*, 1, 350.

75 *JIB*, II, pp. 103-104.

76 *JIB*, II, p. 104: 'Enimverò haec minima filamenti constant ex quibusdam homogeneis, videlicet ex sanguine, bile, melancholiâ et phlegmate, aut saltem horum similibus; haec demum si placet, immediatè ex elementis. Haec homogenea primò ab elementis mutantur. Si igitur calor diutiùs membro adsit, ita ut non solùm poros majores, verùm etiam exiguos inter minima occupet, vel ibi mutat nutrimentum [...], vel à poris ijs pergit ad ipsa minima naturalia eaque penetrat.'

77 *JIB*, II, p. 117: 'Ac jam sciendum est calorem et humorem aut potiùs ignem, aerem, aquam, terram, aut potiùs bilem, sanguinem, pituitam et melancholiam, indesinenter à minimâ constrictae in ipsis minimis haerent, non exerunt vires; imò si nihil perpetuò deflueret, particulâ separari et in poris minimis versari. [...] Hinc necessaria per nutrimentum restauratio particularum, quae jam totae hoc defluxu consumptae sunt, et sua munera jam defuncta evanuerunt.'

78 *JIB*, II, 108: 'Ideò nullo negotio à cibo separatur, cujus omnes particulae inter se sunt connexae, aqueae videlicet cum terreis, adeò ut calor nequeat particulas cibi frangere, sed eas duntaxat ita separat, ut chylus, et chymus legitimus inde existat, unde partes corporis possint nutriri.'

by the pores. As Beeckman explained, this process started with the separation of particles and homogeneous bodies from food by the vital heat.[79] These particles, in turn, bound to the fire particles of the body heat in order to be transformed into 'wind', 'vapour', and other forms of exhalation. From that moment, they passed through the pores of the digestive organs' membrane and were processed into chyle. Thus, for Beeckman, digestive concoction was a process of rarefaction associated with the dilation and contraction of digestive organs. This reasoning was integrated into Beeckman's framework of 'natural *minima*' and pores. Such an approach led him to define nutrition as the renewal of 'useful' food particles, which filled the empty pores of digestive organs.[80]

4 Conclusion

Beeckman's medical theory was the result of a diligent reading of Galen between late 1616 and early 1618 in preparation for his medical degree at the University of Caen in September 1618. It was in this medical context that he developed an atomistic theory of matter in his *Journal*. Beeckman's medical atomism relied on the Galenic framework of elements, qualities, and temperament, but challenged the Aristotelian physics of matter-form. In his theory of matter, he put forward atomic elements with interstitial vacuum, which were each characterized by a particular shape at the origin of their qualities. This atomistic interpretation led Beeckman to abandon the Aristotelian notion of a substantial form in the formation of the body. Nonetheless, he sought to solve the problem of the materialistic tone of his theory by Galenic teleology, which he merged with his Calvinist faith. In his view, it was divine creation that ensured the creation of atoms and determined their organic functioning in nature, including the human body, through their multiple permutations following a regular structure.

79 On homogeneous and heterogeneous humours in Beeckman's conception of pathology, see the chapter 'Physician, Patient, Experimenter, and Observer: Isaac Beeckman's Accounts of Illness and Death' by Dániel Moerman in this volume.

80 *JIB*, II, p. 103: 'Haec minima sunt homogenea respectu ipsorum filamentorum. Omnes enim ejus partes sunt tales, et nutrimentum debet fieri talis pars, antequam possit dici pars corporis nostri. Id autem nutrimentum est ea materia, quae in his exiguis poris continetur fitque talis particula non exeuns è loco suo recipiendo à lateribus suis, id est à minimis his naturalibus, quibus comprehenditur calor, humores, quod in ijs est praecipuum, atque ita antiquum minimum perit, inutili excusso aut exhalante.'

Beeckman integrated his atomistic conception into his Galenic approach to physiology around 1620. Most remarkably, he developed a physiological account of the body as structured into elements, particles, *minima*, and homogeneous bodies. For his account of health and temperament, Beeckman took up the Galenic conception of mixture as a homogeneous union of elemental particles. However, he argued that these particles functioned as atoms that obeyed an accidental concourse and aggregated to form various layers of *minima*. To define the latter, Beeckman proposed an eclectic terminology, of which the notion of homogeneous (*homogenea*) was the most striking example. Drawing on Aristotle's physics, Galenic medicine, Libavius's alchemy, and Keckermann's logic, the notion of *homogenea* designated homogeneous parts whose atomic arrangement was regular and well-disposed. Following this interpretation, the *homogenea* took the meaning of 'homeomerous' parts which were composed of atomic elements and constituted the body's tissues with determined properties.

Beeckman also applied his atomistic account to the physiological phenomenon of digestion. He considered it as a process of contraction and dilation due to the pressure applied to food matter within the digestive organs. The ingested food was concocted and broken down by a vital heat of strictly elemental nature, consisting of fire. Beeckman's interpretation of digestion was nurtured by his expertise in hydraulic engineering, which he combined with Galenic medicine, Lucretian atomism and, presumably, the corpuscular views of Asclepiades and Erasistratus transmitted by Galen. This allowed him to maintain the traditional notions of humours, vital heat, and natural faculties which intervened during food 'concoction'. Nonetheless, it was the atomic arrangement of the bodily substances and the vacuum within their pores that prevailed in Beeckman's explanation of digestion.

In sum, Beeckman's medical theory of matter illustrates how apparently antithetic Galenism and atomism could be combined in the early modern period, as well as the intellectual roots of such a stance in the Renaissance medical tradition. His physiological thinking was part of a classical set of questions on the composition of living bodies in Galenic philosophy, which prompted a corpuscular reinterpretation of elements in the Renaissance. In this context, Beeckman stands as an interesting figure whose views on atoms and physiology were distinct from those of atomist physicians such as Girolamo Fracastoro, Sébastien Basson, and Daniel Sennert. In the early seventeenth century, atomist physicians commonly referred to ancient philosophers in postulating the discrete structure of elements that juxtaposed to form bodies. But Beeckman was remarkable in positing the notions of atomic shape and vacuum within a mechanical framework.

Further studies need to be conducted in order to explore the possible influence of Beeckman's medical atomism on his scholarly network.

About the Author

ELISABETH MOREAU is an FNRS Postdoctoral Fellow at the ULB – Université Libre de Bruxelles (Brussels, Belgium). Trained in the history and philosophy of science, she works on medicine, alchemy, and matter theory in late Renaissance Europe.

8 Physician, Patient, Experimenter and Observer

Isaac Beeckman's Accounts of Illness and Death

Dániel Moerman

Abstract

The received opinion is that Isaac Beeckman never put his medical degree, which he gained in 1618, into practice. His medical interests are, therefore, considered to have been primarily theoretical. However, apart from theoretical treatises on medicine, Beeckman's *Journal* also includes notes on illnesses and ailments which he encountered during his everyday life. These include illnesses that plagued his own body, as well as those of his relatives and friends. Beeckman's notes thus contain a very human aspect, portraying a man who is generally worried about his health and that of others, but they also offer a look into the observant and experimental attitude which he shared with contemporary physicians. This chapter argues that Beeckman practised medicine in a much broader sense than has thus far been considered, which strengthens the established view that he was a practically minded scholar who attributed great value to learning through experience.

Keywords: Isaac Beeckman, medicine, physician, illness, patient

On 6 September 1618, Isaac Beeckman received his doctorate in medicine from the university of Caen. Although Beeckman pursued a number of occupations in his life, such as candle maker, schoolmaster and lens grinder, the received opinion is that he never really put his medical degree into practice and that his medical interests were mainly theoretical. This is illustrated by the many notes in the *Journal* on medical disputations, often

Van Berkel, Klaas, Albert Clement, and Arjan van Dixhoorn (eds.), *Knowledge and Culture in the Early Dutch Republic: Isaac Beeckman in Context*. Taylor & Francis Group 2022
DOI: 10.5117/9789463722537_CH08

in relation to his developing mechanical philosophy of atomism.[1] However, apart from writing about medicine from an exclusively theoretical angle, Beeckman also engaged with illness and bodily ailments he encountered during his everyday life, including a few cases in which he himself was the patient. Considering the fact that he was a trained man of medicine, the question arises: in what manner did Beeckman write about his everyday aches and illnesses? Did he follow the common narrative of many patients at that time, or was he primarily looking at himself, and perhaps even others, through the lens of a trained physician?

The study of lay perceptions of illness and medicine, in addition to the common approach to the physician's view, has become a major subject in the field of medical history since Roy Porter's famous 1985 article 'The Patient's View: Doing Medical History from Below'. Porter's aim was to overturn the exclusive focus of medical history on doctors and their discoveries, as he argued that 'it takes two to make a medical encounter – the sick person as well as the doctor'. Medical history thus needed to be a history of the interaction between the man of medicine and the sick person, not merely the former.[2] The influence of Porter's article cannot be underestimated. His plea to incorporate 'the patient's view' was the starting point of a large number of case studies with regard to the perception of illness in early modern diaries, letters and chronicles.[3] Yet Porter's emphasis on patients and their perception has recently come under scrutiny, as its reliance merely on patient views might disavow the fact that patients are largely constructed by the medical knowledge of a certain period.[4] In certain parts of Beeckman's *Journal* there is an interplay of aspects typical of Porter's 'patient's view' and contemporary medical theories surrounding illness. On the one hand, Beeckman's notes regarding episodes of illness

1 With regard to Beeckman's theoretical approach to atomistic philosophy in his physiological notes, see: Elisabeth Moreau, *Eléments, atomes & physiologie. Le Context medical des théories de la matière (1567-1634)* (PhD diss., Radboud Universiteit, 2018), pp. 291-338, as well as the contribution by Moreau in this volume.

2 Roy Porter, 'The Patient's View: Doing Medical History from Below', *Theory and Society* 14 (1985), pp. 175-198, esp. p. 175.

3 See, for example: Roy Porter, *Patients and Practitioners: Lay Perceptions of Medicine in Pre-Industrial Society* (Cambridge: Cambridge University Press, 1985); Dorothy and Roy Porter, *The Patient's Progress: Doctors and Doctoring in Eighteenth-Century England* (Cambridge: Polity Press, 1989); Michael Stolberg, *Experiencing Illness and the Sick Body in Early Modern Europe*, trans. by Leonhard Unglaub (Basingstoke: Palgrave Macmillan, 2011).

4 See: Flurin Condrau, 'The Patient's View Meets the Clinical Gaze', *Social History of Medicine*, 20 (2007), pp. 535-540, esp. pp. 525-530.

which he, or others around him, experienced, showcase emotions and other aspects common to his medically untrained contemporaries. Yet on the other hand, Beeckman's notes also showcase engagement with experimentation and observation – *experientia* and *observatio* – two epistemic genres which became increasingly popular in medical science throughout the late sixteenth century. Especially the empirical observation of objects and events as a means to gain new insights, became a widely used method among physicians to describe and gain further knowledge about diseases and possible cures.[5] Beeckman, as this chapter will show, was not an exception to this paradigm, although the evidence for this has not received the attention it deserves.

This chapter will apply both the 'patient's view', focusing on certain narrative patterns commonly used by laypeople to describe their experience of illness, as well a focus on observation and experimentation as important aspects in Beeckman's notes on everyday illnesses and death. The primary aim is to highlight the hitherto underrepresented aspect of Beeckman as a man who was a patient and an observant physician at the same time. Although these practical and sometimes experimental approaches only appear in a few passages throughout his *Journal*, they nevertheless present a useful image of Beeckman's combined use of medical training and practical observations – also including aspects of his mechanistic philosophy – to further his knowledge of the human body and its illnesses, even at times when he was suffering badly from rather severe illnesses himself. In order to illustrate this, examples of Beeckman's engagement with various strands of early modern medicine will be discussed as part of his role as a 'patient' or a 'physician'. Yet a distinction between a typical patient or physician's attitude is often difficult to make in Beeckman's case, and should perhaps not be made at all. As this chapter will argue, Beeckman's notes sometimes contain a very human aspect, portraying a man worried about his health and seeking cures to alleviate his suffering. Yet his general view on everyday illnesses and medicine was more akin to that of the observant and experimental practising physicians of his time. This is also testament to his practical philosophical mindset, as a man who tried to discover the world around him through practice and observation.

5 Gianna Pomata, 'Observation Rising: Birth of an Epistemic Genre, 1500-1650', in: Lorraine Daston and Elizabeth Lunbeck, eds., *Histories of Scientific Observation* (Chicago: University of Chicago Press, 2011), pp. 45-80, esp. pp. 47-49, 54-55.

The Perception of Illness in Early Modern Times: Differences between Patients and Physicians

It goes without saying that the perception of illness by early modern medical scholars, physicians, surgeons, and laypeople, was very different from the way we nowadays conceive the workings of illnesses and ailments within our bodies. The experiences of laypeople, which gained more attention after Porter's manifesto regarding the 'patient's view', have already been studied extensively. One of the more encyclopaedic works on the early modern perception of illness by laypeople, Michael Stolberg's *Homo Patiens. Krankheits- und Körpererfahrung in der Frühen Neuzeit* (2003), translated into English as *Experiencing Illness and the Sick Body in Early Modern Europe* (2011), is a fascinating example of 'the patient's view' applied on a large historical scale. Apart from a wide range of personal journals and diaries, Stolberg used around 2,000 letters sent between physicians and patients, to 'explore how people in the early modern period perceived, experienced and interpreted illness and how they dealt with it in everyday life'.[6]

Stolberg, as well as Porter, often perceive the patient and the physician as two different actors, who interact and influence each other, but in the end take different directions with regard to their respective aims. A way to determine whether Beeckman's dealt with his own health and illnesses as a patient or a physician, is to focus on narrative patterns: the overlapping set of discursive themes and styles of description employed by laypeople or physicians to describe illness and pain. The ancient Galenic understanding of illness as the result of a misbalance between the four bodily fluids – blood, yellow bile, black bile, and phlegm – for example, was commonly used by both physicians and patients. Another idea shared by patients and physicians was the thought that illnesses could fundamentally weaken someone's physical constitution, doing permanent damage to his or her well-being. This is not a strange thought given the fact that in early modern times many illnesses, even those considered non-threatening to today's standards, could have fatal consequences. In the case of patients, sensations of pain and symptoms of illness thus triggered anxious reactions that prompted them to seek the help of physicians or, in many cases, alternative methods of healing, and, not to forget, the divine. Patients thus often sought desperate relief from the pain and discomfort caused by their

6 Stolberg, *Experiencing Illness*, p. 1.

illnesses. Physicians, on the other hand, were more pragmatic in their approach as they identified symptoms of illnesses, provided a prognosis, and prescribed treatment.[7]

During the sixteenth century, identifying the symptoms of illness was often done by means of observation. Physicians were trained to carefully observe and describe subtle changes, symptoms, and signs usually overlooked by laypeople.[8] One of the leading scholars who is – rather indirectly – credited with this innovation is the French philosopher Petrus Ramus (1515-1572). Ramus placed a strong emphasis on gaining practical knowledge of the world through experience, in which the writing of detailed case histories – *historia* – played an important part. Ramus's philosophy thus stressed a combined use of *observatio* and *historia*, and his approach to natural philosophy is credited as one of Beeckman's main inspirations for his practical and mechanistic philosophy.[9]

As will be discussed in the following sections, the observant and practical manner by which Beeckman took his notes regarding his mechanistic philosophy is also reflected in many of his notes regarding illness. This observant note taking as an epistemic tool to identify illnesses and possible cures was, however, rather common for physicians during Beeckman's time. A large number of detailed case histories describing such observations and experiments can be found, for example, in the notebooks of Bohemian physician Georg Handsch (1529-1578), and also in the work *Observationes et curationes medicinales* by Dutch physician Pieter van Foreest (1521-1597). A copy of the latter was also part of Beeckman's vast collection of books.[10] Stolberg, in fact, remarked that Handsch and other contemporary physicians often treated their own bodies as reliable sources of knowledge whenever they fell ill.[11] Beeckman was therefore not entirely unique in approaching his own illnesses through a physician's lens. However, as this chapter will illustrate, there are certain aspects within his observations and experiments that remain unique to Beeckman's peculiar manner of looking at the world, including himself.

7 Stolberg, *Experiencing Illness*, pp. 46-49.

8 Michael Stolberg, 'Empiricism in Sixteenth-Century Medical Practice: The Notebooks of Georg Handsch', *Early Science and Medicine* 18 (2013), pp. 487-516, esp. p. 513.

9 Pomata, 'Observation Rising', pp. 66-67; Klaas van Berkel, *Isaac Beeckman (1588-1637) en de mechanisering van het wereldbeeld* (Amsterdam: Rodopi, 1983), pp. 279-290.

10 Stolberg, 'Empiricism', pp. 489-505; Pomata, 'Observation Rising', p. 55; Eugenio Canone, 'Il *Catalogus librorum* di Isaac Beeckman', *Nouvelles de la République des Lettres* (1991), pp. 131-159, esp. p. 146.

11 Stolberg, 'Empiricism', p. 513.

Experimenting with Illness: Beeckman as 'Patient'

An obsession with healthy living is not exclusively something of our present times. During the early modern period people were also, sometimes even crazily, obsessed with their mental and physical well-being, and they were often quite vocal about it. This becomes visible through the copious notes left behind by laypeople who wrote very openly about their mental and physical state: a common narrative pattern in many patient accounts.[12] In the case of Beeckman, this was not very different. In October 1631, he wrote:

> I have in my life never been so out of reason, even when I was very young, that I did something without thinking clearly about the consequences, always thinking whether something would be good [...]. I never did something in my sleep, such as yelling, getting up, hurling or something like that. I have never in my life been drunk, and if I drank a bit too much wine or beer, my head hurt a bit now and then. When I sit at a table drinking with others, they often have to stand up and urinate, but not me, even if a meal lasted 7 or 8 hours, or as long as weddings usually last. This I write down along with other things so that in times of illness, knowing my nature, I can apply the right medicines to myself.[13]

In this section, Beeckman follows a narrative pattern common to many early modern people. The adherence to the Galenic six non-naturals – the six factors outside of the body which contributed to someone's mental and physical well-being – were a common means for early modern people to manage their own health. Keeping these non-naturals balanced was the key to maintaining a healthy lifestyle. Beeckman, however, expresses how he is conscious of his sometimes overindulgent intake of food and drink, but at the same time he is proud to tell that he keeps a good balance between work

12 Stolberg, *Experiencing Illness*, pp. 21-23.

13 *Journal tenu par Isaac Beeckman de 1604 à 1634*, publié avec une introduction et des notes par C. de Waard, 4 vols. (The Hague: Martinus Nijhoff, 1939-1953) [henceforth *JIB*], III, p. 221: 'Ick ben noyt myn leven soo verloopen geweest van sinnen, hoe jonck ick oock was, dat ick yet dede sonder achterdyncken, altyt my bedenckende oft so wel soude syn [...]. Ick en heb noydt yet in myn slaep gedaen, als van roepen, opstaen, smyten of diergelycke. Ick en hebbe noydt myn leven droncken geweest, ende als ick wat te veel wyns of bier gedroncken hadde, so dede myn hooft somtyts wel wat seer. Als ick over tafel sitte ende neffens andere dryncke, d'andere moeten dickwijls opstaen om haer water te maken, doch ick noyt, al duerde de maeltyt 7 of 8 ueren, so langhe als bruyloften gewoon syn te dueren. Dit teyckene ick hier al by malcanderen om in tyt van sieckte, hieruyt myn natuere kennende, de rechte medicine over myselven te doen.'

and rest, sleeping and waking, retention and excretion, and the passions of the mind. A popular physician and contemporary of Beeckman, Johan van Beverwijck (1594-1647), noted: '[T]hey [the six non-naturals] keep a man healthy as they are maintained properly; but can be evil and damaging to the one who does not maintain them correctly.'[14] This explains why Beeckman was rather fond of boasting about his health, and of comparing his own physical and mental capabilities with those of his peers. He also commented, rather proudly, that he was still able to urinate as perfectly and as far away as he was able to do as a child. Even one of the factors less under his control: eating, seemed to have little effect on his health, as he remarked: 'I often eat much, yes more than I think I should; however, I am seldom to never ill, and I have been of the same weight during the last 20 years and longer, namely 125 [pounds] with my clothes on.'[15] Beeckman thus considered himself a physically and mentally stable person, who was taking good care of his health.

Yet a passage from a decade earlier, November 1621, shows that he did not always live up to this standard:

> When I went to Cralingen with Thomas [Cool] on 14 Novemb[er], and arrived at home worn out at noon, I immediately went to eat wholeheartedly, salted meat and roasted meat. But as soon as I left the table, wind came from my stomach into my mouth, that is *ructus*, and my jawbones began to hurt. The following occurred, because the outer parts, that is the muscles, are pulled from the inside to the outside by the heat because of walking, while the stomach [is] not powerful enough to resolve and make the vapours disappear. Wherefore they [the vapours] remained thickened and the muscles of the jawbone irascible, there the heat resolved, extending itself.[16]

14 Johan van Beverwijck, *De schat der gezondheid* (*1636*), ed. Lia van Gemert (Amsterdam: Querido, 1992), p. 16.

15 *JIB*, III, p. 221: 'Ick ete veel, jae meer als my duncke dat behoordt, ende evenwel ben ick selden of noyt sieck, ende ben dese 20 jaeren ende langer al van eenselfde swaerheyt geweest, te weten 125 [pond] met myn kleren.'

16 *JIB*, II, pp. 187-188: 'Also ick den 14en Novemb[er] met THOMAS [Cool] na Cralingen ginck, ende vermoyt snoens thuys quam, ginck ick terstont hertelick eten, gesouten vleesch ende eerst gebraen vleesch. Maer so haest als ick van tafel was, quamender winden uyt de maghe in de mondt, te weten ructus, ende myn kaeckbeenders deden my seer. Twelck was, omdat de uyterste deelen, te weten de musculi, doort gaen de hitte van binnen na buyten getrocken hebbende, de maghe niet machtich en was de dampen te resolveren ende doen verdwynen. Waerdoor sy dick blyvende ende in de musculen van het cakebeen opvlieghende, aldaer door de wermte resolverende, deselvige extendeerden.'

In this entry Beeckman described what he observed with regard to his own body: the cause, name, nature, and the workings of his ailment: *ructus*, a medical term still used to denote 'the voiding of gas or of a small quantity of acidic fluid from the stomach through the mouth'.[17] In Galenic terms the digestive system was understood as a process similar to cooking: food entered the stomach where it became a fluid substance that was turned into 'inner heat', which was then distributed to individual body parts and organs. If the stomach became overburdened, the excessive 'inner heat' (or gas, as we would nowadays refer to it) was voided through the mouth. *Ructus* was therefore a rather common ailment among people who had access to large quantities of food and an esurient attitude. Beeckman also refers to the Galenic 'inner heat' as the cause of his ailment, although he describes it as 'dampen' ('vapours'). The term 'vapours' has to be taken in a literal sense here, as both early modern patients and physicians considered natural gasses produced within the intestines as trails of hot air or smoke moving through the body. These vapours could cause serious discomfort or pain in various areas of the body, ranging from the upper abdomen to sometimes even the head.[18] A German medical councillor in 1665, for example, described his headache to be caused by 'an abundance of vapours [...] bringing about the cruellest pains as well as ceaseless buzzing and droning', which corresponds to the notes of nameless patient who wrapped his head in a scarf as he wrote: 'I believe the wind, if I didn't wrap something around my head, would split it apart by attacking it impetuously.'[19] These descriptive accounts of pain caused by vapours, using terms like 'buzzing' and 'droning', often describe the impact of illness on the emotive state of patients who made detailed descriptions in their letters to physicians, pleading desperately sometimes for advice on what was going on and what to do about it.[20]

Beeckman, however, did not resort to the type of extensive and graphic descriptions. While laypeople were more inclined to think pragmatically in terms of how to deal with sometimes staggering pain and discomfort, Beeckman's approach is characteristic of a physician trying to gain a full understanding and an accurate description of the signs and symptoms of disease, underlining the explicit pathological changes occurring in the

17 *Farlex Partner Medical Dictionary*, s.v. 'ructus', https://medical-dictionary.thefreedictionary.com/ructus (accessed 12 June 2018).

18 Stolberg, *Experiencing Illness*, pp. 123-124, 142.

19 As cited in: Stolberg, *Experiencing Illness*, p. 124.

20 Robert L. Weston, 'Medical Sources', in: Susan Broomhall, ed., *Early Modern Emotions: An Introduction* (London: Routledge, 2017), pp. 105-108, esp. pp. 105-106.

body.[21] Beeckman's observation in this case provides a detailed account of what lay at the cause of his discomfort. His description of the interplay between heat, vapours, and the muscles, displays a mixture of Galenic physiology with his mechanistic philosophy to explain the interconnection of bodily parts as a machine within the process of food digestion. In his more theoretical accounts at the time, Beeckman explained this process to be guided by a suction movement created by the dilation and contraction of organs, such as the liver, which react to the digestive vapours resulting from the concoction of food. Beeckman, however, equalled the working of this suction movement to air pressure caused by a process similar to pump suction, by which he reconciled his knowledge of hydraulic engineering with Galenic medicine and various strands of atomistic thinking.[22] The latter is described in more detail in Elisabeth Moreau's contribution to this volume, which illustrates Beeckman's departure from a purely Galenic to a more mechanical, atomistic explanation of digestion during the early 1620s. This specific example thus displays a typical form of *observation*, which does not merely perceive the symptoms and causes of illness described in medical theory, but extends to an empirical description supporting his ideas regarding the workings of the human body, illustrated by his own discomfort. Many years after Beeckman made these notes, however, another polymath named René Descartes developed a more widely known, fully mechanistic explanation of digestion in terms of matter and motion. Rather similar to Beeckman, Descartes' idea also focused on the motion of food causing movement in the bowels and other parts of the body.[23]

Although Beeckman did not offer a direct remedy for his excessive belching and pain in the above-mentioned case, a more pragmatic, or perhaps even experimental approach to medicine appears in a later case from February 1631. Around the beginning of that month, a large piece of wood accidentally fell on Beeckman's head. What follows is an extensive and lengthy description of the effects this accident had on both his mental and physical well-being. The day-by-day basis in which he recorded the symptoms of his ailments and the effect of certain cures resemble very closely those of a physician observing his patient.[24] One of the first notes

21 Stolberg, *Experiencing Illness*, pp. 79-82.

22 *JIB*, I, pp. 102, 159-160.

23 Antonio Clericuzio, 'Chemical and Mechanical Theories of Digestion in Early Modern Medicine', *Studies in History and Philosophy of Biological and Biomedical Sciences* 43 (2012), pp. 329-337, esp. pp. 333-334.

24 Stolberg, 'Empiricism', p. 498.

after the incident describes how Beeckman felt dazed by the blow and decided to take some rest:

> [I] went to bed, around six o'clock in the evening, and had a vein opened. And the spot on my head, which was a bit red, though not really injured, was chucked with oil of roses and a protective plaster. Neither the next day, nor the day after, did I have any fever.[25]

Beeckman made similar observant notes until the beginning of March, underlining all the examinations regarding his injury and the bodily ailments that followed. He was particularly meticulous in examining his blood and urine, for which he also called the help of a physician and a surgeon. The latter was called to let some blood, which was then tested by a physician who considered it to be of good quality. Yet Beeckman also weighed five ounces of his blood, and set the weight of his blood against a similar quantity of rainwater weighing one 'roosennobel', a medieval English coin. He noted that '[t]he serum, on top of the thick substance, was around a third of the weighed quantity of blood.'[26]

This showcases how Beeckman turned a rather mundane injury into a number of experiments regarding his bodily fluids and their relation to certain ailments. Instead of seeking a cure, which would have been more characteristic for a patient, he was more interested in using his experiences to learn and bolster his medical knowledge. Beeckman therefore continued examining samples of his own bodily fluids. He used small glasses with the width of a thumb and a foot high, in which he observed blood, urine, and other excrements, which allowed him to perceive a proportion of *heterogenea* – heterogeneous substances – which floated on top or at lay still at the bottom. He described his theory behind this as follows:

> [T]he excrement and phlegm, being too thick, might be blended with such liqueur in a way that some of the heterogenea would float upon it, some would sink, and some would mix with it. Doing this one could more accurately judge the quantity of diseases as well as the quality, etc., because the heterogenea, which in the *ordinael* [piss bottle] and pot, etc.,

25 *JIB*, III, p. 191: 'ginck te bedde, also het ontrent ses ueren was savons, ende liet my een ader openen. Ende daer wert op de plaetse des hoofts, die wat root sach, doch gans niet gequetst, gestreken oly van roosen ende een defensif plaester. Hadde 's anderdaechs, noch daeghs daerna, gansch geen gewach van koortse.'

26 *JIB*, III, pp. 191-192: 'Het serum, boven op het dicke staende, was omtrent het derde deel van de heele gewoghene quantiteyt des bloets.'

> are spread out, would drift together in elongated flasks, and could feasibly be measured against each other regarding their proportion, if the glasses are equally wide on both the top and the bottom.[27]

The practice of examining a patients urine with the naked eye, also known by the term 'uroscopy', was very common in early modern medicine. Uroscopy was regarded as a fairly reliable method to recognize symptoms of diseases, and it was usually performed by physicians as part of the central diagnostic procedure by looking at the colour and thickness of the patient's urine.[28] Yet Beeckman also added a new element to uroscopic analysis by using elongated flasks with a cylindrical shape. The preference of elongated flasks over the *ordinael* or *matula* – the commonly used flask with a narrow bottleneck and rounded bottom – came from the fact that Beeckman observed how elongated flasks allowed for a better observation and measurement of phlegm his urine and blood.

In Galenic medicine, phlegm was often specifically associated with patients who suffered from a slow pulse, catarrh, indigestion, or poor appetite.[29] The days following the incident, Beeckman was, perhaps not surprisingly, occasionally struck by fevers and a lack of appetite, which was, as the previous passage from 1621 attests, an anomaly for him. He also suffered from an inflamed left arm, caused by a botched bloodletting. After a troubled night with little sleep and a lot of sweating, he examined his urine on the 19 February 1631, perceiving it to be thick with a reddish colour caused by the blood lying on the bottom of the flask. His urine remained of a similar nature after another test the following day, which he blamed on an excess of phlegm in his blood. He called for a surgeon to perform another bloodletting, but also invited the municipal physician of Dordrecht, Cornelis van Someren (1593-1649), to examine his urine. The latter concluded that Beeckman was in need of extensive purgation, prescribing him *cremoris tartari*, an acid purgative known nowadays as potassium bitartrate. This

27 *JIB*, III, p. 192: 'de excrementa ende fluymen, te dick synde, mocht men met sulcken liqueur menghen dat sommige heterogenea daerop dreven, sommighe daerin soncken, ende sommighe daermede gemeynght werden. So doende soude men veel beschelicker van de quantiteyt van sieckten oock van de qualiteit etc. konnen oordeelen, want de heterogenea, die inden ordinael ende in de pot etc. gespreyt ligghen, soude men in deze lanckworpighe glasen byeen dryven, ende bequamelick konnen gemeten worden wat proportie datse hadden tegen malcanderen, de glasen onder ende boven even wyt synde.'

28 Michael Stolberg, *Uroscopy in Early Modern Europe*, trans. by Leonhard Unglaub and Logan Kennedy (Farnham: Ashgate, 2015), pp. 11, 16-19.

29 William A. Jackson, 'A Short Guide to Humoral Medicine', *TRENDS in Pharmacological Sciences* 22 (September 2001), p. 487.

purgative belonged to the arsenal of physicians adhering to the alchemist ideas of Paracelsus (1493-1541), of which Beeckman owned multiple works, including the *Compendium medicinae & philosophiae Paracelsi*.[30] As a chemical by-product of winemaking, *cremoris tartari*, was conceived in Paracelsist terms as a kind of *tartar*: the acid substance which is also generated as a by-product of digesting food. In case of excess, this substance could cause numerous ailments. The purgative value of *cremoris tartari* was thus to intentionally upset the stomach, which, as we know today, might seriously damage the intestines.[31] Beeckman, however, took the prescribed purgative the following day, and, as a likely result, he started to cough up bloodied slime and suffered from occasional nosebleeds. However, in an unrelenting fashion, he continued to examine his urine the following days, and on 23 February his urine was clear again. According to Beeckman, this indicated that the purgation had worked.[32]

Reflecting back on his ailments, Beeckman concluded that the blow to his head had caused a disturbance in the disposition of the four bodily humours, leading to an excess in certain humoral substances, specifically phlegm.[33] This is largely in line with humoral theory, as the brain was the organ associated with the production of phlegm.[34] The blow to his head must have caused an excessive production of this humoral substance, which was indicated by Beeckman after inspecting his blood and urine. He thus continued to use the purgatives prescribed to him by Van Someren.[35] Around the end of March, however, Beeckman suddenly discontinued his notes on the ailments he had suffered from, which could be an indication that he had finally returned to his old self.

With regard to the descriptions of his own illnesses and bodily ailments, Beeckman's medical training and his observant and experimental attitude are very akin to those of physicians at the time. Although he shared an obsession with healthy living with many of his contemporaries, his detailed descriptions of signs and symptoms, the uroscopic analysis, and use of Latin terminology, indicate that he purposefully approached his own ailments

30 Clericuzio, 'Chemical and Mechanical Theories of Digestion', p. 331; Canone, 'Il *Catalogus librorum*', p. 146.

31 Clericuzio, 'Chemical and Mechanical Theories of Digestion', p. 331; Daniel E. Rusyniak et al., 'Life-Threatening Hyperkalemia from Cream of Tartar Ingestion', *Journal of Medical Toxicology* 9 (2013), pp. 79-81.

32 *JIB*, III, pp. 194-196.

33 *JIB*, III, p. 202.

34 Jackson, 'A Short Guide to Humoral Medicine', p. 487.

35 *JIB*, III, p. 202.

and illnesses from the viewpoint of a medical scholar, seeking a wider understanding of the human body.

Observing the Ill Body: Beeckman as a 'Physician'

Beeckman's practical and experimental physician-like attitude also extended beyond his own body as the object of study. On a few rare occasions, he also performed typical physician's tasks or made careful observations of other people's illnesses. This helps to illustrate his engagement with illness and medical knowledge from a similar practical angle.

One specific case where Beeckman applied an observant approach to a body other than his own, was at a rather difficult moment in his life. It was after the death of his younger brother, Jacob, in reaction to which he wrote:

> Jacob, my brother, has died the 27th of August 1629, in the evening around half past three in Rotterdam of *teeringhe* [tuberculosis]. I had him opened up and found in his lungs many hard grey ulcers, equal to small Turkish beans.[36]

This passage informs us that Beeckman had ordered a post-mortem dissection of his brother's diseased body. Post-mortem dissections, or autopsies, had become a popular form of *observatio* within sixteenth-century medicine, but also outside the medical realm.[37] They were often performed as a means of medical training, but also on the request of family members who wanted know how and why their beloved ones had died. Autopsies were common practice in cases when poisoning or medical malpractice were suspected as the cause of death, but they also helped as part of the grieving process to reconcile with the fate of a beloved one. Yet another incentive for post-mortem dissections was the fear of congenital illnesses, a fear that became more widespread around the fifteenth and sixteenth centuries.[38]

36 *JIB*, IV, 155: 'Jacob, myn broeder, is gestorven den 27en Augusti 1629, t'avons ten half vieren tot Rotterdam van de teeringhe. Hebbe hem doen openen ende in syn longe gevonden veel styve grauwe sweeren, gelyck kleyne Terckse boonen.'

37 Pomata, 'Observation Rising', pp. 65-66.

38 Silvia de Renzi, Marco Bresadola, and Maria Conforti, 'Pathological Dissections in Early Modern Europe: Practice and Knowledge', in: Silvia de Renzi, Marco Bresadola, and Maria Conforti, eds., *Pathology in Practice: Diseases and Dissections in Early Modern Europe* (London: Routledge, 2018), pp. 3-20, esp. pp. 10-15; Annemarie Kinzelbach, 'Dissecting Pain: Patients,

Autopsies were commonly performed by a surgeon, while a physician monitored the process and provided explanations for those present at the scene. In the case of his brother, however, Beeckman made the observations himself. As a medical doctor, he was fully qualified and able to do so. His description is also typical of a straightforward physician's view of the human body, although his motivation for dissecting his brother was also related to concerns regarding his own health. The latter once again highlights the sometimes thin line between a physician-like and 'patient's view' in his writing. The primary reason Beeckman wanted to discover his brother's exact cause of death was related to the increasing belief that *teeringhe* (tuberculosis) was a congenital disease.[39] This fits well within the many indications in his *Journal* that show Beeckman's well-awareness, if not obsession, with symptoms of tuberculosis, such as excessive weight loss and coughing up blood. When he started losing significant weight in 1637, along with coughing up clear drops of blood, he knew very well what had struck him.[40]

However, Beeckman also wished rather explicitly to have his body dissected after death. He requested his friends not to bury him immediately, but to let his dead body decay under supervision for a few days, after which it had to be dissected. This was also related to Beeckman's fear of being buried alive, which was also shared by his daughter Catelyntje, whose frightened testimony he also recorded in his notebook.[41] Although this rather personal testimony regarding the fear of being buried alive is peculiar with regard to his otherwise God-fearing Calvinist attitude, it was not uncommon amongst early modern people. Especially during the 'buried-alive-craze' of the eighteenth century, people often made the similar request to have their bodies placed in a room for two or three days under frequent visitation to prevent being buried alive.[42]

Beeckman's attitude as a careful observant physician also shines through in a few brief segments regarding his deep loathing of what he regarded as untrained healers and imposters who claimed supernatural healing

Families and Medical Expertise in Early Modern Germany', in: De Renzi, Bresadola, and Conforti, eds., *Pathology in Practice*, pp. 170-188, esp. pp. 170-171.

39 Thomas M. Daniel, 'The History of Tuberculosis', *Respiratory Medicine* 100 (2006), pp. 1862-1870, esp. p. 1864.

40 *JIB*, III, p. 431, n. 1.

41 *JIB*, III, p. 215.

42 Václav Grubhoffer, 'Fear of Seeming Death in Eighteenth-Century Europe', in: Albert Classen, ed., *Death in the Middle Ages and Early Modern Times: The Material and Spiritual Conditions of the Culture of Death* (Berlin: De Gruyter, 2016), pp. 491-518, esp. pp. 494-495.

capabilities. In February 1625, for example, he recorded the remarkable case of a woman in Rotterdam, who claimed to relieve people of clavuses with an almost magical ease. In reality, the woman used tiny pieces of cumin cheese to deceive people into believing their skin was contaminated with clavuses. With unreserved pride, Beeckman boasted how he uncovered this scheme by volunteering as a patient. After he had exposed her trickery, the woman was forced by Beeckman to repay the two guilders she had coaxed out of his neighbour's purse. The woman thereafter quickly disappeared without a trace, leaving behind all of her belongings and the materials she had used in her scheme.[43] A similar case described by Beeckman is that of a glass blower named Jacobus Bernardi, who claimed to heal wounds without touching or hurting the patient. This was something Beeckman considered utter nonsense, and he recorded Bernardi's story primarily 'to display the foolishness of such fiddle-faddle'.[44] These reactions to quackery are understandable from Beeckman's practical and observant approach to healing. Yet they are also characteristic from an early modern physician's point of view, as quacks and magical healers were unwelcome competitors on the densely populated medical marketplace.[45] Although Beeckman had little to fear from quacks as he was not a practising physician, he nevertheless felt the need to expose quackery and superstition, as it provided opportunities to boast his critical attitude and philosophy that everything can be explained in practical and natural terms.

However, despite the fact that Beeckman did not actively practise medicine on others for a living, the *Journal* contains a few descriptions which indicate his active participation as an observing, and sometimes even practising physician at the sickbed of others. De Waard collected these notes at the end of Volume 3, labelling them as 'Nosographies', or systematic descriptions of diseases. The appendix contains one very elaborate sickbed description regarding the illness of one of Beeckman's best friends, Gerard van Berckel, who in 1633 informed Beeckman of the fact that he had been diagnosed with 'teeringhe'.[46] Thereupon, Beeckman paid a visit to his ill friend, and took notable time to examine Van Berckel's pulse. He did this repeatedly over a period of a few days, making notes of every examination. As a result, Beeckman noticed that Van Berckel's pulse had slowly

43 *JIB*, II, pp. 321-322.

44 *JIB*, II, pp. 201-202: 'te toonen de sottichheyt van sulcke beuselen'.

45 Christi Sumich, *Divine Doctors and Dreadful Distempers: How Practicing Medicine Became a Respectable Profession* (Amsterdam: Rodopi, 2013), pp. 83-85.

46 *JIB*, III, pp. 443-464.

decreased in speed throughout the period of his visitations.[47] This shows that Beeckman was not just an ordinary visitor at Van Berckel's sickbed. Feeling the pulse was one of the important diagnostic exercises performed by early modern physicians, which allowed them to provide a prognosis regarding the state of illness in which a patient found himself. Doing so, physicians provided an indication of the patient's chances of recovery, and, of course, what kind of medication or therapy could be prescribed. Yet in many cases this was done primarily to ease their patients, who often valued the presence of a well-known physician as both comforting and reassuring.[48] It could very well be possible that Beeckman examined Van Berckel's pulse as a means to comfort his ill friend, but his notes show that he was also generally interested in observing his friend's health while the illness developed.

In January 1634, Beeckman noted that Van Berckel 'could walk much longer and talk more, having tuberculosis, than me being healthy, without alteration of his pulse or fatigue'.[49] Although he noticed that Van Berckel had stopped coughing up bloody phlegm, as he did the year before, this symptom returned after Van Berckel moved from Rotterdam to Amersfoort. The cures prescribed by the local physician in Amersfoort, Theodorus Schut (1595-1672), were noted in detail by Beeckman, including the advice and stories Schut told Van Berckel as a means to comfort him. One of the stories referred to a dog who was given milk to drink, after which the physician who conducted the experiment opened his chest and discovered that the milk had travelled to his lungs. The story of this experiment was meant to prove that orally taken substances, in this case medicaments, could reach the lungs. Beeckman, however, provided his own explanation of this phenomenon in the light of William Harvey's recent theory on the circulation of blood, as a means to explain how substances are transported through the human body via this circulation.[50] Another physician who attended Van Berckel's sickbed was Jacob Dircksz. van den Dael, better known by the Latinized name of Valentius (1570-1644).[51] The latter prescribed a handful of medicines to Van

47 *JIB*, III, pp. 267-268.

48 Sumich, *Divine Doctors*, pp. 33-34.

49 *JIB*, III, p. 446: 'Hy konde oock langher wandelen ende meer spreken, de teeringhe hebbende, dan ick gesont synde, sonder alteratie van syn pols ofte vermoeytheyt.'

50 *JIB*, III, p. 446.

51 Valentius is known primarily as the court physician of stadtholders Maurice and Frederick Henry of Orange. See: H.K. Nagtegaal, 'Dr. Jacob Dircksz. Vallensis (1570-1644)', Hollandse Genealogische Databank, last modified 29 July 2012, http://www.hogenda.nl/wp-content/uploads/2013/02/Vallensis,%20Jacob%20Dircksz.%20(1570-1644).pdf (accessed 2 July 2018).

Berckel, of which Beeckman recorded both the names in Latin, as well as the lengthy Dutch instructions on how these had to be administered. This illustrates a profound interest in the treatment prescribed to his friend, which probably extended to Beeckman's own worries of being struck by the same illness. Doctor Schut, however, also advised Van Berckel to speak as little as possible. This really bothered him, as Beeckman remarked that Van Berckel was a man of many words, who loved nothing more than to converse with others.[52]

Later that year, Valentius advised Van Berckel to undergo multiple purgations, which unfortunately, had little effect. When Beeckman visited his friend in Delft on 29 October 1634, he described the severely weakened Van Berckel as 'doodelick kranck' ('fatally ill'). Although Van Berckel tried to convince him that he was still going strong, his health was noted to be even more dire during Beeckman's following visit. When Van Berckel died a month later, Beeckman was only able to grief the passing of his good friend, as he noted: 'The 1st of Nov[ember] 1634 he died, a man of very good and delicate conscience, the most faithful friend that I had in Holland.'[53]

His relentless day-by-day notations with regard to changes in Van Berckel's health, the observations of other physicians, and the medication prescribed, show a profound interest in the illness and treatment of his friend. These notes can be interpreted in two ways. Firstly, Beeckman was a man who cultivated a lively interest in medicine, keen to make notes of the observations and prescriptions of other physicians, especially prominent ones like Valentius, as a means to learn from them.[54] This extends, however, to the fact that Beeckman was concerned with becoming a tuberculosis sufferer, so perhaps his notes were also of a 'self-help' interest as a means to prepare himself for when the illness would strike his own body. Yet in the case of Van Berckel, it was also not a random patient whom he was observing. Van Berckel was one of his dearest friend, whose physical decay must have been painful for Beeckman to witness. However, this emotional pain, or personal sensation of grief, only becomes visible in his final note regarding Van Berckel's passing, and is in that sense a rarity.

Although detailed descriptions concerning the illness and death of friends and relatives remain a rarity within the *Journal* as a whole, they clearly

52 *JIB*, III, pp. 447-448.

53 *JIB*, III, pp. 448-450: 'den 1en Nov. 1634 is hy gestorven, een man van seer goede ende teere conscientie, myn getrouwste vriendt, die ick in Hollant hadde' (p. 450).

54 Stolberg, 'Empiricism', pp. 502-505.

illustrate Beeckman's interest in illness and death from a practical and observant approach, next to his theoretical knowledge. Apart from the observations and experiments regarding his own body, the accounts of his brother's dissection, his loathing of quackery, and the lengthy sickbed descriptions showcase that Beeckman was not an armchair physician. In fact, he was very keen on extending his practical knowledge with regard to illness and medicine through active observation and the making of copious notes. In the case of tuberculosis, this extended to Beeckman's personal fears regarding his own health, which turned out to true as he succumbed to the illness at the age of 48.

Conclusion

In the history of medicine, a distinction is often made between the 'patient's view' and the medical view. But in the case of Beeckman, such a distinction is very difficult to make, or perhaps should not be made at all. Although his descriptions sometimes show elements that can be associated the 'patient's view', such as the obsession with healthy living and a fear of illnesses such as tuberculosis, the physician and practical philosopher in Beeckman often pervade his descriptions. The manner in which he described his own illnesses and those of friends and relatives was in many ways akin to the observant and experimental approach that had become popular among many physicians during the sixteenth and seventeenth centuries. Beeckman's observations and notes were meant to gain a deeper understanding of the human body through the systematic explanation of the causes of certain illnesses, means of diagnosis, and the effects of certain treatments. His approach to medicine and the human body went beyond theory alone, as the practical observations and notes he made with regard to himself and others helped him to broaden his understanding, confirm theories, and propose new ideas and methods to diagnosis and cure certain illnesses. Taking Beeckman's more practical and observational accounts of illness and death into account, thus showcases very clearly that he practised medicine in a much broader sense than has thus far been considered. However, taking this chapter's conclusion more widely, it confirms even more our understanding of Beeckman as a practically minded scholar, whose lessons of the world were based primarily on what he was able to learn from his own experience.

About the Author

DÁNIEL MOERMAN is a PhD candidate in the NWO-sponsored project 'Coping with Drought: An Environmental History of Drinking Water and Climate Adaptation in the Netherlands, 1550-1850' at the Vrije Universiteit, Amsterdam. He specializes in the sociocultural approaches to crisis and resilience in the early modern period, both on a personal and on a societal level. His most recent publication is a co-authored chapter on the experiences of cross-border nobleman and chronicler Sweder Schele during the Eighty Years' War and the Thirty Years' War in R. Fagel et al., eds., *Early Modern War Narratives and the Revolt in the Low Countries* (Manchester: Manchester University Press, 2020).

9 Beeckman, Descartes, and the Principle of Conservation of Motion*

Samuel Le Gendre

Abstract

Historians of science have always affirmed that Descartes owed his principle of conservation of motion to Beeckman. Beeckman adopted this principle in 1613-1614 and Descartes used this principle after meeting him in 1618. In this chapter it is nonetheless argued that Descartes may actually owe it more to Aristotle and the Coimbrans. While neither Aristotle nor the Coimbrans adopted the principle of conservation of motion, they both discussed it repeatedly. And the Coimbrans provided the element which could lead to the adoption of this principle by arguing in their commentary on Aristotle's *Physics* that motion in a vacuum was possible. Since Descartes studied the Coimbrans at La Flèche in 1612-1613, it is quite possible that he therefore owes the principle of conservation of motion to them.

Keywords: Isaac Beeckman, René Descartes, conservation of motion, principle of inertia, the Coimbrans

* I would like to thank Sophie Roux, Klaas van Berkel, and Arjan van Dixhoorn for giving me the opportunity to write this article. I would also like to thank Sophie Roux for reading and rereading this chapter and for giving me numerous and invaluable advice. It was she who first told me about Beeckman. I would like to thank Klaas van Berkel and Arjan van Dixhoorn for rereading this chapter, for giving me invaluable advice, and for their constructive criticisms. I would like to thank the two anonymous peer reviewers for their valuable criticisms. I would like to thank Hannah Le Gendre for reading and helping me to translate this paper into English. Finally, I would like to thank Giulia Lelli for the days we spent talking about Beeckman and Descartes, for her invaluable advice, for her so relevant remarks, for rereading my paper more than just once and, above all, for her kindness.

Van Berkel, Klaas, Albert Clement, and Arjan van Dixhoorn (eds.), *Knowledge and Culture in the Early Dutch Republic: Isaac Beeckman in Context*. Taylor & Francis Group 2022
DOI: 10.5117/9789463722537_CH09

Introduction

Isaac Newton (1642/1643-1727) owes his principle of inertia to René Descartes (1596-1650), who owes his principle of conservation of motion (PCM for short) to Isaac Beeckman (1588-1637). Thus, if the importance of Descartes for the history of physics is immense, that of Beeckman is no less significant. But, just as Newton hid his debt to Descartes, so Descartes hid his debt to Beeckman.[1] This is the story of the principle of inertia that is currently being told by historians of science.[2] I intend to show that it is possible to tell another one. This chapter will not look at Newton's debt to Descartes. It will look at Descartes' debt to Beeckman. Indeed, after having examined the reasons why historians of science currently think that Descartes owes his PCM to Beeckman, I intend to show why Descartes may rather owe it to the Coimbrans, whose commentaries on Aristotle's *Physics* and *On the Heavens* – in which the PCM is several times explicitly used – he studied at La Flèche in 1612-1613.

1 It has already been pointed out more than once that Newton, who never acknowledged his debt to Descartes, indeed acknowledged his debt to Galileo. Indeed, in the *Principia*, Newton writes in the scholium of the sixth corollary to his *axiomata sive leges motus*: 'The principles I have set forth are accepted by mathematicians and confirmed by experiments of many kinds. By means of the first two laws and the first two corollaries, Galileo found that the descent of heavy bodies is in the squared ratio of the time and that the motion of projectiles occurs in a parabola, as experiment confirms, except insofar as these motions are somewhat retarded by the resistance of the air.' Isaac Newton, *The Principia: Mathematical Principles of Natural Philosophy* (Berkeley: University of California Press, 1999, p. 424). In Latin: 'Hactenus principia tradidi a mathematicis recepta & experientia multiplici confirmata. Per leges duas primas & corollaria duo prima Galilaeus invenit descensum gravium esse in duplicata ratione temporis, & motum projectilium fieri in parabola; conspirante experientia, nisi quatenus motus illi per aeris resistentiam aliquantulum retardantur.' *Isaac Newton's Philosophiae naturalis principia mathematica*, ed. Alexandre Koyré and I. Bernard Cohen (Cambridge, Mass.: Harvard University Press, 1972), p. 64. For Newton's debt to Descartes, see also, for example: J. Bruce Brackenridge, *The Key to Newton's Dynamics* (Berkeley: University of California Press, 1995), p. 22; I.B. Cohen, '"Quantum in se est": Newton's Concept of Inertia in Relation to Descartes and Lucretius', *Notes and Records of the Royal Society of London* 19:2 (1964), pp. 131-155, esp. pp. 131 and 136; I. B. Cohen, *The Birth of a New Physics* (London: Penguin Books, 1992), p. 211; John Herivel, *The Background to Newton's* Principia (Oxford: Clarendon Press, 1965), pp. 42-53; John Herivel, 'L'Influence de Descartes sur Newton en dynamique', *Revue Philosophique de Louvain*, 4th ser., 86:72 (1995), pp. 467-484, esp. p. 472; Alexandre Koyré, *Newtonian Studies* (London: Chapman & Hall, 1965), pp. 64-65. For Descartes' debt, see below.

2 It is generally added that while Descartes was the first to discover the principle of inertia, Gassendi was the first to publish it in his *De motu impresso a motore translato* (1642). Indeed, Descartes abandoned any attempt to publish *Le Monde* (a book he intended to send Mersenne as a New Year gift), after the 1633 condemnation of Galileo. See: Descartes' letter to Mersenne at the end of November 1633, in: *Oeuvres de Descartes*, publiées par Charles Adam et Paul Tannery, 12 vols. (Paris: L. Cerf, 1897-1910) (new ed. in 13 vols., Paris: Vrin, 1974-1986) [henceforth *AT*), I, pp. 270-271. But some historians believe that Gassendi did not clearly adopt the principle of inertia, at least for two reasons: first, he did not,

Definition

Before I begin, I would like to make a few remarks concerning the PCM: 1) by PCM, I do not mean the Cartesian principle according to which God conserves the same quantity of motion in the world; 2) by PCM, I do not mean the principle of inertia: every body perseveres in its state of being at rest or of moving uniformly straight forward, unless it is compelled to change its state by (external) forces; 3) by PCM, I do not mean the idea that the circular motion of celestial bodies can last indefinitely; 4) by PCM, I mean this principle: any body (both celestial bodies and terrestrial bodies), once in motion, continues to move forever, unless it is prevented from doing so by something external.

Thus: 1) the PCM, in my opinion, specifies that the cause that stops the motion cannot be internal to the body but only external; 2) on the other hand, it does not specify the direction of the motion which continues; 3) neither does it specify whether the body continues to move thanks to an internal cause or not.[3]

Why Is It Likely That Descartes Owes His PCM to Beeckman?

As early as 1613-1614, Beeckman, some years after Galileo,[4] and not under his influence, adopts the PCM: *Omnis res, semel mota, nunquam quiescit nisi*

in fact, completely free himself from 'the tyranny of the circle', since he maintained that the circular motion of heavenly bodies was not the effect of an external force; second, the motion of his atoms do not obey the principle of inertia, since they never, strictly speaking, lose their motion. It would therefore be necessary to distinguish two kinds of laws: the laws of motion at a microscopic level and the laws of motion at a macroscopic level. See, on this subject, for instance: Peter Anton Pav, 'Gassendi's Statement of the principle of Inertia', *Isis* 57 (1966), pp. 24-34.

3 As we shall see: in 1618, Beeckman believed that a body once set in motion continues to move without any cause, neither external nor internal; on the contrary, Descartes, in 1618, believed that a body once in motion continues to move without an external cause, but thanks to an internal cause.

4 Galileo is often said to have discovered the PCM before 1607, as it is shown by a letter written by Castelli (*Le Opere di Galileo Galilei*, edizione nazionale, ed. Antonio Favaro and Isodoro Del Lungo, 20 vols. (Florence: Barbèra, 1890-1909), X, p. 170) and to have first published it in 1613, in his *Istoria e dimostrazioni intorno alle macchie solari*, that is to say in his *Letters on Sunspots* (*Opere de Galileo*, V, pp. 134-135). Alan Chalmers and Richard Nicholas claim that Galileo discovered the PCM before 1603-1604 when he came to think of downwards motion as naturally accelerated (and upwards motion as decelerated). Indeed, Galileo was at this time able to characterize neutral motions: they do not accelerate nor decelerate. In other words, they continue at the same speed ('Galileo on the Dissipative Effect of a Rotating Earth', *Studies*

propter externum impedimentum.[5] Or, in Dutch (April 1614-January 1615): 'Dat eens roert, roert altyt, soot niet belet en wort.'[6] Beeckman then provides the PCM to Descartes, his new friend, at their first meeting in Breda in November 1618. Descartes later clarifies the PCM given to him by Beeckman,

in History and Philosophy of Science 14 (1983), pp. 315-340, esp. p. 335). On Galileo and the PCM, see also, for instance: Marshall Clagett, *The Science of Mechanics in the Middle Ages* (Madison: University of Wisconsin Press, 1959), pp. 67-71; Maurice Clavelin, *The Natural Philosophy of Galileo: Essay on the Origin and Formation of Classical Mechanics* (Cambridge, Mass.: Harvard University Press, 1974), pp. 257-259; Maurice Clavelin, 'Galilée et Descartes sur la conservation du movement acquis', *Dix-septième siècle* 242 (2009), p. 33; Maurice Clavelin, *Galilée, cosmologie et science du mouvement. Suivi de Regards sur l'empirisme au XXe siècle* (Paris: CNRS, 2016), pp. 22/139, 5/25, 10/135; Stillman Drake, 'Galileo and the Law of Inertia', *American Journal of Physics* 32 (1964), pp. 601-608; Stillman Drake, *Galileo Studies* (Ann Arbor: University of Michigan Press, 1970), pp. 240-277; Stillman Drake, 'Galileo Gleanings – XVII: The Question of Circular Inertia', in: Drake, *Essays on Galileo and the History and Philosophy of Science*, selected and introduced by Noel M. Swerdlow and Trevor H. Levere (Toronto: University of Toronto Press, 1999), II, pp. 69-85; Anneliese Maier, *On the Threshold of Exact Science: Selected Writings of Anneliese Maier on Late Medieval Natural Philosophy*, ed. and trans. by Steven D. Sargent (Philadelphia: University of Philadelphia Press, 1982), p. 106.

5 *Journal tenu par Isaac Beeckman de 1604 à 1634*, publié avec une introduction et des notes par C. de Waard, 4 vols. (The Hague: Martinus Nijhoff, 1939-1953) [henceforth *JIB*], I, p. 24. Which means: 'Everything, once in motion, never comes to rest, unless impeded by something external.' Except for H.F. Cohen, *The Rise of Modern Science Explained: A Comparative History* (Cambridge: Cambridge University Press, 2015), p. 129, all historians interested in Beeckman's PCM seem to think that he gradually built his principle from a critical reading of Scaliger. See, for example: Klaas van Berkel, *Isaac Beeckman on Matter and Motion: Mechanical Philosophy in the Making* (Baltimore: Johns Hopkins University Press, 2013), pp. 108-109, 147-148, 160; Jean Bernhardt, 'Le Rôle des conceptions d'Isaac Beeckman dans la formation de Thomas Hobbes et dans l'élaboration de son *Short Tract*', *Revue d'histoire des sciences* 40 (1987), pp. 203-215, esp. pp. 207; Dennis Des Chene, *Physiologia: Natural Philosophy in Late Aristotelian and Cartesian Thought* (Ithaca: Cornell University Press, 1996), pp. 276-277. Historians of science usually distinguish five stages in Beeckman's construction of the PCM: 1) Beeckman explains (18 July 1612) that the conservation of the motion of the heavenly spheres is not due to the celestial intelligences nor to the continual command of God but because this motion cannot stop *per se* (*JIB*, I, p. 10); 2) he adopts (July 1613-April 1614) the PCM, i.e., he talks about the conservation of the motion of all bodies (*omnis res*), both celestial and terrestrial (*JIB*, I, p. 24); 3) he insists upon (July 1613-April 1614) the uselessness of the concept of *impetus*: 'Quod verò Philosophi dicunt vim lapidi imprimi, absque ratione videtur' (*JIB*, I, pp. 24-25); 4) he explicitly states (23 November-26 December 1618) that the PCM is valid for circular (on its own circle but also around another centre) and rectilinear motions: '*Id, quod semel movetur, in vacuo semper movetur*, sive secundum lineam rectam seu circularem, tam super centro suo, qualis est motus diurnus Terrae <quam circa centrum, qualis est motus> annuus' (*JIB*, I, p. 253); 5) he explicitly states (5 January-10 February 1631) that the speed is conserved: '*quod movetur in vacuo semper movetur ea motûs celeritate qua coepit moveri*' (*JIB*, III, p. 185). Is is not within the scope of this chapter to discuss this chronology in detail.

6 *JIB*, I, p. 44. Which means: 'What moves once, always moves, unless it is impeded.' I use the translation given by Klaas van Berkel, *Isaac Beeckman on Matter and Motion*, p. 111.

stating the principle of inertia in the form of two distinct laws of nature, which would become a single law of motion for Newton. This is, in broad terms, the story currently being told by historians of science studying the origin of Descartes' PCM[7] and having at their disposal Beeckman's *Journal*.[8]

7 Some historians had at their disposal Beeckman's *Journal* but did not attempt to determine the origin of Descartes' PCM. Two groups can be distinguished. On the one hand, those who do not compare Descartes' PCM with another PCM. See: Desmond M. Clarke, *Descartes' Philosophy of Science* (University Park: Penn State University Press, 1982), pp. 77-107; Tad M. Schmaltz, *Descartes on Causation* (Oxford: Oxford University Press, 2008), p. 106. Frankfurt quotes *Le Monde* and then the *Principia*, pointing out that Gassendi was the first to publish the principle of inertia: Harry Frankfurt, 'Création continuée, inertie ontologique et discontinuité temporelle', *Revue de métaphysique et de morale* 4 (1987), pp. 455-472, esp. p. 458. Others, more numerous, on the contrary, propose comparisons. For a comparison between Descartes' PCM and several other PCMs, see: Michael Elazar, *Honoré Fabri and the Concept of Impetus: A Bridge between Conceptual Frameworks* (Dordrecht: Reidel, 2011), p. 127; Amos Funkenstein, *Theology and the Scientific Imagination: From the Middle Ages to the Seventeenth Century* (Princeton: Princeton University Press, 2018), pp. 13, 74-75, 121, 188, 220, 331, 333, 337; Douglas Jesseph, 'Inertia', in: Roger Ariew et al., *Historical Dictionary of Descartes and Cartesian Philosophy* (New York: Rowman & Littlefield, 2015), p. 185; Michio Kobayashi, *La philosophie naturelle de Descartes* (Paris: Vrin, 1993), p. 65; Gaston Milhaud, *Descartes savant* (Paris: Félix Alcan, 1921), p. 243; G. Rodis-Lewis, *Descartes: Biographie* (Paris: Calmann-Levy, 1995), p. 50; Edward Slowik, *Cartesian Spacetime: Descartes' Physics and the Relational Theory of Space and Motion* (Dordrecht: Springer, 2002), pp. 63-71; James A. Weisheipl, 'The Relationship of Medieval Natural Philosophy to Modern Science: The Contribution of Thomas Aquinas to Its Understanding', *Manuscripta* 20 (1976), pp. 181-196, esp. pp. 187, 190, 191; Richard S. Westfall, *Forces in Newton's Physics* (London: Macdonald, 1971), pp. 58-59, 63-64. For a comparison between Descartes' and Newton's PCM, see: Richard J. Blackwell, 'Descartes' Laws of Motion', *Isis* 57 (1966), pp. 220-234; Cohen, '"Quantum is se est"', pp. 131-155; A. Gabbey, 'Force and Inertia in the Seventeenth Century: Descartes and Newton', in: Stephen Gaukroger, ed., *Descartes: Philosophy, Mathematics, and Physics* (Brighton: Harvester Press, 1980), pp. 286-297; Koyré, *Newtonian Studies*, pp. 65-77; Thomas J. McLaughlin, 'Nature and Inertia', *The Review of Metaphysics* 62 (2008), p. 279. For a comparison between Descartes' and Beeckman's PCM, see, for instance: Frédéric de Buzon and Vincent Carraud, *Descartes et les 'Principia' II. Corps et mouvement* (Paris: Presses Universitaires de France, 1994), p. 101. For a comparison between Descartes and Galileo's PCM, see, for instance: Daniel Garber, 'Laws of Nature and the Mathematics of Motion', in: Geoffrey Gorham, Benjamin Hill, Edward Slowik, and C. Kenneth Waters, eds., *The Language of Nature: Reassessing the Mathematization of Natural Philosophy in the Seventeenth Century* (Minneapolis: University of Minnesota Press, 2016), pp. 134-159, esp. p. 143.

8 It should be remembered that Beeckman's *Journal* (four volumes) was published between 1939 and 1953 by Cornelis de Waard, who discovered the manuscript in 1905. That same year, Cornelis de Waard gave some fragments of the *Journal* to Charles Adam and published others in the Dutch journal *Nieuw Archief voor Wiskunde* under the title 'Eene correspondentie van Descartes uit de jaren 1618 en 1619'. See: 'Descartes et Beeckman' (15 December 1905), in: *AT*, X, pp. 17-18. De Waard also published some fragments of the *Journal* in his edition of Mersenne's *Correspondance*, notably in the second volume, published in 1936. Some of the older historians have taken an interest in Descartes' PCM without having Beeckman's published *Journal* (1939-1953) at their disposal when they wrote their texts. Among them: Emil Wohlwill, Kurd

Lasswitz, Émile Meyerson, Pierre Duhem, and Émile Jouguet. Emil Wohlwill, *Die Entdeckung des Beharrungsgesetzes. Eine Studie zur Geschichte der Physik* (Weimar: Hofbuckdruckerei, 1884), who quotes a text from the *Cogitationes privatae* (1619-1621) – I will call it T3 – does not know Beeckman. He says that the person referred to in that text probably helped Descartes to adopt the PCM: 'Die hier als "Meinung" eines Ungenannten eingeführte Voraussetzung ist offenbar mit dem Beharrungsgesetz identisch. Den Worten, in denen Descartes die gestellte Aufgabe mitteilt, ist ebensowenig wie der unmittelbar folgenden Lösung zu entnehmen, wie er selbst sich jener Zeit zu dieser Lehre verhalten hat. Die Annahme erscheint berechtigt, dass sie ihm neu gewesen ist; es wäre demgemäß der Ursprung der Descartesschen Lehre vom Beharren der Bewegung in der bedeutungsvollen Anregung zu suchen, die ihm – etwa um 1620 – von einem Ungenannten geboten wurde' (pp. 142-143). But Wohlwill is not sure whether 'the unnamed' himself adopted the PCM: 'Ob nun die Auffassungsweise des Ungenannten eine demselben ursprünglich eigene war, ob also ihm etwa zuzuschreiben ist, was später Descartes sich angeeignet hat, lässt sich auf Grund der kurzen Mitteilung nicht entscheiden' (p. 143). Kurd Lasswitz mentions Beeckman, but, like Wohlwill, is unaware that the 'clever man' Descartes is talking about in T3 is Beeckman. And he does not say either that Beeckman adopted the PCM. See: Kurd Lasswitz, *Geschichte der Atomistik vom Mittelalter bis Newton*, 2 vols. (Hamburg: Verlag Leopold Voss, 1890), II, p. 85: 'Auch die mechanischen Entdeckungen DESCARTES' reichen in die früheste Zeit seines Philosophierens zurück. Im Jahre 1620 erfuhr er gesprächsweise von folgender Aufgabe. Ein Stein fällt von *A* nach *B* hin, wird aber von der Erde stets mit derselben Kraft angezogen und verliert nichts von derjenigen Geschwindigkeit, welche ihm durch die vorangegangen Anziehung mitgeteilt ist. Hierbei wird nämlich angenommen, dass dasjenige, was sich im leeren Raume bewegt, sich immer bewege. In welcher Zeit wird die Strecke *AB* durchlaufen?' Lasswitz only states that Descartes discovered the principle of inertia independently from Galileo: 'In der Voraussetzung, dass die Bewegung beharrt und die neue Wirkung der Schwerkraft zur früheren hinzu kommt, welche DESCARTES ausdrücklich in einem Briefe an MERSENNE vom 18. Dezember 1629 hervorhebt, liegt der Beweis, dass DESCARTES ebenfalls frühzeitig und bevor er GALILEIS Werke kannte, sich über das Beharrungsgesetz klar war, das er in *Le Monde* formulierte, aber allerdings erst in den Principien veröffentlichte' (p. 85). Émile Meyerson, *Identité et réalité* (Paris: Félix Alcan, 1908), does not mention Beeckman in his chapter on the principle of inertia (chap. 3). He states that Descartes probably owes his PCM to Galileo (p. 104). Pierre Duhem did not have at his disposal Beeckman's *Journal* either, but only the texts that can be found in Descartes' works (in particular: *AT*, X). He quotes T1, T3, T4 and T5 in *De l'accélération produite produite par une force constante* (Geneva: Congrès Internationale de Philosophie, IIIe Session, 1905), p. 904, and *Études sur Léonard de Vinci*, série III (Paris: Éditions des archives contemporains, 1984), pp. 570, 572. To my knowledge, Duhem does not indicate to whom Descartes owes his PCM, but rather tends to group together Galileo, Beeckman and Descartes. See, for instance: Pierre Duhem, *Le Système du monde* (Paris: Hermann, 1958), VIII, p. 215. Émile Jouguet mentions Beeckman once in his *Lectures de mécanique. La Mécanique enseignée par les auteurs originaux* (Paris: Gauthier-Villars, 1908), I, p. 81, but he does not say that Descartes owes him the PCM. In fact, he does not say that Beeckman adopted the PCM. Jouguet argues that the letter to Mersenne of November 1629 – I will call it T4 – 'contains the first statement ever given of the principle of inertia' (p. 82). Eduard Jan Dijksterhuis in his *Val en worp. Eene bijdrage tot de geschiedenis der mechanica van Aristoteles tot Newton* (Groningen: P. Noordhoff, 1924), p. 305 sq., was one of the first (after Duhem, De Waard and Adam) to draw public attention to several texts of Beeckman's *Journal* in which

Henceforth, I shall call that story S1.[9]

But not all of them tell, in detail, the same story. Three stories are told. For some, while Descartes does indeed owe the PCM to Beeckman, his principle will never be the same as Beeckman's. Indeed, Descartes will never abandon the notion of *impetus* that Beeckman explicitly rejected.

the PCM is explicitly used. Later, in his *The Mechanization of the World Picture* (Oxford: Oxford University Press, 1961; Dutch original Amsterdam: Meulenhoff, 1950), he examines Beeckman's (pp. 332-333) and Descartes' PCM (pp. 410-411). But he does not say that Descartes owes his PCM to Beeckman. René Dugas (1897-1957) could have used Beeckman's *Journal*. But he does not mention Beeckman's PCM. He quickly examines Descartes' principle of conservation of the quantity of motion. But Dugas does not compare it to his principle of inertia, whose discovery he in fact attributes to Galileo, and the reformulation to Huygens and then Newton (René Dugas, *A History of Mechanics*, trans. by J.R. Maddox (New York: Dover Publications, 1988; original ed., 1955), p. 205). Anneliese Maier, too, could have used Beeckman's *Journal*. But I am not sure that she did. Instead, she seems to rely only on two texts quoted by Dijksterhuis in *Val en worp*. See: Anneliese Maier, 'Die Rolle der Impetustheorie in der Entstehung der neuen Mechanik und Naturphilosophie', in: Anneliese Maier, *Die Impetustheorie der Scholastik* (Rome: Edizione de 'Storia e Letteratura', 1940), pp. 153-176, esp. p. 171. Maier made several assertions concerning the PCM and the principle of inertia. She argues that Descartes was not the first to state the principle of inertia, but that he was the first to make it the basis of his mechanics: Anneliese Maier, 'Die Fallbeschleunigung', in: Anneliese Maier, *An der Grenze von Scholastik und Naturwissenschaft* (Rome: Edizione di 'Storia e Letteratura', 1952), pp. 183-218, esp. p. 211. She argues that the principle of inertia was not stated in its classical form by Galileo, but 'in Descartes' circle': Anneliese Maier, 'Impetustheorie und Trägheitsprinzip', in: Anneliese Maier, *Die Vorläufer Galileis im 14. Jahrhundert* (Rome: Edizione de 'Storia e Letteratura', 1949), pp. 132-154, esp. p. 154, n. 42. She argues that Beeckman may have been the first to state the principle of the conservation of motion: Maier, 'Die Rolle der Impetustheorie', p. 171. Later, she also argues that Galileo was the first, before Beeckman and Descartes, to state the principle of inertia if by this one means that he was the first to change the scholastic conception of motion and its causes, i.e., to reject the axioms of Aristotelian-Scholastic mechanics (Maier, *On the Threshold of Exact Science* [see note 5], p. 106).

9 Some historians add that Beeckman most certainly influenced Gassendi. See, for instance: Van Berkel, *Isaac Beeckman on Matter and Motion*, p. 167: 'On the other hand, Gassendi's principle of inertia, published in 1642, echoes Beeckman's. Although Gassendi may have heard others discuss this idea, Beeckman definitely discussed his principle of inertia with Gassendi in 1629. So, although Gassendi never read Beeckman's notebook, Beeckman clearly influenced his work.' Jean Bernhardt even goes so far as to maintain that he influenced Hobbes. See: Bernhardt, 'Le Rôle des conceptions d'Isaac Beeckman', p. 204: 'Hobbes had never heard of Galileo's works till 1634, but he was acquainted with F. Bacon's scientific circle, as a secretary to the (very likely ex-) Chancellor. Directly or not, he probably learnt the so-called principle of inertia, or, more vaguely, a law of conservation of the state of uniform motion, which was the key to modern mechanics, from I. Beeckman, Bacon's Dutch correspondent. Beeckman's ideas and researches were certainly not kept secret in Northern countries.'

Thus, Descartes' debt to Beeckman, if not zero, is not immense.[10] For others, on the contrary, Descartes and Beeckman seem to share the same PCM from the outset. That is why Descartes' debt to his mentor, which he unfairly hid, is immense.[11] Finally, for the majority of historians, Descartes' PCM will gradually become the same as Beeckman's, since Descartes gradually discarded the notion of *impetus* from *Le Monde*. But, at that time, Descartes not only had a conceptual framework different from Beeckman's, but also succeeded in freeing himself 'from the tyranny of the circle'.[12]

10 Peter Damerow et al., *Exploring the Limits of Pre-Classical Mechanics* (New York: Springer, 1992), pp. 39, 45-46, 52, 54, 90, 265; Stephen Gaukroger, *Descartes: An Intellectual Biography* (Oxford: Clarendon Press, 1995), pp. 73, 84, 450; Nicoletta Sciaccaluga, 'Isaac Beeckman e i paradossi della conservazione del moviemento', *Annali della Scuola Superiore di Pisa. Classe di Lettere e Filosofia*, Serie IV, 3 (1998), p. 396; Nicoletta Sciaccaluga, '*Potentia Naturalis*: Kepler, Beeckman, Descartes', *Alvearium* 1 (2008) pp. 39-64, esp. pp. 60-64; John Schuster, *Descartes-Agonistes: Physico-mathematics, Method & Corpuscular-Mechanism 1618-33* (Dordrecht: Springer, 2013), pp. 109-110, 138-144, 206-208, 409, 445-446.

11 Richard Arthur, 'Beeckman, Descartes and the Force of Motion', *Journal of the History of Philosophy* 45:1 (2007), pp. 1-28, esp. p. 10, 20, 23, 26; Danilo Cappechi, *The Problem of the Motion of Bodies* (Dordrecht: Springer, 2014), p. 200; H.F. Cohen, *How Modern Science Came into the World: Four Civilizations, One 17th-Century Breakthrough* (Amsterdam: Amsterdam University Press, 2010), p. 528; Cohen, *The Rise of Modern Science Explained*, pp. 128-129; Cornelis de Waard, in: *JIB*, IV, pp. 165, 171, 199; Klaas van Berkel, 'Descartes' Debt to Beeckman: Inspiration, Cooperation, Conflict', in: Stephen Gaukroger, John Schuster, and John Sutton, eds., *Descartes' Natural Philosophy* (New York: Routledge, 2000), pp. 46-59, esp. pp. 49-50; Van Berkel, *Isaac Beeckman on Matter and Motion*, pp. 61, 171-172. Even if Van Berkel also writes: 'On the question of whether Descartes' ontology really differed from Beeckman's, and if so, why, historians still differ, but this is a topic that belongs exclusively to the study of Descartes' theory of motion, not Beeckman's' (p. 116). See also: Van Berkel, *Isaac Beeckman on Matter and Motion*, p. 230, n. 38; Bernhardt, 'Le Rôle des conceptions d'Isaac Beeckman', p. 208, n. 23.

12 Gabbey, 'Force and Inertia', p. 291. See also: Alan Gabbey, 'New Doctrines of Motion', in: Daniel Garber, Michael Ayers, and Roger Ariew, eds., *The Cambridge History of Seventeenth-Century Philosophy* (Cambridge: Cambridge University Press, 2008), I, pp. 649-679, esp. pp. 662-663. It should be noted, however, that some historians wish to point out that Descartes may not have been as free from the tyranny of the circle as is usually thought. They rely on two text: 1) Chapter 13 of *Le Monde* (*AT*, XI, p. 85); 2) Descartes' letter to Ciermans of 23 March 1639 (*AT*, II, p. 74). See, for example: Gaukroger, *Descartes: An Intellectual Biography*, pp. 246-247; Stephen Gaukroger, *Descartes' System of Natural Philosophy* (Cambridge: Cambridge University Press, 2003), pp. 14-15; Descartes, *The World and Other Writings*, ed. and trans. by Stephen Gaukroger (Cambridge: Cambridge University Press, 2004), introduction, xix-xx, p. 55, n. 86; Stephen Gaukroger, *The Emergence of a Scientific Culture: Science and the Shaping of Modernity, 1210-1685* (Oxford: Clarendon Press, 2006), pp. 293-294, 411-412; Stephen Gaukroger, 'The Foundational Role of Hydrostatics and Statics in Descartes', in: Gaukroger, Schuster, and Sutton, eds., *Descartes' Natural Philosophy*, pp. 60-80, esp. pp. 61-62; William Shea, *La magia dei numeri e del moto. René Descartes e la scienza del Seicento*, trans. by Nicoletta Sciaccaluga (Torino: Bollati Boringhieri, 1994), p. 215, pp. 222-223, pp. 285-288; Westfall, *Forces in Newton's Physics*, pp. 81-82. They are contradicted by: Patrice Bailhache, 'Mécanique et métaphysique: Descartes et la découverte du

Descartes' debt to Beeckman is thus significant, but not immense.[13]

principe de l'inertie', delivered in 1995 in a conference, during the workshop 'Descartes savant', n. 40; Geoffrey Gorham, 'The Metaphysical Roots of Cartesian Physics: The Law of Rectilinear Motion', *Perspectives on Science* 13 (2005), pp. 431-451, esp. p. 435, n. 11. Edward Slowik thinks there is no circular inertia in *Le Monde* but thinks that there is one in the letter to Ciermans: Gorham, *Cartesian Spacetime*, pp. 62-74, esp. pp. 69-71. This is not the place to take part in this debate. I am only pointing out two things: 1) it seems to me that Bailhache and Gorham are right to deny that we find in *Le Monde* and in the letter to Ciermans an inertial circular motion; 2) but I am not sure that Bailhache is right when he writes that the rotation of light corpuscules is due to internal forces ('Descartes eût dit que cette rotation interne – ce spin a-t-on envie de dire – s'accompagnait de forces intérieures dans la particule'), since I think that Descartes explains that the rotation is due to external constraints, for instance in *La Dioptrique* (*AT*, VI, pp. 90-91) or in *Les Météores* (*AT*, VI, pp. 330-335); and I am not sure, either, that Gorham is right when he writes that Descartes, explaining the rotation of these particules, 'may mean simply that there is no reason why the total quantity of motion in the plenum should be distributed differently than it is', for there is a reason, namely the laws of impact.

13 Frédéric de Buzon, 'Beeckman, Descartes and Physico-mathematics', in: Daniel Garber and Sophie Roux, eds., *The Mechanization of Natural Philosophy* (Dordrecht: Springer, 2013), pp. 148-158, esp. 151-153; Daniel Garber, *Descartes' Metaphysical Physics* (Chicago: University of Chicago Press, 1992), pp. 10-11, 197, 210; Wallace Hooper, 'Inertial Problems in Galileo's Preinertial Framework', in: Peter Machamer, ed., *The Cambridge Companion to Descartes* (Cambridge: Cambridge University Press, 2006), pp. 146-174, esp. pp. 149, 162-163; Vincent Jullien and André Charrak, *Ce que dit Descartes touchant la chute des graves: De 1618 à 1646. Étude d'un indicateur de la philosophie naturelle cartésienne* (Villeneuve d'Ascq: Presses Universitaire du Septentrion, 2002), pp. 19, 107, 117, 118, 126, 130, 193; Alexandre Koyré, *Études galiléennes* (Paris: Hermann, 1966), pp. 108, 113, 118-119, 127-128, 161-162, 327; Bernard Rochot, 'Beeckman, Gassendi et le principe d'inertie', *Archives internationales d'histoire des sciences* 5 (1952), p. 284; G. Rodis-Lewis, 'Le Premier register de Descartes', *Archives de Philosophie* 54 (1991), pp. 639-657, esp. p. 644; Sophie Roux, 'Découvrir de principe d'inertie', *Recherches sur la philosophie et le langage* 24 (2006), pp. 449-511, esp. pp. 469-473; William R. Shea, 'The "Rational" Descartes and the "Empirical" Galileo', in: Carla Rita Palmerino and Hans Thijssen, eds., *The Reception of the Galilean Science of Motion in Seventeeth-Century Europe* (Dordrecht: Kluwer, 2004), pp. 67-82, esp. pp. 71-74; Shea, *La magia dei numeri e del moto*, pp. 31, 33, 38. And presumably also, although they are very careful: Maurice Clavelin, 'Galilée et Descartes sur la conservation du movement acquis', *Dix-septième siècle* 242 (2009), pp. 31-43, esp. p. 42; Des Chene, *Physiologia*, p. 275, n. 22, pp. 278-279. I do not know which group Julian Barbour belongs to. It seems to me that he did not consult Beeckman's *Journal* in detail. What is certain is that he maintains that Descartes abandoned the medieval notion of *impetus* and that he writes in Julian B. Barbour, *The Discovery of Dynamics: A Study from a Machian Point of View of the Discovery and the Structure of Dynamical Theories* (Oxford: Oxford University Press, 2001): 'Like Descartes, Beeckman had a concept of inertial motion and should perhaps be given some credit for the idea. It was, however, Descartes, not Beeckman (who published nothing in his lifetime), who presented the idea of inertial motion, in a form quite close to that finally adopted by Newton, to the world with a great flourish and thereby influenced the course of further developments' (p. 407). And then: 'As regards the priority of discovering and recognizing the importance of the natural persistence of motion, there is no doubt that it belongs to Galileo. He had it almost before Descartes (born 1596) was out of his swaddling clothes and he published the essential idea (in the *Letters on Sunspots*) while Descartes was still at school at

The Texts

Historians who adopt S1 often refer to five texts.[14] I shall refer to those texts as T1, T2, T3, T4, and T5. Five things must be said about those texts. Firstly, those five texts have the same object, namely the problem of falling bodies. Secondly, only one of them is written by Beeckman (T1). Thirdly, they do not all date from the same period. The first three are written shortly after the meeting of November 1618, namely between 23 November and 26 December 1618 (T1, T2) and between 1619 and 1621 (T3). The other two are written in November (T4) and December 1629 (T5), when Beeckman and Descartes began to correspond with Marin Mersenne.[15] Finally, those texts do not all have the same status. Two of them are reports that Beeckman and Descartes wrote for themselves (T1 and T3), one is a text given by Descartes to Beeckman (T2), and the other two are taken from letters sent by Descartes to Mersenne (T4 and T5). Here are these five texts:

> T1. Beeckman (*Journal*, November-December 1618): 'This was demonstrated thus by Mr. Peron [Descartes], when I posed the problem to him, asking whether one could know how much space an object traverses in a single hour if it is known how much it traverses in two hours according to my principles, viz. that in a vacuum, what is once in motion will always be in motion, and supposing that there is a vacuum between the Earth and the falling stone.'[16]

La Flèche. More interesting is the question: what led Descartes to inertia? Was he influenced, perhaps indirectly, by Galileo? [...] A much more direct influence than Galileo's will certainly have been that exerted on Descartes by his Dutch physicist friend Isaac Beeckman, with whom, as we noted, he was in regular contact several years before he wrote *The World*. As early as 1614 Beeckman noted in the margin of his *Journal* that 'a stone thrown in vacuum does not come to rest' and that it therefore 'moves perpetually'. Perhaps Beeckman should be credited with the formulation of the first law of motion and not Descartes. Whatever the truth, the outcome of the Beeckman-Descartes line of development was a set of interrelated ideas that owed little to Galileo' (pp. 432-433).

14 I will use the translation from Damerow et al., *Exploring the Limits*, pp. 279-286.

15 Let us review some facts. Descartes and Beeckman met on 10 November 1618, and left one another on 2 January 1619. They subsequently corresponded for a few months, until late April-early May 1619. Then, their exchanges stopped until 8 October 1628, when Descartes paid a surprise visit to Beeckman. The following year, Beeckman, who received a visit from Gassendi in July, and Descartes both began to correspond with Mersenne. It was also from October of that same year 1629 that their relationship deteriorated dramatically, until the violent break-up of October 1630.

16 *AT*, X, p. 60. See also: *JIB*, I, p. 263: 'Haec ita demonstravit Mr. Du Peron, cum ei ansam praebuissem rogando an possit quis scire quantum spacium res cadendo conficeret unica hora,

T2. Descartes to Beeckman ('Physico-mathematica', November-December 1618): 'This question can be posed in another, more difficult, way in the following manner. Imagine that the stone is at point *a* and that the space between *a* and *b* is a vacuum; and that all at once, say today at nine o'clock, God, should create in *b* a force attracting the stone; and that afterwards he should create in every single moment a new force equal to the one he created in the first moment; and that together with the force created before it would attract the stone ever more strongly, because in a vacuum that which is once in motion moves always; and finally the stone which was at *a* will reach *b* today at ten o'clock.'[17]

T3. Descartes (*Cogitationes privatae*, 1619-1621): 'It happened a few days ago that I came to know an extremely clever man who posed me the following problem: A stone, he said, descends from A to B in one hour; it is perpetually attracted by the earth with the same force without losing any of the speed impressed upon it by the previous attraction. According to him [*existimabat*], that which moves in a vacuum will move always. He asks, in how much time will such a space be traversed?'[18]

T4. Descartes to Mersenne (letter to Mersenne, 13 November 1629): 'First I assume that the motion once impressed into a body, remains there always, unless it is removed from it by some other cause, i.e., that which has once started to move in a vacuum moves always and with equal speed.'[19]

T5. Descartes to Mersenne (letter to Mersenne, 18 December 1629): 'To come back to Mr. Beeckman, although what he told you is false, namely

cim scitur quantum conficiat duabus horis, secundum mea fundamenta, viz. *quod semel movetur, semper movetur, in vacuo*, et supponendo inter Terram et lapidem cadentem esse vacuum.'

17 *JIB*, IV, p. 51; *AT*, X, pp. 77-78: 'Aliter vero potest haec quaestio proponi difficilius, hoc pacto. Imaginetur lapis in puncto *a* manere, spatium inter *a* & *b* vacuum; iamque primum, verbi gratia, hodie hora nona Deus creet in *b* vim attractivam lapidis; & singulis postea momentis novam & novam vim creet, quae aequalis sit illi quam primo momento creavit; quae iuncta cum vi ante creata fortius lapidem trahat & fortius iterum, quia in vacuo quod semel motum est semper movetur; tandemque lapis, qui erat in *a*, perveniat ad *b* hodie hora decima.'

18 *JIB*, I, pp. 360-361; *AT*, X, p. 219: 'Contigit mihi ante paucos dies familiaritate uti ingeniosissimi viri, qui talem mihi quaestionem proposuit: *Lapis*, ajebat, *descendit ab A ad B una hora; attrahitur autem a Terraa perpetuo eadem vi, nec quid deperdit ab illaa celeritate quae illi impressa est priori attractione: quod enim in vacuo movetur, semper moveri* existimabat. *Quaeritur: quo tempore tale spatium percurrat.*'

19 *AT*, I, pp. 71-72; *JIB*, IV, p. 168: 'Premièrement je suppose que le mouvement qui est une fois imprimé en quelque corps y demeure perpétuellement, s'il n'en est ôté par quelque autre cause, c'est-à-dire que *quod in vacuo semel incoepit moveri, semper et aequalis celeritate movetur.*'

that once a falling weight reaches a certain point it always continues to fall at the same speed, it is true that after a certain distance the increase in speed is so slight as to be imperceptible. I shall explain to you what he meant to say, for we have discussed this together in the past. He supposes, as I do [*Supponit, ut ego*], that what has begun to move will persist by itself if it is not impeded by some external force, and that it moves always in a vacuum, but in the air it is impeded somewhat by the resistance of the air.'[20]

Reasons for Adopting S1

Historians adopt S1 for three reasons. First, for chronological reasons: the PCM appears for the first time in a text by Descartes between 23 November and 26 December 1618. Second, because they rely on the statements of Beeckman and Descartes: Descartes himself is said to have acknowledged his debt, which Beeckman is said to have also underlined. Finally, for lexical reasons: Descartes states the PCM, which is for him an 'assumption' in *Le Monde*, in the same words as Beeckman. Let us examine each of these reasons in detail.

First Argument: Chronology

First and foremost, historians note that the PCM appears for the first time in a text by Descartes between the end of November and the end of December 1618, when Descartes had just met Beeckman, who had adopted the PCM at least since 1613-1614.[21]

20 *AT*, I, pp. 90-91; *JIB*, IV, p. 171: 'et pour revenir au Sieur Beeckman, encore que ce qu'il vous a mandé soit faux, à savoir qu'il y ait un lieu auquel un poids qui descend étant parvenu, poursuit par après toujours d'égale vitesse, toutefois il est vrai qu'après certain espace cette vitesse s'augmente de si peu qu'elle peut être jugée insensible, et je m'en vais vous expliquer ce qu'il veut dire, car nous en avons autrefois parlé ensemble. // *Supponit, ut ego, id quod semel moveri cœpit, pergere sua sponte, nisi ab aliqua vi externa impediatur, ac proinde in vacuo semper moveri, in aere vero ab aeris resistentia paulatim impediri.*'

21 See, for instance: Van Berkel, *Isaac Beeckman on Matter and Motion*, pp. 133, 148, 160; Van Berkel, 'Beeckman, Isaac (1588-1637)', in: Lawrence Nolan, ed., *The Cambridge Descartes Lexicon* (Cambridge: Cambridge University Press, 2016), pp. 58-59; Buzon, 'Beeckman, Descartes and Physico-mathematics', p. 153; Gaukroger, *Descartes: An Intellectual Biography*, p. 73; Hooper, 'Inertial Problems in Galileo's Preinertial Framework', p. 163; Koyré, *Études galiléennes*, pp. 108-109, 113. Historians add that in June 1629 Beeckman states in a letter to Mersenne that he has been supporting this principle for 20 years, i.e., since 1609 (*JIB*, IV, p. 147). See, for instance: Van Berkel,

Of course, this is an extremely important clue. However, because there is no mention of the PCM in his texts before 1618, it does not necessarily mean that Descartes did not adopt the PCM before he met Beeckman. All the more so, since there is no text ever written by Descartes before the meeting with Beeckman.

Second Argument: Descartes' and Beeckman's Statements

Then, historians examine the statements of Descartes and Beeckman. They notice four things; (a) Descartes himself is said to have acknowledged his debt to Beeckman concerning the PCM before hiding it; (b) Beeckman is also said to have indicated that Descartes owed him the PCM; (c) Beeckman is said to have even blamed Descartes for hiding his debt concerning the PCM; (d) Descartes, acknowledged his debt to Beeckman by stating that he was 'the first mover of my studies and their first author' (*ut studiorum meorum promotorem et primum authorem*).[22] Let us look at each of those four reasons in detail.

a What did Descartes write about Beeckman's PCM exactly?

Historians argue that Descartes himself acknowledged that he owed the PCM to Beeckman, before he denied it. They oppose the *existimabat* of T3, by which Descartes, according to them, acknowledges his debt, to the *ut ego* of T5, by which, according to them, he denies it.[23]

Isaac Beeckman on Matter and Motion, p. 160, 244, n. 129; Buzon, 'Beeckman, Descartes and Physico-mathematics', p. 147. But, as we have seen, Beeckman adopts the PCM in 1613-1614 or, at the earliest, in 1612 (*JIB*, I, p. 10). Must it be said that Beeckman, in 1629, exaggerates? Should it instead be said that in 1613-1614, he does not adopt but applies the PCM, which he possesses for several years? Unless it is necessary to date some notes of the *Journal* differently, i.e., to correct the chronology. Or, as it has been suggested to me by Klaas van Berkel, 'it may very well be that Beeckman, in 1629, alluded to his discussion in 1609 of Tycho Brahe and his arguments against the motion of the Earth' (see: *JIB*, I, p. 2). In favour of S1, it may also be noticed that after 1618-1619, it was not until the letter of 8 October 1629, the year during which the correspondence between Beeckman and Mersenne began, as has already been said, that the PCM can be found in a text written by Descartes (*AT*, I, pp. 28-29). But Descartes did not write many texts between 1619 and 1629.

22 *JIB*, IV, p. 62 [23 April 1619]; *AT*, X, p. 162. It is Klaas van Berkel who advised me to use this letter. On this letter, see: Van Berkel, *Isaac Beeckman on Matter and Motion*, p. 26.

23 Koyré, *Études galiléennes*, pp. 113, 162. Koyré decides to underline the *ut ego* and to oppose it to the *existimabat*. And he decides to translate *existimabat* as 'd'après lui' (p. 113), as Damerow et al., eds., *Exploring the Limits*, pp. 30, 283, do ('according to him'), but contrary to what Jullien

Of course, it is possible to read those texts this way. But it is also possible, it seems, to read them differently. First, one might not oppose the *existimabat* of T3 to the *ut ego* of T5 and instead oppose the *ut ego* of the letter to Mersenne to the *de suo*, which comes just after in that letter.[24] If one does so, one might think that, in this letter, Descartes does not want to hide his debt. He rather wishes to distinguish a principle he already had when he met Beeckman, i.e., the PCM, from a principle he has just adopted, i.e., that the speed of the fall is inversely proportional to the resistance of the air.

Next, we can point out that the *existimabat* is used by Descartes in a private text, but also that Descartes in T2, i.e., in the text he gives to Beeckman, uses the PCM without attributing it to him. Now, one can first think that if Descartes did indeed owe the PCM to Beeckman, his newest friend, with whom he was anything but angry at the time, he would have attributed it to him in T2. We can then think that Descartes did not use *existimabat*, in a private text, to underline the fact that Beeckman, unlike him, adopted the PCM. On the contrary, Descartes perhaps wanted to stress the fact that he had just met someone who adopted the same principle as he did, at a time when almost no one adopted the PCM and at a time when the two friends were accustomed to putting forward the singularity of both their approach and their thought.[25]

and Charrak, *Ce que dit Descartes*, p. 104, do, since they translate *existimabat* with 'estimait-il' ('he thought'). See also: Arthur, 'Beeckman, Descartes and the Force of Motion', p. 15-16, n. 36: '"*Supponit, ut ego*" – an implicit denial of priority that must rank as one of the most disingenuous ever made in the history of physics.' Van Berkel, *Isaac Beeckman on Matter and Motion*, p. 61: 'For example, Descartes stated that Beeckman had taken the principle of inertia from him when they derived the law of falling bodies in 1618, instead of the other way round.' To support their interpretation, historians also sometimes quote Leibniz: 'It seems to me that people have been unjust to Monsieur Isaac Beeckman, by mistreating him on the basis of the reports in the letters of Monsieur Descartes only, from which I have learned not to boast at the expense of others, because Monsieur Descartes gives a strange twist to things when he was mad at someone.' See, for instance: Van Berkel, *Isaac Beeckman on Matter and Motion*, p. 246, n. 28; Gemelli, *Isaac Beeckman*, p. 127, n. 11.

24 *AT*, I, p. 91.

25 *AT*, X, p. 52 and *JIB*, I, p. 244: 'Physico-mathematicians are extremely few – The man from Poitou has much frequented the Jesuits, and other virtuosi and men of learning. He told me, however, that he had never met a man, apart from myself, who was in the habit of studying in the way in which I am pleased to do, joining, in precise fashion, physics and mathematics. And for my part, I have never spoken to anyone who studies in this way.' I use Chikara Sasaki's translation (*Descartes' Mathematical Thought* (Dordrecht: Springer, 2003), p. 99), slightly modified by Klaas van Berkel.

b What did Beeckman write about the PCM?

Historians not only argue that Descartes acknowledged his debt, but think that Beeckman himself, who considered the PCM his principle, said that he gave it to Descartes. They notice that Beeckman indicated that in 1618 Descartes tried to solve the problem of falling bodies 'according to my principles' (*secundum mea fundamenta*).[26]

We can add, that if Descartes had possessed the PCM before meeting Beeckman (and had told him about it), then Beeckman would not have presented the PCM as '[his] theorem' (*meum theorema*) in his *Journal* (8 October 1628-1 February 1629 and 1634).[27]

Of course, it is possible to interpret Beeckman's statements this way. But it seems that they can also be interpreted differently. First of all, by indicating in his notebook, i.e., privately, that Descartes solved his problem with his *fundamenta*, Beeckman only says that he himself fixed the data, among which is the PCM, for the problem he posed.[28] He does not necessarily mean that Descartes owes him the PCM. Furthermore, in deciding to present the PCM as 'my theorem', Beeckman perhaps only wants to emphasize that he owes the PCM to no one, i.e., he designed it himself. Nor does this imply that he provided it to Descartes.

It may be recalled that Duhem, who did not have Beeckman's *Journal* at his disposal but who quoted T1, T3, T4 and T5, did not think he had to read in these texts that Descartes owed his PCM to Beeckman (cf. n. 9).

c What did Beeckman write about Descartes' PCM?

Some historians go so far as to state that Beeckman reproached Descartes for not acknowledging that he owed him the PCM. They do so for two reasons. First, they point out that Beeckman, in a letter dated October 1630, whose content has now almost entirely been lost, reproached Descartes for having presented himself as the author of some ideas that he owed

26 T1, once again: 'Haec ita demonstravit Mr. Du Peron, cum ei ansam praebuissem rogando an possit quis scire quantum spacium res cadendo conficeret unica hora, cim scitur quantum conficiat duabus horis, secundum mea fundamenta, viz. *quod semel movetur, semper movetur, in vacuo*, et supponendo inter Terram et lapidem cadentem esse vacuum.' For example: Garber, *Descartes' Metaphysical Physics*, p. 312; Milhaud, *Descartes savant*, pp. 28-29

27 *JIB*, III, p. 104, p. 341. See also: *JIB*, I, p. 256: 'hoc theorema'. And: *JIB*, III, p. 123: 'meam sententiam de motu'.

28 Étienne Gilson, *Études sur le rôle de la pensée médiévale dans la formation du système cartésien* (Paris: Vrin, 1967), p. 147.

him: 'When your [friend] Mersenne occupied himself entire days with my handwritten book, and when he saw in it quite a few things that he esteemed to be yours, and, from the dates added to these, rightly began to doubt who was their [true] author, I revealed to him, more freely perhaps than pleased you or him, what the case was.'[29] Then, historians believe either that Descartes alludes to the PCM in his famous response letter of 17 October,[30] or, on the contrary, that he chooses not to talk about it,[31] which, in either case, would indicate that Beeckman's accusation concerned his use of the PCM.

It may not be impossible to interpret these texts this way.[32] However, if one chooses, perhaps naively, to recall what is explicitly referred to in Descartes' letter of 17 October, one can only say that Descartes discusses his alleged debt regarding the vibrating strings and the hyperbola.[33] It is therefore quite possible that Beeckman, in his October letter, did not reproach Descartes for presenting himself as the author of the PCM, but only as the author of some of his ideas about music.[34]

d What did Descartes write about Beeckman's influence?

Some historians, finally, recall that in April 1619, Descartes sent his friend Beeckman a letter in which he thanked him for having brought him out of his idleness and for having helped him to remember what he had learnt, even going so far as to allow him to present himself as the author of his future discoveries:

> If, as I hope, I stop somewhere, I promise you that I shall undertake to put my Mechanics or My Geometry in order, and I shall honour you as the first mover of my studies and their first author. For truly, you alone have roused me from my idleness and recalled to me what I had learnt

29 *JIB*, IV, p. 195. I use the translation from: H.F. Cohen, *Quantifying Music: The Science of Music at the First Stage of the Scientific Revolution, 1580-1650* (Dordrecht: Reidel, 1984), p. 193. See also: *JIB*, IV, p. 202; *AT*, I, p. 171.

30 Édouard Mehl, 'Les Années de formation', in: Frédéric de Buzon, Elodie Cassan, and Denis Kabouchner, eds., *Lectures de Descartes* (Paris: Ellipses, 2015), pp. 41-66, esp. p. 49.

31 Cohen, *Quantifying Music*, p. 195.

32 *AT*, I, p. 159.

33 *AT*, I, p. 162.

34 *JIB*, IV, pp. 141-142, 148; *AT*, I, p. 30. It must be said that there is a link between the problem of the vibrating strings and the PCM, even if, contrary to Beeckman, Descartes does not think that the PCM is enough to explain the vibrating strings. See, for instance: *AT*, I, p. 29; Roux, 'Découvrir le principe d'inertie', p. 471, n. 46.

> and already almost forgotten. When my mind had strayed so far from serious occupations, you led it back to better things. Therefore, if by chance I produce something of merit, you can rightfully claim all of it as yours. As for me, I shall not forget to send it to you, not only so that you can use it, but also so that you can correct it.[35]

It is, of course, useful to rely on this letter to show that Descartes owes much to Beeckman. However, first of all, Descartes does not say what exactly he owes to Beeckman. Second, he writes that Beeckman only helped him to recall what he had already learnt, not that he gave him new ideas. Third, he speaks of 'his Mechanics'. We know that Beeckman is not bothered by this formula since he himself encourages Descartes to write 'his Mechanics'.[36] We also know that this formula is very close to the one used by Descartes in 1618. Indeed, in a text he gave to Beeckman, he spoke of 'its mechanical principles'.[37] All these elements seem to suggest that when Descartes met Beeckman at the end of 1618, he was already firm in his opinion on motion and so, perhaps, also on the PCM.

Third Argument: The Wording of the PCM Statements

Finally, historians adopt S1 for lexical reasons. First, (a) they notice that Descartes lays down the PCM in the same words as Beeckman. Second, (b) they notice that, before *Le Monde*, the PCM is for Descartes an 'assumption' whereas for Beeckman, as has been said, it is a *theorema*. Let us examine those two facts in detail.

a *Quod semel movetur, in vacuo semper movetur*

Historians notice that Descartes, when he states the PCM, uses the same phrases as Beeckman.[38] Indeed, he uses not only *semel* + *semper*,[39] but also

35 *JIB*, IV, p. 62; *AT*, X, pp. 162-163. As I said, it is Klaas van Berkel who advised me to use this letter. See also: *AT*, I, p. 24.

36 *JIB*, IV, p. 65.

37 *JIB*, IV, p. 52.

38 See, for instance: Arthur, 'Beeckman, Descartes and the Force of Motion', p. 15, n. 31; Hooper, 'Inertial Problems in Galileo's Preinertial Framework', p. 166. See also, even if they only underline the similarity: Buzon and Carraud, *Descartes et les 'Principia' II*, p. 101.

39 *JIB*, I, pp. 167, 253, 257, 263; III, pp. 74, 101, 104, 117, 123, 136, 165, 280, 341; T2; T3; T4; T5; *AT*, III, p. 208; VIIIa, p. 62.

nisi + impedio,[40] or *perseverare in motu*.[41] At the time when Beeckman and Descartes were writing, none of these phrases were commonplace. Therefore, each one is a real clue.

But none of them is original enough to be sure that Descartes owes them to Beeckman. Indeed, *semel + semper* can be found in Conrad Gessner's third book of *Historiae animalium* (1555) and Nicholas of Cusa's *De ludo globi* (1463).[42] Then, *nisi + impedio* can be found, for instance, in Franciscus Toletus's (1573) and the Coimbrans' (1592) commentaries of Aristotle's *Physics*.[43] Finally, one finds *in ea motione perpetuo et sine variatione perseverat* in the Disputatio XXIX of Francisco Suarez (1597) and *in eodem statu semper perseverantes* in Giovan Battista Pio's commentary on Lucretius's *De natura rerum* (1511).[44] Above all, it is possible that Descartes borrowed his phrases from Beeckman without owing him the PCM.

b The PCM: An 'assumption' or a *theorema*?

Historians finally notice that Descartes considered the PCM an 'assumption' until *Le Monde* when it became a special case of the first 'law of nature',

40 *JIB*, I, 24, 117, pp. 150, 157; III, p. 74, pp. 136, 341; *AT*, I, p. 91; V, p. 173.

41 *JIB*, I, p. 261; III, p. 274; IV, pp. 44, 184; *AT*, V, p. 173; VIIIa, pp. 62, 63, 67, 194. Even if: 1) Beeckman never uses *in suo motu*, contrary to Descartes (*AT*, VIIIa); 2) Beeckman neither uses *perseverare in statu*, contrary to Descartes (*AT*, VIIIa, p. 62); 3) Descartes does not use *perseverare in motu* before the *Principia* (1644), whereas Beeckman uses this formula as early as July 1613-April 1614 (*JIB*, I, p. 25).

42 Gessner, *Historiae animalium liber III, qui est de avium natura* (Zürich: Christoph Froschover, 1555), pp. 288-289; Nicholas of Cusa, *Dialogus de ludo globi* (Hamburg: Meiner, 2000), p. 22.

43 Franciscus Toletus, *Commentaria una cum quaestionibus. In octo libros Aristotelis De physica ausculatione* (Venice, 1606), p. 125; *Commentarii Collegii Conimbricensis* [...], *in libros Physicorum* (Lyon, 1594), Secunda pars, Lib. IV, cap. VIII, text 69, 53.

44 Francisco Suarez, *Disputationes metaphysicae*, ed. and trans. by Sergio Rábde, Salvador Caballero, and Antonio Puigcerver, 7 vols. (Madrid: Ed. Gredos, 1960-1966), IV, p. 304; Giovanni Battista Pio, *In Carum Lucretium poetam commentarii* (Paris, 1514), fol. XLIII; *AT*, VIIIa, p. 62. It may be recalled that the phrase *perseverare in* was to be popularized by Newton and then, of course, by Spinoza. Spinoza gradually replaced *quantum in se est + perseverare in suo statu* by *quantum in se est + perseverare in suo esse*, probably under the influence of the Fifth objections and replies (from and to Gassendi). See, for instance: *AT*, VII, p. 370. Newton used the phrase *perseverationem in suo esse*. *Unpublished Scientific Papers of Isaac Newton*, chosen, ed. and trans. by A. Rupert Hall and Marie Boas Hall (Cambridge: Cambridge University Press, 1978), p. 103. On the use of the phrase *perseverare in* by Descartes, see, for example: Gabbey, 'Force and Inertia', pp. 278-279, 293-294; Koyré, *Newtonian Studies*, pp. 66, 66, 70; Sophie Roux, 'A Deflationist Solution to the Problem of Force in Descartes', in: Delphine Antoine-Mahut and Sophie Roux, eds., *Physics and Metaphysics in Descartes and Its Reception* (New York: Routledge, 2019), pp. 140-158, esp. p. 154.

whereas for Beeckman the PCM was, as we have just seen, a *theorema* as early as 1618.[45] This would prove two things: on the one hand, that Descartes doubted the PCM until *Le Monde*; on the other hand, that he doubted the PCM until *Le Monde* because it was for him a brand-new principle, recently provided by Beeckman.

This reading is undoubtedly possible. But it seems possible to adopt another one. Indeed, in T5, i.e., in 1629, Descartes indicates that he will demonstrate in his future treatise, i.e., in *Le Monde*, the PCM that he 'assumes'. Thus, if Descartes speaks of an 'assumption', it is not to indicate that he is using a principle that he doubts, but that he is using a principle that he has not yet metaphysically grounded.[46] Thus, it is probably not possible to infer from his use of the word 'assumption' that he owes the PCM to Beeckman.

The Coimbrans as an Alternative Source for Descartes' PCM

The reasons why S1 is adopted have just been discussed in detail. But S1 seems to be based above all on the following presupposition: the PCM is too original, in 1618, when Descartes met Beeckman, for Descartes, given his young age and the studies he had pursued, to have been able to adopt it before then.[47] It is above all this presupposition that I think is possible to dispute. This is what I am going to do now, by showing that Descartes perhaps adopted the PCM as soon as he appeared in La Flèche, in 1612-1613, by working on the commentaries of the *Physics* and of *On the Heavens* written by the Coimbrans.[48] Henceforth, I shall call that alternative story S2.

45 *JIB*, I, pp. 256; T4; T5. For example: Arthur, 'Beeckman, Descartes and the Force of Motion', p. 24 and p. 24, n. 51; Buzon, 'Beeckman, Descartes and Physico-mathematics', pp. 152-153; Jullien and Charrak, *Ce que dit Descartes*, p. 89, p. 118; Koyré, *Études galiléennes*, p. 113.

46 Although Descartes will explain to Mersenne in 1631 that 'we cannot suppose the void without error' (*AT*, I, p. 228).

47 See, for instance: Arthur, 'Beeckman, Descartes, and the Force of Motion', p. 23.

48 'Descartes mentions the commentaries of the Coimbrans only twice in his correspondence [...] anticipating objections by the Jesuits to his *Meditationes*' (Dennis Des Chene, 'Descartes and the Natural Philosophy of the Coimbra Commentaries', in: Gaukroger, Schuster, and Sutton, eds., *Descartes' Natural Philosophy*, p. 29). Twice in 1640, in two different letters to Mersenne: *AT*, III, pp. 185, 250. On the influence of the Coimbrans on Descartes, see, for instance: Roger Ariew, 'Descartes and Scholasticism', in: John Cottingham, ed., *The Cambridge Companion to Descartes* (Cambridge: Cambridge University Press, 1992), p. 74; Ariew, *Descartes among the Scholastics*, pp. 43-44; Roger Ariew, 'What Descartes Read: His Intellectual Background', in: Steven M. Nadler, Tad M. Schmaltz, and Delphine Antoine-Mahut, eds., *The Oxford Handbook of Descartes and Cartesianism* (Oxford: Oxford University Press), pp. 24-39, esp. p. 37; Des Chene, 'Descartes and the Natural Philosophy of the Coimbra Commentaries', in: Gaukroger, Schuster,

The Texts

Aristotle uses and then rejects, at least four times, the PCM. Firstly, in *Physics*, III, 5, 205a12-19. Secondly, in *Physics*, IV, 8, 215a19-22. Thirdly, in *On the Heavens*, III, 2, 301b1-17. Fourthly, in *Metaphysics*, 11 (K), 10, 1067a7-15.[49] This has been noted not only by historians of science, but also, in a way, by Galileo and, above all, by Newton himself.[50] Here are these four texts:

> A1. 'Suppose that the infinite sensible body is homogeneous. Then each [part] will be either immovable or always being carried along. Yet neither is possible. For why downwards rather than upwards or in any other direction? I mean, e.g., if you take a clod, where will it be moved or where will it be at rest? For the place of the body akin to it is infinite. Will it occupy the whole place, then? And how? What then will be the nature of its rest and of its movement, or where will they be? It will either be at rest everywhere – then it will not be moved; or it will be moved everywhere – then it will not come to rest.'[51]

and Sutton, eds., *Descartes' Natural Philosophy*, pp. 29-45: Des Chene, *Physiologia*, p. 10; Gaukroger, *Descartes: An Intellectual Biography*, pp. 52-53, p. 84; Étienne Gilson, *Index scolastico-cartésien*, Paris: Vrin, 1979), especially introduction, IV; Étienne Gilson, 'Météores cartésiens et météores scolastiques', *Revue néo-scolastique de philosophie* 88 (1920), pp. 358-384, esp. p. 361; Shea, *La magia dei numeri e del moto*, pp. 16-17.

49 Henceforth, I shall refer to these texts as A1, A2, A3, and A4 for Aristotle's text nr. 1, etc.

50 Historians of science: Dino Boccaletti, *The Equation of Motion* (Dordrecht: Springer, 2016), p. 93; Clagett, *The Science of Mechanics in the Middle Ages*, p. 426, n. 8; Clavelin, *The Natural Philosophy of Galileo*, pp. 17, 103; Cohen, '"Quantum in se est"', p. 141; Israel E. Drabkin, 'Notes on the Law of Motion in Aristotle', *The American Journal of Philology* 59 (1938), pp. 60-84, esp. p. 67; Dugas, *A History of Dynamics*, p. 22; Elazar, *Honoré Fabri*, p. 151; André Goddu, *The Physics of Ockham* (Leiden: Brill, 1984), p. 200; Edward Grant, 'Motion in the Void and the Principle of Inertia in the Middle Ages', *Isis* 55 (1964), pp. 265-292, esp. p. 265; Edward Grant, *Much Ado about Nothing: Theories of Space and Vacuum from the Middle Ages to the Scientific Revolution* (New York: Cambridge University Press, 2008), pp. 7, 43, 266; Amos Funkenstein, 'Some Remarks on the Concept of Impetus and the Determination of Simple Motion', *Viator* 2 (1971), pp. 329-348, esp. p. 335; Funkenstein, *Theology and the Scientific Imagination*, pp. 14-15, 157-160; Jouguet, *Lectures de mécanique*, I, pp. 4-5, 107; Maier, *On the Threshold of Exact Science*, pp. 116-117. For Galileo, see: Galileo, *De motu*, in: *Opere*, I, p. 309; Newton: *Unpublished Scientific Papers of Isaac Newton*, pp. 310-311. We must add that: 1) neither historians of science nor Galileo or Newton refer to A1; 2) when Galileo refers to A3 in the *De motu*, it is to reject the PCM; 3) Newton commits a reference error when quoting A3; 4) Maier believes that Galileo is misunderstanding A3, which, I believe, is not the case. See below.

51 *The Complete Works of Aristotle*, the Revised Oxford Translation One Volume Digital Edition, ed. Jonathan Barnes, trans. by Robert Purves Hardie and Russell Kerr Gaye (Princeton: Princeton University Press, 1995), p. 772.

A2. 'Further, no one could say why a thing once set in motion should stop anywhere; for why should it stop *here* rather than *here*? So that a thing will either be at rest or must be moved *ad infinitum*, unless something more powerful gets in its way.'[52]

A3. 'Again, a body which is in motion but has neither weight nor lightness, must be moved by constraint, and must continue its constrained movement infinitely. For there will be a force which moves it, and the smaller and lighter a body is, the further will a given force move it. Now let A, the weightless body, be moved the distance CE, and B, which has weight, be moved in the same time the distance CD. Dividing the heavy body in the proportion CE:CD, we subtract from the heavy body a part which will in the same time move the distance CE, since the whole moved CD; for the relative speeds of the two bodies will be in inverse ratio to their respective sizes. Thus the weightless body will move the same distance as the heavy in the same time. But this is impossible. Hence, since the motion of the weightless body will cover a greater distance than any that is suggested, it will continue infinitely. It is therefore obvious that every body must have a definite weight or lightness.'[53]

A4. 'Further, every sensible body is somewhere, and whole and part have the same proper place, e.g. the whole earth and part of the earth. Therefore if the infinite body is homogeneous, it will be unmovable or it will be always moving. But the latter is impossible; for why should it rather move down than up or anywhere else? E.g. if there is a clod, where will this move or rest? The proper place of the body which is homogeneous with it is infinite. Will the clod occupy the whole place, then? And how? When then is its rest or its movement? It will either rest everywhere, and then it cannot move; or it will move everywhere, and then it cannot be still.'[54]

Arguments in Favour of S2

There are three main reasons to suggest that Descartes adopted the PCM as early as 1612-1613. First, (a) Descartes had at his disposal those four

52 *The Complete Works of Aristotle*, p. 807.
53 *The Complete Works of Aristotle*, trans. by John Leoffric Stocks, p. 1082.
54 *The Complete Works of Aristotle*, trans. by William David Ross, p. 362.

texts in which the PCM, even if rejected, is not only explicitly used, but is accompanied each time by an argument. Secondly, and above all, (b) Descartes also had at his disposal one theoretical element, provided by the Coimbrans, which could help him adopt the PCM: motion in a vacuum, according to them, is possible. Thirdly, (c) if it is true that Descartes owes the PCM to the Coimbrans, then it is easier to understand, on the one hand, why he, unlike Beeckman, adopted the notion of *impetus* in 1618 and, on the other hand, why the way Descartes justify the PCM looks more like the theses found in the Coimbrans' commentaries than the justifications given by Beeckman. Let us examine these three reasons in detail.

a Why does Aristotle use and reject the PCM?

In A1, Aristotle introduces the PCM to demonstrate that a sensitive, homogeneous and infinite body is impossible. And in A4, Aristotle explicitly rephrases the demonstration given in A1. In both texts, he offers a demonstration *per impossibile* (which seems to be)[55]: if a homogeneous body were infinite, then (but this is impossible) it would either always be at rest, or always be in motion; because it would have the same motion as its parts which would either always be at rest, or would always be in motion; indeed, since the whole of which they would be parts of would be homogeneous, they would have no reason to move in one direction rather than another or to stop in one place rather than another. (Here the demonstration ends.)

In A2, Aristotle uses the PCM to demonstrate that a body cannot move in a vacuum. He offers the following demonstration *per impossibile*: if a body were to move in a vacuum, then it would have to move indefinitely, unless prevented from doing so by something stronger; indeed, such a body would have no reason to stop anywhere rather than anywhere else. (Here the demonstration ends.)

In A3, Aristotle uses the PCM to demonstrate that a body that has neither gravity nor lightness could not be moved by constraint. For this, he proposes the following demonstration *per impossibile*: if a separate body, moved by constraint upwards or downwards, were neither heavy nor light, it would have to move indefinitely; indeed, being neither heavy nor light, it

55 Some commentators understood the argument differently. Indeed, they considered that Aristotle only talked about the motion of the parts and not about the motion of the parts *and* the motion of the whole (see below). For our purposes, it is not necessary to decide. But that difference deserves to be examined in detail.

would necessarily have to travel a distance greater than any distance that a heavy or light body could travel; but this is impossible and consequently, a separate body cannot move by constraint unless it is heavy or light. (Here the demonstration ends.)

Thus: 1) in A3, Aristotle explicitly indicates that the motion he is talking about is violent, whereas in A1, A2, and A4 he does not. Therefore, the PCM found in A3 seems different from the PCM found in A1, A2, and A4. But Aristotle says nothing about the cause of that violent motion; 2) in A1, A2, and A4, Aristotle uses the principle of sufficient reason, whereas in A3 he only uses a physical argument; 3) in A1, A2, A3, and A4, Aristotle proceeds each time in the same way. He offers a demonstration *per impossibile* in which he explicitly uses the PCM, with an argument, but also rejects that principle, without explaining why.

So why does he reject the PCM? As Sir William David Ross says, the argument given in A1 (and so, I add, in A4) is 'difficult'.[56] In particular, it is difficult to know what, according to Aristotle, should be considered impossible, but also why it should be so. Some commentators think that Aristotle is talking about the impossible motion of the portion of the whole and not about the impossible motion of the whole. Philoponus, who comments on Aristotle's argument in detail, seems the think so. Indeed, he writes: 'Now the words "either it will be immobile or it will go on for ever" are not said by him of the whole of the unlimited, but of the portion of it.'[57] He adds: 'It is therefore necessary that the portion should either be always in motion or always in the same position, but each is absurd.'[58] Philoponus explains why a portion should be at rest. He adds: 'If someone will not grant this, sc. that it is immobile, so as not to deny motion, he will at any rate grant that it will go on forever.'[59] Then he writes:

> He [Aristotle] says that it is 'impossible' to be in motion forever, and then infers why it is impossible. For why will it move upward rather than downward at the same time and in the same place? For if (a) all place is germane to it and (b) there is in every place the above and the below (for this I apprehend from patent fact), then at the same time and in the same place it will no more move up than down, if it is indeed naturally liable

56 *Aristotle's* Physics*: A Revised Text with Introduction and Commentary of William David Ross* (Oxford: Clarendon Press, 1936), p. 550.

57 Philoponus, *On Aristotle Physics 3*, trans. by Mark J. Edwards (London: Bloomsbury, 2014), p. 107.

58 Philoponus, *On Aristotle Physics 3*, p. 107.

59 Philoponus, *On Aristotle Physics 3*, p. 108.

to go everywhere. But it is impossible to move in contrary directions at the same time.[60]

According to Thomas Aquinas, what is impossible is for the homogeneous body to be always at rest or always in motion, because 'in either case the intelligibility of nature is destroyed, for nature is a principle of motion and rest'.[61] Ross, on his side, thinks that what is impossible and even 'absurd' according to Aristotle is for the homogeneous body to be always at rest or always moving, 'in view of his experience of the fact that earth sometimes moves, viz. when it is not as near the centre of the universe as it can get, and sometimes rests, viz. when its is as near as it can get'.[62] Themistius, 'which was probably relying on inherited material',[63] does not mention these two reasons: 'But I am not saying that stability is also no less impossible than change. In fact motion to one place if every place is identically its own is as inconceivable as being at rest in one place that is no more proprietary to it than another.'[64] Oresme refers to Averroes: 'Tertia ratio Aristotelis est quod non posset esse infinitum homogeneum, et est melior, quia sequitur quod tale esset immobile et secundum totum et secundum partes. Primo, videmus quod omnino <idem> est locus et motus totius et partis, sicut totius terre et unius glebe. Etiam propter aliud, quia omnis motus naturalis est de loco innaturali in locum naturalem; modo non posset assignari in tali homogeneo locus naturalis uni parti magis quam alteri, et sic illa pars vel continue moveretur vel continue quiesceret, ut deducit Commentator commento 48°.'[65] Then, commenting on Aristotle's text, he writes (it deserves to be emphasized): 'Tertio dico quod posset in perpetuum moveri motu recto. Verbi gratia ponatur quod aliquid sit infinitum versus solum, et non habeat resistentiam ad motum inferius, et ymaginetur in eo gravias unius libre; tunc descenderet et posset

60 Philoponus, *On Aristotle Physics 3*, p. 109. Francesco Vimercato refers to Philoponus when he explains A1. See: Vimercato, *In octo libros Aristotelis De naturali auscultatione commentarii* (Paris: Apud Vascosanum, 1550), p. 199.

61 Thomas Aquinas, *Commentary on Aristotle's* Physics, trans. by Richard J. Blackwell, Richard J. Spath, and W. Edmund Thirlkel (Notre Dame: Dumb Ox Books, 1995), pp. 180-181.

62 *Aristotle's* Physics, p. 551.

63 Themistius, *On Aristotle Physics 1-3*, trans. by Robert B. Todd (London: Bloomsbury, 2014), pp. 171-172, n. 915.

64 Themistius, *On Aristotle Physics 1-3*, p. 97.

65 Nicolas Oresme, *Questiones super physicam, Books I-VII* (Leiden: Brill, 2013), Lib. III, questio 11, pp. 368-369. Averroes, *Quartum volumen Aristotelis De physico auditu libri octo cum Averrois Cordubensis variis in eosdem commentariis* (Venice: Apud Iunctas, 1562), 106D-E.

ymaginari continue moveri, et per idem unum infinitum versus inferius; et per consequens totum aggregatum.'[66]

The argument in A2 is not very easy to understand either. Themistius develops Aristotle's argument, asking what the impediment might be:

> Furthermore, nobody could say why something once it moves will stop anywhere. Why here rather than there? But if there, why will it not come to rest everywhere? For it must either always stand still, or always be carried along. But perhaps something more powerful and stronger will in some way impede the motion? But the argument will revert to the following: is this more powerful thing in fact stationary, and why here rather than elsewhere? Or it moves – and how does it move, or how will it be an impediment?[67]

Philoponus understands Aristotle's argument in a different way. Indeed, Philoponus links A2, which according to him deals with violent motion, to the text which comes just before it in the *Physics*, which deals with the motion of projectiles. And Philoponus explains A2 by invoking the concept known (since Buridan) under the name of *impetus*, whose first use is often attributed to him:

> This is another argument, that nothing moves unnaturally in the void. I say that just as you think that the cause of unnatural movement is the

66 Oresme, *Questiones super physicam*, Lib. III, questio 11, p. 369. See also: Albert of Saxony, *Expositio et quaestiones in Aristotelis* Physicam *ad Albertum de Saxonia attributae*, vol. 1, édition critique de Benoît Patar (Louvain: Peeters, 1999), Lib. III, tractatus secundus, cap. 3, pp. 144-145. And vol. 2, édition critique de Benoît Patar (Louvain: Peeters, 1999), Lib. III, quaestio 11, p. 552. Jürgen Sarnowsky pointed out that in Albert of Saxony, there is a thought experiment which, according to him, involves the principle of inertia: 'Place and Space in Albert of Saxony's Commentaries on the *Physics*', *Arabic Sciences and Philosophy* 9 (1999), pp. 25-45, esp. p. 39. Albert of Saxony refers to Averroes: 'Possibile est aliquod corpus sit compositum equaliter ex gravitate et levitate, modo tali corpori non poterit assignari locus, ut videtur Commentator innuere 2° (sic) celi, quia vel semper quiesceret vel semper moveretur' (*Physics* II, lib. 4, qu. 5, contra conclusionem, fol. 68rb). I have not read Albert's text directly. According to Sarnowsky, Albert probably refers to this text, from Averroes' commentary on *On the Heavens*: 'Impossibile est enim in corporibus compositis aliqua componi aequaliter. Et nos post declarabimus: Et, si esset aliquod compositum aequaliter, contingeret, quod aliquod corpus non moveretur omnino, sed staret in quocunque loco, poneretur, scil. staret aut superius, aut inferius, aut in medio duorum contrariorum, et moveretur in ceteris, quod non invenitur' (*Averrois Cordubensis commentum magnum super libro* De cele et mundo *Aristotelis* (Louvain: Peeters, 2003), vol. 1, Lib. I, vol. C. 7, p. 17).

67 Themistius, *On Aristotle Physics* 4, trans. by Robert B. Todd (London: Bloomsbury, 2014), p. 43.

> thrust of the air, say that it moves so far until the kinetic power given to the air from what originally pushed it expires, in this way clearly even if something were to move unnaturally, in the void it would move so far until the kinetic power given to it by the original thruster was exhausted.[68]

Simplicius writes:

> Having previously proved that if there is a void there will be no motion, either natural or unnatural, I think that he now proves that, if there is a void, not only is motion abolished, but also natural rest. For the natural rest of bodies in their own places comes about through differences, while a void is undifferentiated. So why should something moved come to rest here rather than there, towards which place and because of which bodies move? So, if there is no difference in the void, even if one posits bodies moving in it, they will inevitably move endlessly unless some stronger body stops them.[69]

Then, as Themistius, he asks what the impediment might be:

> But he omitted to enquire what could there be in a void such as to prevent motion by being more powerful, since it is evident that there is no such thing. For that also would prevent while being either naturally or unnaturally stationary. It cannot do so naturally and, if it were itself prevented, there would be an infinite regress. But even if someone should say that something moving should force the travelling body to stop, this absurdity has already been discussed.[70]

Thomas Aquinas does not seem to understand the argument as Philoponus does:

> If there is motion in a void, no one can explain why that which is moved stops somewhere. For there is no reason why it should be at rest in one part of the void rather than in another: neither in things which are moved

68 Philoponus, *On Aristotle Physics 4. 6-9*, trans. by Pamela Huby (London: Bloomsbury, 2014), pp. 45-46.

69 Simplicius, *On Aristotle* On the Void, trans. by Paul Lettinck and J.O. Ormson (London: Bloomsbury, 2014), p. 194. Francesco Vimercato refers to Simplicius when he explains Aristotle's argument. See: Vimercato, *In octo libros Aristotelis De naturali auscultatione commentarii*, p. 276.

70 Simplicius, *On Aristotle* On the Void, p. 195. Simplicius here refers to *On the Heavens*, III, 2, 300a20-300b8. See: *The Complete Works of Aristotle*, p. 1079.

> naturally, since there is no difference in the parts of a void, as was said above, nor in things which are moved by violence. For we say that violent motion ceases when the rebounding or impulse of air – according to the two assigned causes – ceases. Therefore it will be necessary either that every body is at rest and nothing moves, or if something is moved, that it be moved ad infinitum, unless some larger body, which impedes its violent motion, gets in its way.[71]

Averroes talks about natural motion.[72] Albert of Saxony explains that the violent motion of projectiles in a vacuum is impossible according to Aristotle because their motor, amely air, is lacking. Then he adds: 'Hic ponit tertiam rationem: si vacuum esset, nihil in eo quiesceret; ergo nec in ipso aliquid moveretur. Consequentia tenet ex quo non esset ibi causa naturalis quietis, nec per consequens motus. Antecedens patet, quia non esset ratio quare magis quiesceret hic quam ibi.'[73]

What do the historians of science say? Clavelin writes: 'To Aristotle the idea of a motion capable of infinite duration was bound to appear just as untenable as the idea of the movement of movement.'[74] Grant writes: 'Since a perpetual motion without apparent cause was considered unintelligible, Aristotle's inertial consequence was probably an effective argument.'[75] Elazar explains that, according to Aristotle, 'violent motion can never be eternal' since, as Aristotle says 'nothing contrary to nature is eternal'. And he adds: 'And finally, straight motion – whether natural or violent – can never be, by definition, ad infinitum: it must have both a *terminus a quo* and a *terminus ad quem*.'[76] Gabbey, who, unlike the other three, does not comment on A2, writes here: 'Rectilinear motion cannot be infinite in Aristotle's cosmos, so as finite it is either composite (should the mobile retrace its path), or incomplete and perishable; that is, it has *termini ab quo* and *ad quem*, and for the same reason it cannot be eternal.'[77] Indeed, it's quite possible that Aristotle rejects the PCM because, for him, 'a change

71 Thomas Aquinas, *Commentary on Aristotle's* Physics, p. 253.

72 Averroes, *Quartum volumen Aristotelis De physico auditu libri octo*, 157C-E.

73 Albert of Saxony, *Expositio et quaestiones in Aristotelis* Physicam, t. 1., Lib. IV, tractatus secundus, cap. 3, p. 197.

74 Clavelin, *The Natural Philosophy of Galileo*, p. 17.

75 Grant, *Much Ado*, p. 7.

76 Elazar, *Honoré Fabri*, pp. 151-152.

77 Alan Gabbey, 'New Doctrines of Motion', in: Daniel Garber, Michael Ayers, and Roger Ariew, eds., *The Cambridge History of Seventeenth-Century Philosophy* (Cambridge: Cambridge University Press, 2008), I, pp. 649-679, esp. pp. 660-661.

cannot be infinite in the sense that it is not defined by limits'.[78] Aristotle adds that 'every motion is from something to something and is not infinite in respect of its extreme points'.[79] Indeed, what is in locomotion 'will not be in locomotion over an infinite distance; for it cannot have traversed such a distance'.[80] Or: 'For the line traversed in rectilinear motion cannot be infinite; for there is no such thing as an infinite straight line; and even if there were, it would not be traversed by anything in motion; for the impossible does not happen and it is impossible to traverse an infinite distance'.[81] Or: 'It could not move infinitely; for to traverse an infinite is impossible, and impossibilities do not happen.'[82]

The argument given in A3 is also difficult. As I said (cf. n. 51), there is even an explicit disagreement about the manner in which it is to be understood and translated. Some (Thomas Aquinas, Buridan, and Oresme, for instance)[83] argue that Aristotle talks about an infinite speed; others (Simplicius and Galileo, for instance)[84] argue that he talks about an infinite

78 Aristotle, *Physics*, VI, 10, 241b10, in: *The Complete Works of Aristotle*, p. 897.

79 Aristotle, *Physics*, VII, 1, 242a65, in: *The Complete Works of Aristotle*, p. 899.

80 Aristotle, *Physics*, VI, 10, 241b9-10, in: *The Complete Works of Aristotle*, p. 896.

81 Aristotle, *Physics*, VIII, 9, 265a17-20, in: *The Complete Works of Aristotle*, p. 969.

82 Aristotle, *On the Heavens*, III, 2, 300b4-5, in: *The Complete Works of Aristotle*, p. 1078.

83 Thomas Aquinas, *Opera omnia, III: Commentaria in libros Aristotelis De caelo et mundo, de generatione et corruptione et meteorologicorum* (Rome: Ex typographia polyglotta S.C. de Propaganda Fide, 1886), pp. 250-251; Buridan, *Ioannis Buridan expositio et quaestiones in Aristotelis* De caelo, édition, étude critique et doctrinale de Benoît Patar (Louvain: Peeters, 1996), Lib. III, tractatus primus, p. 174; Oresme, *Le Livre du ciel et du monde*, ed. by Albert D. Menut and Alexander J. Denomy, trans. with an introduction by Albert D. Menut (Madison: University of Wisconsin Press, 1968), bk III, chap. 6, 165d, pp. 606-609.

84 'Simplicius, *On Aristotle* On the Heavens *3.1-7*, 593, 15-20, p. 69: 'Furthermore, if there were a moving body [which had neither lightness nor weight], it would be necessary that this be moved by constraint, and being moved by constraint would make the motion infinite. For since what causes motion is a certain power and the lesser or lighter, if moved by the same power, will move farther, let the weightless A have been moved through CE and B, which has weight, have been moved through CD in an equal time. If the body having weight is divided in the ratio of CE to CD, it will follow that what is subtracted from the body having weight moves through CE in an equal time since the whole moved through CD. For the speed of the lesser body will be to that of the greater as the greater body is to the lesser. So the weightless body and the body having weight will move through an equal distance in the same time. But this is impossible. Consequently, since the weightless body will move a distance greater than any assigned, it will move through an infinite distance.' And Simplicius comments: 'Having proved that it is impossible that a body having neither heaviness nor lightness move either up or down naturally, he now proves that it cannot move unnaturally and by constraint either, since it follows that it is moved ad infinitum by the power which moves it.' Galileo (quoted by Maier, *On the Threshold of Exact Science*, pp. 116-117): 'In 3rd Caelo text 27 he says: if something that is moved is neither heavy nor light, it will be moved violently; and whatever is moved violently and possesses no

distance. Indeed, Aristotle's argument can be understood in two ways: either the variable is time; or the variable is distance. But it seems that one must favour Simplicius's and Galileo's reading, confirmed by the translation I have chosen. Indeed, it seems that it fits better with Aristotle's text. If one thinks that he talks about an infinite speed, then one can easily understand the impossibility that Aristotle had in mind, since Aristotle thinks that local motion cannot take place in an instant. It would be a *mutatio*. On the other hand, if one thinks that he talks about an infinite distance, then the impossibility Aristotle thinks of is more difficult to determine. Maybe it is the impossibility he thinks of in A2. Simplicius writes: '[B]ut it is not possible to move an infinite distance, since every motion is from one place to another, and what cannot have come to be cannot be coming to be, as he said previously, therefore a body which does not have either weight or lightness cannot move.'[85] But it is possible to develop Aristotle's argument by recalling, as I also said, that in A3, Aristotle speaks explicitly of a violent motion. So it would seem that the body whose motion Aristotle imagines is moved by something external and in contact with it. Now this thing that moves the body which has to move forever must probably itself be in motion and, in order to avoid infinite regress, must move naturally. It would seem, therefore, that we can say: if the forced motion cannot continue indefinitely, it is because it implies a natural motion, also continuing indefinitely and not necessarily, it seems, in a circle. In which case, we would have there as an equivalent, within the framework of Aristotelian physics, of the principle of inertia. But, of course, Aristotle does not make the above remarks. They are only, I think, implied by what he writes in A3.

To sum up: 1) Aristotle repeatedly used the PCM, which he rejected, without ever explaining why; 2) Aristotle's commentators had those texts at their disposal[86]; 3) some of them were led by those texts to catch a glimpse

resistance in the form of heaviness or lightness moves forever'. In Latin: 'Nam 3° Caeli text. 27 inquit: si quod movetur neque grave neque leve fuerit, vi movebitur; et quod vi movetur nullam gravitatis aut levitatis resistentiam habens, in infinitum movetur.'

85 Simplicius, *On Aristotle* On the Heavens, 3, 1-7, p. 70. See also pp. 57-58.

86 Grant is right when he writes: 'This important consequence failed to emerge as a possible problem in the *Questiones* literature on Aristotle's *Physics*. Indeed, it is not even mentioned in the treatises which I have examined. In *Commentaries* on the *Physics* it is mentioned or recapitulated only because the text is itself included' (Grant, 'Motion in the Void', p. 266, n. 3). However, Grant focuses on A2, leaving out A1, A3 and A4. But A1 is, as I said, mentioned in Oresme's and Albert of Saxony's question 11 (Book III). Above all, the combination of these four passages and their explanations seems to me to change the status of what Grant calls 'the inertial consequence'. Indeed, it is no longer an impossibility just mentioned in passing, although it is true, of course, that it is not examined as carefully (far from it) as the impossibility of motion at infinite speed.

of the PCM, or of the principle of inertia[87]; 4) but none of them seem to have adopted the PCM.

So why might the commentaries of the Coimbrans have led Descartes to adopt the PCM?

b Why did Descartes maybe know and even adopt the PCM before 1618?

Descartes had at his disposal the *Physics*, *On the Heavens*, and the *Metaphysics* before he met Beeckman. Indeed, while at La Flèche, in 1612-1613, he studied the commentaries of the *Physics* and of *On the Heavens* of the Coimbrans, which he did not forget years later, in 1640.[88] And it would seem that he also worked on Book 11 of *Metaphysics*, but not on the commentary of the Coimbrans, since they did not write one.[89]

In their commentaries, as we said, the Coimbrans copy, translate and then explain Aristotle's texts (in this case A1, A2, and A3) (for the English translation, see above pp. 000):

> Translation of A1: 'Omnino autem sensile corpus infinitum esse non posse, ex his constat: omne enim corpus sensile alicubi esse aptum est, & uniuscuiusque locus est aliquis, atque etiam partis, & totius est idem, ut totius terrae, & glebae unius, ignis & scintillae: quare si unius sit forma, aut immobile erit, aut semper feretur; quod fieri non potest. Cur enim deorsum magis, quam sursum, aut ad quemuis locum? Dico autem, si sit gleba, quonam ipsa movebitur? aut ubinam quiescet? Corporis enim, quod ei cognatum est, infinitus est locus. Utrum igitur totum locum occupabit? & quomodo? quis igitur, & ubi est ipsius status & motus? An ubique quiescet? non ergo movebitur. An vero ad omnem locum movebitur? non igitur stabit.'[90]

87 It was while commenting on the *Physics* and *On the Heavens* that Ockham, Buridan and Oresme were all led to catch a glimpse of the PCM. See: Funkenstein, 'Some Remarks on the Concept of Impetus', p. 335: 'Aristotle's formulation comes nearer to the principle of inertia than even Buridan's indefatigable celestial *impetus* – and Buridan might indeed have been inspired by this remark of the Stagirite.' See also: Duhem, *Le Système du monde* [note 9], VIII, p. 196, p. 328; Goddu, *The Physics of Ockham*, pp. 200-203; Maier, *On the Threshold of Exact Science*, pp. 83, 85, 88, 98.

88 See note 49.

89 Ariew, *Descartes among the Scholastics*, p. 15.

90 *Commentarii Collegii Conimbricensis* [...], *in libros Physicorum*, Prima pars, Lib. III, cap. V, text 48, 361 (Lyon, 1594) and 510 (Lyon, 1602).

Explanation: 'Si detur corpus aliquod simplex, uniusque naturae infinita mole praeditum, id aut immobile erit, aut perpetuo ciebitur, imo nec movebitur, nec quiescet. Haec autem absurda sunt, ergo & caet. Probatur Maior. Si enim terrae gleba sumatur, ea nec movebitur, nec quiescet, ad quem enim feretur locum, aut in quo quiescet? Nam cum idem, ut antea sumpsimus, & totius & partis locus sit, & totum infinitorum locorum, versus omnem differentiam, obtineat capacitatem; utique non erit maior ratio cur in unam potius, quam in aliam partem commeet, nec cur in una magis, quam in alia quiescat. Proindeque cum in omnes partes simul ferri nequeat, necessario quiescet: cum item in omnibus partibus eiusdem spatii simul quiescere non possit, necessario movebitur. Quare semper quiescet, & semper movebitur; & numquam movebitur aut quiescet.'[91]

Translation of A2: 'Nullus praterea potest dicere, cur id quod motum est, alicubi stabit. Cur enim hic potius, quam ibi? quare aut quiescet, aut infinite ferri necesse est, nisi potentius aliquid impediat.'[92]

Explanation: 'Vacuum omni ex parte uniforme est, sibique simile. Nulla igitur causa erit cur potius in una, quam in alia eius parte mobile consistat. Quare aut semper quiescet, aut si motum semel arripuerit, perpetuo fluctuabit: quod absurdum est.'[93]

Translation of A3: 'Praeterea si quippiam corpus & ponderis expers & levitatis motum subibit, vi moveatur necesse est : si vero vi moveatur, infinitum efficit motum. Nam cum potentia quaedam sit ea, quae mouet, id autem, quod est minus ac levius, plus ab eadem potentia moveatur:

91 *Commentarii Collegii Conimbricensis* [...], *in libros Physicorum*, Prima pars, Lib. III, cap. V, explanation of text 48, 361-362 (Lyon, 1594). This text is quoted in *Progressum in infinitum* by Johann Andrea Schmidt (Jena, 1686).

92 *Commentarii Collegii Conimbricensis* [...], *in libros Physicorum*, Secunda pars, Lib. IV, cap. VIII, text 69, 53 (Lyon, 1594) and 74 (Lyon, 1602).

93 *Commentarii Collegii Conimbricensis* [...], *in libros Physicorum*, Secunda pars, Lib. IV, cap. VIII, explanation of text 69, 54 (Lyon, 1594) and 76 (Lyon, 1602). Partially quoted by: Elazar, *Fabri and the Concept of Impetus*, p. 152, n. 69. See also: Lib. IV, cap. IX, quaestio IV, articulus II, 73 (Lyon, 1594) and 93-94 (Lyon, 1602): 'Ad rationes Aristotelis, dicendum eas tantum probare contra veterum placita, vacuum ex se potius lationem tollere, quam illius causa esse. Deinde non concludere in nullo prorsus vacuo motum dari; sed in eo, quod omnes locorum differentias confunderet; quale vacuum foret si neque coelum, neque sublunaria corpora extarent; tunc enim cum nulla major esset ratio cur elementa in hanc potius, quam illam partem vergerent, omnis eorum motio cessaret. Quod vero attinet ad caetera argumenta, quibus motum in vacuo refellit, quia oporteret per ipsum traiectione corporum in instanti fieri; ea quidnam momenti habeant, proxima sequenti quaestione disceptandum erit.'

sit motum A. quidem pondere carens, per C.E. spatium, B. vero pondus aequali in tempore per C.D. Si igitur diuisum fuerit id, quod pondus habet, ea ratione, quam C.E. ad C.D habet fiet, ut id, quod auferetur ad eo, quod pondus habet, aequali in tempore C.E feratur, quoniam totum per C.D. movebatur. Celeritas enim minoris ad celeritatem maioris ita sese habebit, ut maius corpus se habet ad minus. Per aequale ergo spatium, & id, quo pondere vacat, & id, quod est ponderis particeps, tempore eodem feretur, quod quidem fieri nequit. Quare si omni quouis proposito per spatium maius id mouebitur, quod pondere caret, per infinitum utique ferri potest. Patet igitur omne infinitum corpus pondus, aut levitatem habere.'[94]

Explanation: 'Ostendit fieri non posse, ut ullum corpus sublunare violentum motum habeat: quia gravitatem aut levitatem habeat : quia cum omne corpus grave & leve, quo minus resistit, eo, caeteris paribus, celerius moveatur, & in eodem spatio decurrendo minus temporis consumat, si daretur corpus omnis expers gravitatis levitatisque, oporteret id accepto impetu celeritate in infinitum maiori deferri, atque ita momenta concitari, quod tamen est contra naturam motus. Adverte rationem hanc probabilem tantum esse, non necessariam. Ut enim quamvis omni gravitate & levitate vacet, tamen circumeundo motam trahit, nec nisi tempore gyrum voluitur, ita licet corpus sublunare neque gravitatem, neque levitatem ullam obtineret: adhuc tempore moveretur. Nimirum velocitas, non sola ponderum vacuitate metienda est, sed ipsa etiam intensione impetus a proiectore impressi, simulque habenda ratio spatii quod traiicitur.'[95]

Three remarks: 1) It can be seen that the explanation of A1 proposed by the Coimbrans is not the same as that of Ross and not, either, the same as that of Thomas Aquinas, but looks like the explanation given by Philoponus and, also, in a way, the explanation given to justify the PCM in Hobbes' *De corpore*.[96] 2) It can be seen, moreover, that the Coimbrans use, to explain A1

94 *Commentarii* [...] *in 4 libros de caelo*, lib. III cap. 2, text 27, 409-411 (Lyon, 1597) and 347-348 (Lisbon, 1593).

95 *Commentarii* [...] *in 4 libros de caelo*, lib. III cap. 2, explanation of text 27, 409-410 (Lyon, 1597) and 347-38 (Lisbon, 1593).

96 Thomas Hobbes, *Elementorum philosophiae sectio prima de corpore* (London: Andrea Crook, 1655), II, chap. 8, 19, p. 70. Even if, in Hobbes' argument, it is not only a question of space, but also of time: 'Quod quiescit, semper quiescere intelligitur, nisi sit aliud aliquod corpus praeter ipsum, quo supposito quiescere amplius non possit. Supponamus enim corpus aliquod finitum existere et quiescere, ita ut reliquum omne spatium intelligatur vacuum. Si jam corpus illud

and A2, the word 'absurd' whereas Aristotle does not use it. 3) Finally, it can been seen that they seem to think (as Thomas Aquinas, Buridan, and Oresme, but contrary to Simplicius and Galileo) that in A3, Aristotle is talking not about an infinite distance but about an infinite speed. If one of Aristotle's texts may have helped Descartes to adopt the PCM, it is therefore perhaps less that one than the others and, in particular, as we shall see now, A2.

Indeed, the Coimbrans, in their commentaries, not only copy, translate and explain Aristotle's text. Deviating from dogma, they also examine and sometimes defend theses that we do not find in Aristotle or that he explicitly rejects. It seems to me that we must insist on one of these theses in particular.

The Coimbrans think that motion in a vacuum is possible. Indeed, according to them motion in a vacuum can be successive and not instantaneous.[97] In other words, it does not necessarily have to be a *mutatio*.

Why is this thesis so important? Because, for Aristotle, indefinite motion (in a straight line) is, as we have seen, an impossibility that is directly implied by this other impossibility, namely motion in a vacuum. But, according to the *Prior Analytics* (I, 15, 34a20) but also to the *Physics* (VII, 1, 242b70-243a30), it seems that as soon as an impossibility ceases to be one (becoming a possibility), then the impossibility it implied also ceases, by the same token, to be one.[98] So, since A (motion in the void) directly implies B (indefinite

ceperit moveri, movebitur sane per aliquam viam. Quoniam igitur, quicquid in ipso corpore erat, disponebat ipsum ad quietem, ratio, quare movetur per hanc viam, est extra ipsum similiter si per aliam viam quamcunque motu esset, ratio quoque motus per illam viam esset extra ipsum. Cum autem suppositum sit extra ipsum nihil esse, ratio motus per unam viam eadem esset quae ratio motus per omnem aliam viam; ergo aeque motum esset per omnes vias simul, quod est impossibile. // Similiter, quod movetur, semper moveri intelligitur, nis aliud sit extra ipsum, propter quod quiescit. Nam si supponamus nihil extra esse, nulla ratio erit, quare nunc quiescere debeat potius quam alio tempore; itaque motus ejus in omni simul temporis puncto desineret, quod non est intelligibile.' See: Aristote, *On the Heavens*, II, 13, 295b11-15: 'Motion upward and downward and sideways were all, they thought, equally inappropriate to that which is set at the centre and indifferently related to every extreme point; and to move in contrary directions at the same time was impossible: so it must needs remain still. This view is ingenious but not true' (*The Complete Works of Aristotle*, pp. 1065-1066).

97 *Commentarii Collegii Conimbricensis* [...] *in libros Physicorum* (Lyon, 1594), Secunda pars, Lib. IV, cap. IX, quaestio IV, articulus I and articulus II, and quaestio V, articulus I and articulus II, 72-73, 73-76 and 92-98 (Lyon, 1602). Des Chene (*Physiologia*, p. 275) insisted on this thesis. But it seems to me that it can be given even more importance than he does, as I hope to show. The Coimbrans are, of course, not the first. Apart from the atomists (Democritus, Epicurus, Lucretius in particular), there are Philoponus, Scotus, Bonetus, etc.

98 *The Complete Works of Aristotle*, p. 901: 'and the assumption of a possible case ought not to give rise to any impossible result'. *The Complete Works of Aristotle*, trans. by A.J. Jenkinson, p. 136: 'If then, for example, one should indicate the propositions by A, and the conclusion by B, it would

– probably rectilinear – motion), if A is possible (and the Coimbrans do think that it is possible), B has to be possible, too. To my knowledge, the Coimbrans do not say so. But it seems to me that this is implied by their thesis concerning motion in a vacuum.

So, thanks to the commentaries of the Coimbrans studied in 1612-1613, Descartes had not only several statements of the PCM, but also, I think, one major theoretical element that should lead to adopt it.[99] It is therefore quite possible that he adopted it at that time.[100]

not only result that if A is necessary, B is necessary, but also that if A is possible, B is possible.' On this subject, see, for instance: Taneli Kukkonen, 'Alternatives to Alternatives: Approaches to Aristotle's Arguments *per impossibile*', *Vivarium* 40 (2002), pp. 137-173; Simo Knuutila and Taneli Kukkonen, 'Thought Experiments and Indirect Proofs in Averroes, Aquinas, and Buridan', in: Katerina Ierodiakonou and Sophie Roux, eds., *Thought Experiments in Methodological and Historical Contexts* (Leiden: Brill, 2011), pp. 83-100, esp. p. 88. I would like to thank Sophie Roux who advised me to read these two texts.

99 It may be added that the Coimbrans remind us that Duns Scotus, to show that an angel can move in a continuous motion, propose to imagine a sphere on a flat surface. This sphere, they say, should move *per se* in a continuous motion: *Commentarii Collegii Conimbricensis* [...] *in libros Physicorum* (Lyon, 1594), II, p. 215. This is a thought experiment which looks like (even if it is not the same) the one proposed by Nicholas of Cusa and from which some historians of science have thought it is quite possible to deduce the PCM. See: Cusa, *Dialogus*, p. 22: 'Therefore, since a perfectly round bowling ball's outermost tip would also be what is innermost and would be an atom, after the ball began to be moved, it would never stop moving by itself, because it could never behave differently. For that which is once moved would never stop moving unless it behaved differently at one time and another. And so, a sphere that behaved always in the same way, on a flat and even surface, would always be moved, once it began to be moved.' On this text, see, for instance: Hans Blumenberg, 'Die Kopernikanische Konzequenz für den Zeitbegriff', in: Jerzy Dobrzycki, ed., *The Reception of Copernicus' Heliocentric Theory* (Dordrecht: Springer, 1973), pp. 57-77, esp. p. 68; Boccaletti, *The Equation of Motion*, p. 96; Pierre Duhem, *Études sur Léonard de Vinci. Série 2* (Paris: Ediutions des archives Contemporaines, 1984), p. 186; Maurice de Gandillac, 'Symbolismes ludiques chez Nicolas de Cues (De la toupie et du jeu de boules au jeu de la sagesse)', in: Philippe Ariès and Jean-Claude Margolin, *Les Jeux de la Renaissance. Actes du XXIIIe colloque international d'études humanistes, Tours, juillet 1980* (Paris: Vrin, 1982), pp. 345-365, esp. p. 360, n. 30; Alexandre Koyré, 'Nicolaus von Cues, *Vom Globusspiel* (*De ludo globi*)', *Revue d'histoire des science et leurs applications* 8 (1955), pp. 83-86, esp. pp. 84-85; Meyerson, *Identité et réalité*, pp. 100-101. Maier thinks that thinkers in the seventeenth century discovered the principle of inertia because they paid attention to the fact that a ball on a flat surface keeps moving. Although it is true that Nicholas of Cusa, who also paid attention to this fact before Galileo, did not discover the principle of inertia: Maier *Die Vorläufer Galileis*, pp. 152-153. Koyré thinks that if seventeenth-century thinkers discovered the principle of inertia, it is not because they paid attention to this fact (as Nicholas of Cusa's position shows), but, to put it briefly, because they stopped conceiving of motion as a process. See: 'Compte rendu du livre d'Anneliese Maier, *Die Vorläufer Galileis im 14. Jahrhundert*, Rome, 1949', *Archives Internationales d'Histoire des Sciences* 30 (1951), pp. 769-783, esp. p. 782.

100 Descartes maybe had at his disposal, in addition to the commentaries of the Coimbrans, also the commentaries of Rubius, Toletus, and Thomas Aquinas, where the texts of Aristotle that we

c Advantages of S2

Firstly, S2 could perhaps explain why, in 1618, as most historians think, Beeckman and Descartes did not use the same PCM, that is to say that Descartes uses the notion of *impetus* whereas Beeckman does not use it.[101] Indeed, like Descartes in 1618, the Coimbrans use the notion of *impetus*.[102]

have quoted are more or less precisely commented upon: Rubius, *Commentarii*, 381-382; Thomas Aquinas, *Opera omnia, II: Commentaria in octo libros Physicorum Aristotelis* (Rome: Ex typograhia polyglotta S.C. de Propaganda Fide, 1884), pp. 127-129, pp. 181-182; Toletus, *Commentaria*, p. 125. This would reinforce Des Chene's judgement: 'Descartes' first law was no novelty in the physics of the period, not even in Aristotelian physics. Though there certainly were Aristotelians, notably Suarez, who would have disagreed with it, there was, especially in the treatment of preternatural motion, ample precedent' (Des Chene, *Physiologia*, p. 279). See also: Ariew, *Descartes among the Scholastics*, p. 169: 'Unlike what might have been expected, late Aristotelianism countenances a kind of corpuscularianism. Its theory of motion also includes a feature that looks similar to the principle of inertia.' And, paradoxically: Arthur, 'Beeckman, Descartes and the Force of Motion', p. 11: 'Still, Beeckman's principle of the conservation of motion in a vacuum ("what is once moved in a vacuum moves always") is clearly contrary to standardly accepted Aristotelian principles, and whatever precedents for it Descartes might have encountered in his Jesuit education at La Flèche, it is reasonable to suppose that he would not have proceeded until he had understood it.'

101 For a contrary interpretation, see, for instance: Arthur, 'Beeckman, Descartes and the Force of Motion', pp. 11-14, 23.

102 *Commentarii Collegii Conimbricensis* [...] *in libros Physicorum* (Lyon, 1594), Secunda pars, Lib. VII, cap. II, quaestio I, articulus VIII, 246 (Lyon, 1594) and 368-369 (Lyon, 1602): 'Contraria tamen sententia arbitrantium principem causam eius motus esse proiectorem ipsum, instrumentariam vero non esse medium corpus, etsi ad id aliquantulum adiumenti conferat, sed vim quandam, seu impetum a iaciente rei motae impressum, eique inhaerentem, verisimilior est. Comprobatur autem primum ex eo, quia non videtur negandum dari in rebus huiusmodi virtutem, sive impulsum, qui non sit motus ipse, sed qualitas realis, qua motus proxime administretur.' On this text, see: Des Chene, *Physiologia*, p. 275: 'The Coimbrans argue that the continuing motion of a projectile is owed to a *vis* impressed upon it by the projector, and not, as Aristotle thought, to the impulsion of the medium (6c2q1a8, 2:246f).' It is not the sixth but the seventh book. See above all: Maier, *Die Impetusheorie*, p. 164: 'Eine eingehende Diskussion des Problems der Wurfbewegung gibt der Kommentar der Conimbricenses zur Physik; auf sie verweist die einschlägige Steile aus dem Kommentar zu De caelo. Die Lösung besteht auch hier in einem Kompromiß, der diesmal freilich die einfache Form hat, daß sowohl Aristoteles wie die Impetushypothese Recht haben sollen: das Medium ist zwar nicht der ausschließliche Träger der bewegenden Kraft, aber es leistet doch eine gewisse Beihilfe bei der Erzeugung der Bewegung. Die eigentliche Ursache für die Weiterdauer ist vielmehr eine vis, s., impetus a iaciente rei motae impressus eique inhaerens. Der Versuch, die Lehre von der dem Körper mitgeteilten vis impressa selbst als aristotelisch nachzuweisen, wird nicht gemacht.' Certainly, it is not impossible that Descartes, having misunderstood it, immediately transformed Beeckman's PCM, as Koyré, for example, argues. Indeed, according to him, in 1618, Descartes 'simply throws away Beeckman's intellectual prize, the principle of conservation of *motion* [and] substitutes the conservation of *force*' (Koyré, *Études galiléennes*, p. 118). Certainly, it is not impossible that Descartes, without misunderstanding Beeckman's principle, but given the conceptual framework he was familiar with at the time,

Let us remember what Gaukroger writes. But also, and perhaps above all, what Des Chene writes:

> (Gaukroger:) Beeckman's principle [...] stated that reiterated applications of a tractive force resulted in added increments of motions which were then conserved. Descartes thinks of this situation in terms of the reiterated addition of internal moving forces, where these forces are the causes, not just of the continued acceleration, but of the continued motion as such. Here Descartes appears to be following the version of *impetus* theory defended in Toletus and the Coimbra commentaries on Aristotle's *Physics* that he would have studied at La Flèche.[103]

> (Des Chene:) It would seem to follow, though the Coimbrans do not say this, that if a projectile moves in a vacuum, then the *vis impressa*, since it is opposed by nothing and since it will not tend to its own destruction, should not decrease. The projectile should therefore continue to move indefinitely.[104]

Secondly, S2 may help explain why Descartes did not establish the PCM using the same principles as Beeckman, namely the parsimony principle ('what can be done with a few means is said to have been done badly with many') and the principle of causality ('nothing changes without some cause of change'),[105] but with the help of principles that he seems to owe, in this context, to Aristotle, the Coimbrans, and Toletus, namely the principle of sufficient reason, a metaphysical principle (divine immutability), an ontological principle ('nothing tends by its own nature towards its own destruction'), and a logical principle ('nothing tends

immediately transformed the PCM provided by his new friend, re-using the notion of *impetus* that Beeckman had explicitly deemed unnecessary (Damerow et al., *Exploring the Limits*, pp. 38-39). But neither of these two hypotheses is very likely. Firstly, we can think that if Beeckman had indeed provided Descartes with the PCM, then Descartes would have asked him to explain this brand-new principle, so as not to misunderstand it. Secondly, one can think that if Beeckman had indeed provided Descartes with the PCM, then Descartes would not have modified it from the outset, that is to say, when he used it for the very first time.

103 Gaukroger, *Descartes: An Intellectual Biography*, p. 84.

104 Des Chene, *Physiologia*, p. 275.

105 *JIB*, I, pp. 10, 24-25, IV, p. 184; III, p. 104. Certainly, in 1613-1614, Beeckman uses the principle of sufficient reason. But this was to rule out the notion of *impetus*, not to establish the PCM directly (*JIB*, I, pp. 24-25). Certainly, in 1623, Beeckman uses a metaphysical principle, the impossibility of passing from being to nothingness, to establish the PCM (*JIB*, II, p. 246). But Descartes does not use this metaphysical principle to establish the PCM.

by its own nature towards its contrary').[106] Let us recall, for example, thanks to Des Chene, that the Coimbrans speak of 'desiring one's own destruction [as] most alien to the laws of nature' and write: 'To have an inborn propensity not to exist is contrary to the nature of created things, since all incline to the opposite, and have [...] an ingrained desire to perpetuate themselves, insofar as that can happen.'[107] And that Toletus writes: '*Motus* tends by itself towards *quies*, *terminus ad quem*, where he stays and rests. *Motus* can be called a kind of quieting, that is, a way toward *quies*, which partly coexists with the *motus* itself. What moves is partly coming to rest with respect to part of the *terminus ad quem* and partly moving toward the remaining parts: it is therefore not contrary to *quies* itself: for a contrary does not tend to its contrary.'[108] As we can see, what the Coimbrans and Toletus write looks very much like what Descartes writes.

106 *AT*, I, pp. 89-90, XI, p. 40, p. 43, III, p. 649, VIIIa, pp. 62-64. Descartes deduces the PCM from divine immutability, unlike Beeckman, for whom, moreover, the relationship between God and motion is not at all the same as it is for him. See: Alan Gabbey, 'Essay Review of W.L. Scott, *The Conflict between Atomism and Conservation Theory, 1644-1680*', *Studies in History and Philosophy of Science* 3 (1973), 373-385, esp. pp. 379-385; Alan Gabbey, 'The Mechanical Philosophy and Its Problems: Mechanical Explanations, Impenetrability, and Perpetual Motion', in: Joseph C. Pitt, ed., *Change and Progress in Modern Science* (Dordrecht: Reidel, 1985), pp. 9-84, esp., pp. 38-41. See also: Van Berkel, *Isaac Beeckman on Matter and Motion*, p. 127, p. 227, n.7; Garber, *Descartes' Metaphysical Physics*, p. 312, n. 35; Slowik, *Cartesian Spacetime*, p. 129, n. 6. And, for a contrary interpretation: Arthur, 'Beeckman, Descartes, and the Force of Motion', p. 23.

107 *Commentarii Collegii Conimbricensis* [...] *in libros Physicorum* (Lyon, 1594), Prima pars, Lib. I, cap. IX, explanatio, 149 and 196 (Lyon, 1602): 'Posterior ratio haec est. Si materia, & privatio formaliter idem essent, sequeretur aliquid sui ipsius interitum appetere, quod a naturae legibus quam maxime est alienum.' Lib. II, cap. VII, quaestio X, articulus III, 269 (Lyon, 1594) and 356-357 (Lyon, 1602): 'Ad tertium, tendere in non esse, idest, habere ingenitam propensionem ut non sint, esse contra naturam creaturarum; cum omnes ad oppositum inclinentur, habeantque ut in argumento recte assumitur, insitum desiderium sese, quoad fieri possit perpetuandi; tendere vero in non esse, id est, non sibi sufficere; sed egere auxiolio Dei sustentantis, ut sint; id non esse contra earum naturam, sed eam potius consequi, & comitari; licet hoc peculiare habeant res corruptioni obnoxiae, quod ad aliis etiam in interitum vocentur.' See: Des Chene, *Physiologia*, p. 274, and p. 278. See also: prima pars, Lib. II, cap. VII, quaestio X, articulus I, 266 (Lyon, 1594) and 352-353 (Lyon, 1602): 'Tendere in non esse est contra rei naturam; cum, unicuique insitum sit perpetuitas desiderium: igitur nulla creatura ex se tendit in non esse proindeque non interibit, nisi aliunde in interitum compellatur: atqui multa sunt ita firma & stabilia, ut nequeant aliunde in interitum compelli, veluti materia prima, anima rationalis, aliaeq; res corruptionis experts. Ergo quae huiusmodi sunt, conservationis beneficium non requirunt.'

108 Toletus, *Commentaria*, p. 169. See: Damerow et al., *Exploring the Limits*, p. 90; Des Chene, *Physiologia*, p. 278. I added the first sentence to Des Chene's quotation.

Conclusion

This chapter began by considering why historians of science who examine the origin of Descartes' PCM argue that he owes it to Beeckman. It then went on to show that he may rather owe it to the Coimbrans, whose commentaries on the *Physics* and *On the Heavens* he studied in 1612-1613 at La Flèche. Indeed, Descartes had at his disposal not only several texts in which the PCM is explicitly used, but also one theoretical element which should lead to adopt it.

Indeed, in arguing, against Aristotle, that motion in a vacuum is possible, because it can be successive and not instantaneous, the Coimbrans, who, moreover, adopted the notion of *impetus* that Aristotle did not, seem to me to have converted the impossibility found in A2 into a possibility.

If S2 were to be adopted, then the following lessons would have to be drawn. First, we should review the often-proposed portrait of Descartes as it is necessary to stop blaming him for hiding his debt to Beeckman. Second, we should review the often-proposed portrait of Beeckman as we should stop considering that he gave the PCM to Descartes.[109] Third, we should review the often-proposed portrait of Aristotle by no longer only considering him the one because of whom it took so long to adopt the PCM and then the principle of inertia, but by also depicting him as the one thanks to whom these two principles were finally adopted. S2 would also prove Funkenstein right: 'Aristotle's theory of motion may be said to have paved the way toward the principle of inertia more than any of its alleged forerunners, including the *impetus* theory.'[110]

About the Author

Samuel Le Gendre is a PhD student under the supervision of Sophie Roux at l'Ecole Normale supérieure-Université Paris Sciences et Lettres (ENS-PSL). He is studying the link between the discovery of the principle of inertia and the idea that there is a right to live in the seventeenth century.

109 This does not prevent us, if we refuse to deny Beeckman's influence on Descartes concerning the PCM and still want to rely on Descartes' letter of April 1619 that we have already discussed, from considering that Beeckman helped Descartes to remember the PCM he had come across when he was at La Flèche. It is a possibility that has been suggested to me by Klaas van Berkel.

110 Funkenstein, *Theology and the Scientific Imagination*, pp. 14-15. See also: Drabkin, 'Notes on the Laws of Motion', p. 67: 'a principle of inertia [...] he formulated precisely enough, only to reject it; from this a sound deductive science of dynamics would not have been a far step.'

10 Beeckman's Corpuscular Study of Plants

*Fabrizio Baldassarri**

Abstract

In his *Journal*, Isaac Beeckman investigated plants by means of his corpuscular and atomistic natural philosophy. These few notes specify Beeckman's interest in the vegetal realm, which was not natural historical nor connected to botanical catalogues, but which concerned the inner structures and processes of vegetal bodies. This chapter explores Beeckman's physico-mathematical approach to plants: his interest in the Touch-me-not plant, his work on medicinal simples, and his investigation of plant formation. Additionally, these notes posit a connection between Beeckman and Bacon, as he comments on a couple of the latter's experiments on vegetal bodies, and Descartes, who discussed similar vegetal features. Beeckman's corpuscular framework sparked the early modern approach to botany as a science.

Keywords: Beeckman, touch-me-not herb, Dutch Baconianism, René Descartes, early modern botany

Isaac Beeckman played a significant role in the history of science, since he devised a *physico-mathematical* philosophy to investigate nature that influenced, if not inspired, René Descartes amongst others. Yet, Beeckman's role in the history of science should not be restricted to his precarious

* Support for this research project was kindly provided by the Kristeller-Popkin JHP Fellowship, the Herzog August Bibliothek Fellowship, by a grant of the Romanian National Authority for Scientific Research and Innovation (CNCS – UEFISCDI), project no. PN-III-P1-1.1-PD-2016-1496, and the European Union's Horizon 2020 research and innovation programme under the Marie Skłodowska-Curie Grant Agreement no. 890770, 'VegSciLif'. I would like to thank the editors of this volume, and also Igor Agostini, Vlad Alexandrescu, and especially Theo Verbeek. Unless otherwise indicated, all translations are mine.

Van Berkel, Klaas, Albert Clement, and Arjan van Dixhoorn (eds.), *Knowledge and Culture in the Early Dutch Republic: Isaac Beeckman in Context*. Taylor & Francis Group 2022
DOI: 10.5117/9789463722537_CH10

relationship with Descartes.[1] On the contrary, he held a pivotal position that lays bare an important attempt to account for natural phenomena and bodies within a systematic theory of mathematical physics. His *Journal* (written between 1604 and 1637, and published in its entirety only in 1939-1953) is a useful source for unearthing the attempt to apply a systematic theory of mathematical physics to the study of nature.[2] In other words, his natural philosophy combines mechanical ingenuity and mathematical methodology with a theoretical view. By means of his method, he fostered, if not anticipated, the modern approach to nature. In this sense, Beeckman was a son of his country. The Dutch Provinces of the time were a laboratory of practices and ideas and a crossroads between cultures, systems, and knowledge, as Delphine Antoine-Mahut and Catherine Secretan have recently shown.[3]

Moving from these premises, in this chapter I explore Beeckman's focus on plants, one of the less-studied subjects of his broad range of interests. Although Beeckman may not be defined as a botanist nor as a botanical *virtuoso* in a strict sense, his attempt to deal with vegetation within his corpuscular, atomistic, and mechanical theory importantly surfaces in a few notes in his *Journal* and significantly anticipates a modern understanding of

1 Klaas van Berkel, 'Descartes' Debt to Beeckman: Inspiration, Cooperation, Conflict', in: Stephen Gaukroger, John Schuster, and John Sutton, eds., *Descartes' Natural Philosophy* (London: Routledge, 2000), pp. 46-60; Richard Arthur, 'Beeckman, Descartes and the Force of Motion', *Journal of the History of Philosophy* 45 (2007), pp. 1-28; Richard Arthur, 'Beeckman's Discrete Moments and Descartes' Disdain', *Intellectual History Review* 22 (2011), pp. 69-90; Frédéric de Buzon, 'Beeckman, Descartes and Physico-mathematics', in: Daniel Garber and Sophie Roux, eds., *The Mechanization of Natural Philosophy* (Dordrecht: Springer, 2013), pp. 143-158; Fabrizio Baldassarri, '"[P]er experientiam scilicet, vel deductionem": Descartes' Battle for *Scientia* in the Early 1630s', *Historia Philosophica* 15 (2017), pp. 115-133.

2 See: Benedino Gemelli, *Isaac Beeckman. Atomista e lettore critico di Lucrezio* (Florence: Olschki, 2002); Peter Damerow, Gideon Freudenthal, Peter McLaughlin, and Jürgen Renn, *Exploring the Limits of Preclassical Mechanics: A Study of Conceptual Development in Early Modern Science: Free Fall and Compound Motion in the Work of Descartes, Galileo, and Beeckman* (Boston: Springer, 2004 [1991]); Natacha Fabbri, *'De l'utilité de l'harmonie'. Filosofia, scienza e musica in Mersenne, Descartes e Galileo* (Pisa: Edizioni della Normale, 2008); Elisabeth Moreau, 'Le Substrat galénique des idées médicales d'Isaac Beeckman (1616-1627)', *Studium* 3 (2011), pp. 137-151; Klaas van Berkel, *Isaac Beeckman on Matter and Motion: Mechanical Philosophy in the Making* (Baltimore: Johns Hopkins University Press, 2013), esp. p. 48.

3 Catherine Secretan and Delphine Antoine-Mahut, eds., *Les Pays Bas aux XVIIe et XVIIIe siècles. Nouveaux regards* (Paris: Honoré Champion, 2015). See also: Paul Dibon, *Regards sur la Hollande du siècle d'or* (Naples: Vivarium, 1990); Klaas van Berkel, Albert van Helden, and Lodewijk Palm, eds., *A History of Science in the Netherlands: Survey, Themes and Reference* (Leiden: Brill, 1999). The notion of the Dutch Republic as a laboratory of early modern science is also explored in: Klaas van Berkel, 'The Dutch Republic: Laboratory of the Scientific Revolution', *Bijdragen en Mededelingen betreffende de Geschiedenis der Nederlanden/Low Countries Historical Review* 125 (2010), pp. 81-105.

the vegetal realm of nature. Beeckman indeed focuses on the arrangement and shape of particles. Additionally, these notes reveal a connection with both Francis Bacon's (1561-1626) experimentation, as Beeckman synthesizes two botanical observations of Bacon's *Sylva Sylvarum* (1627), and somehow anticipates Descartes' natural philosophical study of botany, contained in the *Excerpta anatomica* (posthumously published in 1859-1860).[4] Yet, while positioning Beeckman's work with plants in relation to the scholars of his time is difficult, it is however possible to grasp his contribution to the early modern science of botany as he studied the structure of plants, and reduced their functioning to a corpuscular-mechanical framework.[5]

In the first section of this chapter, I shed light on the Dutch culture in the study of botany and plants, especially focusing on Middelburg, Beeckman's hometown. While looking into Beeckman's *Journal*, in the second section I investigate Beeckman's comment on Bacon's experimentation with plants in his *Sylva Sylvarum*. In the third section I present his interest in the herb Touch-me-not (*Noli me tangere*). In the fourth section I explore Beeckman's study of the internal structure and formation of plants. In the fifth section I present a few notes on the uses of simples for remedies. Since these notes reveal Beeckman's account of plants within his corpuscular framework, in the last section I investigate how far this approach influenced Descartes' study of vegetation, ultimately positioning Beeckman in the early modern Dutch experimental context.

Middelburg: A Centre of Botanical Expertise

In the seventeenth century, Middelburg was an important centre of exchange and one of the busiest cities in the Dutch Republic. Its proximity to Antwerp and its location on the Scheldt estuary allowed it to occupy a central position in Dutch trade, and act as 'a crossroads for economic and cultural

4 On Francis Bacon's study of plants, see: Dana Jalobeanu, 'Bacon's Apple: A Case Study of Baconian Experimentation', in: Guido Giglioni et al., eds., *Motion and Power in Francis Bacon's Philosophy* (Dordrecht: Springer, 2016), pp. 83-113; Dana Jalobeanu, 'Spirits Coming Alive: The Subtle Alchemy of Francis Bacon's *Sylva Sylvarum*', *Early Science and Medicine* 23 (2018), pp. 459-486. On Descartes' study of plants, see: Fabrizio Baldassari, 'Descartes' Bio-Medical Study of Plants: Vegetative Activities, Soul, and Power', *Early Science and Medicine* 23 (2018), pp. 509-529; Fabrizio Baldassarri, 'The Mechanical Life of Plants: Descartes on Botany', *British Journal for the History of Science* 52 (2019), pp. 41-63.

5 On experimentation with plants in the early modern Dutch Republic, see: Fabrizio Baldassarri, 'Descartes and the Dutch: Botanical Experimentation in the Early Modern Period', *Perspectives on Science* 28:6 (2020), pp. 657-683.

exchanges between the southern and northern Netherlands'.[6] Significantly, this trade network stimulated technological innovation and manufacturing production, which goes hand-in-hand with higher education and scientific interest. Towards the end of the sixteenth century, according to Klaas van Berkel, the 'inhabitants of Middelburg developed a keen interest in natural history, particularly in exotic plants and beautiful flowers'.[7]

Apothecaries, doctors, clergymen, amongst others, collected rare specimens and exchanged fruits, seeds, and plants with a botanist in Leiden, Carolus Clusius (1526-1609), a pioneering botanist, who corresponded with all these sorts of people in Middelburg.[8] The apothecary Willem Jasperse Parduyn (1550-1602) introduced himself to Clusius in November 1593, after a month of Clusius's arrival in Leiden. With his letter, Parduyn sent botanical gifts, such as oranges, lemons, and pomegranates: indeed, he used his shipping contacts, as he was the brother of a merchant, to transport letters, plants, seeds, and so on, generally obtaining exotic and rare plants from the Gold Coast of West Africa and the East Indies.[9] It should be remembered that Middelburg was one of the most important centres for the Verenigde Oostindische Compagnie (VOC, the Dutch East India Company), founded in 1602. The port provided exotic plants to the nobility in the Republic, as the case of Yzabeau van Arkel (1536-1617), who received from Middelburg a ginger plant for her garden at Castle Merkenborch, reveals.[10]

The circle of *naturalia* lovers and garden enthusiasts in Middelburg was not limited to Willem Parduyn, but included Tobias Roels (1565-1602), the town physician of Middelburg, the minister Johannes de Jonghe, the apothecaries Johan Somer, Reynier van de Putte, and Thomas de la Fosse, and

6 Van Berkel, *Isaac Beeckman on Matter and Motion*, p. 8. Cf. Klaas van Berkel, 'The City of Middelburg, Cradle of the Telescope', in: Albert Van Helden et al. eds., *The Origins of the Telescope* (Amsterdam: KNAW Press, 2010), pp. 45-71. See also: Maarten Prak, *The Dutch Republic in the Seventeenth Century* (Cambridge: Cambridge University Press, 2005); J. Parmentier, ed., *Noord-Zuid in Oost-Indisch perspectief* (Zwolle: Walburg Pers, 2005); K. Heyning, *Return to Zealand: Masterpieces from the 16th and 17th Centuries* (Middelburg: Zeeuws Museum, 2008).

7 Van Berkel, 'The City of Middelburg', p. 61.

8 See Huib J. Zuidervaart's contribution in this volume, as well as: Huib J. Zuidervaart, 'The Middelburg *Theatrum Anatomicum*: A Location of Knowledge and Culture in an Early Urban Context', in: F.J. Dijksterhuis, A. Weber, and H.J. Zuidervaart, eds., *Locations of Knowledge in Dutch Contexts* (Leiden: Brill, 2019), pp. 64-104.

9 On this correspondence, see: F.W.T. Hunger, *Acht Brieven van Middelburgers aan Carolus Clusius* (Middelburg: J.C. & W. Altorffer, 1925). On Clusius, see: Florike Egmond, Paul Hoftijzer, and Robert Visser, eds., *Carolus Clusius: Towards a Cultural History of a Renaissance Naturalist* (Amsterdam: KNAW, 2007); Florike Egmond, *The World of Carolus Clusius: Natural History in the Making, 1550-1610* (London: Pickering & Chatto, 2010), chap. 9.

10 Egmond, *The World of Carolus Clusius*, p. 45.

the merchants Gerard Staels, Jacques Noirot, and Pieter Courten, whose wife, Hortensia del Prato, indulged herself in gardening. It is relevant to name the two artists Balthasar van der Ast (1593/1594-1657) and Ambrosius Bosschaert Sr (1573-1621), whose still lifes and, in the case of Bosschaert, exotic plants defined an important part of the pictorial art of the Dutch Golden Age. Matthias de L'Obel (1538-1616), a member of the sixteenth-century Flemish school of botany, was city physician in Middelburg, before returning to England and becoming Royal Botanist to King James I. However, he would be intermittently found in Middelburg, where he frequently returned.[11] In general, Middelburg was a hub for a broad network of botanists spreading from Antwerp to London.

The diversity of the people in this list reveals the wide attraction to botany and rare plants of the elites in this epoch. Roels, for example, was highly educated and was fluent in Latin, while Parduyn had no Latin and no university training. Their collaboration follows the difference in their professional status. Roels's interest was principally medical, as he claimed nobody could develop a successful medicine without a knowledge of simples, but he also had sufficient knowledge to state whether a plant was newly discovered or unknown. For the others, their interest was largely in creating gardens. However, as Florike Egmond noted, 'none of the *curiosi* in Middelburg seems to have actually created a private botanical collection-garden in Middelburg that could have stood comparison with those of [...] Antwerp and Bourdeaux.'[12]

In 1610, Caspar Pelletier (d. 1639) published a work on plant names and their equivalents in various languages, listing around 1,600 plants.[13] Anything that grew on the island of Walcheren, from native to foreign and exotic flora, is collected in this book, showing the variety and richness in the botanical exchanges and knowledge in Middelburg at the beginning of the seventeenth century. In this context, Beeckman grew up. His family was connected to the cultivated citizens of Middelburg: as Van Berkel has shown, his father belonged to the merchant group and was a highly cultured man with an acute mind.[14] These connections shaped the climate in which the young

11 See: A. Louis, *Mathieu de L'Obel, 1538-1616. Épisode de l'histoire de la botanique* (Ghent: Story-Scientia, 1980).

12 Egmond, *The World of Carolus Clusius*, p. 150. See: Anne Goldgar, *Tulipmania: Money, Honor, and Knowledge in the Dutch Golden Age* (Chicago: University of Chicago Press, 2007), p. 26.

13 Caspar Pelletier, *Plantarum tum patriarum tum exoticarum in Walachria, Zeelandiae insula, nascentium synonymia* (Middelburg: Schilders, 1610). See: P. Eldering, 'Caspar Pelletier. Ein Botaniker in den Anfängen des 17. Jahrhunderts', *Janus* 64 (1977), pp. 263-278.

14 Van Berkel, *Isaac Beeckman on Matter and Motion*, p. 11. See: Van Berkel, 'The City of Middelburg', pp. 58-60.

Isaac Beeckman was raised, although it is very likely he did not mingle in these circles, as Huib Zuidervaart argues in his contribution to this volume, and his botanical interests followed a different line and reveals a much more theoretical approach.

Beeckman and Bacon: A Theoretical Study of Plants

Caspar Parduyn, son of Willem Parduyn, was a close friend of Isaac Beeckman.[15] Yet, the latter did not share the same obsession with botanical rarities and floral bodies of his fellow-citizens. Beeckman's lack of interest in collecting flora was due to physical limitations, since he was short-sighted, but was not unrelated to a different theoretical approach to nature. As Beeckman claims, his myopia prevented him 'from exercising the senses' on vegetation, or from observing green nature, as the study of plants required.[16] While he 'could not reach an exact knowledge of visible things [of herbs], [he] thus ought to confine himself to knowing things only *animo et mente* [soul and mind]', which means from a more theoretical point of view.[17] This is an important claim: Beeckman's physical handicap forced him to drop any observational interest and pursue a more theoretical perspective.

An example of this theoretical approach to the study of plants surfaces in a 1628 entry on two experiments of Bacon's *Sylva Sylvarum*, namely Experiment 601 and Experiment 607, which deal with the difference between animals, plants, and metals. This is what Beeckman writes in the *Journal*:

> 601 and 607. He [Bacon] says that the difference between animals, plants and metals or stone and things alike is, that the animals and plants have 'spiritus' or spirit in little channels, but the rest has spirits that are secluded here and there in little holes, so that the spirits can and will not come together. Second, [he says] that the spirits of animals and plants will ignite and burn immediately, but the others won't, even though the substance of heat is often greater in it, once it is ignited, as in naphtha and spices.

15 *Journal tenu par Isaac Beeckman de 1604 à 1634*, publié avec une introduction et des notes par C. de Waard, 4 vols. (The Hague: Martinus Nijhoff, 1939-1953) [henceforth *JIB*], I, p. xxii.
16 *JIB*, I, p. 181.
17 *JIB*, II, p. 84: 'mihi vero contigit myopia, deficitque sensus praecipuus, nec possum herbas accurate videre nisi inclinato capite, quod, cum non liceat facere perpetua et res frequenti duntaxat intuitu accurate cognosci possint, sequitur me ad exactam rerum visibilium cognitionem pervenire non posse. Quae cum ita sint, domi me contineo, solis animo et mente comprehensibilibus intentus.'

> The difference between animals and plants is, that the spirits of animals also gather in hollow spaces next to their veins and channels and that they burn and ignite much more than the plants do. I have written about this burning in more detail before. With regard to the spirit in stones it should be researched, because it could be there, and [there could also be] a vacuum, because spirit is often much less dense than a stiff body, because water (which can also be called spirit) is more dense than wood.[18]

This significant commentary reveals Beeckman's haste in reading the recently published *Sylva Sylvarum*, as he was not making detailed comments. Beeckman was indeed one of the first Dutch readers of Bacon's work. This note is part of a larger section of the *Journal* in which he discusses Bacon's experiments on sound and light in the *Sylva Sylvarum*, in which, as Benedino Gemelli has remarked, Beeckman insists on an atomistic-corpuscular interpretation of these phenomena, and this note is related to this interest.[19]

Within this corpuscular framework, in this note on plants Beeckman deals with spirits endowing natural bodies, metals, plants, and animals, and considers these spirits to be material.[20] More than a note on plants in their own right, or a note on botanical experimentation, this is a comment on what differentiates inert and living bodies, a crucial question at the time. Beeckman's note concerns the presence of spirits that ignite within bodies. This is connected to his claim that life corresponds to a fire in bodies,

18 *JIB*, III, 64: '601 en 607 seght hy, dat het verschil tusschen dieren, planten ende metallen of steen ende diergelycke is, dat de dieren ende planten spiritus of geest hebben in gootkens, maer de reste hebben geest, die besloten is hier ende daer in gaetjens, also dat de geesten byeen niet kommen en konnen. Ten tweeden, dat de geesten van dieren ende planten gelyck ontsteken syn ende branden, maer de andere niet, al ist dat de substantie van hitte daerin dickwils grooter is, soo sy ontsteken wort, als in naphta ende aromatibus. Het verschil tusschen dieren ende planten is, dat der dieren geesten noch bovendien in seker hollicheden vergaderen behalven haer aeren of gootkens, ende dat sy meer branden ende ontsteken syn dan der planten. Van dit branden hebbe ick vooren ergens breeder geschreven. Van de geest in de steenen is te ondersoecken, want die kan daer wel in syn, ende vacuum oock, dewyle dat geest veeltyts ondichter is dan styf lichaem, want water (twelck oock geest kan genoempt worden) is dichter dan houdt.' In the margin: 'Animalium, plantarum et metallorum differentia.' (Translated with the help of Suzanne Kooloos and Robert Vinkesteijn.)

19 See: Paul Dibon, 'Sur la réception de l'œuvre de F. Bacon en Hollande dans la première moitié du XVIIe siècle', in: Dibon, *Regards sur la Hollande*, pp. 191-220, esp. 193-199; Elena Alberto, 'Baconianism in the Seventeenth-Century Netherlands: A Preliminary Survey', *Nuncius: annali di storia della scienza* 6:1 (1991), pp. 33-47; Benedino Gemelli, 'Isaac Beeckman as a Reader of Francis Bacon's *Sylva Sylvarum*', *Journal of Early Modern Studies* 2:2 (2013), pp. 61-79.

20 See: Gemelli, *Isaac Beeckman*, p. 16.

as he writes one page after this note, in the margin to *Sylva Sylvarum*'s Experiment 704.[21]

In the *Sylva Sylvarum*, these two experiments are titled Experiment 601, 'Experiments in consort touching the affinities and differences between plants and inanimate bodies', and Experiment 607, 'Experiments in consort touching the affinities and differences of plants and living creatures, and the confiners and participles of them.'[22] According to Bacon, what helps differentiate between bodies is the diversity in spirits and their flammability, something that confirms Beeckman's conception of life.

In commenting on Experiment 601, Beeckman synthesizes two ways to differentiate between inanimate and animate bodies. This first resides in the fact that plants and animals have spirits in branches, veins, and little channels that are within the body, thereby moving throughout it, while spirits in metals and stones are in little holes and do not come together nor move. Bacon states that spirits in living bodies 'are continuous with themselves' and communicate, while in inanimate bodies 'are shut [...] and not pervious one to another', or are interrupted. The second difference consists in the fact that the spirits in living bodies are ignited and burn, while spirits in inert matter are more difficult to ignite, even though they possess much hotter spirits, as the cases of naphtha, petroleum, and spices show.[23] Following Bacon, Beeckman stresses that the spirits in living bodies are quicker to ignite, while those in inert bodies are more difficult to ignite, but once they do ignite, they have more heat substance in them and the burning is fiercer.

In commenting on Experiment 607, Beeckman acknowledges Bacon's division between animals and plants. Accordingly, spirits in animals have

21 *JIB*, III, p. 65: 'Vita nostra quomodo sit ignis.'

22 Francis Bacon, *Sylva Sylvarum, or A Natural History in Ten Centuries*, in: James Spedding, Robert Leslie, and Douglas Denon Heath, eds., *The Works of Francis Bacon*, 14 vols. (Stuttgard-Bad Cannstatt: Frommann, 1961-1963), II, pp. 528-529.

23 Bacon, *Sylva Sylvarum*, 528: 'The differences between Animate and Inanimate Bodies, we shall handle fully under the title of Life, and Living Spirits, and Powers. [...] All Bodies have Spirits, and Pneumatical parts within them; but the main differences between Animate and Inanimate are two. The first is, that the Spirits of things animate, are all continued with themselves, and are branched in Veins, and secret canals, as Blood is: And in Living Creatures, the Spirits have not only Branches, but certain Sells or Seats, where the principal Spirits do reside, and whereunto therest do resort: But the Spirits in things Inanimate are shut in, and cut off by the Tangible parts; and are not pervious one to another, as Air is in Snow. The second main difference is, that the Spirits of Animate Bodies are all in some degree (more or less) kindled and in flamed, and have a fine commixture of Flame, and an Ærial substance: But Inanimate Bodies have their Spirits no whit inflamed or kindled.' On the analogy between spirits and flame, see: *JIB*, II, pp. 175, 327-328.

a position, something that Bacon calls 'a Cell or Seat', which is absent in plants, and for this reason ignite much more than in plants. This is what Bacon considers a radical difference between living bodies. Beeckman appears, however, to be dissatisfied with this explanation, and suggests investigating more thoroughly the spirits in stones, since their properties may depend on 'a vacuum' or on the density of these bodies. Beeckman does not mention the eight secondary differences between animals and plants listed by Bacon, and appears therefore less interested in plants in themselves. In contrast, his primary concern relies on the presence of spirits in natural bodies and in the differences in bodily flammability.

Finally, it is to be noted that Beeckman appears not to be attracted to the observation of plants, which is the focus of the following experiments in the *Sylva Sylvarum* conducted or proposed by Bacon.[24] The latter claims that the difference between animate and inanimate beings does not entail different classes of spirits, but only different structures of the same spirit; plants are thus alchemical laboratories for investigating this difference.[25] In contrast, Beeckman focuses less on plants as a means of investigating this differentiation. Rather than a direct observation of plants, Beeckman follows a more theoretical approach to vegetal bodies and uses them as case studies to explore spirits and flammability.

The *Noli Me Tangere* or Bursting Plant

In an earlier note of the *Journal*, dated between December 1616 and March 1618, in which he discusses fire, answering the question 'why does fire produce fire, and a great fire from abundant matter?', Beeckman presents the case of a particular plant. This is the text:

> For the reason that the matter of fire is compressed against the nature of fire itself, in the way the air is compressed, and with little difficulty returns to its nature, as happens in the herb called *noli me tangere* [*krudeken en roert my niet*] that bursts when it is touched. Suddenly, a greater force produces bursting [...], this occurs if several of these plants are put close

24 On Bacon's theory of matter, see: Guido Giglioni, 'Mastering the Appetites of Matter: Francis Bacon's *Sylva Sylvarum*', in: Charles Wolfe and Ofer Gal, eds., *The Body as Object and Instrument of Knowledge: Embodied Empiricism in Early Modern Science* (Dordrecht: Springer, 2010), pp. 149-168.
25 On this, see: Doina-Cristina Rusu, 'Same Spirit, Different Structure: Francis Bacon on Inanimate and Animate Matter', *Early Science and Medicine* 23 (2018), pp. 444-458.

> by, and someone with the lightest touch makes one burst, and so while many burst they do it as though they were a singular thing dissolving many things, therefore increasing the motion until all the herbs will be destroyed. Similarly inflammable matter [...].[26]

First, the reference to a plant in this note is problematic. Beeckman uses a plant called Touch-me-not or *Noli me tangere* to discuss the production of fire. It is to be remembered that early modern scholars conceived plants as cold bodies, without an internal source of heat activating their functions. In a note to the text, De Waard claims that this plant is 'the sensitive herb (*Mimosa pudica*), recently imported from Brazil or the West Indies', which was indeed called Touch-me-not after the phenomenon it generates.[27] This interpretation, however, is incorrect. Indeed, there is no correspondence between the phenomenon of the *Mimosa pudica*, or sensitive herb, which, as Guido Giglioni pointed out recently, consists of 'its tendency to fold its leaves when they are touched by an external agent', and the phenomenon described by Beeckman.[28] He writes of the explosive dehiscence or of a process of spontaneous bursting of plant structures, and thus refers to another plant designated by the common locution Touch-me-not, which is known today by its Latin term *impatiens*. This was a renowned plant at the time: Caspar Bauhin (1550-1624) includes this plant among the *balsaminae* in his *Pinax theatri botanici* (1623), and naturalists such as Andrea Cesalpino (1519-1603), Conrad Gessner (1516-1565), Mathias de L'Obel, and Rembert Dodoens (1517-1585), amongst others, discuss it. The latter also gives its Flemish name, which Beeckman uses (fig. 10.1).[29]

26 *JIB*, I, p. 124: 'Cur ignis ignem producit et ex abundante materiâ ignem abundantem? Respondeo: Quia ignis materia a naturâ compressa videtur contra ignis naturam, eo modo ac si quis aerem comprimat, qui minimo negotio ad suam naturam redit, ut fit in herbâ quam vocamus *krudeken en roert my niet*; dissilit enim si quis eam tangat. Subitò majoremque vim dissiliendo facit, quae adhibita est ad ejus dissolutionem, unde fit, si multae tales herbae prope invicem collocarentur, atque aliquis levissimo tactu unam dissolveret,<ut> haec dissiliendo plures dissolveret quae singulae rerum quaeque plures dissolveret, atque ita motus incresceret, dum omnis herba dissoluta foret. Hoc pacto quoque orditur materia inflammabilis compacta, cujus aliquam particulam si quis dissolvat, id est incendat, incensa proximam perpetuò dissolvet. Atque ita flamma increscit.' (Italics in the text.)

27 *JIB*, I, p. 124, n. 1: 'L'Herbe sensitive (*Mimosa pudica*)'.

28 Guido Giglioni, 'Touch Me Not: Sense and Sensibility in Early Modern Botany', *Early Science and Medicine* 23 (2018), pp. 420-433.

29 Caspar Bauhin, *Pinax theatri botanici ... sive Index in Theophrasti Dioscoridis Plinii et botanicorum* (Basel: sumptibus et typis Ludovoci Regis, 1623), pp. 306-307: 'Noli me tangere & Impatiens herba [...] quae ubi maturae levissimo contactu dissiliunt.' Rembertus Dodonaeus,

Figure 10.1 ***Impatiens herba*** **or** ***Balsaminae***

From: Rembertus Dodoens, *Stirpium historiae pemptades sex* (Antwerp: Christopher Plantin, 1583), p. 33

Moreover, difficulties surface. The first regards Beeckman's botanical concerns. He focuses neither on the nature of this plant nor on its unique phenomenon, but uses it as a means to exemplify the functioning of the particles of heat when stored in bodies, and appears to be more attracted by the corpuscular nature of fire. This lack of interest in the plant makes his reference to it harder to understand, because he could have used the example of the natural fire that ferments in stored hay more profitably, as Descartes will do a few years later in the *Principia philosophiae*, in order to explain the spontaneous combustion of natural bodies due to the motion of particles.[30] In another note dated January-February 1625, Beeckman again uses the *herba impatiens* as an example to describe the ways fire operates within bodies (wood), and thus to clarify the corpuscular nature of bodies.[31]

Second, it is unclear whether Beeckman observed the phenomenon of this plant personally or just reported something he heard. Apparently, although this plant was known and discussed, it was not easy to find in the Dutch Republic at the time.[32] Presumably, Beeckman did not achieve any direct observation of this plant, but was aware of its phenomenon and reduced it to the corpuscular structure of light, conceiving it as an apt example to describe natural bodies. We are thus confronted with a theoretical approach to plants.

The Formation of Plants

In another note of September 1626, Beeckman claims that 'when one saws through a tree, one can tell by the rings how many years old it is. Every year there grows a ring from the inside of the heart [of the tree].'[33] While focusing on the internal parts of trees, this note reveals Beeckman's interest in the structure and formation of plants. This is a much more important feature of the mechanical study of plants.

Stirpium historiae pemptades sex (Antwerp: Chr. Plantin, 1583), pp. 648-649: 'Belgae Crudeken en ruert my niet: vulgo Noli me tangere…' Clusius translated Dodoens' work into French.

30 Descartes, *Principia philosophiae*, IV, art. 92. In: *Oeuvres de Descartes*, publiées par Charles Adam et Paul Tannery, 12 vols. (Paris: L. Cerf, 1897-1910; new ed. in 13 vols., Paris: Vrin, 1974-1986) [henceforth *AT*], VIII-1, p. 256. See: Baldassarri, 'The Mechanical Life of Plants', pp. 42-43.

31 *JIB*, II, p. 319: 'Nam et lignum ab igni solvitur et solutum dissilit eo modo, quo dissilit herba *Noli me tangere*.'

32 Egmond, *The World of Carolus Clusius*, p. 36.

33 *JIB*, II, p. 431: 'Alsmen eenen boom doorsaeght, soo siet men aende ryngen hoeveel jaer hy out is. Elck jaar groeyter eenen rinck aen van binnen het herte af, soomen seght.' [During his youth, the city sawmill in Middelburg was just across the street where Beeckman lived (see fig. 11.17) – editor's note.]

This is the subject of a large note of April 1619, in which he deals with the ways living bodies, and more specifically trees, grow. This is the text:

> Trees grow in this way: the rays of Sun enter in, and enlarge their pores, making their humour more tenacious and fluid. […] As pores are widened, the more fluid humour is attracted, and every part [of the plant] draws humour as the inferior parts [do] from the roots. Spread in the earth, the roots take in [the humour] by the force of vacuum [*vi vacui*]. This humour is cooked in these pores by the returning heat, and by the nature of these parts [the humour is] converted into a similar substance […] that passes in the near parts by the rays of the Sun. Superior parts, which are more recent and humid, and smaller, receive earlier and more easily the rays of the Sun, and are widened. Since they do not have anything above, when the rays leave the inferior parts [of the tree], these remain filled with humour and press it toward the superior parts, these latter transfer it between them due to the compression of the tree, and they press it out, making the tree taller.
>
> Due to the tenacity of the humour and the weakness of the inner heat, trees have vigour during the summer; animals in contrast, due to influence of the heat of the heart and the internal perpetual motion, perpetually grow, since the soft parts can release heat.[34]

Regrettably, I cannot deal with every issue raised in this note. Beeckman claims that trees grow by means of the movement of particles of humours drawn from the soil, by a sort of attraction, and by the heat of sunrays that enlarges the pores of the plant and prepares, or cooks, the humour. As these particles move towards the top of the plant, they make it grow upwards. It is to be noted that, since plants are cold bodies, they mostly grow during the

34 *JIB*, I, p. 284: 'Arbores crescunt hoc pacto: Solis radij ingrediuntur ejus poros eosque dilatant atque humorem tenaciorem magis fluidum reddunt. […] Poris igitur latioribus redditis, attrahitur humor fluidior factus, idque omnes partes praestant, ita ut inferiores humorem ex radice. Radix verò ex terrâ propagata sugit vi vacui.
Hic humor in his poris a calore redeunte excoquitur, et per naturam partis in similem substantiam convertitur, […] quod vicinas partes passas esse diximus a radijs solaribus. Superiores partes cùm sint recentiores, ideòque humidiores et minores, primum et faciliùs Solis radios admittentes, dilatantur, cùmque supra se nihil humoris habeant, radijs exeuntibus partes inferiores, humore plenae subsidentes, exprimunt suum humorem in superiores; superiores verò, cùm nequeant exprimere infra se propter totius arboris compressionem, exprimunt extra se, atque ita arbor fit altior. Arbores autem, propter humoris tenacitatem et caloris nativi imbecillitatem, aestate dumtaxat vigent; animalia verò, propter cordis calorem influentem et motum perpetuum actionum, perpetuò crescunt, quamdiù pars flexibilis calori cedere potis est.'

summer, while animals grow perpetually thanks to their internal source of heat. More precisely, we shall note that Beeckman acknowledges a sort of digestion in plants, that is a conversion of humour into the parts of the plant, activated by what he calls a returning heat, which is probably the heat of the sunrays that make the humours melt. At the end of the note, in contrast, he speaks of a *calido innato* endowing living bodies. Moreover, what makes humours arise within the plant is a sort of attraction presumably produced by the fact that the sunrays widen the pores of the plants, and this enlarged space attracts the humours from the bottom parts of the plant and from the earth. Beeckman claims this enlargement of pores creates a vacuum that makes humours arise as happens in air pumps.[35]

Beeckman confirms this motion in another note of June 1619, in which he explains why trees grow taller in regions with fresh water. The reason is that the sunrays make the water rise within the tree, while they cannot lift salt. As a result, in those regions where the water is mixed with salt, trees grow less, whereas plants such as wheat and *Gramineae*, which only grow larger and not taller, grow abundantly in those regions.[36] It is to be noted that Beeckman conceives the heat and water to be sufficient to produce bodies.[37]

In a note of July 1619, in which he deals with the ways the earth grows, that is the transformation of natural bodies in earth matter, Beeckman repeats that heat and water make natural bodies grow. He moves from the experience that ditches in Holland that have been cleared of mud and herbs after a certain time are refilled with the same material. This experience reveals that corpuscles change and assume a different arrangement [*positura*] in bodies, converting them to herbs, mud or possibly earth.[38] As such, the

35 *JIB*, I, p. 23. Cf. Van Berkel, *Isaac Beeckman on Matter and Motion*, p. 85.

36 *JIB*, I, p. 130: 'In regionibus ubi aqua dulcis abundat, plurimae proveniunt arbores, quia sal non elevatur satis commodè a Solis radijs. Ascendit igitur arborum materia copiosior in altum. Praeterea, ubi sal miscetur cum aquâ, minus aquae est eodem loco. Frumenti et graminis plus in locis humilioribus quàm si dulcis aqua sale non inficeretur: sal enim in aquâ existens particulas aquae aliquantulum distendit. Sic in altissimis montibus altissimae arbores crescunt. Frumentum verò, gramen etc., quae non sunt alta, in salsis regionibus abundantiùs proveniunt, quia sal fit materia pinguedinis, cùm non sit necessè ut altè ascendat.'

37 *JIB*, I, p. 150. See: Gemelli, *Isaac Beeckman*, p. 86.

38 *JIB*, I, pp. 326-327: 'Terra reverâ videtur crescere his argumentis quia in Hollandiâ eae fossae, unde cespites exceptae sunt, successu temporis repleri dicuntur, quod manifestissimè in nostris fossis videmus. Haec enim, profundae factae, post aliquot annos iterum implentur per excressentias herbarum etc. ex ipso fundo pullulantium, quae herbae ibidem putrescendo in terram vertuntur. Quin igitur eo modo materia cespitum non cresceret? Sic sylvae densantur arboribus quae, si omnes terrae conderentur, eam valdè augerent altioremque redderent. Quare credendum est, cùm nihil augeatur cui nihil accedit, cùmque sola pluvia agris incidat, aquam fieri terram atque aquae atomos per calorem coeli itadisponi ut non ampliùs fluant, sed firmiter consistant atque in terram

movement and change of atoms assume different positions and shapes, forming natural bodies. By dealing with the transformation of water and earth into plants, Beeckman appears to be aware of the ways plants grow by the addition and internal transformation of particles, although he does not provide a philosophical explanation or a framework for it.

He also explains this in a note on grafting of May 1626. This observation reveals that 'elements change in everything, and the humours of the trunk or rootstock are converted in their nature by the scion or grafted branch, thanks to the different *positura* of the particles of the scion'.[39] While passing through one plant to the other, the particles of humours take different arrangement, what he again calls *positura*, and constitute the plant, according to Beeckman. The mechanical motion and change of particles that take different shapes, and their arrangement throughout the body explain the formation of plants in Beeckman's mechanical philosophy of nature, thereby detailing about several cases, such as the growth of plants, herbs and shrubs, and the case of grafting. It is to be noted that the heat of the Sun and fresh water favour the motions of humours and particles deriving from the earth, passing through the channels and pores of plants, and taking different shapes and positions, thereby resulting in the formation of vegetation. Still, Beeckman does not observe this change in the form of particles.

The Uses of Vegetal Simples for Treating the Body

For Beeckman, plants were an interesting subject also for exploring the fabrication and the action of therapeutics. Although he graduated in medicine at the University of Caen, Beeckman only practised on himself and some friends and relatives; thus, in dealing with simples, his goal was not just to verify their effectiveness, but especially to investigate the corpuscular and atomic structure of natural bodies and their interrelations.[40] In a note of June 1618,

vertantur. Idque fieri non in arenâ quia ibi non fit mixtio nec infractio partium per Solis radios, quia nimis crassis arena constat particulis, sed in terrâ viscosâ ac aptâ herbis producendis. Haec enim eum aquâ et igni per minima potest misceri: in eâque aquae particulae per mutuam actionem in suas atomos resolvuntur atque ita alium situm positumve ob ingressionem particularum terrae et ignis sortiuntur, vertunturque tùm in herbas, tùm in cespitum materiam, tùm in ipsam fortassis terram.'

39 *JIB*, II, pp. 341-342: 'Ipse verò antè de mutatione harum rerum scripsi elementa in omnia mutari, et humorem trunci ab insititio ramo in suam naturam verti propter diversam insititij particularum intrinsecarum posituram. Quomodo enim ramus in summo arboris maximi insitus, tantas radices deorsum mitteret? Non videtur author vidisse semen avenae ex pendulo cucumere, sui generis herbam producere etc.'

40 On Beeckman's medical practice, see Dániel Moerman's contribution in this volume.

Beeckman refers to *senna*, a plant of the tribe *Cassieae*, described by Dodoens as a plant used for purgative and laxative purposes. Accordingly, the vapours of this plant free the pores of the body, providing relief from constipation.[41]

As Beeckman grounds his medicine in the evacuation of humours, the relief is due to the interrelations between humours in the body, ruled by the impact between particles and the actions of the particles of the plant. As particles of different sizes and shapes collide with one another, they eliminated the parts that obstruct the body.[42] Together with these purgative effects, Beeckman uses therapeutics as astringents, in order to close the veins.[43] In a section made up of several notes, while discussing several passages from Galen, Beeckman speaks of the therapeutic virtues of herbs such as *myrrha* or *smyrna*, which helps evacuate the body, or other astringents or laxatives, such as cinnamon.[44] Additionally, eating unripe fruit eases the evacuation of *cacochymia*, or the black bile, as such purgatives divide the humours in the body.[45]

Drawing on the experience of Galen and several physicians and naturalists in the sixteenth century, Beeckman explains the functioning of these vegetal remedies, focusing on the corpuscular structure of matter and natural bodies, therefore claiming to combine medicine with *mathesis*.[46] His comments on the efficacy of therapeutics confirm his mechanical understanding of nature, however embedded in the Galenic theory of the humours.

Beeckman's Corpuscular Botany: Conclusion

The few notes discussed above reveal Beeckman's narrow focus on plants. He never exchanged botanical specimens, as was common in Middelburg,

41 *JIB*, I, p. 192: 'In corpore verò nostro, quod poros majores habet, non semper ventus fit, nec rarefactus ibidem perpetuò coercetur, sed plerumque ejus partes attenuatae tam parvae sunt, ut insensibiliter poros corporis pertranseant, quantumque rarefactum est, exitum invenit nisi ubi cutis est astricta etc. Praeterea cruda ventos excitant, ut scribit Dodoneus de sennâ immaturâ. Est enim vapor eorum tenax et crassa, ignisque eum ingrediens, dilatat quidem eum, sed nequit secare in tam parvas partes, quae poris conveniunt. Neque existimandum est vaporem constare ex solo igni et aquâ, sed terrea substantia etiam plerumque admixta est, quae – cùm non sufficienter est extenuata et dissecta – coercet intrinsecum humorem ab igni dilatatum, fitque hoc pacto vapor crassissimus et maximarum partium.'

42 See: *JIB*, II, pp. 75-76.

43 See: *JIB*, II, pp. 55-56, 79, 133.

44 *JIB*, II, pp. 109-113.

45 *JIB*, II, p. 114; see also *JIB*, II, p. 109, 'στοματικὸν διὰ καρύων'.

46 *JIB*, II, p. 56: 'Medicina requirit mathesim.'

his hometown, and rarely observed vegetal bodies per se, although he studied plants in their own right, that is, focusing on the formation and structure of vegetation, and on their virtues. He mostly conceives plants as objects to provide examples or to buttress his reasoning and study of nature. This is evident in his double references to the exotic *Noli me tangere*, or the *herba impatiens*, whose behaviour is analogous to explaining the presence of fire in natural bodies. Still, Beeckman may have used the case of cut hay more effectively, since this latter was a common occurrence, while the behaviour of this herb required a more direct observation, something that fails to attract his attention. Similarly, when commenting on Century VII of Bacon's *Sylva Sylvarum*, a collection of experiments also dealing with plants, Beeckman focuses on a very restricted subject, the motions of spirits within natural bodies, rather than using this text as a source to perform botanical experimentation as Henricus Reneri (1593-1639) does later.[47] It is therefore clear that Beeckman's role in drawing attention to Bacon's work remains a general claim for the case of plants, as he did not develop a more specific experimental interest in vegetation, but mostly proposed a more theoretical approach to plants.[48]

Yet, something more significant surfaces in those notes in which Beeckman discusses the formation of plants in mechanical terms, as he describes the internal structure of plants and the ways particles shape and arrange within the body, also detailing the role of the Sun in opening pores, raising water and the earth matter constituting the plant. This approach was not different from that of Descartes and characterized the early modern botanical knowledge in the second half of the seventeenth century.

Yet, drawing a direct connection between Beeckman and Descartes, or between Beeckman and other scholars in this field, is difficult. Although Beeckman and Descartes deal with similar botanical issues, such as the presence of heat and fire in plants, the attraction to exotic plants, and the interpretation of grafting, and both investigate the formation of plants, discuss the motion, shape, and arrangement of particles, and unearth the differences arising from planting trees in different places, their notes and their focus do not overlap.[49] For example, both acknowledge the importance of plant simples to heal and unclog bodies, and speak of *senna* as a purgative,

47 See: Baldassarri, 'The Mechanical Life of Plants', pp. 45-46.
48 See: Baldassarri, 'Descartes and the Dutch', pp. 659-662.
49 Baldassarri, 'The Mechanical Life of Plants', pp. 48-61.

but many differences arise.[50] Both focus on the *Noli me tangere*, but deal with different plants. Both discuss the formation of plants, but their notes are different. Many dissimilarities surface, making it difficult to know how much Descartes appropriated from Beeckman, if indeed he ever read these notes. Still, if not in the specific content of these notes, a theoretical link between them emerges in connection to the mechanical framework of their interpretations of vegetal bodies.

In some ways, however, Beeckman's mechanistic-corpuscular framework operated as a source of inspiration for Descartes' natural philosophy, and was also part of the intellectual milieu in which early modern Dutch scholars such as Henricus Regius (1598-1679) and Florentius Schuyl (1619-1669), amongst others, were trained. Indeed, their investigations of vegetal bodies reflect a theoretical, corpuscular, and atomistic perspective, far from the botanical collections of the time. Moreover, although no clear connection between them and Beeckman ever arises, they provided a mechanistic interpretation of green nature not far from the latter's. In the end, if Beeckman was unable to promote the contents of his notebook and make an authority of his natural philosophy as Descartes did, his mathematical physics was at work even in the case of plants, and a corpuscular-mechanical approach to plants in Holland during the 1620s ultimately surfaces. While his interest in plants remains marginal as the few notes show, Beeckman's botany is consistent with a new investigation of nature that spread in Holland during the 1630s and 1640s.[51] If he failed to bring this section to a more complete condition, a corpuscular study of plants, however, develops as a compelling piece of his research at the heart of the Scientific Revolution.[52]

50 See: René Descartes, *Remedia et vires medicamentorum*, in: *AT*, XI, p. 642: 'Hinc et facile reddi ratio poterit multorum astringentium, ut *ver de gris*, acerbi omnes fructus, sorba, mespili, etc. Certum est meatus istos occludere, contra vero, quae frigida atque humida, ut pruna, cassia, poma, etc. illos laxare; ideoque esse purgantia.' Cf. Fabrizio Baldassarri, 'Seeking Intellectual Evidence in Sciences: The Role of Botany in Descartes' Therapeutics', in: J.A.T. Lancaster and R. Raiswell, eds., *Evidence in the Age of the New Sciences* (Cham: Springer, 2018), pp. 47-75.

51 See: Baldassarri, 'Descartes and the Dutch', pp. 671-677.

52 On the seventeenth-century natural philosophical study of plants, see: Fabrizio Baldassarri and Oana Matei, 'Manipulating Flora: Seventeenth-Century Botanical Practices and Natural Philosophy', *Early Science and Medicine* 23 (2018), pp. 413-583.

About the Author

FABRIZIO BALDASSARRI (Ca' Foscari University of Venice/Indiana University Bloomington) is an intellectual historian, working on the natural philosophical studies of plants in the early modern period. He currently holds a Marie Skłodowska-Curie fellowship. He recently published articles on the study of plants in seventeenth-century Dutch culture in the *British Journal for the History of Science*, *Early Science and Medicine*, and *Perspectives on Science*, a monograph on Descartes' medical studies, and has edited (with Andreas Blank) *Vegetative Powers: The Roots of Life in Ancient, Medieval and Early Modern Natural History* (Cham: Springer, 2021).

Colour Section

1 Image projection with a camera obscura, from Johan van Beverwijck's *Schat der Ongesontheyt*

From: Johan van Beverwijck, *Schat der Ongesontheyt* (Amsterdam, 1656). Rijksmuseum Boerhaave, Leiden. The lens is placed in a shutter to the left.

2 Joos Lambrechtse and his family

Design for a glass window, 1654. Atlas van Stolk, Rotterdam. Joos Lambrechtse (1597-1669) was Isaac Beeckman's assistant and successor as chandler in Zierikzee. He was married to Susanna Langenes (1597-1658), a daughter of the Middelburg bookseller Barent Langenes.

3 'Reaal van Achten', issued by the Verenigde Zeeuwsche Compagnie, 1602

Nationale Numismatische Collectie – De Nederlandsche Bank, Amsterdam. This coin (42 mm) was minted in the Middelburg Abbey by the mint master Melchior Wyntgens. For more information, see this volume, p. 268, note 22.

4 Portrait of Symon Jasperse Parduyn, by an unknown painter

Zeeuws Museum, Middelburg. Symon Jasperse Parduyn (d. 1612) was a merchant and botanical enthusiast in Middelburg. He is portrayed as the representative of the 'First Noble' in the States of Zeeland (1596-1612). The 'First Noble' was Count Maurits van Nassau, whose portrait is in the golden chain.

5 Signboard of the workshop of the Middelburg printer Jan Pietersz van de Venne, 1623

Daily Mail and General Trust, London. The signboard of the workshop, located on the Korte Burg in Middelburg (no. 3 on the map shown in fig. 11.10, p. 290) was painted by the printer's brother Adriaen. The windows on the right were looking onto the Koopmans- or Heerenbeurs (no. 4 on the map on p. 290). In the garden in the background a column can be seen with the statue of Mercury. This statue also appears in the drawing of De L'Obel's Lauwerhof (see fig. 11.6 on p. 276).

6A The Nieuwe Kerk (New Church) in Middelburg, with the house of Hans Lipperhey

Sketch by J. Tuyter, 1848, just before the demolition of the houses in the Kapoenstraat, built against the church. Zeeuws Archief, Zelandia Illustrata, II-549. In 1609, Hans Lipperhey bought the house on the right (no. 9 in figure 11.10 on p. 290), which he called *De Drie Vare Gesichten*, after the three telescopes he had supplied to the States General. Before 1609, he rented the adjoining house on the right (not depicted here).

6B Network of Middelburg scholars, printers and amateurs, around 1600

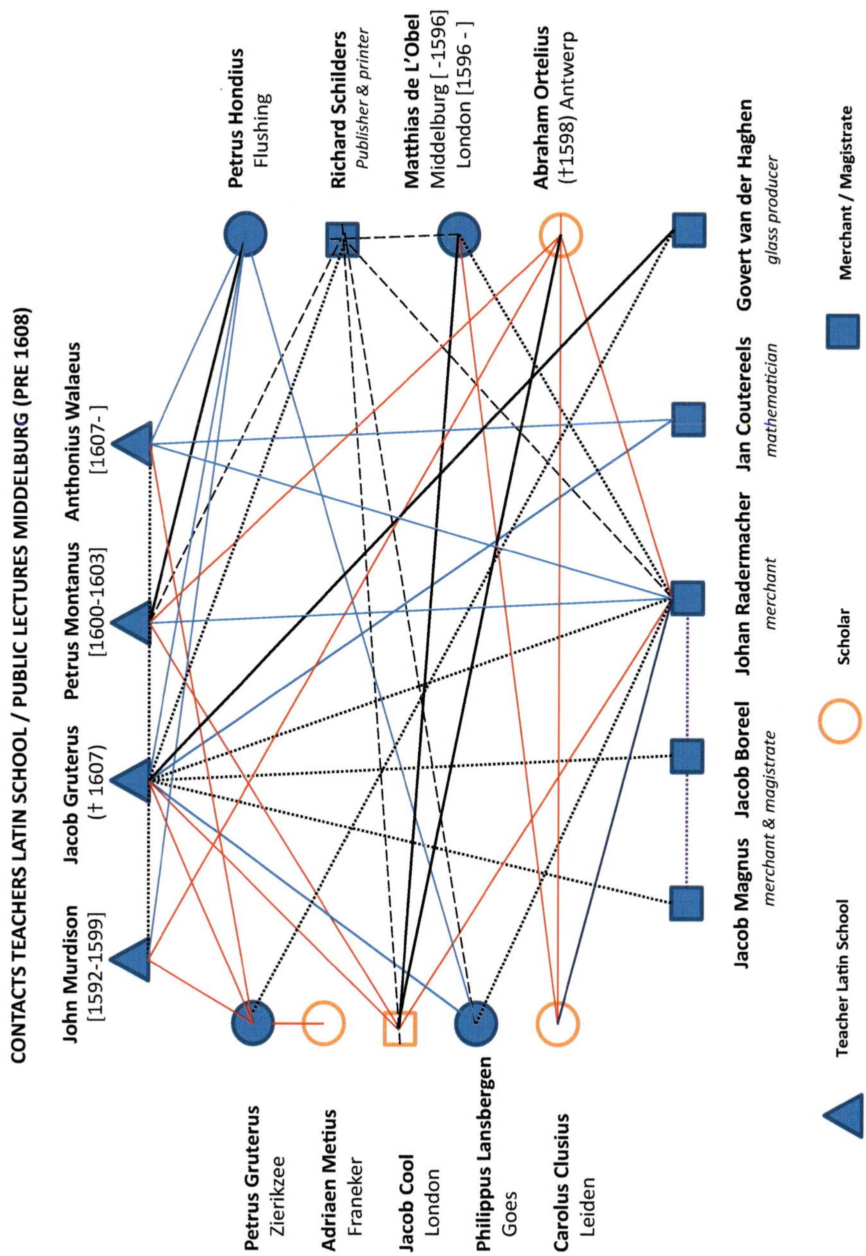

This network represents the contacts between teachers at the Latin School in Middelburg and other learned persons between 1591 and 1608. Filled icons: persons living in Zeeland; empty ones: persons living elsewhere. Black line: family relationship; red line: known mutual correspondence; blue line contact documented via poem or *album amicorum*; dotted line: direct personal contact, known from letters or other documents; striped line: books published by Richard Schilders. For sources: Zuidervaart, 'Scientia', pp. 97-98.

7 The use of Lansbergen's quadrant explained, 1620

Title engraving after a drawing by Adriaen van de Venne in: Philippus Lansbergen, *Verclaringhe van 't ghebruyck des Astronomischen ende Geometrischen Quadrants* (Middelburg, 1620). In the background we see the city of Middelburg.

8 Title page and first page of the auction catalogue of Antonius Biesius

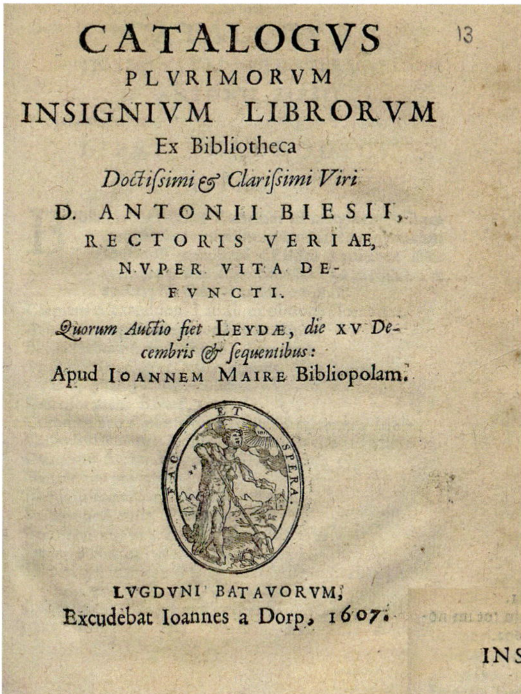

CATALOGVS
PLVRIMORVM
INSIGNIVM LIBRORVM
Ex Bibliotheca
Doctiſſimi & Clariſſimi Viri
D. ANTONII BIESII,
RECTORIS VERIAE,
NVPER VITA DEFVNCTI.

Quorum Auctio fiet LEYDÆ, *die* XV *Decembris & ſequentibus:*
Apud IOANNEM MAIRE Bibliopolam.

LVGDVNI BATAVORVM,
Excudebat Ioannes a Dorp, 1607.

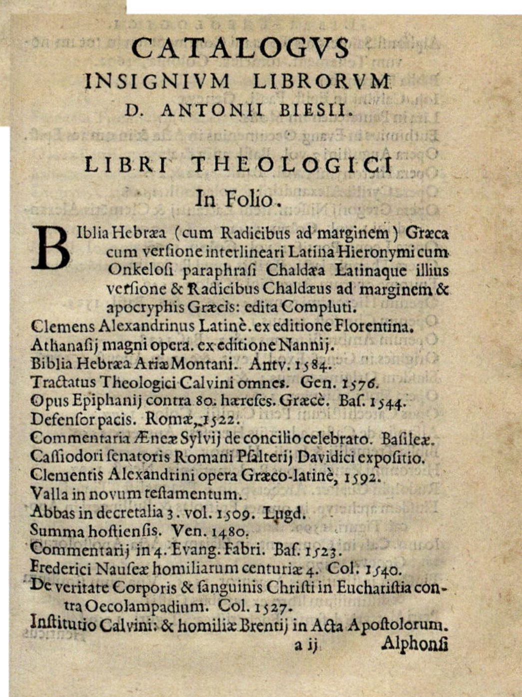

CATALOGVS
INSIGNIVM LIBRORVM
D. ANTONII BIESII.

LIBRI THEOLOGICI
In Folio.

Biblia Hebræa (cum Radicibus ad marginem) Græca cum verſione interlineari Latina Hieronymi cum Onkeloſi paraphraſi Chaldæa Latinaque illius verſione & Radicibus Chaldæus ad marginem & apocryphis Græcis: edita Compluti.
Clemens Alexandrinus Latinè. ex editione Florentina.
Athanaſij magni opera. ex editione Nannij.
Biblia Hebræa Ariæ Montani. Antv. 1584.
Tractatus Theologici Calvini omnes. Gen. 1576.
Opus Epiphanij contra 80. hæreſes. Græcè. Baſ. 1544.
Defenſor pacis. Romæ, 1522.
Commentaria Æneæ Sylvij de concilio celebrato. Baſileæ.
Caſſiodori ſenatoris Romani Pſalterij Davidici expoſitio.
Clementis Alexandrini opera Græco-latinè, 1592.
Valla in novum teſtamentum.
Abbas in decretalia 3. vol. 1509. Lugd.
Summa hoſtienſis. Ven. 1480.
Commentarij in 4. Evang. Fabri. Baſ. 1523.
Frederici Nauſeæ homiliarum centuriæ 4. Col. 1540.
De veritate Corporis & ſanguinis Chriſti in Euchariſtia contra Oecolampadium. Col. 1527.
Inſtitutio Calvini: & homiliæ Brentij in Acta Apoſtolorum.
a ij Alphonſi

University Library Heidelberg, F 9801. The library of Biesius, rector of the Latin Schools of Arnemuiden and Veere, and teacher of Beeckman, was auctioned in Leiden in December 1607.

9 The encounter of a mathematician, a lawyer, a painter, and an engraver, with a sculptor in the background

Watercolour by Adriaen van de Venne, British Museum, London. The figures possibly represent Philippus Lansbergen (the mathematician), Jacob Cats (the lawyer), Adriaen van de Venne himself (the painter), François Schillemans (the engraver) and Pieter Roman senior (the sculptor in the background).

10 The house on the Beestenmarkt (now Varkensmarkt no. 11) in Middelburg where Beeckman was born

Photograph: Huib Zuidervaart

11 The Beeckman residence in the Hoogstraat in Middelburg and neighbouring houses

Photograph: Huib Zuidervaart. Isaac Beeckman's father settled in the house called *De Twee Haentgens* (The Two Roosters) in the Hoogstraat in 1593. At some moment in time the house was split into two parts (today nos. 19 and 21-23). The left part (less high) now bears the date 'Anno 1735' on top of the facade. The right part, with a large door, still seems to contain elements of an older edifice. The wider house to the left of *De Twee Haentgens* still bears the name *De Olyfberch* (The Mount of Olives), as in Beeckman's time. In 1626 this house was also owned by Abraham Beeckman. The house to the right has the date '1737' on top of the facade, and is called *De Oude Bakkery* (The Old Bakery). Between 1599 and 1606 the baker Tobias Antheunisse lived her.

12 Identification of the houses around the Beesten- or Varkensmarkt in Middelburg in 2020

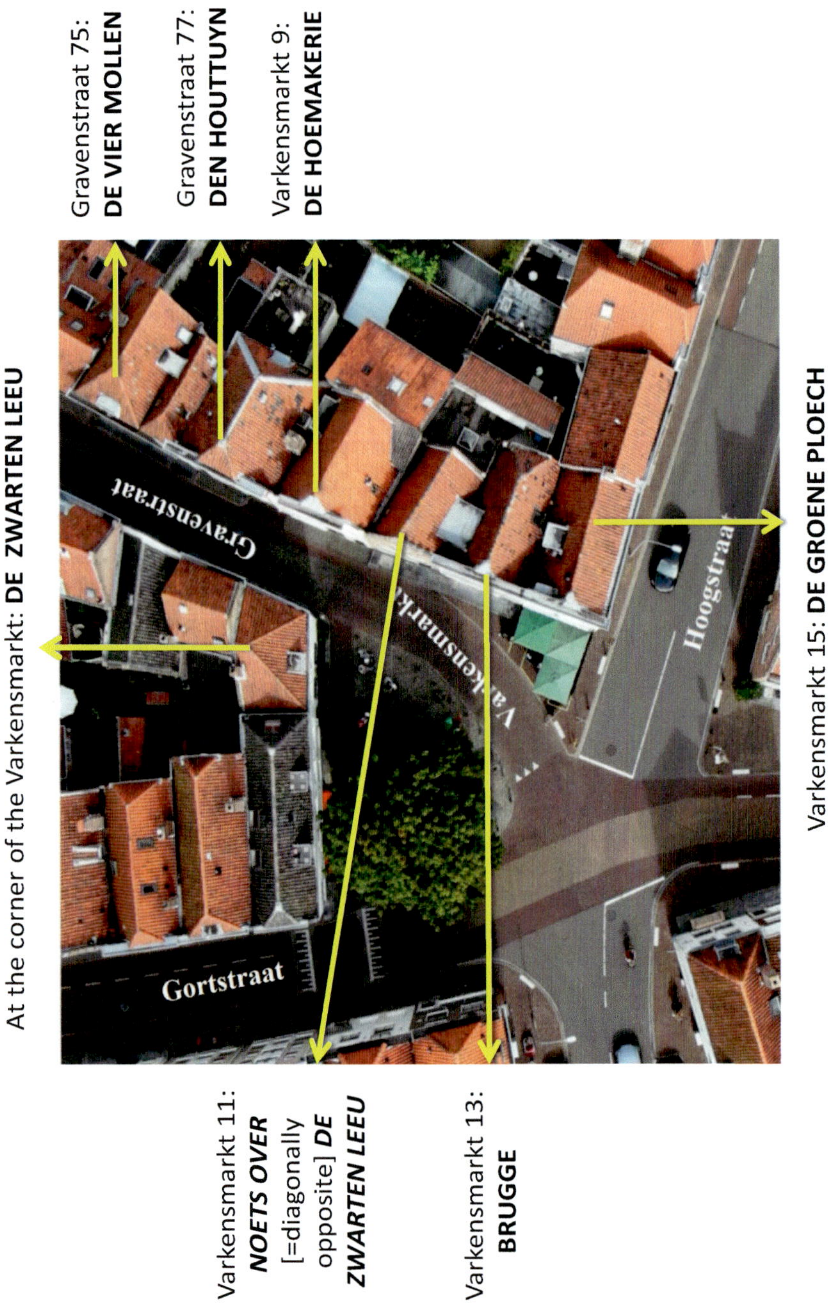

Drone photograph by Hans Zuidervaart

13 Portrait of Janneken van Ryckegem, sister-in-law of Isaac Beeckman, 1632

Oil painting on oak panel, attributed to the Rotterdam painter Jan Daemen Cool (1589-1660). Städel Museum, Frankfurt am Main, inv. no. 716. Janneken van Ryckegem (1595-1639) was the second wife of Isaac Beeckman's brother Jacob Beeckman (1590-1629). She married Jacob Beeckman in 1619. After the death of her husband, she moved to Dordrecht and for a few years lived with Isaac Beeckman and his family. Janneken is painted here on the occasion of her remarriage with the wealthy widower Thomas Vergrue, a manufacturer of salpetre in Middelburg. This portrait is the only surviving painting with a link to the Middelburg Beeckman family and it nicely illustrates the rise in prosperity and status of the group of Flemish immigrants to which they belonged. In 1776 this painting was in the possession of the Middelburg magistrate Daniël Radermacher (1722-1803), a great-great-grandchild of Jacob Beeckman and Janneken van Ryckegem. After Radermacher's death the painting was bought by Johann Justinian Georg, Freiherr von Holzhausen (1771-1846). In 1818 it was purchased by the Städel Museum in the Holzhausen auction.

14 Organ of the Nieuwe Kerk (New Church) in Middelburg

Drawing, 64 x 46 cm, Middelburg, Zeeuws Archief, Zelandia Illustrata, II-557. The organ was made by Jan Roose and completed by Johann Morlett in 1603.

15 Double virginal, made by Lodewijck Grouwels, 1600

The Metropolitan Museum of Art, New York (Crosby Brown Collection of Musical Instruments), www.metmuseum.org/art/collection/search/501767. The original instrument (190.5 x 50.8 cm) was made by Lodewijck Grouwels in Middelburg in 1600. The octave keyboard is by A. Dolmetsch.

16 Portrait of Jacob Pergens, by Salomon Mesdach, 1619

Rijksmuseum Amsterdam, SK-A-918

Part III

Situating Beeckman

11 Networks of Knowledge in Middelburg around 1600

The Context of Isaac Beeckman as a Young Man

*Huib Zuidervaart**

Abstract

Isaac Beeckman's innovative attitude to the study of nature has been attributed to his mixed training as both a craftsman and a scholar. More generally, the Dutch contribution to early modern science has been ascribed to three factors: (1) the easy mingling of scholars, merchants and craftsmen in the early Dutch Republic; (2) the vital role of the Dutch academic institutions as centres of both teaching and innovative research; and (3) a congruence of early scientific and mercantile activities and values in the early modern Dutch trading communities. Against this background, this chapter examines the question of which persons and circumstances have contributed to Beeckman's early education in his birthplace Middelburg. Although it appeared possible to identify early modern Middelburg as a fertile melting pot of mercantile, artisanal and learned contacts, this study underpins Van Berkel's earlier conclusion that the life of the young Beeckman unfolded for the largest part in a milieu of Flemish immigrants, with no demonstrable connection to the Middelburg learned community.

Keywords: Isaac Beeckman, networks of knowledge, Zilsel thesis, Middelburg, telescope

* This chapter builds on my Dutch paper '"Scientia" in Middelburg rond 1600. De wisselwerking tussen academici, kooplieden en handwerkers in het vroegmoderne Middelburg', *Archief KZGW* (2018), pp. 43-100, which for this occasion is adapted to the case of Isaac Beeckman. The reader is referred to that paper for more detailed references.

Van Berkel, Klaas, Albert Clement, and Arjan van Dixhoorn (eds.), *Knowledge and Culture in the Early Dutch Republic: Isaac Beeckman in Context*. Taylor & Francis Group 2022
DOI: 10.5117/9789463722537_CH11

Introduction: Beeckman and the Zilsel Thesis

Isaac Beeckman was both a craftsman and a scholar. He was trained by his father as a candle maker and a constructor of waterworks, but he also studied philosophy, mathematics, theology and medicine. This mixed education seems to be the key to Beeckman's innovative attitude towards the study of nature. In his 2013 monograph on Beeckman his biographer Klaas van Berkel claims that this early modern researcher 'was the first to devise a completely mechanical philosophy of nature, and thus introduced an approach that would become a cornerstone of the new science'.[1] Van Berkel even goes further to stipulate that 'Beeckman played a crucial but not always recognized role in the early stage of the Scientific Revolution', even in such a way that Beeckman could be seen as 'the missing link between artisanal knowledge and mathematical science'.[2] With this latter statement Van Berkel refers to the thesis formulated in 1942 by Edgar Zilsel, namely that the new science emerged from the empirical work of artisans and from the interaction between craftsmen and scholars.[3] Hands-on knowledge of materials and craftsmanship, combined with theoretical academic knowledge, or, succinctly put, 'the union of hand and mind', had resulted in the empirical and experimental methodology that formed the core of the new science of the seventeenth century.[4]

In his 2010 article 'The Dutch Republic: Laboratory of the Scientific Revolution', Van Berkel elaborated on – what he then called – the 'Zilsel-Rossi thesis',

1 Klaas van Berkel, *Isaac Beeckman on Matter and Motion: Mechanical Philosophy in the Making* (Baltimore: Johns Hopkins University Press, 2013), p. 1.

2 Van Berkel, *Isaac Beeckman on Matter and Motion*, p. 4. Earlier, H. Floris Cohen, *The Scientific Revolution: A Historical Inquiry* (Chicago: University of Chicago Press, 1994), p. 362, mentioned Beeckman as a perfect example of the merger of a scholar and an artisan.

3 Edgar Zilsel, 'The Sociological Roots of Science', *American Journal of Sociology* 47 (1942), pp. 544-562. See also: Zilsel, *Die sozialen Ursprünge der neuzeitlichen Wissenschaft*, ed. by W. Krohn (Frankfurt am Main: Suhrkamp, 1976; Diederick Raven, W. Krohn, and Robert S. Cohen, eds., *The Social Origins of Science* (Dordrecht: Kluwer, 2000).

4 Van Berkel, *Isaac Beeckman on Matter and Motion*, pp. 2-3, 253. See also: Klaas van Berkel, 'The Dutch Republic: Laboratory of the Scientific Revolution', *Bijdragen en Mededelingen betreffende de Geschiedenis der Nederlanden/Low Countries Historical Review* 125 (2010), pp. 81-105, esp. p. 89; Pamela H. Smith, *The Body of the Artisan: Art and Experience in the Scientific Revolution* (Chicago: University of Chicago Press, 2004); Pamela Long, *Artisan/Practitioners and the Rise of the New Sciences, 1400-1600* (Corvallis: Oregon State University Press, 2011); L.B. Cormack, Steven A. Walton, and John A. Schuster, eds., *Mathematical Practitioners and the Transformation of Natural Knowledge in Early Modern Europa* (Cham: Springer, 2017).

focusing on the Dutch context.[5] He attributed the remarkable contribution of the Dutch Republic to early modern science to three factors: firstly, in the early Dutch Republic there was an easy mingling of scholars, merchants and craftsmen, which was facilitated by a lack of social stratification in this period, characterized by turmoil and high immigration numbers; secondly, there was the vital role of the Dutch academic institutions as centres of both teaching and innovative research; and thirdly, there was a congruence of early scientific and mercantile activities and values in the early modern Dutch trading communities.

In my contribution, I want to investigate these three factors of the 'knowledge culture' in the city of Middelburg around the year 1600, and apply these findings to the question: can we determine what this 'culture' contributed to Isaac Beeckman's formation as an ingenious scholar? Therefore, I first present a short outline of Isaac Beeckman's life. Subsequently, I discuss the vibrant intellectual and artisanal climate of the city of Middelburg during Beeckman's youth, in which context some attention will also be given to the invention of the telescope (an instrument to which Beeckman would devote much energy later in life). As my conclusion will be that Beeckman had no demonstrable contacts with the Middelburg circle of *curiosi*, I conclude with an investigation of the local artisanal milieu in which he grew up.

Short Outline of Isaac Beeckman's Life

Let us first ascertain *when* and *where* Isaac Beeckman resided in Middelburg, the capital of the province of Zeeland, in the centre of the island of Walcheren. Isaac was born here in December 1588, as the eldest son of the candle maker Abraham Heijndricksz Beeckman (1563-1626) and his wife Suzanna Pieters van Rhee (1568-1629).[6] They had been married in Middelburg

5 Van Berkel, 'The Dutch Republic: Laboratory of the Scientific Revolution', p. 89. Van Berkel added Paolo Rossi, because his work did much to give this interpretation credibility. See: Paolo Rossi, *Philosophy, Technology, and the Arts in the Early Modern Era* (New York: Harper & Row, 1970).

6 For the determination of the location of the various people mentioned, the Middelburg Haardstedenregisters (tax registers based on the number of fireplaces) were used, dated 1601 and 1606 and preserved in the Zeeuws Archief, Rekenkamer van Zeeland, 'Rekenkamer D', inv. nrs. 47270a (1601) and 47270b (1606). For 1599, see also: *Journal tenu par Isaac Beeckman de 1604 à 1634*, publié avec une introduction et des notes par C. de Waard, 4 vols. (The Hague: Martinus Nijhoff, 1939-1953) [henceforth *JIB*], IV, p. 9.

in January 1588. As Van Dixhoorn observes, both families Beeckman and Van Rhee were 'by no means ordinary artisans', being embedded in what he calls a 'vernacular *consten*-culture', brought with them from their area of origin, Flanders.[7] Both families were rather well-to-do, and surviving letters by Isaac's father, Abraham, demonstrate that he was indeed an educated intellectual.[8] Isaac was named after a brother of his father who had died in England at a young age.[9] In 1593, when he was four years old, his family moved from their small house in the Gravenstraat to a larger house around the corner, called *De Twee Haentgens* (The Two Roosters – see the Appendix).[10] From there, Isaac went to a local primary school at the age of seven. Middelburg stayed important in his life even when in 1601, at the age of twelve, he went to the Latin School in nearby Arnemuiden. When a year later its rector, Antonius Biesius, moved to the Latin School of Veere, Isaac went with him. Both Arnemuiden and Veere were small towns on Walcheren, less than two hours' walking distance from Middelburg. Isaac left the island in April 1607 at the age of eighteen, when he enrolled at Leiden University as a student in letters and philosophy.[11] That same year he went to Rotterdam for three months to receive instruction in arithmetic, geometry and navigation, visiting the school of his distant relative, Jan van den Broecke.[12] This was

7 See the chapter by Van Dixhoorn in this volume.

8 Zeeuwse Bibliotheek, Middelburg, handschriften, nos. 5065-5091: letters by Abraham Beeckman to various persons (1597-1608).

9 Isaac Beeckman, original notebook 'Loci communes', fol. 48r, end of second column (Zeeuwse Bibliotheek, manuscript no. 6471).

10 Beeckman, 'Loci communes', fol. 49r, column 1. See also: Haardstedenregister 1601: Hoogstraat: 'Thuijs *tHaentkens*, aencomende Abraham Bekemans, kersmaker, ende heeft bij den selven bewoont, 8 schouwen'; idem, 1606: ''tHuys genaempt *De Twee Hanen* aencommende Abraham Vekemans ende wert bij hem bewoont ende heeft zeven schouwen.' The spelling 'Bekemans' or 'Vekemans' is a mistake by the person who made the list. Names were only transmitted orally, so these registers contain many misspelled surnames. In 1606 a Hans Vekemans lived in a house on the Nieuwe Haven, the extension of the Hoogstraat, where Abraham Beeckman resided. Vekemans also possessed *De Cleyne Soutkeete* in the nearby Gravenstraat, a house he rented out. Later in life Isaac Beeckman made a last will at a notary in Dordrecht called 'Vekemans'. *JIB*, I, p. xx.

11 Leiden University Library, ASF 8: *Volumen inscriptorium*, 21 May 1607: 'Isaacus Beeckmannus, Middelburgensis, annorum XVIII, linguarum et philosophiae studiosus, in de Sonneveltssteech.' Beeckman enrolled together with Abraham Teerling from Arnemuiden, who most likely was one of his fellow students at the Latin School led by rector Antonius Biesius. After completing his studies, Teerling settled in Dordrecht, where he became collector of taxes relating to the slaughter of cattle ('impost van het bestiaal').

12 Jan van den Broecke was probably a grandchild of Aelken Beeckman, the aunt of Isaac's grandfather Hendrick. Earlier he lived in Flushing (Vlissingen). In Rotterdam he had a school in the 'Nyeu Poort'. His brother Cornelis van den Broecke lived in Middelburg, in the brewery *De*

followed in the spring of 1608 by a course of Hebrew in Amsterdam given by Henry Ainsworth, the leader of the local Brownist congregation.[13] Isaac attended these lessons together with his younger brother, Jacob (1590-1629), who had entered Leiden University in October 1607, at the age of sixteen.[14] In the fall of 1608 Isaac experienced a mental breakdown as a result of studying too intensely, so he returned to his parents' home in Middelburg. After his recovery, Isaac and Jacob Beeckman registered again in Leiden in September 1609, now both as students in the *artes liberales*.[15] In August 1610 Isaac finished his education in Leiden without a formal degree and although his brother Jacob went to Franeker for further study of Hebrew, Isaac now returned to Middelburg, where he – as the eldest son – would be trained in the profession of his father: making candles and constructing waterworks.

Wolsack, at the corner of the Lange Delft and the Kerkstraat, where Isaac Beeckman renewed the waterworks in 1621. *JIB*, I, p. 174; II, p. 173; III, p. 78. Notary Archive Rotterdam, several deeds 1610-1612. See also: Jan vanden Brouck, *Instructie der Zee-Vaert* (Rotterdam: Abraham Migoen, 1609; 1610), which gives an impression of his mathematical teaching.

13 The Brownists were separatists of the Church of England, opposed to all remaining Catholic rituals and dogmas in the Anglican Church. They were named after their leader, Christian Browne (c. 1550-1633), who with his followers lived in Middelburg between 1581 and 1584. There are strong indications that the Beeckman family was very much inclined to the Puritan ideas of the Brownists. At Ainsworth's request Isaac Beeckman would preach once for the Amsterdam Brownist congregation in 1608. See also the conclusion of this chapter.

14 UB Leiden, ASF 8: *Volumen inscriptorium*, 1 October 1609: 'Jacobus Beekmannus, Middelburgensis. Annorum XVI, studious literarum, apud Jacobum Brutsaert, int Noordende'. The Noordeinde is a street in Leiden leading to the Rapenburg canal where the university was (and still is). Jacob Broetsart was a *trapier* (cloth weaver), who made a testament in 1595 (Erfgoed Leiden, notary archive 506, nos. 59, 11 March 1595). He was the father of Pieter Broetsart, mentioned in the next footnote.

15 UB Leiden, ASF 8: *Volumen inscriptorium*, 29 September 1609: 'Isaacus Beeckmannus, Middelburg. ann. 20, studios. bonar[um] litterar[um], at the house of Pieter Broetzart. Jacobus Beeckmannus, Middelburg, ann. 18, stud. bonar. litterar. apud eundem'. The Broetsart family, where the two brothers found their lodgings, were refugees from from Hontschoten in Flanders. On 15 March 1599 Pieter Broetsart became a *poorter* (citizen) of Leiden, still being a minor. At that time he worked as a *pletsverkoper* (wool trader). Later he was mentioned as a *cruidenier* (grocer) (Erfgoed Leiden, Register van poorterinschrijvingen E, inv. no. 1266, fol. 183; Register van poorterinschrijvingen F, inventarisnummer 1267, fol. 102 (1617); DTB Trouwen, archive 1004, no. 3 (10 April 1593) and no. 87 (27 and 30 March 1602)). Isaac and Jacob Beeckman lived there together with a third student, Carolus Borbonius from the French city of Angoulême. Borbonius matriculated on 27 October 1609 as a student of medicine, living 'apud Petr. Brotzardum'. His name suggests that he was a (Protestant) member of a minor branch of the French royal family, the princes of Bourbon-Angoulême, probably related to the Hugenot leader Prince Henri I de Bourbon, second Prince of Condé (1552-1588), but I have not been able to find more details about this Charles de Bourbon, who probably was born in 1588.

Being approved by the Middelburg guild as a master chandler in March 1611, Isaac settled in Zierikzee, to be joined there in September by his brother Jacob, who after his stay at Franeker University had been appointed vice-principal at the Zierikzee Latin School. In that town the two brothers – still unmarried – probably lived together.

The intellectual mind of Isaac, however, desired more than just producing candles. In Zierikzee, for instance, Isaac started to record daily observations of the weather, the first of this kind of activity in the Netherlands.[16] It was also in Zierikzee that he formulated for the first time his own version of the – well known – principle of inertia, nowadays attributed to Galileo Galilei and René Descartes.[17] But in Zierikzee the Beeckman brothers were also influenced by the Pietist movement in the Dutch Reformed Church, of which Zierikzee at the time was the centre. It was here that Jacob Beeckman began to prepare himself for a position as a Reformed minister. Isaac decided to follow his example. In June 1612 Isaac travelled to France for a summer course in theology at the Reformed University of Saumur. After his return to Zierikzee in November 1612, Isaac continued this studies and he passed the theological exam before the classis of Walcheren in July 1613, only a few months after his brother Jacob had done the same. However, in the next two and a half years, no church congregation could be found that desired to nominate these two aspirants as their minister. Maybe the deviating theological views of their father were partly to blame for this.

So, in September 1615 Isaac accepted that his career would go differently than anticipated. From then on he spent most of his time in Middelburg again, leaving his workshop in Zierikzee in the hands of a deputy. This was a distant cousin, Joos Lambrechtse, who had also been trained as a candle maker by Isaac's father. Isaac now focused his activities more and more on the construction of waterworks and fountains, a craft that his father Abraham Beeckman had practised before. When in April 1616 his brother Jacob was appointed rector of the Latin School in Veere,[18] Isaac decided to

16 In Zierikzee Beeckman recorded observations of the weather from March 1612 until 17 March 1615. See: Klaas van Berkel, 'Vruchtbaar isolement. Isaac Beeckman in Zierikzee (1611-1616)', *Kroniek van het land van de zeemeermin (Schouwen-Duiveland)* 11 (1986), pp. 59-74, esp. pp. 64-65.

17 See the chapter by Samuel Le Gendre elsewhere in this volume.

18 Although Jacob Beeckman (1590-1629) was appointed in April 1616 as rector of the Latin School in Veere, he only moved to that small town in May 1619, living in the Middelburg Wagenaarstraat during the years 1616-1619 (Zeeuws Archief, DTBL Veere 2: NG Lidmatenregister 1610-1700). As Abraham Beeckman Jr., the youngest of the Beeckman brothers, noted in the original *Loci communes* (fol. 296recto) that Isaac Beeckman – when in Zeeland during these years – lived with his brother Jacob, this means that he too resided in the Middelburg Wagenaarstraat in that period. This in contrast to De Waard in: *JIB*, I, p. ix.

leave Zierikzee for good. He sold his chandler workshop to Lambrechtse, who in the following decades managed to expand this business into a very prosperous enterprise (colour ill. 2).[19]

In May 1616 Isaac returned to Middelburg, from where he made several business trips, among others to England and the Southern Netherlands. In the mean time his inquiring mind was caught by another subject: the human body. In 1614 Isaac started to record his heartbeat and he began to consult various books on medicine.[20] With the intention of obtaining an official degree in medicine, Isaac went to France in the summer of 1618. His printed thesis at the University of Caen, discussing three-day fever – well known on the island of Walcheren – was dedicated to his father, his brother Jacob and his 'old friend' Jacques Schouten.[21] However, being home again, Isaac decided not to become a physician after all, but instead to pursue a career in education, just as his brother Jacob had done. To explore this possibility he made trips to Gorinchem, Rotterdam, Delft, Brielle and Breda in the summer of 1619. With success: in November he was appointed vice-principal of the Latin School of Utrecht.

When on 20 April 1620 Isaac married Cateline (also Catherina) de Cerf (fig. 11.1), a Flemish girl who had arrived in Middelburg in 1615, it seemed he was really ready to settle. However, when in October 1620 his brother Jacob, who had become rector of the Latin School in Rotterdam, asked Isaac to assist him there without a formal appointment, Isaac did not hesitate for a moment. He quit his job in Utrecht and joined his brother and his newly-wed second wife, Janneken van Ryckegem, in Rotterdam. Still, Walcheren remained important. To increase the one income his brother shared with him, Isaac once again carried out some pipe-laying work in Middelburg and Veere in the summer of 1621. This extra work became no longer necessary after he was officially appointed as the Rotterdam vice-principal in 1624. Finally, in May 1627 Isaac was invited to become rector of the Latin School of Dordrecht, where he stayed for the rest of his life. Nevertheless, in all these years, even after the death of his parents, Beeckman frequently returned to Middelburg for family visits, business affairs and (in the 1630s) for training in lens grinding, arising from his optical studies.

19 *JIB*, I, p. 217; IV, p. 33. Lambrechtse became a prosperous merchant in Zierikzee, who traded in flax, cheese, beans and other foodstuffs and also received profits from a shop in linen, silk and other manufactures. See: Katie Heyning, *Zeeuwse meesters uit Gouden Eeuw. Een selectie uit de Goedaert Collectie* (Zwolle: Wbooks, 2018), pp. 8-9.

20 *JIB*, I, p. 34 [12 April 1614].

21 Isaac Beeckman, *Theses de febre tertiana intermittente* (Thesis, Caen, 1618). See: *JIB*, IV, pp. 43-44.

Figure 11.1 Registration of Beeckman's engagement with Catharina de C(h)erf, 1620

Utrecht Archive, Nederduits gereformeerde gemeente, inv. nr. 93, p. 44. The engagement was registered by the Reformed Church in Utrecht on 19 March 1620 (o.s.) and the marriage itself took place in Middelburg in 20 April 1620 (n.s.). See: *JIB*, I, xiii and IV, 68.

So, we may conclude that the city of Middelburg was pivotal in Isaac Beeckman's life. Therefore, let us return to our questions: what may Isaac Beeckman have learned in this city of his youth? What do we know about the knowledge culture in this city in his formative years, and can we establish what this culture contributed to Isaac Beeckman's education and further formation as a scholar?

The City of Middelburg around 1600

Middelburg is located in the centre of Walcheren, an island in the estuary of the Scheldt river. In Beeckman's youth the city was a melting pot of people with different backgrounds. Its population tripled in 30 years, from around 5,500 inhabitants in 1569 to around 18,000 in 1600. The conquest of Antwerp by Spanish troops in 1585 had caused a huge influx of Protestant refugees from the Southern Netherlands, some of whom had first fled to German regions and England and then moved to Middelburg. These circumstances resulted in a booming economy. Middelburg was the centre of the Dutch trade in wine and after 1582, when the English Company of Merchant Adventurers moved its stable to Middelburg, the city succeeded in becoming, at least for a short while, also important for the trade in English woollen cloth. In the late 1590s Middelburg merchants (many of southern origin) organized the first successful trading expeditions to the Indies. In 1602, the Verenigde Zeeuwsche Compagnie (United Merchant Company of Zeeland) even had its own coins minted, with the recently conceived proud Zeeland logo 'Luctor et emergo' ('I struggle and emerge') in the circumscription (colour ill. 3).[22] Later that year this merchant company merged into the

22 Early in 1601, the Amsterdamsche Compagnie van Verre (Amsterdam trading company for the Far East) received permission from the States of Holland to mint a series of coins that could

newly founded Verenigde Oostindische Compagnie (VOC, the Dutch East India Company), of which the Middelburg chamber was the second largest party involved.

The fall of Antwerp, however, also had its downside. The subsequent restrictions on the trade to Antwerp made the commercial relations with the Southern Netherlands more difficult. When after the Twelve Years' Truce (1609-1621) it became clear that the southern provinces were lost for the Dutch Republic for good, many merchants and artisans moved to Holland, where cities such as Amsterdam, Haarlem and Rotterdam had become the leading merchant centres. Then Middelburg's heydays were over. So, during Beeckman's youth, when he grew to adulthood, the intellectual setting in Middelburg was favourable for scholarly curiosity. The city's prosperity offered time and space for a knowledge culture to flourish. So let's look more closely to the three factors mentioned in the introduction: (1) the easy mingling of scholars, merchants and craftsmen; (2) the vital role of the academic institutions; and (3) the congruence of early scientific and mercantile activities and values in the trading communities.

I Academic Lectures in the Former Abbey

In 1575, when the first university in the Northern Netherlands was established, two cities had been candidates for the establishment of this new

remedy the lack of silver *reaals* (originally a Spanish coin) in the Indonesian archipelago. This impelled their Zeeland counterpart, the Verenigde Zeeuwsche Compagnie (VZC), to request the same to the States of Zeeland on 15 November 1601. After hearing the Zeeland mint master Melchior Wyntgens, this permission was granted on 28 November. At that time, Wyntgens was just producing the new Zeeland *daalder van 60 groot* (dollar of 60 *greats* or 30 *stuiver*), a currency type introduced in 1601 by the States of Zeeland on its own authority. Given the haste with which the Zeeland *reaal van achten* had to be made (the production of 1,200 silver marks (that is, 10,800 coins) had to be completed within a week), it was decided to give the obverse of these trade coins an effigy and legend identical as on the new *daalder*: a combination of the coat of arms of the Zeeland cities and those of the First Noble (count Maurits of Nassau) with the legend 'MONE[ta] ARG[entea] ORD[inum] · ZELANDIAE' (silver coin of the States of Zeeland). The reverse shows the Zeeland coat of arms with the swimming lion, and the value '8 R[eaal]' besides it. The legend displays the Zeeland motto 'LVCTOR ET EMERGO' ('I struggle and emerge'), issued for the first time in 1601 on the Zeeland *roosschelling* (a coin with a value of 6 *stuivers*). Today the coin ordered by the VZC is of the utmost rarity, because eventually almost all of the coins were melted down in Asia. See: *Notulen van de edel mogende heeren Staten van Zeeland, d'anno 1601* (n.p., 1602), pp. 376-378, 400; E. Netscher and J.A. van der Chijs, *De munten van Nederlandsch Indië, beschreven en afgebeeld* (Batavia: Lange, 1863), p. 3; C. Scholten, *De Munten van de Nederlandsche Gebiedsdeelen overzee, 1601-1948* (Amsterdam: Schulman, 1951), p. 32.

institution: Middelburg and Leiden. As we know, Middelburg was overtaken by Leiden. But the hope of a form of higher education in Middelburg was never lost. In 1593 an annually recurring series of academic lectures in natural philosophy and theology started in an auditorium below the Abbey Tower.[23] The founding of a university had always been the privilege of the sovereign. Leiden University was indeed founded in the name of King Philip II of Spain, as the rightful Count of Holland. But after the abjuration of Philip in 1581, eventually the various provinces constituting the Dutch Republic seized that sovereignty for themselves. Thus the establishment of a university was now a prerogative of the provincial governments. Friesland became the first province where in 1585 by sovereign provincial initiative a university was founded. This also rekindled the desire to have some sort of higher education in Zeeland. In June 1590, the States of Zeeland took the first step to establish a so-called 'Illustrious School' or Athenaeum. In this academic institution, modelled after an example that had already existed in Ghent and elsewhere, only the first years of university training would be offered. In the spring of 1592 indeed a first lecturer was appointed. The Scot John Murdison (c. 1568-1605) became *praeceptor* (tutor) at the Middelburg Latin School, with the additional assignment to provide public lessons in natural philosophy; this in anticipation of the actual founding of an Illustrious School. From 1593 onwards these public lectures were given, not only by Murdison, but also by the rector of the Latin School, Jacobus Gruterus, a learned man who excelled in the classical languages. Disputations by students were also introduced and academic theses were published by the Middelburg city printer Richard Schilders (figs. 11.2 and 11.3). After Murdison's departure to Leiden in 1599, his lectures on philosophy were continued by Gruterus. After Gruterus's death in 1607 he was succeeded by one of his former students, Antonius Walaeus (1573-1639), who lectured in ethics, philosophy and astronomy.[24] Towards the turn of the century, Murdison and Gruterus were supplemented by the local pastor Johannes Isenbach (also known as Hitzenbach), who provided public lessons in theology. Still, Middelburg had to wait until 1610 before an Illustrious School was formally founded.

23 Huib J. Zuidervaart, 'Zeeuws pre-academisch erfgoed. Een filosofische disputatie uit Middelburg uit de late zestiende eeuw', *Zeeland* 18 (2009), pp. 50-56.

24 In 1619 Antonius Walaeus would become a professor at Leiden University. See: Willem Frijhoff, 'Zeelands universiteit. Hoe vaak het mislukte en waarom', *Archief KZGW* (1987), pp. 7-41, esp. p. 32, and Johannes Polyander, 'Oratio Funebris', in: Antonius Walaeus, *Opera omnia, I* (Leiden: Franciscus Hackius, 1643).

Figure 11.2 Abbey church in Middelburg

Drawing after Matthias Smallegange, Zeeuws Archief, Historisch-Topografische Atlas Middelburg, 284. The Abbey church consisted of the Nieuwe Kerk (New Church) and the Koorkerk (Choir Church). In 1592 an auditorium for public lectures was established in the part of the church between the Nieuwe Kerk and the Koorkerk, almost under the tower.

Figure 11.3 Title page of a disputation held in the Abbey church in Middelburg, 1597

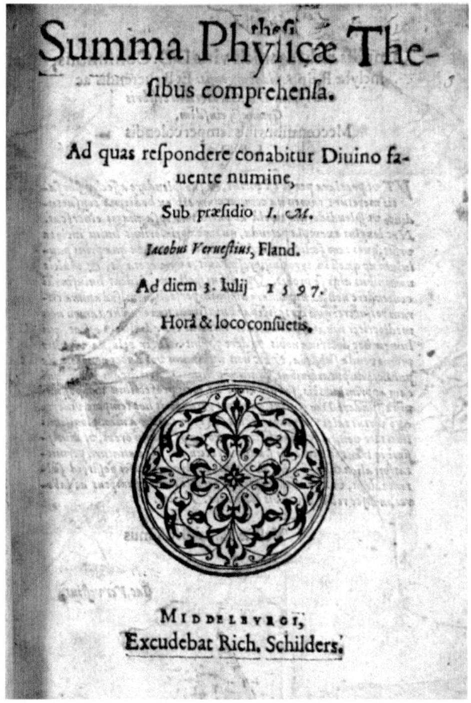

Summa Phyſicæ The-
ſibus comprehenſa.
Ad quas reſpondere conabitur Diuino fa-
uente numine,
Sub præſidio I. M.
Iacobus Verueſtius, Fland.
Ad diem 3. Iulij 1597.
Horâ & loco conſuetis.
MIDDELBVRGI,
Excudebat Rich. Schilders.

Trinity College, Dublin. The disputation was held by the student Jacob Vervestius on 3 July 1597. The presiding lecturer was John Murdison.

II A Fruitful Place for Scholarly Curiosity

Inquisitive Merchants and Their International Network

The Middelburg academic lectures of the 1590s were also attended by 'inquisitive merchants', as we learn from a letter by a brother of rector Gruterus.[25] Who these merchants were is not mentioned, but the mathematician Zacharias van Hoorebeeke in his *L'Art de tenir livre de comptes* (1599)

25 Reinier Gruterus to his brother Petrus Gruterus, 18 June 1592, in: Petrus Gruterus, *Epistolarum centuria secunda* (Amsterdam: Paulus Arnoldus, 1629), p. 218.

provides us with the names of the main merchants active in Middelburg.[26] Some of them – especially those who mastered the internationally used Latin language – will have attended these academic lectures. In Van Hoorebeeke's book we find the names of renowned Zeeland immigrant families, such as Adriaansen, Boreel, Braem, Cornelisse, De Hase, De la Palma, De la Rue, De Lobel, Franc, Le Clerc, Munnix, Sanders, Thibaut, Van der Goes, Verhouven and Vermeyden. These Middelburg merchants traded all over Europe.[27] In 1602 the Dutch East India Company was founded, with the important participation of Middelburg. As a consequence Middelburg became a place in which many natural 'rarities' could be obtained. For example, in the shop of Willem Jasperse Parduyn (c. 1550-1602), pharmacist in *Den Gouden Mortier* on the Middelburg Market (fig. 11.4),[28] items were sold such as alligators, corals, rare stones and 'unicorn' horns.[29] We know from his letters that Parduyn personally collected his rarities from ships that entered the Middelburg port. In some cases he also received curiosities through his brother Symon Jasperse Parduyn (d. 1612) (colour ill. 4), a very wealthy merchant and partner of the trading company of Balthasar de Moucheron, whose ships were among the first to explore the Far East.[30]

26 Zacharias van Hoorebeeke, *L'Art de tenir livre de comptes ou de raison, contenant le train de marchandise par divers pais et villes capitales de l'Europe.* [...] *Ensemble la correspondance des changes de Middelbourg* (Middelburg: Simon Moulert, 1599).

27 Van Hoorebeeke mentions the following places with which Middelburg merchants traded: Amsterdam, Antwerp, Bordeaux, Castillia, Danzig, Frankfort, Genoa, Hamburg, Lisbon, Livorno, London, Mechelen, Rome, Rouen, Seville and Venice.

28 Willem Jasperse Parduyn mentions the location 'Markt' in a letter to Carolus Clusius, dated 1593. *De Gouden Mortier* is also mentioned on that location in the Middelburg Haardstedenregister of 1601, but then occupied by his successor, the apothecary Daniel Snellaert. In this same register Willem Jasperse Parduyn is mentioned as living in the Korte Delft (nowadays Damplein). In 1606, when Parduyn's orphans lived in this house, the name *De Gouden Mortier* appears to have moved to this site. Today the house there still bears this name. In my earlier Dutch article on this topic, I made the mistake of identifying the former apothecary *De Morinne* in the Korte Delft, where today a gilt mortar can still be seen, with Parduyn's 1606 location. For that building, which was a grocer's shop in the 1600s, see: *Nehalennia* 33 (autumn 1979), pp. 9-15.

29 Matthias de L'Obel, *Den Leytsman ende Onderwijser der Medicijnen, oft ordentlicke uytdeylinghe ende Bereydingh-boeck vande Medicamenten* (Amsterdam: Hendrick Laurensz., 1614), 144. This book was originally published in Middelburg in 1596 (no publisher indicated), with a dedication to the Middelburg city council. Only one copy of this first printing has survived. See: Andrew Pettegree and Malcolm Walsby, eds., *Netherlandish Books: Books Published in the Low Countries and Dutch Books Printed Abroad before 1601*, 2 vols. (Leiden: Brill, 2010), no. 8899. See also: Peter M.N. Eldering, 'Middelburgs biologisch onderzoek in de zeventiende eeuw', *Archief KZGW* (1987), pp. 87-102

30 Symon Jasperse Parduyn was one of the most influential merchants in Middelburg. He was also mayor of the city. Just like his brother, the pharmacist, he too had a 'flower garden'. See

Figure 11.4 North-eastern part of the Middelburg Market in 1605

Detail of a drawing made in 1783 after a now lost painting by N. de Bast in 1605. Zeeuws Archief. Zelandia Illustrata II-229. The house with the arrow is *De Gouden Mortier* (The Golden Mortar), the apothecary shop of Willem Jasperse Parduyn.

Merchants as Botanical Enthusiasts

The Parduyn brothers owned a garden in which they cultivated exotic flowers and herbs. Especially Willem Parduyn manifested himself in the 1590s as the spider in a web of botanical enthusiasts from Middelburg, consisting of pharmacists, physicians and merchants. We know their names thanks to numerous letters that were exchanged with Carolus Clusius, the founder of the Leiden *hortus botanicus*. The Middelburg enthusiasts supplemented the

Cornelis de Waard, entries 'Willem Jasperse Parduyn' and 'Simon Jasperse Parduyn', in: *Nieuw Nederlandsch Biografisch Woordenboek*, 10 vols. (Leiden: Sijthoff, 1911-1937) [henceforth *NNBW*], III (1914), cols. 958-959; F. Hunger, 'Acht brieven van Middelburgers aan Carolus Clusius', *Archief ZGW* (1925), pp. 110-133, esp. p. 112.

Leiden botanical garden with numerous exotica from East and West.[31] Apart from Parduyn, Clusius corresponded with several others in Middelburg, for instance, the pharmacist Thomas de la Fosse,[32] the reverend Johannes de Jonghe,[33] and the banker Louis de la Bistrate.[34] The importance of the international mercantile network is nicely illustrated by the example of the Middelburg shipowner Jacques Noirot, who introduced himself to Clusius as a 'young lover of flowers'. Noirot lived in Middelburg in a large mansion at the Balans, near the Koopmansbeurs. He had brothers living in Antwerp, Naples and Venice.[35] At Parduyn's request, Noirot arranged for the dispatch of Clusius's letters to Seville in Spain, a city difficult to reach, as being in enemy territory. The Middelburg merchant Arnout Verhouven (one of the founding merchants of the Dutch East India Company) played a similar role.[36] Remarkable, too, is the role of Johan Somer, the son of a former Middelburg bailiff, who in 1597 sent Clusius the seeds and a 'contrefeytsel' ('image') of a flower that had bloomed in his garden.[37] Somer enjoyed a certain fame as the result of his journey to the Middle East in the years 1590-1592, in

31 All letters mentioned can be found online. See: Esther van Gelder, ed., *The Clusius Correspondence: A Digital Edition-in-progress* (2015), http://clusiuscorrespondence.huygens.knaw.nl. See also: Hunger, 'Acht brieven'. Eldering, 'Middelburgs biologisch onderzoek'; Anne Goldgar, *Tulipmania: Money, Honor, and Knowledge in the Dutch Golden Age* (Chicago: University of Chicago Press, 2008), pp. 20-29; Florike Egmond, *The World of Carolus Clusius: Natural History in the Making, 1550-1610* (London: Pickering & Chatto, 2010), pp. 143-155; Sylvia van Zanen, *'Een buitengewone verscheidenheid'. Uitwisseling van kennis en materiaal in het netwerk van Carolus Clusius* (PhD diss., Leiden University, 2016).

32 De la Fosse to Clusius, 31 August 1589; De la Fosse to Clusius, 12 July 1596. De la Fosse lived in the house on the corner of the Giststraat (today Dam) and the Molstraat. In 1601 this house was called *Den Gouwen Pot*, and in 1606 *Den Blompot*. His first letter to Clusius dates from 1589, when he still lived in Tournay (Doornick).

33 De Jonghe to Clusius, 14 May 1596. Johannes de Jonghe was a minister in Middelburg since 1583. In 1601 and 1606 he lived just outside the city, near the Seispoort.

34 Bistrate to Clusius, 2 August 1600. He is also mentioned in: Thomas Chauvijn to Clusius, 30 June 1599. Louis de la Bistrate (Van der Bistraeten (1578-c. 1610) from Valenciennes was married to Suzanne de Malapert. He was a cousin of the Middelburg merchant Samuel Godin. Their houses could not be located in the registers of 1601 and 1606. In the latter year an Anthony Godin lived in a house at the corner of the Noordstraat and the Lombardstraat.

35 Jacques Noirot to Clusius, 6 February 1601. At that time Noirot lived in *De Vergulde Peere*, still the largest house at the Balans. In 1606 he left Middelburg. Noirot was probably a relative of Jean Noirot, who in 1580 became the first master of the Zeeland Mint.

36 Willem Jasperse Parduyn to Clusius, 1 November 1596. See also: Enrique Hoons from Antwerp to Clusius, 15 May 1596. Arnout Verhouven lived in the mansion *Leeuwenburch*, in the Lange Noordstraat at the corner of the Blindenhoek. This large house was partly destroyed during riots in 1787, then owned by the physician Lucas van Steveninck.

37 Jehan (or Johan) Davidsz Somer to Clusius, 8 May 1597. In 1601 he lived in the Nieuwstraat, adjacent to *Den Voghel Phenix*. After his death in 1605, his widow Elisabeth Joosdr van de Vivere (d. 1620) married the physician Petrus Gruterus, brother of Jacob Gruterus, the late rector of

which period he had been imprisoned for some time.[38] On his return to Middelburg, Somer was one of the first in the Netherlands to import a variety of tulip bulbs and other plants, reason why he frequently is mentioned by Clusius in his *Rariorum plantarum historiae*, published in 1601.[39] Clusius's correspondence with Tobias Roels (c. 1565-1605), city doctor of Middelburg from 1591 to 1602, was also very intensive.[40] One of their letters, discussing plants from South America, is published in full in Clusius's *Rariorum*.[41]

Botanical enthusiasts are also mentioned in the books of the renowned physician Matthias de L'Obel (1538-1616), also known by his scholarly name Lobelius (fig. 11.5).[42] De L'Obel was the former court physician of Prince William of Orange, and was a city doctor in Middelburg from 1585 to 1596. He lived at the Korte Burg since 1590 (fig. 11.10, location no. 3), in one of the houses he built on the empty land of the former Bogarden monastery.[43] In later years the beautiful Renaissance garden behind that building, laid out by De L'Obel, was drawn several times by the painter Adriaen van de Venne (fig. 11.6). In Middelburg De L'Obel made his acquaintance with many *curiosi* on the island of Walcheren. He not only described plants of several botanical enthusiasts, such as the merchants Michiel de Lannoy,[44] Cornelis

the Middelburg Latin School. See: C. de Waard, entry 'Petrus Gruterus', in: *NNBW*, III (1914), col. 508-509.

38 See for this journey: Jan Somer, *Beschrijvinge van een zee ende landt reyse naer de Levante, als Italien, Candyen, Cypres, Egypten, Rhodes, Archipelago, Turckyen, en wederom door Duytslant* (Amsterdam: Joost Hartgers, 1649).

39 Carolus Clusius, *Rariorum plantarum historiae* (Antwerp: Johannes Moretus, 1601), pp. 133, 153, 257. See also: Goldgar, *Tulipmania*, pp. 20-22.

40 Tobias Roels to Clusius, 6 Sept. 1590; 19 Feb. 1591; 1592; 10 April 1593; 2 Jan. 1594; 12 June 1596; May 1597. In 1601 Roels (d. 1605) lived in the house *De Sonnewijser* in the Lombardstraat, at the side of the Korte Burg, where he owned another house. His Middelburg relative 'the young' Johannes Roels is mentioned in a letter from Johan Somer to Clusius, 8 May 1597.

41 Clusius, *Rariorum plantarum historiae*, pp. cccxv-cccxx.

42 Petrus Pena and Matthias de L'Obel, *Dilucidae simplicium medicamenorum explicationes* (London: Thomas Purfoot, 1605), pp. 483-485, 505; Matthias de L'Obel, *Medici Insulanj* [...] *botanographi, sive plantarum historiae physicae, tam indigenarum & Britanniae inquilinarum, quam exoticarum scriptoris* (London: Thomas Purfoot, 1605), pp. 67, 137; Matthias de L'Obel, *Stirpium illustrationes* [ed. by William How] (London: Thomas Warren, 1655), pp. 83, 102, 156. See also: A. Louis, *Mathieu de L'Obel, 1538-1616. Épisode de l'histoire de la botanique* (Ghent: Story-Scientia, 1980); Eldering, 'Middelburgs biologisch onderzoek'. The genus *Lobelia* is named after De L'Obel.

43 According to the Middelburg Haardstedenregister, De L'Obel owned three houses on the Korte Burg in 1601. In 1606 one of these was sold. His former house *Het Blauw Lacken* was owned by bookseller Adriaen van de Vivere since 1599, who renamed it *The Bible*.

44 Pena and De L'Obel, *Dilucidae simplicium medicamentorum*, p. 503. In 1601 Michiel de Lannoy owned the house *Het Groene Wout* in the Giststraat. However, he himself lived in the narrow

Figure 11.5 Matthias de L'Obel

Engraving by Francis Delaram, 1615. London, National Portrait Gallery. De L'Obel was city physician of Middelburg between 1585 and 1596. He then moved to London to become court physician to King James I.

Figure 11.6 Matthias de L'Obel's garden, the Lauwer-hof

Drawing from Adriaen van de Venne, *Tafereel van Sinne-Mal* (Middelburg, 1623), p. 1. The garden, behind the shop of Van de Venne brothers on the Korte Burg and adjacent to the Koopmansbeurs (Stock Exchange), was designed by De L'Obel.

Coorne,[45] or the goldsmith Johannes Knibbe,[46] he also had a special eye for the go-betweens who provided the exotic plants and rarities. Of course, De L'Obel mentions Willem Parduyn, but he points also to fellow pharmacist Thomas de la Fosse, living in *Den Gouwen Pot* at the Dam at the corner of the Molstraat. A few houses further on the Dam was the shop *St Andries Cruyse* of the chemist Cornelis Bouwens, where one could admire curiosities such as the spiral horn of the mythical unicorn, already correctly interpreted by De L'Obel as belonging to a 'fish with [a] great tooth' (that is, a narwhal).[47]

These botanical enthusiasts also had an impact on the Middelburg painters. Reverend Johannes de Jonghe, for instance, provided Clusius

Pijpstraat. His Middelburg relative 'the young' Jacques de Lannoy is mentioned in a letter from the Tournay botanist Jacques Plateau to Clusius, dated 8 February 1604.

45 Pena and De L'Obel, *Dilucidae simplicium medicamentorum*, p. 492. In 1601-1606 Cornelis Coorne lived at the east side of the Koepoortstraat.

46 Pena and De L'Obel, *Dilucidae simplicium medicamentorum*, pp. 485, 503, 505. The goldsmith Johannes (or Hans) Knibbe became member of the St. Luke guild in 1586. In 1601-1606 he lived on the west side of the Molstraat, in *De Gouden Hamer*. This was almost adjacent to the pharmacist-botanist Thomas de la Fosse, who lived on the corner of the Molstraat. See also: A. Bredius, 'Gildeboeken St. Lucas Middelburg', *Archief voor Nederlandsche Kunstgeschiedenis* 6 (1884-1887), pp. 106-264, esp. p. 159.

47 De L'Obel, *Den Leytsman*, 144. For De la Fosse, see above, notes 32 and 46.

with a drawing of a 'Tulip, depicted by the painter with a bare bulb'.[48] The unnamed painter was probably the then 22-year-old Ambrosius Bosschaert (1573-1621). He is the first in a long row of Middelburg painters of flower still lifes, called *blompots*.[49] These paintings bear witness to both the current culture of botanical collecting and those of accurate observation. The latter possibly even with a magnifying glass. We know that in later years Johannes Goedaert, one of the last Middelburg painters of *blompots*, used such lenses. In his insect study published in the 1660s, Goedaert writes that – as a rule – he draws the larva of an insect with the help of 'a magnifying glass'.[50] This practice could well be an old tried and tested habit.[51]

Lovers of Alchemy and Astronomy

The intellectual interest in Middelburg extended also to other fields of knowledge. Around 1600, the Middelburg publisher Richard Schilders issued two remarkable books on alchemy; books that clearly originated from a local setting. Until recently 'alchemy' had a rather negative connotation, but in the last decade historians have realized that the hands-on knowledge which accompanied alchemy and other crafts was decisive for the development of modern science. And indeed, in 1600 alchemy was still cutting edge knowledge.

The first alchemical book published by Schilders (in 1597) was the *Apologia chymica* by the old Middelburg physician Josephus Michelius (or Joseph Michels), born in Lucca (Italy). After his training in Paris in 1561, Michelius had settled in Antwerp, but, like so many Protestants, he fled to England after the fall of that city in 1585. Eventually he went to Middelburg, because his son worked there as a merchant.[52] In 1600, he would be appointed court

48 'met den bolle ontbloot, opdat hem de schilder mochte sien'. De Jonghe to Clusius, 14 May 1596.

49 Laurens J. Bol, 'Een Middelburgse Brueghel-groep', *Oud-Holland* 70 (1955), pp. 1-20, 96-109; Laurens J. Bol, *The Bosschaert Dynasty: Painters of Flowers and Fruit* (Leigh-on-Sea: Lewis, 1960); Noortje Bakker, I. Bergström, and G. Jansen, *Masters of Middelburg: Exhibition in the Honour of Laurens J. Bol* (Amsterdam: Kunsthandel K. & V. Waterman, 1984).

50 J. Goedaert, *Metamorphosis naturalis* (Middelburg: Jacobus Fierensius, [1662]), II, pp. 236, 272.

51 See also: Van Berkel, 'The City of Middelburg', p. 70: 'It is hard to imagine a painter like Bosschaert studying a fly or a caterpillar without the help of a magnifying glass. In the sixteenth and seventeenth century such an instrument was not uncommon in the textile industry.'

52 Biographical details on Michelius can be found in two printed letters to the Heidelberg professor Henricus Smetius, one dated 1 March 1601, written by Daniël Miverius, Michelius's former colleague as city physician of Middelburg, and the other, dated 8 February 1601, by Carolus Battus, city physician of Dordrecht. See: *Miscellanea Henrici Smetii* (Frankfurt: Jonas Rhodius, 1611), pp. 721-725.

physician to stadtholder Maurits van Nassau, so in that year he left for The Hague, where he died six months later.

In his *Apologia chymica* Michelius contradicted some of the views of the German physician Andreas Libau, better known as Libavius (1555-1616).[53] Michelius was a follower of the medical theories of Philippus von Hohenheim (1493/1494-1541), also known as Paracelsus, who based his ideas about illness and health on the need to maintain a balance between micro- and macrocosm. In this theory (attributed to the mythical Egyptian priest-deity Hermes Trismegistus) all kinds of physical, chemical and astronomical processes played a role. But Libavius was critical of those who were adherents of astrology and alchemy. For Libavius medication was above all else pure chemistry: the art of preparing and separating matter. In an extensive tract, dedicated to the Middelburg city council, Michelius opposed Libavius's disenchantment with the supernatural and also attacked the lofty tone of his writings. Libavius, however, returned this attack with a vigorous defence, which was also directed at the Middelburg city council.[54]

Michelius's example of writing an apology was followed in 1601 by the former city physician of Middelburg, Daniël Miverius, who had been Michelius's colleague in 1597.[55] His book, too, was published by Schilders in Middelburg and was again dedicated to its city council.[56] But Miverius's apology was foremost astronomical in character. After moving to Goes in 1598, Miverius

53 Michelius dedicated his *Apologia chymica* to Count Maurits van Nassau, the stadtholder of Holland and Zeeland. His preface to the Middelburg city council is dated 'Middelburg, 2 July 1597'. See: Bruce T. Moran, 'Eloquence in the Marketplace: Erudition and Pragmatic Humanism in the Restoration of Chymia', *Osiris* (2014), pp. 49-62, esp. pp. 52-53, and Bruce T. Moran, *Andreas Libavius and the Transformation of Alchemy: Separating Chemical Cultures with Polemical Fire* (Sagamore Beach, Mass.: Science History Publications, 2007), pp. 43-49.

54 Andreas Libavius, *Novus de medicina veterum tam Hippocratica, quam Hermetica tractatus. In cuius priore parte dogmata plaeraque inter utriusque professores recentes controvesa, adversus ultimum per Iosephum Michelium Paracelsitarum conatum discutiuntur* (Frankfurt am Main: Petrus Kopff, 1599). Preface dated 18 April 1598. Michelius was also attacked for his views by the French medical chymist Bernard Penot (c. 1522-c. 1617) in his *Apologia Bernardi G. Penoti,* [...] *ad Iosephi Michelii Middelburgensis medici scriptum* (Frankfurt: Jonas Rhodius, n.d. [1600]).

55 Daniël Miverius (d. 1602) was an physician of Flemish descent, born in London, who had studied in Heidelberg (1589), Leiden (1591-1593) and Padua (1594). In 1595 he practised in Vlissingen (Flushing). Being appointed as one of Middelburg's city physicians in 1597, he settled in the local Kerkstraat. In May 1598 he moved to Goes, where he obtained a similar appointment, but with a salary four times higher. See: C. de Waard, entry 'Daniel Miverius', *NNBW*, II (1912), col. 925-926, and Zeeuws Archief, Middelburg, lidmatenregister Vlissingen, K 481, fol. 11 (1595).

56 For this dedication to the Middelburg magistrate, dated Goes, 4 August 1602, Miverius's widow received a small reward at the end of 1602. See: H.M. Kesteloo, 'De stadsrekeningen van Middelburg V. 1600-1625', *Archief ZGW* 8 (1902), pp. 41-200, esp. pp. 33-34.

had met the pastor-astronomer Philippus Lansbergen there. In addition to medicine, Miverius had also studied mathematics, both in Heidelberg and in Leiden. When in 1601 his former Heidelberg teacher, the orientalist Jakob Christmann (1554-1613), criticized one of Lansbergen's theorems about spherical trigonometry published in Lansbergen's *Triangulorum geometriae libri IV* (Amsterdam 1591), Miverius decided (perhaps guided by Lansbergen) to refute him with an elaborate statement of defence.[57] Miverius's book also contained Lansbergen's first published astronomical observations, as well as a discussion of the 'Nova Zembla effect', witnessed in 1597 during the expedition of Willem Barentz. This reappearance of the Sun before it was expected according to astronomical calculations was correctly interpreted by Miverius as an optical phenomenon.[58]

However, the most interesting of the scholarly books published by Richard Schilders in these years is the *Opera mineralia*, a Latin translation of a Low German manuscript by the mysterious alchemist Isaac Hollandus, published in 1600 (fig. 11.7).[59] Annelies van Gijsen recently argued that Hollandus never existed and that this text is a convolute of several manuscripts that circulated in the middle of the sixteenth century, presumably in Flanders.[60] Nevertheless, Hollandus's text is of particular importance because it is the first to deal with the process of making enamel and artificial gems.

How this alchemical manuscript ended up in Walcheren and who took the initiative to translate it into Latin is unknown. Contacts with the Hoghelande family could have played a role. The Middelburg-born Theobald van Hoghelande (c. 1560-c. 1608), who lived in Cologne, was a well-known alchemist. Hoghelande had collected much alchemical information from all over Europe. Theobald was in regular contact with his brother Johan van Hoghelande in Leiden, a close friend of Clusius. From the correspondence between Clusius and the Middelburg city doctor Roels, we know that in the 1590s Johan van

57 Lansbergen's theorem was used in the description of the trajectory of the Sun. See Jacob Christmann, *Observationum solarium libri tres, in quibus explicatur verus motus solis in zodiaco* (Basel: L. Zetner, 1601), pp. 86-92 and Daniël Miverius, *Apologia pro Philippo Lansbergio ad Iacobum Christmannum* (Middelburg: Richard Schilders, 1602). In the preface Miverius tells his readers that he studied mathematics in Heidelberg with Professor Hermann Witekind [= Wilken, 1522-1603) and in Leiden with Professor Rudolph Snellius (1546-1613).

58 Isaac Beeckman refers to the optical explanation of both Daniël Miverius and Johannes Kepler in 1616. *JIB*, I, p. 99.

59 Joannes Isaac Hollandus, *Opera mineralia, sive de Lapide Philosophico, omnia, duobus libris comprehensa* (Middelburg: Richard Schilders, 1600).

60 Annelies van Gijsen, 'Isaac Hollandus Revisited', in: Miguel López Pérez, Didier Kahn, and Mar Rey Bueno, eds., *Chymia: Science and Nature in Medieval and Early Modern Europe* (Newcastle: Cambridge Scholars Publishing, 2010), pp. 310-330.

Figure 11.7 Joannes Isaac Hollandus, *Opera mineralia*, 160

MAGISTRI
IOANNIS ISAACI
HOLLANDI, VIRI IN
Philoſophia, potiſſimùm
verò in Chymia cele-
berrimi,

Opera Mineralia, ſiue de Lapide Phi-
loſophico, omnia, duobus libris
comprehenſa.

Nunquam antehac edita, ac nunc primùm ex
optimis manu-ſcriptis Teutonicis exem-
plaribus fideliſſimè in Latinum
ſermonem tranſlata, à
P. M. G.

MIDDELBVRGI,
Excudebat Richardus Schilders, Ordinum
Zelandiæ Typographus.
1600.

54 OPERVM MINERALIVM

Fixatio lapidis. Imbibitio lapidis cùm ſuo ſpiritu. Mercurio videlicet, qui in ſiccando aſcédebat, pro vt latius de ſale metallorum. Reductio lapidis ad perfectiorem virtutem per ſublimationé.

ſecreta donec omnino fixum ſit, ac penetret candentem laimnã æream, vt laminæ ſimiles fiant puræ lunæ: ac ſi tum etiàmnum perfectiorem vis reddere, ſuo ipſius ſpiritu immerge, & fixa ſuo ipſius ſpiritu, ac facto projectioné ad album. Sique filius meus hunc lapidé conficere vult ad album, poterit materiam ſublimandam ponere in hunc modum.

Sume ollam firmam amplam vitream quæ infra latum habeat fundum, eique materiam tuam inde, vt corám tibi facilè dicam, ac ollæ magnum impone alembicum cum amplo roſtro, receptaculumq; ei luto affixum. Erit autem inferior olla vitrea admodum humilis, plana & admodum lata in inferiore parte, vt hæc figura tibi demonſtrat, huic tuas materias præparatas inde, & alembico ſuperaddito, luto probè obtura, & fornaci ſublimationis impone, cum teſta cinerum, luto receptaculum roſtro adglutina, ac luto ſicco, paratus es ad ſublimandum. Qua de re quidam Philoſophus ſic ait: Terra eſt infimus mundus, & aqua quæ ſurſum aſcendit habet in ſe animã: nec ſunt plura quam 2 co-

Hollandus's *Opera mineralia* is a Latin translation of a manuscript on minerals and the philosopher's stone, with various illustrations, including this distillery apparatus, called alambic.

Hoghelande frequently visited Middelburg, where his sister Magdalena van Hoghelande lived in the impressive family mansion *De Witte Hazewind* (The White Greyhound) in the Lange Delft, one of the city's prominent streets.[61]

However it may be, Hollandus's *Opera mineralia* offers further insight into the knowledge community of Middelburg of around 1600. The book contains poems by three young men, all of whom in life would become rather well known. They are (1) Petrus Hondius (c. 1578-1621), who would become famous as a botanist and as the author of the poem *De Moufe-schans* (1619), describing a botanical garden near Terneuzen[62]; (2) Enoch Sterthemius (c. 1576-1626), at

61 The house *De Witte Hazewind* in the Lange Delft, was situated diagonally opposite the Kerkstraat (see fig. 11.10, no. 21). Magdalena van Hoghelande, widow of Jan Matens, bequeathed the house to the Middelburg hospital, the Gasthuis, in 1612. See: C. de Waard, entries 'Theobald van Hoghelande' and 'Johannes van Hoghelande', in: *NNBW*, II (1912), cols. 595-596.

62 J.G. Frederiks, 'Petrus Hondius, geboren te Vlissingen, omstreeks 1578, overleden te Neuzen in 1621', *Tijdschrift voor Nederlandsche Taal- en Letterkunde* 6 (1885), pp. 103-159; Ronald Rijkse, 'De Moffenschans van Petrus Hondius, 1-6', *Nehalennia*, no. 103 (1995), pp. 25-28; no. 105 (1995),

that time (1599-1600) *praeceptor* at the Middelburg Latin School, where he himself had been educated, and who in 1615 would become an influential minister, not only in Middelburg, but also overseas (in 1623 he was the first Calvinist minister sent with the Dutch fleet to the recently conquered territories in Brazil)[63]; and (3) Cornelis Herls (d. 1625), who called himself a student in medicine, becoming a member of the Middelburg surgeons' guild two years later (afterwards Herls wrote an authoritative textbook for surgeons that would be reprinted for more than a century).[64] The question of the identity of the translator, marked on the title page as 'P.M.G.' is solved by an entry in Petrus Hondius's *album amicorum*.[65] This translator was 'Petrus Montanus Gandavus' (or Piet van den Berghe from Ghent). He was Hondius's former tutor at the Latin School in Flushing (Vlissingen), but in 1600 he worked as vice-principal of the Latin School in Middelburg. In 1603 Montanus would move to Amsterdam where he continued his work as the translator of many more books.[66]

A last question to be solved is the identity of the signatory 'L.D.' who wrote the dedication of the *Opera mineralia* to George Eberhard, count of Solms (1563-1602), the commander of the Zeeland military. Name combinations with the initials 'L.D.' are very rare in the Middelburg Haardstedenregister of 1601. I only found one such combination, pointing to 'Louys Doutrelau', the Walloon minister of Middelburg since 1574. Louis D'Outreleau (d. 1606) was a former army pastor, so the dedication to the commander of the Zeeland army would fit him well. His son David D'Outreleau would become one of the four Middelburg city doctors in 1604. Another son, Jacob D'Outreleau, was a pharmacist.[67] (Al-)

pp. 3-6; no. 107 (1996), pp. 4-7; no. 108 (1996), pp. 16-20; no. 112 (1997), pp. 14-19; no. 115 (1997), pp. 18-24.

63 Huib Uil, 'Repertorium van onderwijsgevenden in Zeeland, 1578-1801', appendix in: Huib Uil, *'De scholen zijn planthoven van de gemeente'. Het onderwijs in Zeeland en Staats-Vlaanderen 1578-1801* (Bergschenhoek: Marberg Media, 2015), entry 'Sterthemius'.

64 See: A.A. Fokker, 'Philippus Lansbergen en zijne zonen Pieter en Jacob', *Archief ZWG* 1:5 (1863), pp. 52-100; Cornelis Herls, *Examen der chyrurgie* (Middelburg, c. 1625). Subsequent editions were published in Amsterdam in 1645, 1660, 1663, 1680 and 1723). The early editions of Herls's book were dedicated to the Middelburg directors of the East and West India Companies.

65 A. Meerkamp van Embden, 'Het album amicorum van Petrus Hondius, 1578-1621, predikant te Terneuzen, 1604-1621', *Archief ZGW* (1934), pp. 45-62, esp. p. 52. See also: L.A. Burgersdijk Jr., 'Speurtocht tusschen de bladen van het album amicorum van Petrus Hondius', *Archief ZGW* (1934), pp. 63-109.

66 W.M.C. Regt, entry 'Petrus Montanus', in: *NNBW*, X (1937), cols. 645-646; Uil, 'Repertorium', entry 'Montanus'.

67 In 1601 the French minister Louis D'Outreleau lived in de St. Pieterstraat adjacent to the Walloon church. His son, the physician David D'Outreleau, lived in the Lombardstraat in 1606, possibly in the former house of his predecessor as city physician Tobias Roels (d. 1605). The other son, the pharmacist

chemical interest was evidently strong in this family. An additional argument why Louis D'Outreleau could be this 'L.D.' is that it is most likely his library that was auctioned anonymously at Middelburg in February 1607.[68] The catalogue lists 925 books, the majority of which consisted of theological and philosophical works (63 per cent), but among the 81 books classified as 'medical' several alchemical classics can be found. Along with similar surviving book auction catalogues, this in itself bears witness to a clear local interest in such learned topics.[69] In summary, it is sufficiently clear from these examples that in Middelburg of around 1600 there was indeed a flourishing culture of scholarly curiosity, based on a network of international correspondents.

The Emergence of the Telescope in 1608

Middelburg around 1600 was not only a centre of botanical and alchemical interest, it was also the cradle of the telescope.[70] In September 1609 the Middelburg spectacle maker Hans Lipperhey (d. 1619) demonstrated in The Hague a spyglass that is generally considered to be the first operational telescope. This demonstration happened during the peace negotiations between the Dutch Republic and the Spanish king that led to the Twelve Years' Truce, so several foreign diplomats who attended the peace conference spread the news of this new device. Due to this dissemination of information within a year after Lipperhey's demonstration similar telescopes were in the hands of almost every monarch and scholar in Europe.

For quite some time it was a mystery why a useful telescope did not appear long before 1608. After all, it was already known for almost a century that two consecutive spectacle lenses could produce a magnified image.[71] The optical research of the Swiss historian of optics Rolf Willach

Jacob D'Outreleau, owned houses at the Balans and the Wagenaarstraat (1606). A probable third son, Louis D'Outreleau, is noted as a Middelburg schoolmaster in 1629. In 1633 he became a notary. A Steven D'Outreleau lived in the house *Den Olijfboom* in de Lange Delft (1601 and 1606). Uil, 'Repertorium'.

68 *Catalogus librorum theologicorum, iuridicorum, medicorum, mathematicorum, historicorum, & philosophicorum, qui vendentur Middelburgi ad diem 26 Februarij 1607* (University Library Heidelberg, F 9801:7). Although the catalogue is issued anonymously (which was a rather common habit), the contents point to a provenance of a minister. Louis D'Outreleau is the only Middelburg reverend who died in 1606, which makes him a good candidate for being the original owner of these books.

69 The above-mentioned Heidelberg convolute also contains auction catalogues of the Middelburg libraries of Pieter Herquebout [= Erckebout] (1605) and Anthonio Taymont (1606). The latter catalogue in particular contains a relatively large number of scientific works, including various botanical manuals.

70 Van Berkel, 'The City of Middelburg'.

71 Rolf Willach, 'The Long Road to the Invention of the Telescope', in: Albert Van Helden et al., eds., *The Origins of the Telescope* (Amsterdam: KNAW, 2010), pp. 93-127. See also: Peter Louwman and Huib J. Zuidervaart, *A Certain Instrument for Seeing Far: Four Centuries of Styling*

Figure 11.8 Telescope with a diaphragm

Rijksmuseum Amsterdam. Detail of the print *Besiet u selven* (Look at yourself), published in Amsterdam by Johannes de Ram, c. 1670, after an earlier drawing by Adriaen van de Venne, who lived in Middelburg until 1625. The small aperture, the diaphragma, at the front of the telescope, is clearly recognizable.

has brought forward a plausible explanation. Willach's optical measurements of early spectacle lenses in many European museum collections revealed that the shape of these early modern lenses was always faulty at the edges. A combination of two of these early lenses can only produce a sharp image when the front lens is covered with a diaphragm, leaving only a small aperture around the centre of the lens (fig. 11.8). Even though with such a diaphragm the intensity of the light was reduced, the resulting image became sharper and therefore useful. So, according to Willach, it is not so much the discovery of the telescope, but rather the discovery of the effect of a diaphragm that enabled the instrument to become a success.[72] This also explains the quick spread of the telescope across Europe. After all, spectacle makers lived everywhere, and one only needed to know the trick of the small aperture in order to copy the instrument.

But what brought Lipperhey to the step of applying a diaphragm? Was it mere chance? Or, was it the result of a series of experiments? One has to

the Telescope, Illustrated by a Selection of Treasures from the Louwman Collection of Historic Telescopes (Wassenaar: Louwman Collection of Historic Telescopes, 2013), p. 12.

72 See also: Huib J. Zuidervaart, 'The "True Inventor" of the Telescope: A Survey of 400 Years of Debate', in: Van Helden et al., eds., *The Origins of the Telescope*, pp. 9-44.

realize that the application of a diaphragm is completely counterintuitive. By partially covering the lens, the image becomes darker, but strangely enough it becomes also sharper! As long as the optics of a lens was not properly understood, this phenomenon seemed supernatural. This is precisely the context in which we have to place the origin of the Zeeland telescope. Initially it was foremost a magical instrument. In the first years no one understood how the instrument worked. There was hardly any knowledge about the refraction of light in glass. The behaviour of light was almost exclusively investigated in the context of hollow (burning) mirrors. It is not surprising that Arjen Dijkstra, in his study of early modern mathematics at Franeker University, concluded that the early telescope received foremost attention in alchemical circles.[73] In Franeker, this probably concerned the instrument made by the Alkmaar optician Jacob Adriaensz Metius, brother of the Franeker professor of mathematics Adriaen Metius, who also claimed to have invented the telescope.[74] Related to this Franeker alchemical interest was Isaac Beeckman's first attempt to explain the operation of the telescope. Tiemen Cocquyt has shown that Beeckman, in 1618, also assumed that light rays in a telescope were deflected from their straight path, using a principle borrowed from the metaphysics of light. This is in accordance with the illustration published in Girolamo Sirtori's book *Telescopium*, published that very year (see fig. 5.2).[75]

This focus on magic and exotic phenomena is not at all surprising. In the second volume of the recent *History of Zeeland*, it is described how intensely society in Zeeland in the seventeenth century was coloured by religious motives and how the natural and the supernatural world were constantly intertwined.[76]

73 Arjen Dijkstra, *Between Academics and Idiots: A Cultural History of Mathematics in the Dutch Province of Friesland (1600-1700)* (PhD diss., University of Twente, 2012), pp. 151-154. See also: Fokko Jan Dijksterhuis, 'Magi from the North: Instruments of Fire and Light in the Early Seventeenth Century', in: Arianna Borelli, Giora Hon, and Yaakov Zik, eds., *The Optics of Giambattista Della Porta (ca. 1535-1615): A Reassessment* (Cham: Springer, 2017), pp. 125-143.

74 In October 1608 Jacob Metius also made a claim to have invented the telescope. However, at the time he was unable to provide a properly working spyglass. Nevertheless, thanks to the influence of his brother, the Franeker professor, Jacob Metius was long (but erroneously) regarded as the prime inventor. For more detail, see: Zuidervaart, 'The "True Inventor" of the Telescope', pp. 19-21, and idem, 'The Invisible Technician Made Visible: Telescope Making in the Seventeenth and Early Eighteenth-Century Dutch Republic', in: Alison D. Morrison-Low et al., eds., *From Earth-Bound to Satellite: Telescopes, Skills and Networks* (Leiden: Brill, 2012), pp. 41-102, esp. pp. 60-64.

75 See the chapter by Tiemen Cocquyt elsewhere in this volume. See also: Girolamo Sirtori, *Telescopium, sive ars perficiendi novum illud Galilaei visorium instrumentum ad sydera* (Frankfurt: Apud Pauli Jacobi, 1618), and *JIB*, I, p. 208.

76 Arno Neele, Katie Heyning, and Clasien Rooze-Stouthamer, 'Religie en cultuur', in: Paul Brusse and Wijnand Mijnhardt, eds., *Geschiedenis van Zeeland, Vol. 2: 1550-1700* (Zwolle: Wbooks,

Moreover, the telescope did not originate from scratch. The search for an instrument that could provide magnification had already been going on for decades.[77] In the sixteenth century, a legendary story (wrongly supposed to come from Antiquity) was used as a guideline. According to this myth, large concave mirrors had been used at the lighthouse in Pharos, the port of Alexandria in Egypt, to locate ships from afar. These ships even were set on fire by these huge mirrors. So, from sixteenth-century Europe, there are several reports about attempts to make such magnifying devices. But almost always these attempts concerned a combination of a concave mirror and one or more lenses.

The first known references to the promise of a magnifying instrument can be found in the writings of magi, such as those of the Englishman John Dee (1527-1608) or the Italian writer Giambattista della Porta (c. 1535-1615). In his *Magia naturalis* of 1589 Della Porta described the enlarging effect of combinations of concave mirrors and lenses. He treated this topic more extensively in his *De refractione optices*, published in 1593.[78] These ideas were so common, that when Galileo first heard of the Dutch telescope, he assumed that it concerned a combination of a mirror and some lenses. Galilei had already experimented with such combinations, just as Della Porta and other Italian opticians.[79]

The Convex Mirrors of Wine Merchant Johan Radermacher

With this knowledge in mind, it is fascinating to note that some of Della Porta's books, including his *De refractione optices*, were present in the library of Johan Radermacher (1538-1617), also known as Rotarius (fig. 11.9).

Radermacher was a Flemish wine merchant who, after running an influential business as well as having administrative career in Antwerp and London, had settled in Middelburg in 1599, after having lived in Aachen in the first years after the fall of Antwerp.[80] Radermacher's fame

2012), pp. 199-295.

77 Eileen Reeves, *Galileo's Glassworks: The Telescope and the Mirror* (Cambridge, Mass.: Harvard University Press, 2008).

78 Borrelli, Hon, and Zik, eds., *The Optics of Della Porta*.

79 Reeves, *Galileo's Glassworks*.

80 Karel Bostoen, *Bonis in Bonum. Johan Radermacher de Oude (1538-1617), humanist en koopman* (Hilversum: Verloren, 1998), p. 16. C. de Waard, entry 'Johan Radermacher', in: *NNBW*, II (1912), col. 1154. According to the Haardstedenregister of 1606 Radermacher lived in Middelburg on the Brakstraat, in the still existing house *De Schelpe* (The Shell). Karel Bostoen et al., *Het album J. Rotarii. Tekstuitgave van het werk van Johan Radermacher de Oude (1538-1617) in het Album J. Rotarii, handschrift 2465 van de Centrale Bibliotheek van de Rijksuniversiteit te Gent* (Hilversum: Verloren, 1999), erroneously places Radermacher in the house *De Globe* at the Rotterdamsche

as a humanist scholar reached far beyond the national borders. When the French scholar Nicolas Claude Fabri de Peiresc (1580-1637) learned that his brother would visit the Netherlands, he insisted that he should visit this 'homme docte' in Middelburg to admire his numismatic collection.[81] The learned Radermacher was also interested in astronomy and optics. The Zeeland astronomer Philippus Lansbergen calls Radermacher a 'great lover of astronomy'.[82] In one of his works Lansbergen published Radermacher's measurements of a solar eclipse, performed with a quadrant in 1605.[83] A surviving letter by Radermacher, written in August 1606, shows that he, too, longed for a concave mirror that 'makes things infinitely bigger than they really are'.[84] At the time, Radermacher expected to receive two such mirrors from his friend, the London silk trader of Flemish descent Jacob Cool (1563-1628) (written in English as 'Jacob Cole', or in Latin as 'Jacobus Colius Ortelianus' – a reference to his famous uncle, the Antwerp geographer Abraham Ortelius). Deborah Harkness has shown that this *mercator sapiens* was the pivotal figure in a large network of scholarly enthusiasts in London, near his house in Lime Street.[85] Cole also played a key role in contacts with kindred spirits elsewhere in Europe. Radermacher was highly regarded by Cole, as is evident from his dedication to Radermacher of his book *Den staet van London in hare groote peste* (a description of the London plague of 1604).[86] Cole visited Middelburg frequently.[87] At the end of 1606 he married

Kaay, built by his descendants in 1661 (destroyed 1940). See for his library: Anna E.C. Simoni, *The Radermacher Sale Catalogue* (2007), http://www.johanradermacher.net.

81 Philippe Tamizey de Larroque, ed., *Lettres de Peiresc* (Paris: Imprimerie nationale, 1896), VI, p. 678.

82 Philippus Lansbergen, *Verclaringhe vande platte Sphaere van Ptolemaeus anders Astrolabium genaemt* (Middelburg: Zacharias Rooman, 1628), p. 39. See also: Bostoen et al., *Het album Rotarii*, p. 103r.

83 Philippus Lansbergen, *Observationum astronomicarum thesaurus* (Middelburg: Zacharias Rooman, 1632), p. 118; French edition: *Tresor d'observations astronomiques* (Middelburg: Zacharias Rooman, 1633), p. 48. In 1606 Lansbergen composed a poem for Radermacher on a new star. See: Bostoen et al., *Het album Rotarii*, p. 103r.

84 Radermacher to Cole, Middelburg, 5 August 1606. See also: 14 August 1603, in: J.H. Hessels, ed., *Abrahami Ortelii (geographi Antverpiensis) et ad Jacobum Colium Ortelianum epistulae* (Cambridge: Typis Academiae, 1887), no. 331, 780-783; no. 335, 792-795.

85 Deborah E. Harkness, *The Jewel House: Elizabethan London and the Scientific Revolution* (New Haven: Yale University Press, 2007), pp. 19-24.

86 [Jacob Cool], *Den staet van London in hare groote peste in het jaer des Heeren MDCIII* (Middelburg: Richard Schilders, 1606). The book was edited by the London-based minister Johannes Regius, who had married a sister of Johan Radermacher. Modern edition: Jan A. Dorsten and K. Schaap, eds., *Jacob Cool, Den staet van London in hare groote peste* (Leiden: Brill, 1962).

87 See, for example, his inscription, signed 'Middelburg, 1602', in the *album amicorum* of Samuel Radermacher Johanzn (University Library Leiden). Samuel Radermacher lived in Hamburg, but

Figure 11.9 Johan Radermacher

Copper engraving. Zeeuws Archief, Zelandia Illustrata, IV-735. Johan Rademacher (1538-1617) was a wine merchant in Middelburg since 1599.

Louise de L'Obel, daughter of the former Middelburg physician Matthias de L'Obel (then living in London).

Radermacher knew De L'Obel very well. His Middelburg neighbour was a widow De L'Obel, probably a relative of the botanist.[88] It is telling that at the occasion of Cole's marriage, Radermacher did not congratulate Cole so much for his wife, but rather for his acquisition of a father-in-law with whom Cole could constantly converse about his botanical studies.[89] De L'Obel could well be the contact through whom Cole obtained the above-mentioned concave mirrors. A close friend of De L'Obel was Richard Forster, president of the famous London College of Physicians.[90] Forster was well acquainted with the Italian mathematician Giovanni Antonio Magini (1555-1617) in Bologna,

stayed often with his father in Middelburg.

88 The 'widow Lobel' – probably the widow of the merchant Jean de L'Obel, mentioned by Van Hoorebeeke in 1599 – lived in *De Robijn* at the west side of the Middelburg Brakstraat, adjacent to Radermacher's *De Schelpe*. One of their neighbours across the street, in *'t Gulden Brantijsser*, was the merchant Richard Tulleken, mentioned in a letter from the canon-botanist Joachim Levenier (Venerius) to Clusius, dated 18 February 1600.

89 Radermacher to Cole, 1 March 1607. Hessels, *Ortelianum epistulae*, no. 338.

90 De L'Obel, *Stirpium illustrationes*, p. 38. See also: Charles E. Raven, *English Naturalists from Neckam to Ray: A Study of the Making of the Modern World* (Cambridge: Cambridge University Press, 1947; 3rd ed., Cambridge: Cambridge University Press, 2010), p. 247.

famous as constructor of concave magnifying mirrors.[91] However it may be, it is rather remarkable that shortly before the invention of the telescope in Middelburg, Cole provided Radermacher with these intriguing magnifying mirrors. Knowing this new fact, a testimonial of the Italian scholar Girolamo Sirtori, written down in 1612, may be seen in a different light. After the invention of the telescope in 1608, Sirtori tried to discover the 'secret' of this new instrument. Therefore, Sirtori also visited the Netherlands. He even claims to have examined Lipperhey's first telescope, which he found 'better and more suitable' than many others he had seen. About the invention itself Sirtori wrote:

> In the year 1609 [read 1608] some creative spirit [Lat., *genius*] appeared in Middelburg in Zeeland. I do not know who, but one from the Low Countries. He visited Johannes Lippersein [sic!], a spectacle maker and a man distinguished from others by his remarkable appearance. There was no other spectacle maker in that city, and the unknown man ordered several lenses with him, both concave and convex. On the agreed day the man returned, eager for the finished work, and as soon as he had these [lenses] before him, he took two of them, namely a concave and a convex one, and he kept one and the other in front of his eye and moved them slowly back and forth, either to test the assembly point or the craftsmanship, and then he left, after he had paid the maker. The craftsman, who was anything but stupid, was curious about the novelty and began to do the same by imitating the customer. His ingenuity suggested that these lenses should be put together in a tube. And as soon as he had completed one, he rushed to the court of Prince Maurice and showed him the invention.[92]

What if this story about the *genius* that came to Lipperhey is true? It is tempting to identify Johan Radermacher as this 'genius'. If Radermacher was engaged in experiments with mirror-lens combinations – just as other contemporary scholars – then such an order of 'several lenses, both concave and convex' would make sense. Unfortunately, we can only establish that Radermacher had the interest, the knowledge and the possible intention of carrying out such experiments. Moreover, in Della Porta's *De refractione*

91 Sven Dupré, 'Ausonio's Mirrors and Galileo's Lenses: The Telescope and Sixteenth-Century Practical Optical Knowledge', *Galilaeana* 2 (2005), pp. 145-180.

92 Sirtori, *Telescopium*, pp. 23-24. English translation: Albert Van Helden, 'The Invention of the Telescope', *Transactions of the American Philosophical Society* 67 (1977), pp. 1-67, esp. pp. 48-51. Although Sirtori's book was published in 1618, his manuscript already circulated in 1612. See: Frederico Cesi to Galileo, 28 October 1612, in: A. Favaro, ed., *Le Opere di Galileo Galilei*, edizione nazionale (Florence: Edizione nazionale, 1901), XI [1611-1613], p. 420.

optices not only mirror-lens combinations were discussed; there was also an early example of the camera obscura, first as the well-known pinhole camera and later one provided with a convex lens.[93] This device comes really close to the concept of the diaphragm! So, could it be that the knowledge of a pinhole camera was also transferred to Lipperhey by means of this visiting Dutch 'genius'? Of course, we will never know. We do know, however, that in 1609 Lipperhey managed to meet the difficult request of the States General's delegates to make several telescopes suitable for two eyes. A later visitor also testified that Lipperhey improved his first telescopes with a magnification four times larger than his first spyglass.[94] So, Lipperhey, in his own right, must have been a rather clever artisan. In the light of the foregoing, the assumption that he was able to lean on a network of scholarly interested people seems not too daring.[95]

III A Location Favourable for an Easy Mingling of Scholars, Merchants and Craftsmen

With the aforementioned examples, it is convincingly demonstrated that in Middelburg, around 1600, an interaction between merchants, scholars, physicians, surgeons and other artisans was amply present. When several persons from this 'inquisitive' group of scholarly interested people are plotted on a map of Middelburg with the street plan from before World War II (fig. 11.10), it is striking how close together the locations were where they could meet each other. This was definitely the case in the area around the former Abbey: this site was a real meeting place for merchants, scholars and artisans.

On this map we can point out the locations of: (no. 1) the Latin School, where the rector, Jacobus Gruterus, lived and where the academic public lectures were prepared. In the adjacent Lombardstraat lived (no. 2) city doctor Tobias Roels, Clusius's main correspondent, in the house *De Sonnewijser*, near the Korte Burg.[96] Roels also owned a house around the corner, almost adjacent

93 Reeves, *Galileo's Glassworks*, pp. 85-86.
94 This message was transferred to Galileo through a befriended physician, Alfonso Strozzi from Venice. See: Filippo Salviati (Verona) to Galileo Galilei (Florence), 13 November 1613, in: Favaro, *Le Opere di Galileo Galilei*, XI, p. 595. See also: Cornelis de Waard, *De uitvinding der verrekijkers. Eene bijdrage tot de beschavingsgeschiedenis* (The Hague: Nederlandse Boek- en Steendrukkerij, 1906), p. 273.
95 On Lipperhey's personal network, see: Zuidervaart, '"Scientia" in Middelburg', p. 75 (see note * on p. 261).
96 Roels had inherited this site from his father, the pensionary Willem Roels. See for details: P.K. Dommisse, 'Onderzoek naar de eerste omwalling en omgeving der stad Middelburg', *Archief ZGW* (1904), pp. 1-209, esp. pp. 97-98.

Figure 11.10 Map of the inner city of Middelburg (area around the Abbey), 1873

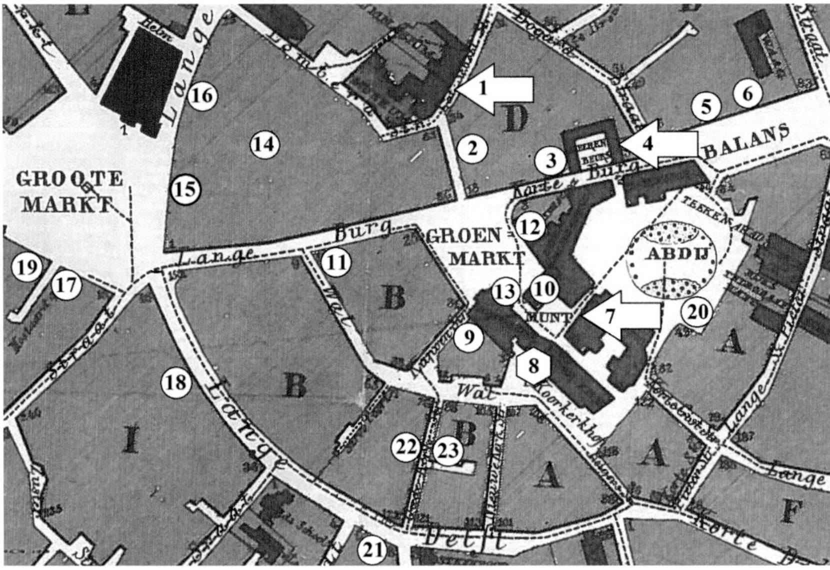

Detail from Nieuwe plattegrond der stad Middelburg by T.P. Roest, 1873. Zeeuws Archief, Historisch-Topografische Atlas Middelburg 15-I. The late nineteenth-century street plan matches the situation around 1600. The numbers indicate the various locations of houses and institutions discussed in the text. For this purpose, the tax registers of 1601 and 1606 were used.

to (no. 3) the home of his colleague city doctor Matthias de L'Obel. After De L'Obel's departure for England in 1596, this house was transformed by Adriaen van de Vivere into the bookshop *In de vergulde Bijbel*,[97] continued after his death in 1618, in the left part by his heirs, and in the right part by the brothers Jan and Adriaen Pietersz van de Venne, who exploited here a paintings shop (*schilderijenwinckel*), complemented by a printing press in 1622 (colour ill. 5).

Next to this shop was (4) the *Koopmansbeurs*, built in 1592, where the Zeeland merchants did their business (fig. 11.11). On the adjacent Balans was (5) *De Vergulde Peere*, the still impressive home and garden of the botanical enthusiast Jacques Noirot.[98] In 1601 the mathematician Johan Coutereels, known for his many textbooks on accounting, lived two houses further

97 Adriaen van de Vivere 'van Duijsburch bouckbinder' became *poorter* (citizen) of Middelburg in December 1592. H.M. Kesteloo, 'De stadsrekeningen van Middelburg, IV, 1550-1600', *Archief ZGW* 7 (1894), no. 1, pp. 1-182, esp. p. 95.

98 Hunger, 'Acht brieven', p. 130. Noiret still lived here in 1601. In 1606 the house was owned by the merchant Pyeter van Peenen, from Roeselaere. See: Haardstedenregister 1601, fol. 79r.; Haardstedenregister 1606, fol. 53r.

Figure 11.11 The Middelburg Stock Exchange or Koopmansbeurs

From: Matthias Smallegange, *Nieuwe Cronyk van Zeeland* (Middelburg, 1696). The Middelburg Stock Exchange (no. 4 in figure 11.10) was built in 1592 on the Korte Burg. On the left side, a small part of the printing shop of Pieter Janz van de Venne (no. 3 in figure 11.10 and colour ill. 5) can be seen, whose windows overlook the stock exchange.

(6).[99] From the *Koopmansbeurs* one could walk to the Groenmarkt via the courtyard in the Abbey and the adjacent cloister corridor. In that corridor, trade was not only carried out in stalls, but here, too, (7) the Zeeland Mint was located, where metalworking was combined with artistic design. In the buildings around the cloister corridors were also the melting furnaces for making alloys. Next to this is the Mint Square, where artisans such as essayers, mint cutters and balance makers had their domicile.

Through a still existing door one had direct access from the cloister corridor (no. 8 in fig. 11.10) to the auditorium in the Nieuwe Kerk (later the Illustrious School), where the academic lectures were held. On the other side, in the Kapoenstraat, there was (no. 9 in fig. 11.10) the house and workshop of spectacle maker Hans Lipperhey (fig. 11.12 and colour ill. 6A). On the adjacent Groenmarkt was (no. 10 in fig. 11.10) the home of the Zeeland mint master (until 1601 Jacob Boreel and from 1601 to 1612 his successor, the art collector Melchior Wyntgens).[100] Boreel, in 1602 one of the founders of the Dutch East

99 This was his address in 1601. Two years later Coutereels lived at the Market, in the house *De Hooge Deure*. In 1606 he then moved to the large house *Koning Solomon* in the Gravenstraat (in Isaac Beeckman's back yard). In 1610 Coutereels moved again, this time to the Latijnsche Schoolstraat. In 1613 he moved his French school to Arnemuiden. See: C. de Waard, entry 'Johan Coutereels', in: *NNBW*, I (1911), col. 644, and Klaas Hoogendoorn, *Bibliography of the Exact Sciences in the Low Countries from ca. 1470 to the Golden Age (1700)* (Leiden: Brill, 2018), p. 260.

100 M.G.A. de Man, 'De verpachting der Zeeuwsche munt in 1601', *Jaarboek voor Munt- en Penningkunde* (1916), pp. 178-182; M.G.A. de Man, 'De grafelijke Munt van Zeeland en de balansmeesters die er voor hebben gewerkt', *Jaarboek voor Munt- en Penningkunde* (1922), pp. 41-69, esp. pp. 46-47.

Figure 11.12 Hans Lipperhey

Engraving in: Pierre Borel, *De vero telescopii inventore* (The Hague, 1655). Zeeuws Archief, Zelandia Illustrata, IV-597

India Company, lived then in *Het Pater Noster* (no. 11 in fig. 11.10), a house on the south side of the Lange Burg, towards De Wal. In 1655 it was Jacob Boreel's son Willem – then ambassador of the Dutch Republic in France – who would initiate an investigation into the invention of the telescope.[101] On the Groenmarkt was also 's Lands Giethuys (the present Statenzaal) (no. 12 in fig. 11.10), where cannons and church bells were produced with metal furnaces operated by the Burgerhuys family (fig. 10.13).[102] It is telling that this place was referred to as the 'Hell'.[103]

On the corner of the Groenmarkt also lived the silversmith Fredrick Muntynchx, who in March 1604 claimed to have witnessed in his house the alchemical transfer of lead into gold, performed by a Scottish nobleman.[104] Opposite the Giethuys, next to a gate to the Mint Square, was (no. 13 in fig. 11.10) the tiny house – built against the wall of the church – where Sacharias Jansen lived in his childhood, the man who in 1655 was posthumously proclaimed as being the first inventor of the telescope (fig. 11.14).

101 In his younger years, Willem Boreel enrolled at Leiden University on 12 October 1609, only two weeks after Isaac and Jacob Beeckman.

102 Jan (= Hans) Burgerhuys, the founder of the bronze foundry, also settled in the Kapoenstraat in 1593, to the left of the entrance to the Nieuwe Kerk. The watercolour by Tuyter from 1848 (colour ill. 6A) only shows a shed on that place. An earlier house can be seen on the engraving from 1796 (fig. 11.14). Cf. J.W. Enschedé, 'Het geslacht Burgerhuys, klok- en geschutgieters te Middelburg', *Bulletin Nederlandsche Oudheidkundigen Bond*, second series, 9 (1916), pp. 221-222, and W.S. Unger, 'Het klokkengietersgeslacht Burgerhuys', *Archief ZGW* (1926), pp. 19-29.

103 De Man, 'De grafelijke Munt van Zeeland', p. 47.

104 E. Wiersum, 'Alchemie te Middelburg in 1604', *Archief ZGW* (1918), pp. 119-120. See also Van Dixhoorn, elsewhere in this volume.

Figure 11.13 's Lands Giethuys with the adjacent home of the Zeeland Master of the Mint

Engraving from: Matthias Smallegange, *Nieuwe Cronyk van Zeeland* (Middelburg 1696). 's Lands Giethuys (today the Statenzaal) is no. 12 in figure 11.10 and the home of the Zeeland Master of the Mint is no. 10 in figure 11.10.

Figure 11.14 Groenmarkt with the Nieuwe Kerk (New Church)

Engraving from: *Zeeuwsche Chronyk Almanach* voor 1797, Zelandia Illustrata, II-548. In the houses against the wall lived Sacharias Jansen (no. 13 in figure 11.10) and Hans Lipperhey (no. 9 in figure 11.10; see also colour ill. 6A).

This claim was made by his son, the lens grinder Johannes Sachariassen, whose statements, however, have proven to be very unreliable.[105] Today, a nineteenth-century memorial stone in the church wall still bears witness to this serious falsehood. Since then, archival research has revealed that Jansen was a spectacle maker only since 1615, after having worked first as a peddler, and later as a counterfeiter of copper coins.[106] With the Zeeland

105 Johannes Sachariassen's claims are scrutinized in: Zuidervaart, 'The "True Inventor" of the Telescope', pp. 43-44. In 1601 and 1606 this house against the wall of the church was owned and occupied by the widow of Jan Claesse. Given the fact that this house was the shared property of Sacharias Jansen and his sister Sara in 1622, Jan Claesse's widow can be identified as their mother Maeyken Meertens (d. after 1610). Previously she had been married to Hans Martens, who died in 1592, and of whom is said that he lived 'between the pillars' of the Nieuwe Kerk. So, Sacharias Jansen indeed will have lived here as a child. De Waard assumes that Sacharias left his mother's tiny house after his marriage in 1610 and then moved to a house on the Koorkerkhof. De Waard, *De uitvinding der verrekijkers*, pp. 178, 322, 326.
106 Sacharias Jansen was convicted and charged a fine of '400 Carolus guldens' for the counterfeiting of coins in 1613. He returned to this criminal craft on a larger scale in 1619, after he had moved to the nearby city of Arnemuiden. There he was also accused of these criminal activities,

Mint as his neighbour Jansen evidently had well observed how coins could be made. Perhaps he did the same around the corner, at Lipperhey's workshop, spying how spectacle lenses were made.

In the vicinity was also the seat of the Middelburg chamber of rhetoric, a convergence point for literary culture.[107] Its meeting place (no. 14 in fig. 11.10) was in a house behind an inn on the Market, which could be reached through an alley near the homes of both the pharmacist Willem Jasperse Parduyn (no. 15 in fig. 11.10) and the flower painter Ambrosius Bosschaert (no. 16 in fig. 11.10).[108] On the other side of the Market was also the house *De Galeye* (17), used by the publisher and bookseller Richard Schilders as his workshop between 1593 and 1600 (fig. 11.15). Behind the adjacent tavern *De Groote Soutkeete* lived Petrus Montanus, the vice-principal of the Latin School and the translator of Hollandus's book on alchemy, the printing of which had occurred in *De Galeye* in the year 1600.[109] At the end that year, Schilders moved to *Den Olyphant* in the Lange Delft (no. 18 in fig. 11.10), where he remained active until the end of his career in 1618.

but Jansen escaped prosecution by fleeing outside Arnemuiden's jurisdiction, being helped by the local bailiff, who seems to have been one of Jansen's accomplices. Jansen returned to Middelburg in 1621, where he bought the house *Den Swarten Leeuw* in the Schuitvlotstraat in 1622. He moved to Amsterdam in November 1626, where he worked as a spectacle maker on Dam Square until his bankruptcy in May 1628. He died before 1632. His son Johannes Sachariassen (1611-after 1655) returned to Middelburg, where he was active as a *brilmaker* from 1630 onwards. See: De Waard, *De uitvinding der verrekijkers*, p. 120 (living at the gate of the Mint up to 1610); pp. 121-122 (counterfeiter in Middelburg); pp. 123-136 (counterfeiter in Arnemuiden). See also: Huib J. Zuidervaart, '"Uit vaderlandsliefde". Pierre Borel, *De vero telescopii inventore* (The Hague, 1655) en het negentiende-eeuwse streven naar een gedenkteken voor de "ware uitvinder" van de verrekijker', *Archief KZGW* (2008), pp. 5-58, esp. p. 51, n. 6, and p. 52, n. 20.

107 According to *Zeelands chronyk almanach* (1786), 1020-1024, the Prinsenzaal was built or refurbished on this location in 1675, but surviving notes by Abraham Bredius, made in the Middelburg archives before 1940, reveal that the chamber of rhetoric was already housed at this location in 1619. See Bredius's notes at the RKD-Nederlands Instituut voor Kunstgeschiedenis in The Hague. See also: M.G.A. de Man, 'De voormalige Middelburgsche rederijkerskamer Het Bloemken Jesse onder de kenspreuk "In minnen groeiende" en hare gildepenningen', *Jaarboek voor Munt- en Penningkunde* (1917), pp. 1-40, esp. pp. 19-20, and Arjan van Dixhoorn, *Lustige Geesten. Rederijkers in de Noordelijke Nederlanden (1480-1650)* (Amsterdam: Amsterdam University Press, 2009).

108 Bosschaert lived in Middelburg until around 1615, first with his father in the Noordstraat 'across from the town hall'. After 1611 he moved to the south side of the St. Pieterskerk. Cf. A. Bredius, 'De bloemschilders Bosschaert', *Oud-Holland* (1913), pp. 137-140, esp. p. 138.

109 The Haardstedenregister of 1601 places 'Pieter Monthanes' behind the tavern *De Groote Soutkeete*, adjacent to *De Galeye* at the Middelburg Market, the house where the publisher and bookseller Richard Schilders resided between 1593 and 1600 (see fig. 11.4). At the end of 1600 Schilders moved to *Den Olyphant* in the Lange Delft. In my article '"Scientia" in Middelburg' I misread the Haardstedenregister, concluding erroneously that Montanus also lived in *De Galeye*. See also: Paul J. Aarsen, 'Richard Schilders, de eerste (?) drukker in Middelburg', *Nehalennia* 174 (2011), pp. 2-7.

Figure 11.15 South-eastern part of the Middelburg Market in 1605

Detail of a drawing made in 1783 after a now lost painting by N. de Bast in 1605. Zeeuws Archief, Zelandia Illustrata, II-229. The house with the arrow is *De Galeye* (no. 19 in figure 11.10), later called *De Franse Galeye*. Between 1593 and 1600 the publisher Richard Schilders had his workshop here. The vice-principal of the Latin School, Petrus Montanus, the translator of the alchemical book by Hollandus, lived behind the adjacent tavern on the right, called *De Groote Soutkeete*. Two houses closer lived Jacob Simonsz. Magnus (1563-1625), lord mayor of Middelburg and several times chairman of the States General. When in 1616 Cateline de Cerf, Isaac Beeckman's future wife, settled in Middelburg, she found accomodation in *De Galeye*, at that time the home of her brother-in-law Pierre Osel, a grain merchant from Nieppe, who had married a sister of hers.

The Market was a posh neighbourhood. Only two houses away from *De Galeye* lived (no. 19 in fig. 11.10) the influential Lord Mayor Jacob Simonsz Magnus (1563-1625), in the house *De Sonne*. He was Zeeland's long-serving representative to the States General. In December 1608 it was Magnus who – in that capacity – tested Lipperhey's telescope after his application for a patent on this invention. Lipperhey's letter of introduction to the States General was composed by Bonifacius de Jonge, the secretary of the Gecommitteerde Raden (Executive Committee of the States of Zeeland), who lived nearby, in one of the quarters of the Abbey (no. 20 in fig. 11.10).

Middelburg Networks

This all concerned only the direct neighbourhood of the Abbey. The mutual proximity of all those people and locations is striking. Rather remarkable also is the network of persons connected to the Middelburg Latin School, especially from those who lectured in the Abbey for the general public. The Latin School clearly formed one of the nodes of knowledge in the fabric of contacts between the learned world and the Zeeland merchants, magistrates, physicians, mathematicians and botanical enthusiasts (colour ill. 6B).[110]

Rector Jacob Gruterus, for example, was close friends with both Johan Radermacher and Philippus Lansbergen. He corresponded with various (mostly Leiden) scholars, including Bonaventura Vulcanius, professor in Greek, and Petrus Scriverius, independent philologist. Poems by (or devoted to) Gruterus feature in the works of yet others, such as those of his Dordrecht colleague Franciscus Nansius or the Leiden professors Dominicus Baudius (chair of eloquence), Daniel Heinsius (poetics), Johannes Meursius (classicist and historian) and Antonius Walaeus (theology). Poems by Gruterus were also included in the works of some authors living in Zeeland, such as twice in a book by the astronomer Philippus Lansbergen and once in a publication by the mathematician Johan Coutereels.[111] Gruterus was also brother-in-law of the Middelburg glass manufacturer Govert van der Haghen, whose crystalline glass Lipperhey may have used to construct his first telescope.[112] Moreover,

110 See my '"Scientia" in Middelburg', pp. 97-98, for the sources used to compose the graph of this network.

111 Lansbergen, *Triangulorum geometriae*; Philippus Lansbergen, *Catechesis religionis christianae quae in Belgii et Palatinatus ecclesijs docetur, sermonibus LII explicata* (Middelburg: Richard Schilders, 1594); Johan Coutereels, *Den vasten stijl van boeckhouden* (Middelburg: Symon Moulert, 1603; 2nd ed., 1623). Poems devoted to Jacob Gruterus are made by Dominicus Baudius and Antonius Walaeus. See my '"Scientia" in Middelburg' for more details.

112 In 1601 Govert van der Haghen (d. c. 1605) lived in *Het Nieu Glashuys* on the Cousteense dijck. Both he and Jacob Gruterus (d. c. 1607) were married to a daughter of the Middelburg merchant

Gruterus was also familiar with Jacob Simonsz Magnus, the man who, as representative of the States General, investigated Lipperhey's first telescope. As president-curator of the Latin School, Magnus was in fact Gruterus's supervisor. In December 1608, Magnus was succeeded as Zeeland's representative in the States General by Jacob Boreel, the man who in 1601 presented Gruterus with an astronomical quadrant, to be used in the Latin School.[113] In short, when on 13 February 1609 Lipperhey delivered his three telescopes with binocular view in The Hague, Jacob Boreel was among the deputies of the States General who had to approve the payment of the remainder of the agreed 900 guilders.[114]

Isaac Beeckman's Middelburg Network

As we have outlined above, in the melting pot of immigrants and local inhabitants who constituted Middelburg around 1600, the young Isaac Beeckman had every opportunity to acquaint himself both with artisanal work as well as with scholarly knowledge. But, of course, the question is: did he indeed profit from these circumstances? Can we prove an inspiring link between the young Beeckman and the Middelburg circle of *curiosi* mentioned above? This is a difficult task, given the fact that Isaac Beeckman started his notebook in the years between 1611 and 1616 when he had already settled in Zierikzee as a chandler. Nevertheless, I will make an educated guess.

Beeckman's biographer Klaas van Berkel has argued that the published documents (mostly from archives that were lost during the Second World War) indicate that the life of the young Beeckman mainly unfolded in the artisanal Middelburg milieu of Flemish immigrants.[115] In his *Journal* Isaac Beeckman reveals only very little about his youth. He informs us that he went to a primary school when he was seven years old. But further details are missing.[116] He also left a remark that he had composed a theatre play

Carel Verhasselt (d. c. 1600) and his wife, Petronelle Thielens. This couple lived in the – still existing – house *Het [Groot] Paradijs* in the Giststraat (today Damplein). Verhasselt was deacon of the church in 1590-1594 and elder in 1598. De Waard, *De uitvinding der verrekijkers*, p. 309; De Waard, entry 'Gruterus'.

113 Kesteloo, 'De stadsrekeningen van Middelburg V, 1600-1625', *Archief ZGW* 8 (1902), p. 118; H.H.P. Rijperman and H. Japikse, eds., *Resolutiën der Staten-Generaal van 1576 tot 1609*, vol. 14 [1607-1609] (The Hague: Martinus Nijhoff, 1970), p. 324.

114 Van Helden, *The Invention of the Telescope*, pp. 43-44; Rijperman and Japikse, eds., *Resolutiën Staten-Generaal*, p. 914.

115 Van Berkel, *Isaac Beeckman on Matter and Motion*, p. 11.

116 *JIB*, I, p. 217: 'Ic was 7 jaer out doen ic schole ginck.' Which primary schoolmaster Isaac Beeckman attended is unknown. Huib Uil lists some 20 schoolmasters working in Middelburg

at the age of eleven. And he tells that in 1601 he went to the Latin School in Arnemuiden, remarkably *not* to the one in Middelburg, led by rector Gruterus, but the much smaller institution headed by the less famous Antonius Biesius.[117]

Father Abraham Heijndricksz Beeckman (1563-1626)

Both De Waard and Van Berkel ascribe this remarkable choice for the Latin School in Arnemuiden to the fact that Isaac's father, the chandler Abraham Beeckman, was involved in a fierce conflict with the Middelburg church council.[118] The dispute was about the question whether or not it was allowed to baptize in the Calvinist church the children of parents who still adhered to the Catholic faith. This affair started in 1597 after Beeckman had heard a sermon by the reverend Johannes Isenbach (the same who gave public lectures in the Abbey, together with Jacob Gruterus). The conflict more or less ended in 1611, with a short period of reconciliation in the years 1607-1608. Abraham's stubborn refusal to recognize in this matter the authority of the Middelburg consistory even led him to be excluded from the Lord's table between December 1602 and 1607. As the Middelburg clergy was closely connected to the Middelburg Latin School, this rules out any influence on the young Beeckman of those attached to this semi-academic institution that – as explained above – played a vital role in the circle of Middelburg *curiosi*.

Abraham Beeckman's renunciation of the authority of the Calvinist consistory also explains the absence in Isaac's *Journal* of almost any reference to the merchant-scholar Johan Radermacher. Not only was Radermacher a close friend of the rector, Gruterus, but in these years of Abraham Beeckman's conflict with the Middelburg church council, Radermacher served twice as one of its elders.[119] Only in 1628, when one of Radermacher's sons accidently met Isaac Beeckman in Dordrecht, did Beeckman note down in

in the years between 1596 and 1601. See: Uil, *'De scholen syn planthoven van de gemeente'*, p. 781.

117 *JIB*, I, p. 217: '12 jaer, twee weecken voor Paesschen (denwelcken quam) den 2en April [1601] – ginc ick t'Armuyen by Antonius Biesius schole liggen om latyn te leeren.'

118 Abraham's co-leader in this conflict was Hans de Swaef, a baker from Antwerp, who was married to Josijntje Panneel, daughter of the Middelburg minister Michiel Panneel (d. 1604). In 1601 Hans de Swaef lived on the Lange Burg, next to Jacob Boreel's house *Het Pater Noster*. After settling the conflict with the church, in September 1607 Hans de Swaef himself became (as an elder) member of the Middelburg church council. His son Daniël de Swaef (b. 1594) became a minister in Middelburg in 1638. In 1626 Daniël married Margriete Coene, daughter of Isaac Beeckman's uncle Hans Coene and his aunt Elisabeth Beeckman.

119 Frederik Nagtglas, *De algemeene kerkeraad der Nederduitsch-hervormde gemeente te Middelburg van 1574-1860* (Middelburg: Altorffer, 1860), pp. 71-72: appointment as an elder for two years on 9 September 1600, 2 October 1604 and 14 September 1608.

his *Journal* that his grandfather Hendrick Beeckman (d. 1581) had been 'very familiar' with Johan Radermacher during his stay in London, this apparently in contrast to his – then recently deceased – father in Middelburg.[120]

Therefore, as there are no provable indications of contact between the Beeckman family and the Middelburg circle of *curiosi*, described above, let's concentrate on the direct neighbourhood of Beeckman, his family and known friends during his formative years in Middelburg, as an alternative milieu that may have shaped his mind. So, what can be established about these possible influences on the young Isaac?

First and foremost, the influence of father Beeckman on the young Isaac must have been considerable. Abraham has been described as 'a man well versed in theology as well as languages'.[121] It was Abraham who trained his son in the craft of producing candles and the construction of waterworks. From such a waterworks project, executed in 1601, Abraham had kept an interesting and inspiring contact with the *vernufteling* (proto-engineer) Cornelis Drebbel. That year Drebbel constructed a fountain just outside Middelburg's Northern gate, most likely together with Abraham Beeckman, who owned a small house there.[122] In later years, Isaac Beeckman always received news from Drebbel through his father, for instance, the oldest known sketch of a compound microscope, invented by Drebbel around 1620.[123]

The Neighbourhood of Beeckman's Childhood

A second influence may have come from the direct neighbourhood of the young Isaac. As the tax registers of 1601 and 1606 show, the houses in the street where Isaac grew up (the Hoogstraat and the Nieuwe Haven) were mostly populated by artisans and craftsmen in several fields (fig. 11.16).

120 *JIB*, IV, p. 1, refering to Beeckman's original notebook, fol. 314vs. This contrary to Van Dixhoorn's remark (in this volume) that the Middelburg Beeckman family 'belonged to the Radermacher circle'. In my view this was definitely not the case.

121 Pieter de la Rue, *Geletterd Zeeland* (Middelburg: M. en A. Callenfels, 1741), p. 8.

122 According to the city registers of 1600-1601 Drebbel received £23-6-8 'over de reste en volle betalinghe van alle tgene hy aen de stadt voor date van deze heeft gewrocht ofte verdient soo int maken van de fonteyne buyten de Noortpoorte als anders' ('as the full payment of all that he has wrought for or earned from the city, both for making the fountain outside the Noortpoort, as otherwise'). In 1601 Abraham Beeckman is named as the owner of a small house at the 'waterganck van de Noorderpoort', which was rented out to a certain Jacob de Backer. Kesteloo, 'De stadsrekeningen van Middelburg V', p. 119; De Waard, *De uitvinding der verrekijkers*, p. 147; Haardstedenregister 1601, fol. 97.

123 *JIB*, III, p. 442. The Zeeland diplomat Willem Boreel reported in 1655 to have seen such a microscope at Drebbel around 1619. Leo Nellissen, 'De echte uitvinder van de telescoop. Pierre Borels werk uit het Latijn vertaald', *Archief KZGW* (2007), pp. 61-103, esp. p. 88.

Figure 11.16 Map of the neighbourhood of Beeckman's youth around the Hoogstraat

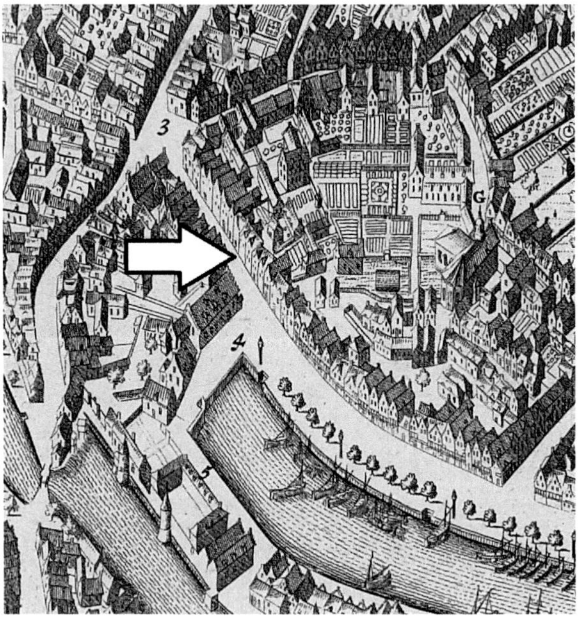

Detail of a map of the surveyor Cornelis Goliath, 1657. At the centre are the Hoogstraat and the Nieuwe Straat, from the Beestenmarkt (no. 3) to the St. Jansstraat (bottom right). The arrow indicates the Beeckman residence, *De Twee Haentgens*. Diagonally opposite is the Stadsschuur (see also figure 11.17).

It seems that this area of the city functioned quite on its own. Isaac, for instance, in his notes never mentions the district around the Middelburg Abbey (discussed above), although this vital part of the city was less than ten minutes' walk from his home. In that respect the Middelburg market in the city centre seems to have functioned a kind of watershed. So, who were the persons young Isaac could meet on a daily basis and what were their occupations? In the years between 1599 and 1606 the following professions were present in Isaac's direct neighbourhood: a furrier, a hatter, a baker, a baker of pies, a woodworker, a glass maker, a drum butcher, a belt butcher, a skipper, a merchant in clothes and various carpenters (including the upper dean of the guild and member of the chamber of rhetoric, Laureys Willemsen Verpoorte [d. 1628]).[124] Unfortunately not every person in the

124 Residents of houses around the Beeckman residence (the Hoogstraat and Nieuwe Haven, between the corner of the Beestenmarkt and the St Jansstraat), mentioned in tax registers with their profession (in Dutch): *1599*: Aelbrecht Janssen (*), *bontwercker*; Tobias Antheunissen (*), *backer*; Loys de Bosscher (*), *pasteybacker*; Franchois van Damme, *schrynwercker*; Joos

Figure 11.17 The prison, a water mill, and the Stadsschuur (with a sawmill) in front of the Beeckman residence

Engraving from: Matthias Smallegange, *Nieuwe Cronyk van Middelburg* (Middelburg, 1696). From left to right we see: the Rasp- en Spinhuys (a prison for men, resp. women), a water mill, and the Stadsschuur.

registers is mentioned with his (or her) profession. Some neighbours most likely were merchants, such as Jan Baltenssz Perduyn from Oostkapelle. He lived with his wife, Neelken Cornelis from Zierikzee in *De Olyfberch* (The Mount of Olives), a large house next to the Beeckman family (which house at some point before 1626 was also acquired by Abraham Beeckman).[125] On

Quackelbeen, 'd'oude meester'; Silvester Romboutssen, *schipper. 1601*: Cornelis van Dale (*), *glaesmaker*; Landereijs Willemsz (Verpoorte) (*), *timmerman*; Jan van Steendamme, *steenhauwer*; Wynant van Toore, *tromslager*; Jan Coucke, *schrijnwercker*; Abraham Bekemans (*), *kersmaker*; Bastiaen Gillis, *cleermacker*. Around the corner at the Beestenmarkt: Lucas (unreadable), 'oudt cleervercoper'; Pieter Tellier (*), *hoeymaker*; Benedictus 'de timmerman' (*). *1606*: Jacob Marinissen, *calfslager*; Reynier Mulders, *riemslager*. Persons with an asterik are also named in later tax registers.

125 Jan Baltenssz Perduyn was a distant relative of the brothers Willem and Symon Jasperse Parduyn, mentioned above. He must have died before 1606, when his son Balten Jansz (Parduyn) is listed as the owner and resident of *De Olyfberch*. See also the appendix.

the other side of the road, Isaac could witness all the activities going on at the Stadsschuur, the large repository of the city and the centre of several technically oriented activities (fig. 11.17). Adjacent to the Stadsschuur in the Hoogstraat was a soap factory and, at the other side, a sluice with a water mill.[126] Not far from there, on the 'Cousteense dijck', adjacent to the Rasp- en Spinhuys at the other side of the Nieuwe Haven, was the Middelburg glass factory, founded by Govert van der Haghen in 1581. This factory was continued in 1607 by Antonio Miotto and his foreman Simon Fabri, both from Venice.[127] All in all, it is evident from this tour that, from an early age on, Isaac was familiar with all sorts of technical practices.

Family Members

Uncle Jan Pietersz van Rhee (1570-c. 1634)

Practical craftsmanship was also what the young Isaac Beeckman would see when he visited some of his Middelburg relatives. Of his four uncles, two were skilled artisans who lived nearby. First there was Jan Pietersz van Rhee (1570-c. 1634), a brother of Isaac's mother, born in Sandwich (England). Van Rhee's father (Isaac's grandfather), had been the entrepreneur who in 1586 had bought the empty plot at the Middelburg Hoogstraat, on which in 1588 the house *De Twee Hanen* was built where Isaac would live with his parents from 1593 onwards. Uncle Jan Pietersz van Rhee was a wheelwright, just as his father had been. But he was also a candle maker and a constructor of waterworks. As such he was a direct colleague of Abraham Beeckman. Van Rhee had settled in Rotterdam after his marriage in April 1595, but when in 1597 the plague had robbed him of his parents and two of his brothers, he returned to Middelburg to take over the affairs of his late father.[128] But Van Rhee soon rented out his father's *kersmackerie* (candle factory) outside the Lange Viele gate (in the western part of the city, a ten minutes' walk from his house at the Beestenmarkt),[129] so that he could be engaged with other

126 C. Sanderse, 'De sluis bij de stadsschuur te Middelburg', *Zeeland* 5:2 (1996), pp. 61-70.

127 Anthonio Miotto, 'M[eeste]r van de Kristalijne Glasen Oven binnen Middelburg', took over the business of Govert van der Haghen in December 1607. The patent expired in 1615 and was extended that year by seven years. See: *Notulen van de edel mogende heeren Staten van Zeeland, d'anno 1607* (n.p., [1608]), p. 242; ibidem, *d'anno 1614* (n.p., [1615]), pp. 115-116, 237-238; ibidem, *d'anno 1615* (n.p., [1616]), p. 63. De Waard, *De uitvinding der verrekijkers,* pp. 110, 315-319.

128 *JIB*, IV, p. 8.

129 Haardstedenregister 1601: 'Comende buyten de Lange Viele poort [...] het Huis genaempt *De Kersmackerie*, huerder Hans Wylhemsen, eygenaer Jan Pietersen van Ree, 7 hs [haardsteden] | Op de sijde van den waterganck, huerder Adriaen Bouve, eygenaer Jan Pietersz van Ree, 4 hs.'

business, such as being the collector of the tax on textiles.[130] But Van Rhee was also an inventor. In 1616 he moved to Amsterdam, where he obtained a patent for some improvements relating to 'water pipes, mill works and the deepening of canals'.[131] The fact that one of Isaac's earliest surviving notes (from 1612) mentions experiments that he has carried out with this 'Jan-oom', indicate that this uncle may have been one of the early practical educators of the young Isaac Beeckman.[132] That this 'Jan-oom' accompanied Isaac to the University of Caen in 1618, underpins their close relationship. In Caen both men obtained an academic degree; Isaac in medicine (on 6 September 1618) and Jan Pietersz van Rhee remarkably in law (at the end of October). For that occasion Isaac even had to lend his uncle the sum of 25 guilders.[133]

Uncle Hans Coene (1576-1634)

Another uncle with a technical inclination was Hans Coene (1576-1634), born in Roeselare (Flanders). In 1600 he had married Elisabeth Beeckman, the half-sister of Isaac's father Abraham. Coene settled in Middelburg in 1594 as an apprentice at a *borstelmaecker* (brush maker), but he soon became a skilled carpenter. Several times he was *beleeder* (member of the managing board) and *deken* (dean) of the carpenter's guild (fig. 11.18). He lived close to the Beeckman family, in the same row of houses at the end of the Nieuwe Haven (the extension of the Hoogstraat).[134] Coene was a well respected citizen. In 1604 the Middelburg church council used him as mediator in the theological conflict between Abraham Beeckman and the consistory. In 1606 Coene was already 'captain' of the city quarter in which he – and the Beeckmans – lived. Isaac Beeckman respected this uncle. In 1618 he remembered – and sketched – in full admiration Coene's skilfully designed

130 *Notulen van de edel mogende heeren Staten van Zeeland, d'anno 1612* (n.p., 1613), p. 183: request by Jan Pietersen van Rhee for compensation for the fact that his job as tax collector of 'Lynwaet, Breynaet, etc.', a profession he had excercised for more than ten years, was discontinued without notification of the reasons. This request was denied.

131 *JIB*, IV, pp. 35-37. It concerned 'inventien van verscheyden manieren van waterleydingen, molenwercken ende verdiepingen van grachten'. The patent was granted by a commission led by Symon Magnus from Middelburg, but later withdrawn.

132 *JIB*, I, p. 15.

133 *JIB*, IV, pp. 47-48.

134 According to De Waard, Coene settled in the *De Dry Halve* [*Witte*] *Manen*, with the adjacent *Huys metten Houttuyn* at the corner of the Jansstraat and the Nieuwe Haven. However, the tax register of 1606 mentions Coene as the tenant of a house further on the Nieuwe Haven, near the Vismarkt (Fish Market), between the houses – called in 1601 – *Vogelsanck* and *De Goude Cramer*. In 1601 this third house from the corner of the Vismarkt and the Nieuwe Haven did not yet exist.

Figure 11.18 Medal of the Middelburg Carpenters' Guild

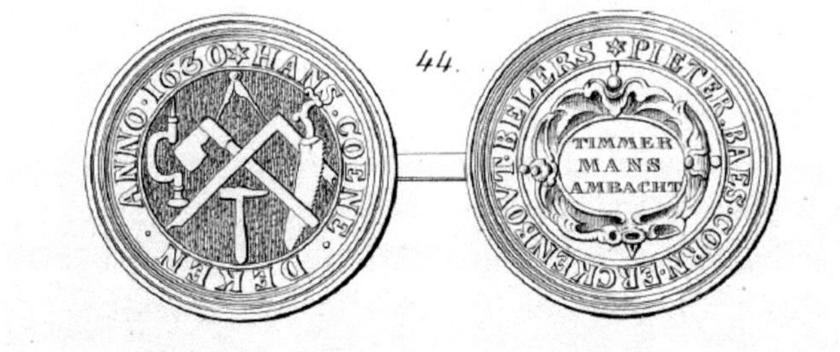

From: J. Dirks, *Atlas van platen, behoorende bij het 2de deel (Nieuwe Reeks) van de Verhandelingen uitgegeven door Teylers Tweede Genootschap (De Noord-Nederlandse Gildepenningen)* (Haarlem, 1879). Isaac Beeckman's uncle Hans Coene was dean of this guild when this medal, depicting all sorts of carpenter's tools, was made by the silversmith Lowys in 1630.

doors that could open in two directions.[135] And when in 1620 Isaac Beeckman participated in a technical discussion on the question how to maintain the canal from Middelburg to the sea at the required depth, Coene was among the 'seventeen or eighteen of the most distinguished enthusiasts from Middelburg' who attended this meeting. It was in part thanks to Coene's advice that the project was cancelled.[136]

Coene was also a well-respected member of the Middelburg Chamber of Rhetoric, of which he became treasurer in 1619.[137] In the carpenter's guild Coene was also a colleague of the rhetoricians Laureys Willemsen Verpoorte and Pieter Joossen 'altijd Rechthout'. With the latter Coene shared the office of *beleeder* of the guild.[138] Coene's activity as a rhetorician makes one wonder whether it was his influence that stimulated the young Isaac to compose a playlet, in the very year (1600) in which Coene became a member of the Beeckman family.[139] In 1628 Isaac remembers the following of this event:

135 *JIB*, I, p. 181.
136 *JIB*, II, p. 39, n. 1. Charles van den Heuvel, 'Tot meerder bewijs. Een kaart en een model van Daniel Note van 1620 ter demonstratie van een nieuwe uitvinding om de haven van Middelburg zandvrij te schuren', in: Paul Hoftijzer et al., eds., *Bronnen van kennis, wetenschap, kunst en cultuur in de collecties van de Leidse Universiteitsbibliotheek* (Leiden: Primavera Pers, 2006), pp. 100-108.
137 De Man, 'De voormalige Middelburgsche rederijkerskamer het Bloemken Jesse', p. 33.
138 Robert Fruin, 'De kroniek van Pieter Joossen Altijt Recht Hout', *Archief ZGW* (1909), pp. 65-96, esp. p. 94. Pieter Joossen had started the new carpenter's guild book with a poem in 1594 and entered another one, praising – among others – Coene, in 1620. See also Arjan Dixhoorn's database of rhetorians in the Northern Netherlands (1480-1650): http: //www.lustigegeesten.nl/index.html.
139 During the editoral review of this chapter, Klaas van Berkel suggested that young Isaac composed this play on the occasion of the wedding of Coene and his aunt Elisabeth Beeckman.

> When I was eleven years old, I made various songs, and I also invented a history by heart, similar to the one I frequently had read, about Valentijn and Ourrson etc.[140] And I made a comedy for four persons, of about 500 verses in rhyme, and it was played by me, my brother Jacob, Jacques Schouten – who now is our brother-in-law[141] – and another boy, on our courtyard, on a scaffold, hung with curtains, played in the presence of our friends and neighbours. We played it four times, because at the time I didn't know anything about act or scene.[142]

Such an influence of his uncle, the rhetorician, is quite conceivable and is in line with Van Dixhoorn, who assumes a strong influence of the rhetorician culture on the young Beeckman.[143] However, it is also possible that this initiative stemmed from something he was used to do at the Latin School with his teacher Antonius Biesius.[144]

140 *Valentijn and Oursson* is a late medieval adventure novel, set in the days of Pepin the Short (714-768). The text of this play has not survived in a sixteenth-century printing. The oldest known copy kept in a Dutch library dates from 1698: *Een schoone ende wonderlijcke historie van Valentyn en Oursson, de twee edele vroome ridders, soonen van den mogenden keyser van Griecken, en neven van den edelen coning Pepijn, doen ter tijd coningh van Vranckrijck. Uyt de Francoysche in onse Nederlandsche Sprake over-geset* (Utrecht: J. van Poolsum, n.d.).
141 Beeckman's oldest friend Jacques Schouten (1588-1655) was a son of Nicolaes Schouten (d. 1610) and Agatha Pieters (d. 1604). Jacques Schouten matriculated at Leiden University on 16 April 1608 as a student in the arts, together with his brother Pieter. In August 1610 Isaac Beeckman and his friend returned together to Middelburg. Schouten would travel with Beeckman to the University of Saumur in 1612. At Isaac's request he would perform some meteorological observations during his further travels in France. In 1615 he married Isaac's sister Janneken Beeckman (d. 1633). Schouten became a Calvinist minister, first in Noordgouwe (near Zierikzee) and later in 's-Heer Arendskerke and Baarsdorp (near Goes). See: *JIB*, I, pp. vii, xxxvi, and III, p. 4 (Beeckman's 1627 recollection of their dangerous trip to France).
142 Beeckman, 'Loci communes', fol. 315vs. 'Als ick elf jaer oudt was maeckte ick verscheyde lydekens, ende versierde oock een historie ut mijn hooft gelijck die ick veel geleesen hadde van Valentijn ende Oersen &c. Ende maeckte daer wt een commedie van vier persoonen van omtrent 500 verssen al op rijm ende wert van mij, mijn broeder Jacob, Jaques Schouten, die nu ons swager is, ende noch een jonghen openbaerlick op ons plaetse op een stellage met gordijnen behanghen in presentie van ons vrienden ende ge- bueren gespeelt. Men ginck viermael wt, want ick en wist doen van geen actus en schene' (my transcription, assisted by Klaas van Berkel, as being not published by De Waard). At the time Beeckman was not yet aware of the classical rules of three unities of drama that were ascribed to Aristotle.
143 Van Dixhoorn, this volume.
144 Beeckman's school teacher Biesius used to practise theatrical plays with his pupils in the Latin School of Arnemuiden. See H.W. Fortgens, 'De Latijnsche scholen te Veere en te Arnemuiden', *Archief ZGW* (1944), pp. 51-73, esp. p. 71.

Other Uncles

Isaac's other uncles seems to have had less influence, probably because they had left Middelburg rather soon after their marriages. Uncle Anthony van Alderwerelt, for instance, who married Sara Beeckman (half-sister of Isaac's father) in 1595, left Middelburg somewhere between 1601 and 1606,[145] most likely to go to Flushing (Vlissingen), where until 1913 his eponymous son's gravestone (from 1628) could be found.[146] Another uncle was Pieter Cools, a *huyvetter* (tanner) from Blanckeberch in Flanders, who had married Elisabeth Pieters van Rhee – a sister of Isaac's mother – in 1595.[147] At that time he lived on the other side of the city, in the Waegenaerstraete, next to the *Samaritaensche Vrouwe*. But as he left Middelburg before 1601, probably for Breda where he became *poorter* (citizen) in 1613, his influence on young Isaac can be discounted.[148] However, in 1618 this 'Pieteroom' was the reason why Isaac went to Breda, to assist him for a short time in his craft of tanner. It was during that visit that Beeckman made his acquaintance to René Descartes, possibly in the local chamber of rhetoric Het Vreuchdendal.[149]

Philippus Lansbergen

Finally there is only one person left in the network of the Beeckman family who can be connected with the group of Middelburg *curiosi* described above. This is the Calvinist minister and astronomer Philippus Lansbergen, who until the end of 1613 lived in Goes on the nearby island of Zuid-Beveland. In the years 1600-1604 Abraham Beeckman and Lansbergen exchanged many letters about Beeckman's theological controversy. Although Lansbergen did not agree with Beeckman in this matter, the tone of their letters is always very friendly. Abraham, for instance, often provided Lansbergen with a box of his best

145 In 1595 Anthony van Alderwerelt from Antwerp lived in the house *De Witte Pluyme*, in the Lange Delft, one of the city's prominent streets. In the Haardstedenregister of 1601, he is mentioned as living in *Het Scharlaecken*, also in the Lange Delft. As the name *De Witte Pluyme* does not appear in the tax register, this probably was the same house, renamed by Van Alderwerelt after his trade in wool. In 1606 this house is occupied by a certain Abraham Maertensz. As in that year the name Alderwerelt is not mentioned elsewhere in the register, he must have died or moved out of the city. The last child of the couple was born in 1605.

146 P.K. Dommisse, 'Enige grafschriften uit de afgebrande St. Jacobskerk te Vlissingen, met archivalische toelichtingen', *Archief ZGW* (1913), pp. 1-146, esp. pp. 84-85.

147 About 'uncle' Pieter Pietersen, a widower who in July 1595 married Isaac's aunt Sara Pieters van Rhee, no further details are known. *JIB*, IV, p. 9.

148 Van Berkel, *Isaac Beeckman on Matter and Motion*, p. 196, n. 80.

149 *JIB*, I, p. 228: 'Voor de slachtyt des jaers 1618 ben ic te Breda gecomen om Pieteroom te helpen wercken en te vryen oock.' For the chamber of rhetoric, see: *JIB*, I, p. 257. See also: Van Berkel, *Isaac Beeckman on Matter and Motion*, p. 24, and the chapter by Arjan van Dixhoorn elsewhere in this volume.

candles.[150] So it is possible that, although Lansbergen only moved to Middelburg when Isaac Beeckman already worked in Zierikzee, he may have had some influence on Isaac's formation. For instance, in the period around 1613 when Isaac tried to become a physician. After all, Lansbergen's son Jacob, who Isaac probably knew from his earlier stay in Leiden, had obtained his medical degree earlier that very year. In any case, they must have been in contact at least since 23 December 1616, when Isaac – shortly after his return to Middelburg – made an astronomical observation with Lansbergen's quadrant (colour ill. 7). This observation – or a similar one made at that date – is also printed in Lansbergen's description of that quadrant.[151] In 1622 it was also Philippus Lansbergen who stimulated Isaac to obtain a telescope 'like the one Galileo used'.[152]

The Latin Schools of Arnemuiden and Veere

As so little is known about Isaac's younger years, this question is relevant: what kind of scholar was Antonius Biesius, Beeckman's teacher at the Latin School? What was his background? Biesius (d. 1606) was born in Delft, in a family with its origin in the city of Ghent in Flanders. He was probably related to the Leuven professor of medicine Nicolaus Biesius (1516-1572), former court physician of the Emperor of the Holy Roman Empire Maximilian II.[153] Antonius had not followed in his relative's footsteps, for he had studied law at Leiden University from 1592 until 1594.[154] But instead of seeking a career as a lawyer or notary, he had found a job as a teacher, first in Arnemuiden and thereafter in Veere.[155] By a lucky coincidence the catalogue of Biesius's library (auctioned in Leiden in 1607) has been preserved (colour ill. 8), and that document gives us a glimpse of what kind of knowledge Biesius can have offered to his pupils.[156]

150 Zeeuwse Bibliotheek, Middelburg, handschriften, nos. 5070, 5078-5079. See: Van Berkel, *Isaac Beeckman on Matter and Motion*, chap. 1, n. 9, 14, 15. Lansbergen lived in London between 1566 and 1579. Van Berkel supposes that the friendship between Lansbergen and Abraham Beeckman dates back to that time.

151 *JIB*, II, p. 151; Philippus Lansbergen, *Verclaringhe van 't ghebruyck des astronomischen ende geometrischen quadrants, ghesneden ende uytghegheven door F. Schillemans* (Middelburg: Zacharias Rooman, 1620), p. 14.

152 *JIB*, II, p. 294.

153 His library (see footnote 156) contained three books by Nicolaus Biesius, dated 1558, 1564 and 1573.

154 Biesius matriculated in Leiden on 27 December 1592. His *Theses de acquirendo rerum dominio* (Leiden: Fr. Raphelengius, 1594) has been preserved in Leiden University Library, sign. ASF 353: 23.

155 Fortgens, 'De Latijnsche scholen te Veere en te Arnemuiden', pp. 55, 71.

156 *Catalogus plurimorum insignium librorum ex bibliotheca* [...] *D. Antonii Biesii* [...] *quorum auctio fiet Leydæ* [...] *apud Ioannem Maire* (Leiden: Ioannes a Dorp, 1607).

The catalogue contains 34 pages, listing some 700 books from several fields. There are 27 pages with books in Latin, of which nine pages are books on theology, four and a half pages on law, one small page on medical literature, and the rest of the books categorized as 'Libri Miscellanei'. The remaining seven pages list books in French, Spanish and Dutch. Recently published books were scarce. Most editions dated from the middle of the sixteenth century, which suggests that Antonius Biesius acquired them through inheritance. Nevertheless, in the Miscellaneous section there were some books well worth mentioning, such as an edition of Sebastian Munster's *Cosmographia* (1559) – one of the earliest descriptions of the world – but also a nicely illustrated edition of Ptolemy's *Geographia*, published in Basel in 1540. Books in the field of natural philosophy were almost entirely absent in Biesius's library. One that was present was Lambert Daneau's *Physique françoise* (1581), in which the classical views about the phenomena in the natural world are summarized. And, of course, Biesius had some copies of the medical and natural philosophical works of his relative Nicolaus Biesius. Remarkably the catalogue mentions only one textbook on mathematics: Simon Stevin's *l'Arithmetique* (Leyden, 1585). Intriguing is, of course, the one manuscript offered in auction: a 'libellus nuptialis' or a marriage manual. Was it intended for private or public use? Finally, the catalogue mentions only two books published in nearby Middelburg, both first editions: Lansbergen's Latin *Catechimus* (1594) and Michiel Panneel's Dutch translation of Napier's explanation of the Revelation of Saint John (1600).[157] All in all, this gives the impression that Biesius was not a very frequent visitor to the Middelburg book shops, and that his interest was not very focused on questions relating to nature or natural philosophy. Only one link between Biesius and a person from the Middelburg network of *curiosi* can be established: in 1603 he stood witness at a baptism together with the reverend-botanist Johannes de Jonghe, one of Clusius's correspondents – but it seems plausible that this incidental meeting has little meaning.[158]

157 Lansbergen, *Catechesis religionis christianae*; John Napier, *Een duydelijcke verclaringhe vande gantse Openbaringe Joannis des Apostels*, trans. by M. Panneel (Middelburg, 1600). The edition of 1613 was also in Isaac Beeckman's library. See: Eugenio Canone, 'Il *Catalogus librorum* di Isaac Beeckman', *Nouvelles de la République des Lettres* (1991), pp. 131-159.

158 Baptism in the Church of Arnemuiden of Johanna Ryckelem, daughter of Joannis Ryckelem, on 5 March 1603. Witnesses: 'Johannes de Jonghe, Mr. Predicant tot Middelburch', and Anthonius Biese, 'Mr. Rector ter Veere', Zeeuws Archief, DTB, Arn-1, consulted via the website 'Zeeuwen gezocht'.

Conclusions

From this overview of the scientific climate in Middelburg during Isaac Beeckman's youth we can draw the following conclusions.

(1) In the years around 1600 the city of Middelburg was a fertile melting pot of mercantile, artisanal and learned contacts, where special relationships and networks existed, extending both inside and outside the Dutch Republic. With its *curiosi* the Zeeland capital was unmistakably part of what sometimes is referred to as the European 'Republic of Letters'. The constant exchange of letters, news, ideas and objects, combined with visits to and from abroad, contributed step by step to the accumulation of natural knowledge, leading to what often is called the Scientific Revolution of the seventeenth century. In that respect, the Middelburg situation is comparable to what, for example, Deborah Harkness has outlined for early-seventeenth-century London, a city with which Middelburg maintained many close contacts.[159]

(2) Moreover, with this local case study, Van Berkel's hypothesis about the factors that contributed to the flourishing of early modern science in the Dutch Republic, has obtained a concrete interpretation, in line with his assumptions. This encounter between the artisanal, administrative and learned world of Middelburg is beautifully depicted by Adriaen van de Venne, who lived in this city until 1625. In his watercolour he sketched an encounter of a mathematician, a lawyer, a painter and an engraver, with a sculptor in the background (colour ill. 9).[160] This watercolour is iconic for the limited social stratification that characterized Dutch society in the early seventeenth century. Van de Venne literally portrays the interaction between academics, magistrates, merchants and artisans that is considered so fruitful for the development of early modern science.

(3) Nevertheless, almost no proof could be found that relates the young Isaac Beeckman to the network of Middelburg scholarly enthusiasts identified in this article. Van Berkel's conclusion that the life of the young Beeckman unfolded for the largest part in the Middelburg milieu of Flemish immigrants, is underpinned by our research. Yet, although Beeckman hardly mentions any of these Middelburg *curiosi*, it is almost inevitable that during his upbringing, young Isaac has encountered

159 Harkness, *The Jewel House*.

160 Martin Royalton-Kisch, *Adriaen van de Venne's Album in the Department of Prints and Drawings in the British Museum* (London: British Museum, 1988), plate nr. 38 (The arts). In this picture Van de Venne has portrayed himself as the painter. The lawyer can be identified as Jacob Cats. The mathematician is probably Philippus Lansbergen and the engraver François Schillemans. The sculptor may refer to Pieter Roman the elder.

some persons or activities related to these scholarly circles of interest. In his *Journal*, for example, Beeckman mentions his habit of always making the same round through the city centre.[161] Remarkably, in this tour he avoids the area around the Abbey with its merchant centre and artisanal workshops, designated earlier in this chapter as a probable point of contacts with the Middelburg circle of *curiosi*.[162] Still, during his wanderings through Middelburg Isaac still must have seen, smelled, or experienced things that we have no knowledge of, but which will most likely have stimulated his inquisitive mind. The only meagre clue of some influence from the Middelburg *curiosi* is perhaps the fact that at the end of his life, in Isaac's library, three books could be found that were linked to the Middelburg enthusiasts for natural knowledge, all three published *before* Beeckman began his studies at Leiden University. These were Hollandus's *Opera mineralia* (1600), Daniël Miverius's, *Apologia pro Philippo Lansbergio* (1602), and De L'Obel's adapted Dutch version of Cordus's and Coudenberghe's *Den Leytsman der Medicijnen* (1596). This last book has several references to scholarly enthusiasts in Zeeland. But did he own these Middelburg books before he left the city in 1607? This, of course, we will never know.

(4) Interestingly, there is another book in Beeckman's library which confirms a Puritan theological influence on the young Isaac. The catalogue of his books lists an extremely rare copy of *A Plaine Refutation of Mr. Giffard's Book*.[163] This publication by the Brownists Henry Barrow and John Greenwood was printed anonymously in 1591, but at the request of the English ambassador Sir Robert Sidney the entire edition of this book was seized by the Dutch authorities. Sidney desired to have this separatist's book

161 *JIB*, I, pp. 279-280: 'Als ic uyt de Hoochstraete na de Wal gaen wil langs de Haven, soo gae ic gemeynlick de S^t. Jansstrate voorby en also door de Segeerstrate en het Kerckstraetken; maer van de Wal comende door het Kercstraetken, soo gae ic langs den Langen Delft door de St Jansstrate nae de Hoochstrate, omdat men van naturen altyt geern naerdert de plaetse, daer men wesen wil, en men wyct niet geern uyt den naesten wech, die men voor ons siet leggen, tensy dat de wegen sóó veel verschillen, dat men by reden afwyckt' (note from 1619).

162 Was this perhaps because the abbey was also the seat of the consistory that had cut off his father from the sacrament? It is tempting to search for such a reason.

163 Henry Barrow and John Greenwood, *A plaine refutation of M.G. Giffardes reprochful booke, intituled a short treatise against the Donatists of England. Wherein is discouered the forgery of the whole ministrie, the confusion, false worship, and antichristian disorder of these parish assemblies, called the Church of England. Here also is prefixed a summe of the causes of our seperation* [...] *by Henrie Barrovve. Here is furder annexed a briefe refutation of M. Giff. supposed consimilituda betwixt the Donatists and vs* [...] *by I. Gren. Here are also inserted a fewe obseruations of M. Giff. his cauills about read prayer & deuised leitourgies* (n.p. [probably Dordrecht], 1591). See: Canone, 'Il *Catalogus librorum* di Isaac Beeckman', p. 142, nr. 28.

destroyed, because the ideas of the very Puritan Brownists (who fiercely opposed all Catholic remainders in the Church of England) were seen as a threat to the unity of this church. Therefore, the stock of confiscated books was brought to Middelburg where the preacher of the local English church, Francis Johnson (1562-1618), ensured that the edition was destroyed. Only very few copies of the book were saved from destruction.[164] Ironically, Johnson would later become a Brownist himself, according to the rumours, because he had read one of the confiscated copies of this very book.[165] So, the fact that Isaac Beeckman possessed one of the few surviving copies – possibly the one formerly owned by his father – is again an indication of the influence on the Beeckman family of the Puritan Brownist community in Middelburg.[166]

(5) All in all, the final conclusion can be that, whereas the universities of Leiden, Saumur, and Caen introduced Isaac Beeckman into the world of scholarship, the Middelburg setting of craftsmanship and artisanal skills provided Beeckman with the practical outlook onto the problems we encounter in his *Journal*. As far as we can ascertain, this was an essential combination for Beeckman to build his ideas relating to natural philosophy.

Appendix: The Houses in Which Isaac Beeckman Was Raised

In his notebook Isaac Beeckman states that he was born 'on the Beestenmarkt, diagonally opposite the corner house of the 's Gravenstraet, in a small house' ('op de Beestenmarckt, noes over het hoeckhuys van de 's Gravenstraete, in een kleyn huysken') (see colour ill. 10).[167]

In the past the identification of Isaac's birth house has been problematic (today the wrong house bears a plaque stating Isaac Beeckman was born there). This confusion has been caused by the fact that at the end of the

164 This Anglican church in Middelburg was founded by the London-based Company of Merchant Adventurers, a group of English merchants who used Middelburg as their Continental seat between 1587 and 1611.

165 Leland H. Carlson, ed., *The Writings of Henry Barrow, 1590-1591* (London: Allen & Unwin, 1966; reissued 2004), pp. 261-264.

166 The presence in the catalogue of John Penry, *The Historie of Corah, Dathan, and Abiram etc. Numb. 16. Chap. Applied to the Prelacy, Ministerie and Church-assemblies of England* (n.p., 1609), also bears witness to Beeckman's early interest in the English Puritans. See: Canone, 'Il *Catalogus librorum* di Isaac Beeckman', p. 142, nr. 13. Beeckman's copy was probably printed in Middelburg by Richard Schilders. John Penry (1559-1593) was a Welsh Brownist who was executed in 1593 for his alleged rebellion. He was considered a martyr for the Brownists.

167 *JIB*, I, p. iv, n. 1.

sixteenth century several plots in the neighbourhood were not (or no longer) filled with houses. The Middelburg Beestenmarkt (Animal Market), today called Varkensmarkt (Pig Market) has been constructed in or shortly before 1576, after the demolition of several medieval houses.[168] Various streets, such as the Koestraat, Gortstraat, Gravenstraat, Hoogstraat, and Vlissingsestraat, all lead to this small triangular market (no 3 in fig. 11.16). As a result, houses facing the Beesten- or Varkensmarkt have nowadays very different addresses. Problematic also is that De Waard, in his marvellous edition of Beeckman's *Journal*, distributed the available information on the Beeckman family homes across the four volumes, which has led to some minor errors in the interpretation of the facts.

When in July 1578 Isaac's grandfather Pieter Janssen van Rhee and his wife, Janneken van Rentergem, came from Sandwich (England) to Middelburg, they settled in what would become Isaac's birth house, then described as 'the corner house on the Beestenmarkt, diagonally opposite the house, called The Black Lion' ('int houckhuys op de Beestemarct, noets over *De Zwarten Leeu*').[169] The Haardstedenregisters of 1601 and 1606 show that *De Zwarten Leeu* was a house at the other side of the street, at the western corner of the Gravenstraat and the Beestenmarkt.[170] Van Rhee rented his house until March 1580, when he was able to buy this property, limited north by feudatory land owned by the city of Middelburg and south by the (former) house *Brugge*.[171]

However, this house called *Brugge* was demolished in 1576, and a new house on this plot would only be built in the 1590s. De Waard was probably unaware of this. That's why Van Rhee's house could be called a 'corner house' in 1578. Next to it was only an empty plot 'where once has stood the house

168 1576: Bought from Cornelia Pieter de Hortersdochter 'een vervallen huijs en de materialen' located 'in de Goorstraete [= Gortstraat] daer nu de beestenmerct aff gemaect is' (£5-2). *Archief ZGW* (1894), p. 31.

169 *JIB*, IV, p. 2.

170 Haardstedenregister 1606, fol. 126vs: "*t Swart Leeuwken*. Aencommende ende bewoont door Jan Bosschaert ende wesende het houckhuijs, 4hs'. In 1601 the name of this house was *not* recorded. The tax collector only stated that the house on the corner of the Gravenstraat and the Beestenmarkt was inhabited by a *backer* called 'Jane' (probably Jan Bosschaert, the same man as in 1606). However, that year another tax collector – who recorded the data of houses in the Gortstraat – called the neighbouring house at the corner of the Beestenmarkt 'Thuys achter *Het Swerte Leeuwken*', which confirms the name of the other house also in 1601. See Haardstedenregister 1601, fol. 104vs and 15vs. The name was still used in the eighteenth century, when the *De Zwarte Leeuw* 'op de Beestenmarkt' was an inn. See: https://middelburgdronk.nl.wiki/De_Zwarte_Leeuw (accessed 10 July 2020).

171 *JIB*, IV, 4, n. 6.

called *Brugge*'.[172] As years went by, Van Rhee's business as a wheelwright could afford him a larger house, so on 25 April 1586, he took out a mortgage for the purchase of an empty plot at the Hoogstraat, across the 'Stadtsschuere', where he built a new house, called *De Twee Haentgens* (The Two Roosters) (colour ill. 11).[173] Van Rhee took up residence there, at least before January 1588 when his former house was occupied by his newly-wed daughter Suzanna Pieters van Rhee and her husband, Abraham Beeckman, together with his half-sisters Sara and Elisabeth Beeckman.[174] It is in this house that Isaac Beeckman was born in December 1588.

However, only a few years later, in August 1592, Van Rhee succeeded in buying from the merchant Lieven van Winckel (who owned several plots in the Gravenstraat) two newly built houses adjacent to the house of the Beeckman family, one called *Brugge* (just like the earlier property) and the other one *De Groene Ploech*. The latter house constitutes the new corner of the Beesten- or Varkensmarkt until today (see colour ill. 12).[175] Van Rhee moved to his new residence sometime in 1592 or early 1593. Now Van Rhee's house *De Twee Haentgens* became available for the Beeckman family, who settled there at some moment after the birth of Daniël Beeckman in April 1593.[176] In that house Isaac Beeckman would grow up to adulthood. The small house on the Beestenmarkt was sold, probably to an old cloth merchant, called Lucas, who owned the property in 1601.[177]

172 1576: Bought from Simon Brouckhoven 'een ledighe erve met alle de materialen daerop liggende daer eertijts op gestaen heeft het huijs genaempt *Brugge*' (£17). *Archief ZGW* (1894), p. 31.

173 *JIB*, IV, 3, 96, n. 4. The land was about 72 feet deep and measured at the front about 30 feet and at the back about 24 feet. The plot was bought from Matthys Augustynsz (Crompvliet) (*JIB*, IV, p. 3), who also provided the mortgage. He was also the owner of the adjacent mansion *Coninck Salomon* in the Gravenstraat, the remainder of the Middelburg seat of the Counts of Holland. In 1601 this mansion was rented by the Middelburg mayor Adriaen van de Hooghe and in 1606 by the *rekenmeester* (mathematician) Johan Coutereels.

174 *JIB*, I, p. 3. In note 6 and also in *JIB*, IV, p. 6 note 1, De Waard makes the mistake that the small Beeckman house was later divided into the adjacent houses *De Hoemakerie* (The Millinery) and *Den Houttuyn* (The Timber Yard), today Varkensmarkt 9 and Gravenstraat 77, respectively. In 1592 both plots were owned by the merchant Lieven van Winckel, who sold the newly built houses to the *huydevetter* Nicolaes de Buck on 4 February 1593. The names of these houses were given at a later date. In 1601 and 1606 the *De Hoemakerie* was occupied by the English *hoeymaker* (hatter) Pieter Tellior (written as 'Pieter Tailor' in 1606). The adjacent house *Den Houttuyn* was in these years occupied by Benedictus 'de Timmerman' (Benedict 'the Carpenter'). The next house, called *De Vier Mollen* (built in 1593), was owned and inhabited by the merchant Jan de Mol (1601).

175 *JIB*, IV, p. 5, n. 8.

176 *JIB*, IV, p. 6, n. 1.

177 Haardstedenregister 1601, fol. 104vs. The family name of this 'Lucas' is difficult to decipher. It resembles 'Hernycop'. In 1606 the property was owned and inhabited by Lucas Hoorne, perhaps

In the spring of 1597 the plague ravaged the city of Middelburg, causing the death of both Pieter Janssen van Rhee, his wife, Janneken van Rentergem, and two of their children.[178] Later that year, both houses *Brugge* and *De Groene Ploech* were sold by their surviving two children, Jan Pieters van Rhee and his sister Suzanna Pieters van Rhee, the wife of Abraham Beeckman.[179] Thanks to this inheritance Abraham Beeckman could obtain a large part of the neighbouring terrain in January 1599, in this way greatly enlarging the space for his chandler workshop.[180]

After the death of Abraham Beeckman, his heirs sold several buildings in the immediate vicinity of the parental home on 30 January 1626.[181] From this deed it appears that Abraham Beeckman at some moment in time also had acquired the large adjacent house known from other documents as *De Olyfberch* (The Mount of Olives). This property was bought by the grain merchant Coolaert Osel, a brother of Abraham Beeckman's brother-in-law Pierre Osel (married to Jacquemine de Cerf).[182] In 1630 it seems that the parental home itself was split into two. One half was obtained by Isaac's youngest sister, Hesther Beeckman, and her husband, the chandler Louys Vergrue, a widower from Bruges, who continued the candle-making business of the family.[183] Isaac Beeckman and the widow of his brother Jacob, Janneken van Ryckegem (colour ill. 13), sold the other half of the property to an old cloth merchant, Pieter Bartholomeeusen.[184] So the division of the property, visible today, is perhaps the result of one of Isaac's actions.

the same man.

178 *JIB*, IV, p. 8.

179 *JIB*, IV, p. 9. *De Groene Plouch* was sold on 13 August 1597 and *Brugge* on 29 September 1597. The buyers are not mentioned. In 1599 the house *De Groene Plouch* was owned by Ieman Jobssen, but two years later it was the property of Alexander de Keyser and inhabited by Gillis van der Straaten. In 1606 the house was owned by Alexander de Keyser's widow and rented by Isau Baute. Alexander de Keyser and his wife also owned the house behind the *De Groene Plouch*, in the Hoogstraat, rented in 1601 by Bastiaen Gillis 'de Cleermacker' and in 1606 by Jan Mout.

180 *JIB*, IV, p. 9.

181 *JIB*, IV, p. 96. See also Stadsarchief Rotterdam, toegang 18, inv. nr. 183 (notary Jacob Cornelisz van der Swan), akte 134 (12 January 1626): Authorization to sell a house and homestead located at the Hoochstraet in Middelburg by Jacob, Isaac and Hesther Beeckman, as well as by Abraham Jansz du Bois, widower of their sister Maria Beeckman.

182 *JIB*, IV, p. 34, n. 2.

183 *JIB*, IV, pp. 108, 121, 168. Hesther Beeckman still lived in the Hoochstraete in 1661. Her daughter Catelintje Vergrue died there in May 1679, so the house *De Twee Haentgens* probably stayed in the family at least until then. See website 'Zeeuwen gezocht'.

184 *JIB*, IV, p. 191. In 1632 Janneken van Ryckegem got remarried – to the wealthy widower Thomas Vergrue, a manufacturer of saltpetre near the Vlissingsche Poort (Flushing Gate) in Middelburg. The groom's brother, Louys Vergrue, had married Hesther Beeckman, Isaac's youngest sister, in

About the Author

HUIB ZUIDERVAART is a retired senior researcher at the Huygens Institute of the Royal Netherlands Academy of Arts and Sciences in Amsterdam. He published widely on early modern science, especially with regard to the role of scientific instruments, scholarly institutions, and the Dutch province of Zeeland (where he lived for 23 years).

1626. After her marriage Janneken continued to care about the Beeckman family. In 1636, when a sister and other relatives of Isaac Beeckman's wife, Cateline de Cerf, moved from Calais to Middelburg, the couple offered shelter to them. See: *JIB*, IV, p. 248. Her portrait (colour ill. 13) was recently rediscovered. An inscription on the frontside of the painting 'Aetatis 36, A° 1636' and a specific mark (*huismerk*) on the back, described in a list made in 1776, fit with those of a hitherto unidentified female portrait in a German museum. See: Daniël Radermacher, 'Memorieboek van portretten [1776]', *Wapenheraut* (1916), pp. 261-274, esp. 265. Some of Radermacher's paintings were auctioned together with his library in June 1803. See: Marjam Bogers et al., *Veilingcatalogi Zeeland 1731-1925* (Middelburg: Zeeuwse Bibliotheek, 1987), no. 22. See further: Mirjam Neumeister, 'Bildnis einer Dame', in: *Holländische Gemälde im Städel 1550-1800, Band 1: Künstler geboren vor 1615* (Frankfurt am Main: Städelsches Kunstinstitut, 2005), pp. 245-254.

12 Musical Culture in Middelburg in the Times of Isaac Beeckman

Albert Clement

Abstract

This chapter gives a first, overall impression of musical culture in Isaac Beeckman's hometown Middelburg and its environment. Middelburg's long musical tradition, the Reformation, the explicit presence of musical instruments in Beeckman's times, the activities of important instrument builders, including the Grouwels and Burgerhuys families, domestic music making, and several individuals, including Jacob Cats and Adriaan Valerius, are discussed. Public as well as domestic music making are described. Music appears to have been omnipresent in Beeckman's time in Middelburg and the developments as described in this essay must have made a lasting impression on him.

Keywords: Isaac Beeckman, Middelburg, music, keyboard instruments, cultural history

Isaac Beeckman has paid much attention to music all his life, and it seems likely that the seed for this fascination was laid during his youth. Middelburg appears to have had a lively musical culture in Beeckman's times; yet, it has never been mapped out.[1] This essay aims to give an impression of a number

1 Several periods in time have been described in issues of the Exempla Musica Zelandica series (a series of scholarly editions of sheet music, related to Zeeland; general editor: Albert Clement), published by the Royal Zeeland Society of Arts and Sciences (Koninklijk Zeeuwsch Genootschap der Wetenschappen). An essay describing Middelburg's musical life in the second half of the eighteenth century was published on the occasion of this society's 250th anniversary: Albert Clement, 'Het muziekleven in Middelburg ten tijde van de oprichting van het Zeeuws Genootschap: een nadere verkenning', in: Arjan van Dixhoorn, Henk Nellen, and Francien Petiet, eds., *Een hoger streven. Bouwstenen voor een geschiedenis van het Zeeuws Genootschap, 1769-2019*

Van Berkel, Klaas, Albert Clement, and Arjan van Dixhoorn (eds.), *Knowledge and Culture in the Early Dutch Republic: Isaac Beeckman in Context*. Taylor & Francis Group 2022
DOI: 10.5117/9789463722537_CH12

of aspects related to music in Beeckman's hometown and its environment, and in doing so, to sketch a background that may have influenced Beeckman in his earlier years.

Although it is difficult to determine with certainty to what extent Beeckman possessed musical talent,[2] the many remarks related to music in his *Journal* reveal a profound interest that is beyond any doubt. Beeckman does not discuss polyphonic music from the Renaissance in his *Journal*.[3] By contrast, much attention is paid to Genevan Psalms, in particular relating to questions of modality, intonation, the practice and notation of leading tones (*musica ficta*), correct harmonization, etc. Born in 1588 in the Reformed milieu of a city in which many Protestants had settled down after the fall of Antwerp in 1585, this cannot come as a surprise. The explicit presence of keyboard instruments (organs, harpsichords) and bells (the introduction of the carillon) in Middelburg as well as the tuning of keyboard instruments seem to have aroused Beeckman's serious interest: he discusses matters of tuning (and the differences between organs and harpsichords related to this) rather in depth. The 'floating' of a pitch which is out of tune (against a properly tuned tone) is described by him with a fine term, *wywauwen* – in Dutch serving as an onomatopoeic word.[4]

Early Evidence of Music in Middelburg: The 1364 Procession

Music has played a major role in Middelburg throughout the centuries. An early example testifying to the fact that Middelburg can boast on a very long musical tradition indeed, is an event that took place in 1364. On 23 May of that year a crowd of musicians poured into Middelburg to add to the pomp

[*Archief KZGW* (2019)] (Middelburg: Koninklijk Zeeuwsch Genootschap der Wetenschappen, 2019), pp. 241-264.

2 The very fact, for instance, that Beeckman did not have a great voice (see below), cannot serve as an indication: it is a well-known fact that no lesser composer than Wolfgang Amadé Mozart (1756-1791) had difficulty singing in tune. Also, the fact that Beeckman's skills in harpsichord playing may have been limited does not necessarily provide information on his musical talents, but possibly more about his lack of time to practice.

3 One might regard Beeckman's reference to Andreas Speth, *Psalmorum Davidis* [...] *paraphrasis* (Heidelberg: Petrus Mareschal, 1596), as the exception. However, these settings in four parts of the Genevan Psalter on German texts by Ambrosius Lobwasser are in fact homophonic rather than polyphonic.

4 *Journal tenu par Isaac Beeckman de 1604 à 1634*, publié avec une introduction et des notes par C. de Waard, 4 vols. (The Hague: Martinus Nijhoff, 1939-1953) [henceforth *JIB*], I, p. 130.

of the annual procession.[5] The performers (see below) were of all kinds; *vedelers* (the *vedel* was a predecessor to the violin), *gitterners* (musicians performing on a small kind of lute), wind players, singers (both male and female), and trumpeters. They were drawn from varied sources, including Middelburg itself, other cities, courts, and the Church. Such a grand display of talent might lead to the conclusion that the city fathers of Middelburg placed a high value on culture in general, and music in particular. Without excluding this possibility indeed, it seems even more likely that the city was making a statement about its ambitions, for in the late Middle Ages most music was tied to ritual, and ritual, in turn, was closely linked with social, political, and economic undercurrents.

In short, while the accounts which record the festival in Middelburg tell us a great deal about instrumental practices in the fourteenth century, they also reveal the outlines of larger forces at work in the town and region. Middelburg was emerging as one of the new, dynamic towns in the Low Countries. It had claimed, though briefly, the staple for English wool (challenging powerful Bruges in the process) and was successful in achieving long-term dominance in the trade in French wines. By pursuing such goals, Middelburg's ambitions came into conflict with those of other centres. Close at hand were the other harbour towns of Zeeland, in particular Veere, Vlissingen, and Zierikzee. In this competition Middelburg was generally quite successful. As the stature of the town rose, however, it then encountered more formidable opponents in the larger cities.[6]

Against this backdrop of courtly and civic ambitions, 51 musicians arrived in Middelburg in the spring of 1364, all mentioned by name, and many with their instruments indicated. The information is recorded in the account book of that year, one of the rare surviving documents of the times; in fact, this specific information has survived only because that the account was

5 If a city at that time had a *jaerlicschen ommeganck* (circumambulation once a year), it was almost always the sacrament procession (in late May or early June), decreed by Pope Urban IV in 1264 with the bull *Transiturus de hoc mundo*. This is also the case here. Although formally an ecclesiastical activity in celebration of the fact that Jesus Christ offered himself to the faithful in the form of bread and wine (*Sanctissimi corporis et sanguinis Christi solemnitas*), this public event required permission from the city authorities. The latter gradually took over the organization.

6 This information can be found in: Keith Polk, 'Instrumenten en muzikanten in een middeleeuwse stad', in: Louis Peter Grijp, ed., *Een muziekgeschiedenis der Nederlanden* (Amsterdam: Amsterdam University Press-Salomé, 2001), pp. 36-43. Bruges, of course, could draw upon a tradition of decades of dominance as the leading commercial city of the north; Antwerp, too, was rapidly achieving enormous popularity with foreign merchants, and would completely overwhelm the competition in the region in the next century. Dordrecht's previous position of strength was beginning to erode.

published in 1926[7]; the original was destroyed in the bombardment of the city in 1940. First mentioned among the instruments was the *vedel*. The *vedelers* were followed by three *gitterners*, six players of psaltery (a small plucked instrument, played on the lap), two of them women, and one player of the rebec (a small bowed instrument). Two wind instrument players, probably playing the shawm (the medieval version of the oboe), were also mentioned, with a separate entry for 'Gillis den piper', who was probably playing a bagpipe. The loud timbres were further represented by at least three pairs of trumpets. Six vocalists, three of them women, are specifically included in the accounts. Five other women were probably also singers. The paymaster in Middelburg recorded both the name and the instrument of many of the musicians, which makes it possible to track down further details about several individuals.[8]

The Middelburg procession accounts of 1364 reveal a mature, sophisticated musical culture. The performance practice was evidently dynamic, showing an interplay between tradition and innovation. Tradition was maintained in the division of instruments between *haut* and *bas*. Innovation is shown in the recent arrival of the lute, and in the rising stature of shawm ensembles. The procession payroll figures also demonstrate that an ambitious city could mount an impressive display, for the forces assembled in Middelburg could rival even court festivities sponsored by Albrecht, Count of Holland and Zeeland, and Duke of Bavaria. The city was thriving, and the town fathers chose to signal their prosperity with visible and aural demonstration of their rising stature.

Two Composers from the Renaissance

From the information that has come down to us, it can be safely assumed that music played an important role in Middelburg and its surroundings in the fifteenth and sixteenth centuries as well. Within this context, the names of two composers born not far from Middelburg before Beeckman may be mentioned briefly.

7 The text giving the extract from the Middelburg accounts for 1364 is available in: W.S. Unger, *Bronnen tot de geschiedenis van Middelburg in den landsheerlijken tijd* (The Hague: Martinus Nijhoff, 1926), II, pp. 110-111.

8 See: Polk, 'Instrumenten en muzikanten', pp. 41 sq. The first *vedeler* mentioned, one of three musicians paid at the highest rate of 56 'groten' (groschen), was Willem den Corten, who must have been a striking personality. He was recorded throughout the Low Countries performing at courts (e.g. in The Hague for Count Albrecht, the highest authority for Zeeland, and for Jan van Blois, his deputy), and in other cities, including Dordrecht and Deventer.

The first name is that of Wulfaert Hellinck, born, as recent research has shown, most likely in Axel in 1493 or 1494.[9] He is one of the truly great composers of the Renaissance, and his works have been spread and published in Italy, France, Germany and the Netherlands. Wulfaert Latinized his name to Lupus in 1523, when he was appointed as succentor at the church of Sint Donaas (Saint Donatian) in Bruges, after having been admitted there as a choirboy in 1506. Next to fifteen Masses for four and five voices, he composed motets, French chansons, and songs in Dutch. He also published eleven German chorales which indicate that he had clear sympathies towards the Lutheran Reformation, although being employed by the Roman Catholic Church. Moreover, he and his singers participated in the play presented by the Bruges chamber of rhetoric 'De Helighe Gheest' (The Holy Spirit) at the contest of such chambers in Ghent in 1539, after which the complete edition of all plays was placed on the Index of the Roman Catholic Church. Hellinck's liberal ideas may be explained through a connection with Savonarola, who was critical of the Church. Hellinck stayed at the court in Ferrara, where Savonarola was born and held in high esteem. There, Hellinck could have been exposed to his influence, and, in turn, have developed sympathetic feelings for the Reformation. At his death in January 1541 Hellinck was praised as 'the prince of all musicians'.[10]

Even before Hellinck was born, Renaissance polyphony sounded in churches in Zeeland. Thus in 1442 the so-called *getijden* (Seven Hours) were sung in Mary Magdalene's church in Goes on all Sundays and holy days, and perhaps even daily between 1443 and 1452; in 1471 it was decided with certainty to have the Hours sung 'perpetually'.[11] Hellinck's works were published by Tielman Susato in Antwerp and four-part books printed by him and used by the *getijden* singers in Goes have survived.[12] This *getijdencollege* was still financially supported by the town of Goes in 1576.[13] Other towns also had such choirs, where liturgical singing could develop into real art. Such choirs were present in Hulst and Zierikzee, but also

9 The latest biographical information on him can be found in Exempla Musica Zelandica IX: *Lupus Hellinck: Three Four-Part Masses* (Middelburg: Koninklijk Zeeuwsch Genootschap der Wetenschappen, 2016), provided with a 'Biographical Sketch' by Bonnie J. Blackburn. The Masses published in this edition are the *Missa Peccata Mea*, the *Missa Intemerata Virgo*, and the *Missa Virgo Mater Salvatoris*. In our times, Axel is a town in the south-west of the Netherlands, located in the municipality of Terneuzen, Zeeland, and close to St. Niklaas, Belgium.

10 *Lupus Hellinck, Three Four-Part Masses*, p. viii.

11 Eric Jas, *Piety and Polyphony in Sixteenth-Century Holland: The Choirbooks of St Peters's Church, Leiden* (Woodbridge: The Boydell Press, 2018), p. 38.

12 Jas, *Piety and Polyphony*, p. 39; P. Scherft, *Een speurtocht door Zeeuws muziekverleden* (Middelburg: Koninklijk Zeeuwsch Genootschap der Wetenschappen, 1984), p. 4.

13 Jas, *Piety and Polyphony*, p. 39.

in the collegiate churches of Middelburg: the Westmonsterkerk (or Sint Maarten) and Noordmonsterkerk (or Sint Pieter). Six singers performed daily in Middelburg, Masses and Vespers (Evensongs) on Sundays and holy days included; however, contrary to the situation in Goes, this tradition was ended in 1574, when Goes and other cities – Middelburg included – turned Protestant and Genevan Psalms were introduced.[14]

The second name is that of Ghiselin Danckerts, born in the city of Tholen in or around 1510. He left Zeeland, and in 1538 he became a singer in the most prestigious vocal ensemble of his time in Europe: the papal choir in the Sistine Chapel in Rome, in which he would be active from 1538 to 1565. The mere fact that he was accepted in that choir testifies of the quality of the musical training he had received in Tholen. Most probably he went to school in Tholen and had developed a love for music there. In Tholen, well-presented liturgical services enlivened by beautiful music (both Gregorian chant and polyphonic compositions) were much appreciated. An agreement between the dean and the chapter of the Church of Our Lady of Tholen and the city council, dating from October 1482, provides some detailed information. Every day, six of the best students would come to the church to strengthen the choir, so that the services could be properly arranged. These choirboys had to appear every evening to sing the praises and served to strengthen the choir at all the Masses of Mary. This was no small effort, but the students received something in return: attendance fees, a *tabbaert* (toga) sheet every year, and no less than six pairs of shoes, six pairs of slippers, and two pairs of socks. A remarkable passage in the agreement requires special attention: it is explicitly mentioned that the choirmaster would teach the choirboys daily in 'simpelen sanck, discant ende conterpont', which means: Gregorian chant, improvised music, and polyphony.[15]

Danckerts's musical reputation has traditionally been connected to his role in a theoretical dispute between Nicola Vicentino and Vincente Lusitano, because little was known about his oeuvre for many years. The five compositions known at the time were published in 2001 in the music series of the Royal Zeeland Society of Arts and Sciences. The edition includes the resolution of his famous and highly sophisticated riddle canon 'Ave Maris Stella' (fig. 12.1) in modern notation.[16]

14 *Bouwsteenen. Eerste jaarboek der Vereniging voor Nederlandsche Muziekgeschiedenis 1869-1872* (n.p., n.d.), pp. 77-78; Scherft, *Een speurtocht door Zeeuws muziekverleden*, pp. 4-5.

15 Derk Buddingh, *Geschiedenis van opvoeding en onderwijs, met betrekking tot het bijbellezen en godsdienstig onderrigt op de scholen, in de Nederlanden* (The Hague: J.M. van 't Haaff, 1843), pp. 134-137.

16 Cf. Exempla Musica Zelandica V: *Ghiselin Danckerts: The Vocal Works*, ed. Eric Jas (Middelburg, Koninklijk Zeeuwsch Genootschap der Wetenschappen, 2001).

Figure 12.1 Ghiselin Danckerts's 'Ave Maris Stella' in the form of a chessboard

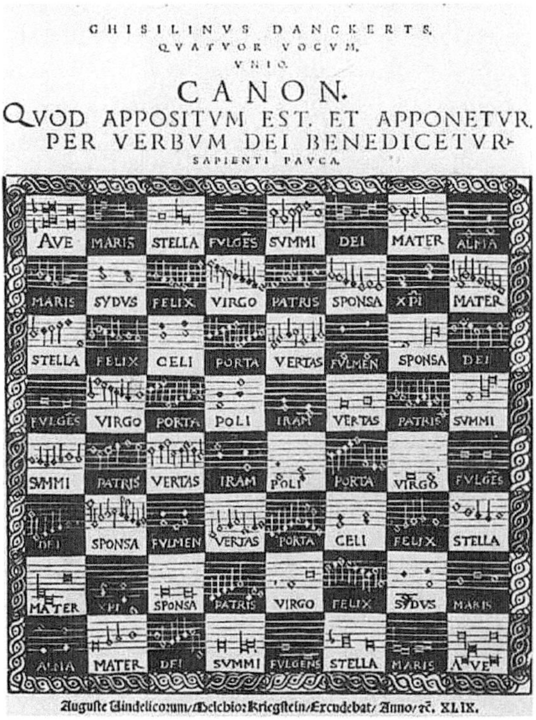

Folio sheet. Herzog August Bibliothek Wolfenbüttel, 186 musica div, fol. (1). Danckerts presented the motet in the form of a riddle with 21 possible solutions. The riddle can be solved as follows. Starting from the 'Ave' at the upper left-hand and lower right-hand squares, one voice has to move from the upper left-hand corner through the horizontal rows to the lower right-hand corner and the other voice has to move in the opposite direction. Similarly, two more voices (3 and 4) can be found, starting from the same squares, but moving through the vertical columns. Just like voices 1 and 2, they are each other's retrogrades. These four voices have the same sequence of words as their text. The squares on the diagonals contain both filled-in and empty notes. The vertical parts have to select empty notes; thus the filled-in notes belong to the horizontal parts.

More recently, Antonio Morelli presented a new source for Danckert's music, which, after further study and transcription, could significantly enrich our knowledge of his work.[17] Although both Hellinck and Danckerts did not stay in the environment where they were born, nowadays called Zeeland, the musical roots of these two composers of major importance must have been sound.

17 Arnaldo Morelli, 'A New Source for the Music of Ghiselin Danckerts, "musico e cantore cappellano della cappella del papa"', *Tijdschrift van de Koninklijke Vereniging voor Nederlandse Muziekgeschiedenis* [henceforth *TVNM*] 64 (2014), pp. 47-75; earlier published in Italian in: *Recercare* 21 (2009), pp. 75-110.

Singing Towers and Silenced Organs

One of Middelburg's long musical traditions is that of playing the bells of various towers. Middelburg was one of the cities of the Northern Netherlands in which a so-called *voorslag* (first stroke) sounded in the fifteenth century. It was located in the tower of the Westmonster Church. In 1526, a chime was hung in the town hall. The construction of a city carillon was started at the end of the century. The creation of this instrument, to be placed in the Abbey Tower (fig. 2),[18] as well as the construction of a clockwork was made around 1595 by master builder Hendrick van Trier and his (former) fellow craftsman Jan Burgerhuys.[19] The new bells were inspected and approved by Simon de la Barre, *musicien* (musician), possibly from Hainaut, who in 1612 received £20 'for his services and work carried out in inspecting the new clocks, serving as clockwork in the Abbey Tower' ('over sijnen dienst ende arbeijt gedaen int accordeeren vande nieuwe clocken, dienende tottet horologie opden Abdijetoren').[20] This same musician had been involved in the clockwork of the Middelburg town hall already in 1580. Together with his father, he was paid £10 for his services 'for giving instructions related to the clockwork of the town hall' ('int geven van ordinantien van het oorlogie op het stadthuijs').[21] Simon de la Barre was also charged with taking care of the tuning and general condition of the clockwork, and he made melodies for the clockwork as well.[22] This clockwork could also be played manually, as the city accounts include payments for the *beijaerden* (chiming) of the bells

18 In later times, this tower became more generally known as the 'Lange Jan' ('Tall John').

19 F.A. Hoefer, 'Aanteekeningen betreffende de klokkenspellen van Middelburg', *Archief ZGW* 8 (1902), pp. 1-40, esp. pp. 21-22. The construction of the 'nyeu orlogie voor de stadt' ('new clockwork for the town') seems to have started in 1593. See H.M. Kesteloo, 'De stadsrekeningen van Middelburg IV. 1550-1600', *Archief ZGW* 7 (1891), pp. 1-143, esp. p. 74. Huibert Martin Kesteloo (1842-1918), who served as town clerk of Domburg for more than 40 years, published – among other things – extensively on the city accounts (and church accounts) of Middelburg. Since the demolition of the city archives during the Second World War, these publications have been a source of inestimable value. They are used in this chapter for the first time to gain a further general insight into musical life in Middelburg in Beeckman's time.

20 H.M. Kesteloo, 'De stadsrekeningen van Middelburg V. 1600-1625', *Archief ZGW* 8 (1902), pp. 41-120, esp. p. 79.

21 Hoefer, 'Aanteekeningen betreffende de klokkenspellen', p. 12. The collaboration with Simon de la Barre seems to have worked out well, and was continued by Jan Burgerhuys's son Michael, e.g. in 1627, when the carillon for Tholen made by him was to be inspected. André Lehr, *Van paardebel tot speelklok. De geschiedenis van de klokgietkunst in de Lage Landen* (Zaltbommel: Europese Bibliotheek, 1981), p. 173.

22 Hoefer, 'Aanteekeningen betreffende de klokkenspellen', p. 12.

Figure 12.2 Carillon in the tower of the Abbey church

Engraving (detail) from: Matthias Smallegange, *Nieuwe Cronyk van Zeeland* (Middelburg, 1696). The carillon was already installed in the tower in the times of Isaac Beeckman. He must therefore have heard the sound of its bells daily.

of the clockwork of the town hall in the years 1592-1645,[23] i.e. throughout Beeckman's life. The first city carillonneurs in a long tradition, still existing today, were appointed, and the instruments of the Abbey Tower and the town hall were both played by carillonneurs, especially appointed for each tower (fig. 12.2).[24] This all happened during Beeckman's youth, and will not have gone unnoticed. The name of one musician of Beeckman's times is known: Gilles Sacharias was responsible for the bells of the town hall from 1592 until his death, and in addition for performing on the Abbey Tower as well as on the organ of the Nieuwe Kerk in the years 1611-1638. He died in May 1638, and was buried in the Nieuwe Kerk.[25]

In addition to carillonneurs, bell ringers were active. A number of examples can be found in the Middelburg city accounts; mostly this was related

23 Hoefer, 'Aanteekeningen betreffende de klokkenspellen', pp. 35-36.

24 Albert Clement, 'Houd de Zeeuwse beiaardtraditie hoog!', *Zeeuws Tijdschrift* 61:5-8 (2011), pp. 60-63; Exempla Musica Zelandica I: *Pierre Bustyn, IX Suittes pour le clavessin*, Amsterdam c. 1712, facsimile ed., ed. Albert Clement (Middelburg: Koninklijk Zeeuwsch Genootschap der Wetenschappen, 2011), Introduction.

25 Hoefer, 'Aanteekeningen betreffende de klokkenspellen', p. 35.

to the conquest of cities. Two examples of the sixteenth century may be mentioned here. In 1574, the year in which the city turned Protestant, bells from three Middelburg towers were ringing on the occasion of the Leidens Ontzet (the siege and relief of Leiden on 3 October 1574, the day on which the Spanish army fled).[26] Another city account mentions a payment to Hendrik Jansz. Bevelander 'for sounding five long peals of the big bell of the Abbey church for the triumph of the Prince of Breda' ('over dat hij geluijt heeft vijff langhe poosen de groote clocke vander Abdie kercke over de tryumphe van prynse van Breda').[27] One of the many other occasions in Beeckman's earliest youth took place in 1600. On 31 May, the city bell ringer was paid to ring the bells with his company in both churches (obviously the Noordmonster and the Nieuwe Kerk) 'because of the happy arrival of the ships called the *De lange Bercke* [The Long Birch] and *De Sonne* [The Sun], coming from the East Indies' ('over de blijde incompste vande schepen genaemt *De lange Bercke* ende *De Sonne*, commende vuijt Oost Indien').[28] Two examples from later times may suffice to make it clear that this tradition was put into practice throughout Beeckman's life. The bell ringer Jacob Stagniete received £1-10 (Flemish) for ringing the big bell in 1628

> at the time of the celebration of the conquest of the silver ship fleet of the King of Spain, conquered by the naval fleet of the registered West Indian Company of the United Netherlands under the command of noble and brave Pieter Pietersen Hein as general of the same fleet, and such in the Bay of Matanzas in the West Indies [tijde van de vieringhe ter cause van de veroveringhe van de silver schips vlote des Conincx van Spaegnien, verovert bij de scheepsvlote der Geoctryeerde West Indische Compagnie der vereenichde Nederlanden onder 't beleijt van E ende manhafte Pieter Pieterss Heijn als Generael der selver vlote ende dat in de Baye van Matanca in West Indie].[29]

In 1629, the bells were ringing because of the conquest by the fleet of the West Indian Company of the city of Olinda in Pernambuco in Brazil, and

26 H.M. Kesteloo, 'De stadsrekeningen van Middelburg IV', p. 136. This *ontzet* is still celebrated in Leiden every year.

27 Kesteloo, 'De stadsrekeningen van Middelburg IV', p. 138. Kesteloo mentions the year 1589, although Breda was actually conquered in March 1590.

28 Kesteloo, 'De stadsrekeningen van Middelburg IV', p. 92. At certain (festive) events, other musicians were also performing, e.g. on trumpets.

29 Kesteloo 'De stadsrekeningen van Middelburg VI. 1626-1650', *Archief ZGW* 8 (1902), pp. 1-98, esp. p. 89. In this chapter, the symbol £ always refers to Flemish pounds.

also for the celebration of the conquest of Wesel and 's-Hertogenbosch and the departure of the enemy from the Veluwe.[30]

The tradition of liturgical singing ended when Middelburg turned Protestant. Not only was Renaissance polyphony replaced by Genevan psalms at this occasion; moreover, organs were put out of use and several were removed, e.g. from the Westmonsterkerk, the large organ of which was bought by the *baljuw* (bailiff) of the city at a rate of £73 (Flemish), and the small organ to a citizen who had settled down in Middelburg only six years earlier, François Valéry.[31] The city accounts show payments to several musicians for the compositions of Psalms, e.g. to Matthijs Mercker, *musicien*, who received £3 for 'the dedication of Psalm 128 in eight parts, set to music' ('de dedicatie van den 128 psalm met acht partijen op musijcke gestelt') in 1606, and to Hendrik Speuij, who received £5 for a 'music book of King David's Psalms, composed by him and presented to the city'('musyckboucken van de psalmen Davidts bij hem gecomponeerd ende de stadt vereert') in 1610.[32]

The period of silenced organs (as required by the synod of Dordrecht in 1574) in Protestant churches did not last long in the Netherlands.[33] Very soon after 1578 – the year in which the National Synod, held in Dordrecht,

30 Kesteloo, 'De stadsrekeningen van Middelburg VI', p. 89. Later occasions giving rise to bells ringing include the conquest of 'Paribo' (Brazil) in 1634, the conquest of Breda in 1637, and the birth of a son of the king of France in 1638 (Kesteloo, 'De stadsrekeningen van Middelburg VI', p. 90). 'Paribo' refers to Paraíba, now Paraíba do Sul (a town in the Brazilian state of Rio de Janeiro).

31 Kesteloo, 'De stadsrekeningen van Middelburg IV', p. 67. Some more information about this Valéry will follow below.

32 H.M. Kesteloo, 'De stadsrekeningen van Middelburg V', pp. 74-75. Matthijs or Matthias Mercker was an organist and composer, probably born in Amsterdam around 1575, and mainly active in the Dutch province of Friesland between 1599 and 1622. His survived works include seven fugues for four instruments (1609) and instrumental dances such as pavanes ('paduanes') and galliardes. Hendrik Joosten Speuy (Brielle, 1575-Dordrecht, 1625) also was an organist (appointed as such in 1595 at the Grote Kerk in Dordrecht) and composer, nowadays still known for his edition *De Psalmen Davids / gestelt op het Tabulatuer van het Orghel ende Clavecymmel / met 2. Partijen* (Dordrecht: Peeter Verhaghen, 1610), possibly the earliest example of printed keyboard music from the Netherlands. It included 21 arrangements of 20 Psalm melodies from the Genevan Psalter, two settings of the Lord's Prayer (*Vader in ons Hemelrijck*), and a setting of the *Magnificat*, all arranged for performance on a house organ or harpsichord, although Speuy will have had church organists in mind as well, who were only a few years after 1574 again required to play Psalm arrangements before and after church services. Speuy himself sent copies of his edition not only to Middelburg, but also to other cities, including Gouda and places abroad (including Great Britain). His Psalm arrangements became well-known in the seventeenth century.

33 The very fact that these organs were at many occasions property of the civil government rather than of church authorities was obviously also of importance here.

decided that organs should be removed from the churches[34] – organists were asked to play before and after the sermon, as well as before and after the church service. Furthermore, organ performances took place regularly. Town organists were appointed by Dutch cities in the seventeenth century and required to give recitals. Middelburg was by no means an exception. Thus, the municipality paid £50 to organ builder Jan Roose for a new organ in 1597, followed by payments of £300 in 1598, and £216-13-4 in 1599. Another payment in 1599 was made for details of the organ case. Later payments reveal that this new organ was placed in the Nieuwe Kerk (i.e. the large church in the centre of the city, which soon became its main church).[35]

Due to blindness, Roose could not complete this instrument himself, but a drawing testifies of his quality in the field of organ architecture (colour ill. 12). The organ was completed by Johan Morlett (from Arnhem) in 1603, and tested and approved in that same year by no less an expert than Jan Pieterszoon Sweelinck (1562-1621), the famous organist and composer from Amsterdam. His verdict must have been favourable, judging from the generous gratuity Morlett received.[36]

Instrument Builders

Middelburg had its own organ builder in the times of Beeckman: Jan Roose (see above).[37] He worked in Utrecht in the middle of the sixteenth century and settled in Middelburg around 1561, probably at the behest of the new bishop of Middelburg, Nicolaas de Castro (1503-1572), who was welcomed in this capacity in Middelburg on 31 December 1561. When Middelburg turned Protestant, organs were put out of use and some were removed. Roose left the city, but the large organs of the Noordmonster (St. Pieter)

34 With regard to this situation in Amsterdam and the position of the most important musician in the Netherlands of the time, see: Albert Clement, 'Jan Pieterszoon Sweelinck: een stadsorganist van wereldfaam tussen calvinisme en katholicisme', in: Grijp, ed., *Een muziekgeschiedenis der Nederlanden*, pp. 182-189.

35 Kesteloo, 'De stadsrekeningen van Middelburg IV', p. 72. Beeckman's father did not attend services in this church and went to church in Arnemuiden with his family after 1603. However, this does not detract from the importance of this instrument, which was undoubtedly played by the organist at other occasions.

36 Cf. Kesteloo, 'De stadsrekeningen van Middelburg V', p. 79.

37 The information provided on Roose here can be found in: J.H. Kluiver, 'Historische orgels in Zeeland I', *Archief ZGW* (1972/1973), pp. 43-116, esp. p. 46; J.H. Kluiver, 'Historische orgels in Zeeland II', *Archief ZGW* (1974), pp. 23-135, esp. pp. 56-58.

and the Westmonster (St. Maarten), found a new destination in the St. Martinkirche and the Ueberwasserkirche in Münster. When the organs were given a function in public musical life in the Netherlands, it was Roose who became involved again. Obviously, his good name was maintained in Zeeland, because he also restored the organ in Brouwershaven in 1588, and worked on other organs in Zeeland as well. At the end of the century, many town councils realized that the sudden departure of all church organs had a negative influence on public music. Thus, organs were reintroduced in many cities. In Middelburg, Roose was commissioned to build a new organ for the Abbey church (Nieuwe Kerk) in Middelburg (see above) (colour ill. 14).

The developments related to the carillon gave rise to the establishment of a bell foundry in Middelburg. Jan Burgerhuys, the former master servant of Hendrick van Trier, originating from Aachen, decided to settle down in Middelburg in 1593. Together they moulded a number of bells for the Abbey Tower, for which Van Trier was paid £333-6-8 (Flemish) in 1594.[38] His helpers and his wife were also paid several amounts; furthermore, the city reimbursed Jan Burgerhuys, who rented a house next to the entrance of the Nieuwe Kerk.[39] The way in which these payments are described – e.g. 'Jan Burgerhuys, bell-founder, for what he created with his people and son in a period of 19 days' ('Jan Borgeroys, clockgieter, over dat hy met zijn volck ende zone gewrocht hebben den tijt van 19 dagen')[40] – shows that he was not merely 'an' employee of Hendrick van Trier, but a fellow on his way to becoming a master himself.

Jan Burgerhuys's activities marked the beginning of a family tradition. He poured not only bells (including a number for Scotland) but also artillery and mortars. Jan was buried in Middelburg on 9 November 1617. His sons Evert (about whom almost nothing is known) and Michael had the same profession and continued the family business. Michael enjoyed several privileges from the city, such as exemption from excise duty as well as the use of the city's warehouses for making moulds. Bells, guns, and other castings are known from him. In 1627 he cast a small carillon of nineteen bells for the town hall of Tholen. Michael also delivered bells to Scotland, of which about 50 are known. The delivery probably took place via a merchant in Middelburg. He died in 1651 and was buried on 5 April 1651 in the 'Choorkerk'

38 Kesteloo, 'De stadsrekeningen van Middelburg IV', p. 75.

39 Cf. Kesteloo, 'De stadsrekeningen van Middelburg IV', p. 75; W.S. Unger, 'Het klokkengietersgeslacht Burgerhuys', *Archief ZGW* (1926), pp. 19-29, esp. pp. 1-2.

40 Unger, 'Het klokkengietersgeslacht Burgerhuys', p. 2.

(nowadays called 'Koorkerk' or the Choral Church, which is the smaller of the two Middelburg Abbey churches).[41]

Joannes Burgerhuys was a son of Michael. He married Judith van Bruynich with whom he had two daughters. He was buried in Middelburg on 3 January 1679. In official documents by the States of Zeeland he was simply referred to as the country's gun- and bell-founder. As such he first asked the city of Middelburg and later the States of Zeeland for an annual wage of 200 guilders, the heavy tools needed for moulding and casting, as well as free housing. In 1658 he obtained a patent for Zeeland, i.e. the exclusive right to supply bells and artillery in this province. His casting house was located in the old refectory (the present Statenzaal) of the Abbey. Like his father and grandfather, he also supplied bells to Scotland. After his death – he was also buried in the Koorkerk on 3 January 1679[42] – the company was dissolved. Over the course of more than 70 years, three successive generations of this family made bells for many churches and towers in Zeeland, but also for other provinces and countries, including Friesland and Scotland.[43]

When Beeckman was aged four, it was not only Jan Burgerhuys who settled down in Middelburg: the municipal financial records of 1593 list that a *clavesimmaker* (harpsichord maker) Johannes Grouwels became a *poorter*,[44] meaning that he had obtained the rights of citizenship.[45] Moreover, on 6 November of that same year, *clavesimmaker* Lodewijck Grouwels – in all likelihood Johannes's son – bought a house in Middelburg.[46] They both came from Antwerp, where they had been taught in the craft of instrument building according to the Flemish tradition of harpsichord building. Only one virginal built by Johannes Grouwels has survived. It is now preserved in

41 Unger, 'Het klokkengietersgeslacht Burgerhuys', p. 24.

42 Unger, 'Het klokkengietersgeslacht Burgerhuys', p. 24.

43 Information on this family can be found in the articles mentioned above by Kesteloo, Hoefer, and Unger, as well as in an earlier contribution: J.W. Enschedé, 'Het geslacht Burgerhuys, klok- en geschutgieters te Middelburg', *Bulletin van den Nederlandschen Oudheidkundigen Bond* 2/9 (1916), pp. 221-222. One of the surviving smaller bells, a so-called table bell, made by Michael Burgerhuys, is now preserved in the museum of Zierikzee: www.stadhuismuseum.nl/collectie/klokken-kanonnen-bellen/.

44 Kesteloo, 'De stadsrekeningen van Middelburg IV', p. 94.

45 At the beginning of the Dutch Revolt, Antwerp remained under Spanish rule. However, on 4 November 1576, Spanish soldiers mutinied and massacred 6,000 people and burned 800 houses in the city. After the departure of the Spanish troops Protestants could openly profess their faith in Antwerp, but after the fall of the city in 1585 Protestant citizens were given two years to arrange their affairs and leave. Keyboard maker Lodewijk Theewes settled down in London, and his colleague Marten van der Biest went to Amsterdam. Johannes Grouwels, appointed master of the Guild of Saint Luke in 1579 despite being Protestant, fled with his family to Middelburg, a nearby stronghold of the Dutch Republic.

46 Alan Curtis, 'Dutch Harpsichord Makers', *TVNM* 19 (1960/1961), pp. 44-66, esp. p. 51.

the Musical Instruments Museum, Brussels. The instrument has a compass of three octaves and a sixth. Also one virginal by Lodewijck Grouwels is still known. It is kept in the Metropolitan Museum of Art in New York and may well be the earliest surviving virginal made in the Northern Netherlands.[47] Its case shows various images (see below) and the soundboards are painted with flowers and fruit, each with a rose bearing the initials L.G., referring to the maker of this double virginal: 'Ludovicus Grovvelus' (colour ill. 15).

With its main keyboard on the left, Lodewijck made another chest on the right that could hold a very small virginal inside, called the *octave virginal* because its range was an octave higher. The eighteenth-century Middelburg notary and author of the first Dutch music dictionary Joos(t) Verschuere Reynvaan (1739-1809) described this very luxurious type of virginal and called it *mother with child* – a term which became widely accepted and used by musicologists later. It conceivable that Verschuere Reynvaan was inspired by a Middelburg instrument made by Grouwels.[48] The octave virginal could be taken out and placed on top of the instrument above the main keyboard. Thus, one could transform the instrument into a double manual virginal. Moreover, the keys of the mother keyboard could be connected to those of the octave keyboard through a slot that the latter had in the bottom of its case.[49]

Besides paintings, the virginal bears the Latin inscription 'LVDOVICVS GROVVELVS ME FECIT 1600' ('Lodewijck Grouwels made me [in] 1600'). A motto is written on the lid of the mother case: 'SCIENCIA NON HABET INIMICVM NISI IGNORANTEM' ('Knowledge has no enemy but the ignorant'). The lid of the child instrument bears the saying: 'ARS VSV IVVANDA' ('Art is to be aided by practice'). These and comparable mottos are also found on other harpsichords from the so-called Flemish tradition. Closer study of this lavishly decorated instrument reveals, among other things, that its lid painting features a combination of three themes, which, in doing so, conveys a message to the observer (colour ill. 15).[50] Entirely on the right, one sees the story of the battle between David with his slingshot and the giant Goliath. In the centre, a procession is depicted, showing David with the head of the slain giant Goliath. On the left the gates of a strong city are visible: Jerusalem (1 Samuel 17:54). The combination of these themes seems to provide a metaphor for

47 Curtis, 'Dutch Harpsichord Makers'; Donald H. Boalch, *Makers of the Harpsichord and Clavichord 1440-1840*, ed. by Charles Mould (Oxford: Clarendon Press, 1995), p. 342.

48 Clement, 'Het muziekleven in Middelburg', pp. 263-264.

49 The original *octave virginal* made by Lodewijck was lost. The current one was made by Arnold Dolmetsch at the end of the nineteenth century.

50 Cf. www.metmuseum.org/toah/works-of-art/89.4.1196/.

Middelburg, Grouwels's new home town, bordering the Catholic Spanish Netherlands and resisting the Spanish giant, who is to be defeated.

Domestic Music Making

In Middelburg, there must also have been a lively tradition of domestic music making at the time of Beeckman. In the Netherlands, the organist/carillonneur was usually also in charge of the *collegium musicum* if there happened to be one in the city. A *collegium musicum* was a group of approximately fifteen music lovers who performed under the direction of a professional musician in each member's house in turn. *Collegia* already existed in Arnhem and Amsterdam in the sixteenth century, and they also appeared in other places soon after. It is known that Middelburg already had a *collegium musicum* in 1622.[51]

Many well-to-do citizens also possessed musical instruments, and with several builders available close by, notably in Antwerp, but also in Middelburg itself, it may come as no surprise that keyboard instruments seem to have been popular in Middelburg. Within the present context, a fine example to be mentioned is how a man named François Valéry, who originated from France, settled down in Middelburg and obtained citizenship there in 1569, bought the small organ of the Westmonsterkerk for £25 (Flemish) in 1575.[52] This François, whose family name was soon changed into Valerius, was the father of the composer and poet (among many other things) Adriaen Valerius, born in that same year. We now know that the latter first and foremost served as a member of the chamber of rhetoric of Veere and author of publications connected to this institution and its literary and musical culture. Adriaen's daughter Catharina (1610-1641) was married in 1629 to Hendrik Somer (1598-1647), alderman of Veere and administrator of the West Indian Company,[53] with whom Beeckman was regularly in touch.[54] One of

51 Cf. Exempla Musica Zelandica I: *Pierre Bustyn, IX Suittes pour le clavessin*, Amsterdam c. 1712, facsimile ed., ed. A. Clement (Middelburg: Koninklijk Zeeuwsch Genootschap der Wetenschappen, 2011), Introduction.

52 Kesteloo, 'De stadsrekeningen van Middelburg IV', p. 67.

53 The parental of the Somer family can be found here: http://hzomer.antenna.nl/some/index.htm (accessed 1 November 2021).

54 *JIB*, I, p. 218. Moreover, Somer was related to the second wife of Beeckman's brother, Jacob, and himself a former pupil of the same. Cf. Klaas van Berkel, *Isaac Beeckman on Matter and Motion: Mechanical Philosophy in the Making* (Baltimore: Johns Hopkins University Press, 2013), p. 195, n. 70.

the indicators of Valerius's mastery was the invitation to participate in a poetry volume entitled *Zeevsche Nachtegael*, published in 1623.[55]

In 1626, one year after his death, Valerius's *Neder-landtsche gedenck-clanck* was published, presenting the most important histories of the Seven Provinces.[56] In between these histories, one finds 76 songs, mostly written by Valerius and set to largely unknown melodies, of which only fourteen were of Dutch origin. The texts focus on trust in God and longing for freedom. The collection gained interest in the second half of the nineteenth century and several songs became rather popular, in particular 'Wilt heden nu treden voor God, den Here', which is known in Germany as 'Das niederländische Dankgebet',[57] and in the United States as 'The Prayer of Thanksgiving'. Valerius also included a melody of the 'Wilhelmus' in a version that would much later become the national anthem of the Netherlands (fig. 12.3), with its added notes to selected syllables, e.g. in its first line (here in italics): 'Wilhelmus *van Nas*souwe ben ick van *Duit*schen bloet', etc.

Speaking of Zeeland composers from Beeckman's times, Valerius may act as *the* example. Another name may perhaps come as a surprise. It is Jacob Cats (1577-1660), whom Beeckman met with on various occasions and whom he befriended.[58] Born in Brouwershaven, Zeeland, Cats is now first and foremost known as poet, statesman, and lawyer. In the years from 1603 to 1623, he worked in several capacities as a man of law in Middelburg, where he lived in the Lange Noordstraat (nr. 31). In 1605 he bought another attractive building, called 't Munnikenhof, in Grijpskerke, completely in line with his motto: 'I shunned the bustle of the city, and chose the lonely field, because that's what my soul longed for at the time' ('Ick schoude stadts gewoel, en koos het eensaam velt, Want daar was toen ter tijt mijn wesen naar gestelt').[59]

Next to poems, Cats wrote about 65 songs. While doing so, he invented new texts on pre-existing, primarily French melodies. Cats's songs became remarkably popular, and like Valerius, Cats was represented in the *Zeevsche Nachtegael* of 1623. Using pre-existing melodies was common practice

55 A digitized copy can be found here: http://objects.library.uu.nl/reader/index.php?obj=1874-39982&lan = en#page//10/66/16/106616853264702711567765062173045344670.jpg/mode/1up.

56 A digitized copy can be found here: https://books.google.be/books?id=QlhhAAAAcAAJ&printsec=front cover&hl =nl&source=gbs_ge_summary_r&cad=0#v=onepage&q&f=false.

57 Later on, it was also used by the Nazis, which is why it is not that known anymore in Germany.

58 *JIB*, I, p. xx.

59 Jacob Cats in his autobiography *Twee en tachtig-jarig leven* (Eighty-two years of life), which he completed in 1657, but which only first appeared in 1700, in the back of his collected works.

Figure 12.3 The Wilhelmus, from Adriaen Valerius's *Neder-landtsche gedenck-clanck* (1626)

Valerius presented the melody of the 'Wilhelmus' in the version that would become the national anthem of the Netherlands.

at the time, but while other poets seem to have been satisfied when the number of notes and syllables matched, Cats made sure that the quality and characteristics of the melody on the one hand and his own text on the other were in accordance.[60] A fine example, testifying to the close attention that Cats paid to the melody, is the song 'Ziel-sucht, gepast op het hoogen en vallen van de Musicq' ('Passions of the soul, set to the rising and falling of the music'), published in the collection *Klagende Maeghden* (Mourning maidens) of 1633, consisting of a series of songs and texts about careless virgins, and published a few years before the death of Beeckman (fig. 12.4).[61]

It is obvious that with each mention of 'high' and 'low', the melody corresponds to the text. Baroque composers applied such word painting

60 This special quality seems to have been insufficiently appreciated by Beeckman when he criticizes the, in his opinion, mismatch between word accent and melody accent in Cats's songs. *JIB*, II, pp. 208-209.

61 Cf. Exempla Musica Zelandica VIII: *Jacob Cats: 'Klagende maegdhen': Mourning Maidens and Other Songs: An Anthology*, ed. Louis Peter Grijp (Middelburg: Koninklijk Zeeuwsch Genootschap der Wetenschappen, 2008).

Figure 12.4 Cats, 'Ziel-sucht', stanza 1

Jacob Cats, 'Ziel-sucht, gepast op het hoogen en vallen van de Musicq', stanza 1

when setting text to music. But in this case, it is the poet, Cats, who carefully follows the already existing melody (of English origin) with his newly composed text, throughout all stanzas.[62] In this respect, Cats did a much better job than most of his Dutch contemporaries.

62 *Jacob Cats, Klagende maegden*, pp. 61-62.

Concluding Remarks

Being born in Middelburg, Beeckman grew up in a city in which music played a very important role. Beeckman was exposed to the sound of bells in towers, performances on keyboard instruments in churches and houses, activities of instrument builders, domestic music making, the publication of collections with songs, and last but not least numerous individual contacts. These contacts were obviously not part of Middelburg's musical culture, and therefore not the subject of this chapter. Yet, they should not go entirely unnoticed here either. The best known and at the same time most important contacts of Beeckman in the field of music, not limited to his years in Middelburg, include those with Marin Mersenne, René Descartes, and Evert Verhaer. Descartes' *Compendium musicae*, his first work, was the first result of their intensive exchange of thoughts; the treatise was offered to Beeckman on 1 January 1619 and copied into Beeckman's *Journal* about two decades later. It is conceivable that through his contacts with Descartes, Mersenne, and others, Beeckman encountered a respectable number of the theoretical writings on music referred to in his *Journal*.[63] It appears that Beeckman took music lessons – in singing (although far from successful) and harpsichord – with Evert Verhaer, a former student of Sweelinck, the famous organist who had been in Middelburg in 1603 in order to approve the new organ of the Nieuwe Kerk. In Utrecht, Verhaer was active as teacher at the Hieronymusschool (Beeckman worked as vice-principal of this Latin School in 1619 himself) and as precentor of the former cathedral (the so-called Dom). He was an obvious figure of authority to Beeckman, who regularly refers to him in his *Journal* ('Mr. Verhaer says…').[64]

63 These include the following works: Jacques Lefèvre d'Étaples [Iacobus Faber Stapulensis], *Musica libris quatuor demonstrata* (Paris: G. Cavellat, 1552); Heinrich Glarean [Henricus Glareanus], *Dodekachordon* [*Δωδεκαχορδον*] (Basle 1547; repr. Hildesheim: Olms, 1969); André de Pape [Andreas Papius], *De consonantiis, sev pro diatessaron libri dvo* (Antwerp: Plantijn, 1581); Gioseffo Zarlino, *Le Istitutioni harmoniche* (Venice: Francesco Senese, 1558); Thomas Morley, *A Plaine and Easie Introdvction to Practicall Mvsicke, Set downe in forme of a dialogue* (London: Humfrey Lownes, 1608); Pierre Maillart, *Les Tons, ov discovers svr les modes de mvsiqve, et les Tons de l'Eglise, et la distinction entre iceux* (Tournai: Ch. Martin, 1610); Johannes Nucius, *Musicae poeticae, sive de Compositione cantûs praeceptiones absolutissimae* (Neisse: Chrispinus Scharffenberg, 1613); Salomon de Caus, *INSTITVTION harmoniqve* (Frankfurt: Jan Norton, 1615); Jacques Vredeman, *Isagoge Musicae. Dat is Corte, perfecte ende grondighe Instructie van de Principale Musijcke* […] (Leeuwarden, 1618); Johannes Kepler, *Harmonices mvndi libri V* (Linz: Godfried Tampach, 1619); Simon Stevin, *Vande Spiegheling der Singconst* (manuscript, c. 1606). Furthermore, one can safely assume that Beeckman also possessed a copy of Marin Mersenne's famous *Traité de l'harmonie universelle* (Paris: G. Baudry, 1627).

64 *JIB*, II, pp. 4-5, 15-19. On 28 February 1620, Beeckman paid him for three months of singing lessons. *JIB*, II, p. 19.

Although Beeckman maintained friendly contact with Jacob Cats, the poetry of the latter seems not to have been to his liking. According to a statement by Beeckman made in 1622, Cats and others only applied the iamb and trochee, and those who did this 'cannot compose a skilful poem on all kinds of musical pieces, not even on all the psalms' ('konnen geen bequamen dicht op alderhande musyckstucken maken, ja selfs niet op alle de psalmen').[65] Although one may argue that Cats's *Klagende Maeghden* are from a later date, Cats paid much closer attention to the character of the already existing melody than his Dutch contemporaries, and tried to write poetry in accordance with the character of that melody. This seems to have escaped Beeckman's attention, either because he was not aware of such songs, or because he did not have the musical knowledge or talent to establish this.

The *clavercyne* (also *klavercyne*, *claversyne*) is mentioned many times related to its tuning and construction; Beeckman played the instrument himself. Organs also had his fond attention, their tuning and technical aspects included. The same applies to bells, and Beeckman appears to have asked both organists and carillonneurs to supply him with information several times. Examples include Bartholomeus 't Fel (1578-c. 1619), organist of the Sint Lievens Monsterkerk in Zierikzee,[66] and the famous blind composer, recorder virtuoso, and carillonneur 'Joncker' (Sir) Jacob van Eyck (c. 1589-1657), carillonneur of the Dom in Utrecht.[67] In 1628, Beeckman received information about the tuning method of the deceased organist of Nijmegen, shown to him by 'the harpsichord maker and bell programmer here in Dordrecht' ('de clavercynmaker ende clockstelder hier te Dort'), Jan Dirckszoon Tegelbergh, who was also carillonneur there.[68]

Beeckman made a large number of remarks related to these instruments and also comments on new inventions, including a perpetuum mobile machine developed by Cornelis Drebbel: 'so it's not strange that Drebbel makes a harpsichord play through the heat of the Sun' ('Ten is niet vrempt dat Drebbel een klavercyne met de hitte van de Sonne doet spelen').[69] The following quotation (c. 1617) provides proof that in his youth Beeckman devoted close attention to practices related to music indeed:

> To make a bell ring when you only move the clapper. In Middelburg, someone did this, but the blows occurred too quickly, one after the other,

65 *JIB*, II, p. 208-209.
66 *JIB*, I, p. 84.
67 *JIB*, III, pp. 310-311.
68 *JIB*, IV, pp. 128-129.
69 *JIB*, II, p. 363.

> and he could not make the clapper move any slower, because then it would not have had enough strength. Therefore, had he been wise enough to make the clapper heavier – as heavy as was necessary – the slow movement would have produced a stronger sound. [Om een klocke te doen luiden als men alleen den klepel roert. Te Middelborgh dedet eene, maer de slagen gingen te ras opeen, ende hi mochte den klepel niet trager doen roeren, want dan soude sy geen kracht gehadt hebben. Daerom, hadde hi soo wys geweest ende den klepel soo veel swaerder gemaeckt als van node was; so soude meteen de traegheit een stercken klanck gehadt hebben.][70]

All in all, it seems fair to conclude that music was omnipresent in Beeckman's times in Middelburg, and that the developments described in this chapter must have made a lasting impression on the young Beeckman. These remarkable events in Zeeland's capital during the late sixteenth and early seventeenth centuries are worthy of attention in our times, and Beeckman's musical utterings provide remarkable, detailed, and at times unique information.

About the Author

ALBERT CLEMENT is Professor of Musicology (Utrecht University), Professor of Theology and Music (Theological University of Apeldoorn), and organist. He was deeply involved in setting up Utrecht University's International Honours College in Middelburg in the years 2003-2004, where he also teaches (University College Roosevelt), and introduced performing arts here into the Dutch university system. He published widely on the history of music from the fifteenth to the twentieth centuries, notably on Johann Sebastian Bach and Felix Mendelssohn Bartholdy, and supervises a large international circle of PhD students.

70 *JIB*, I, p. 117. The quote continues as follows: 'because big ships do exercise more power, sailing slowly against something, than small ships, sailing fast' ('want groote scepen doen meer kracht, tegen iet traech varende, als kleyn sceepkes, ras varende'). This is yet another example demonstrating how easily Beeckman moves to another field whilst addressing physical aspects of music.

13 *Consten*-Culture

Beeckman, the Rhetoricians, and a New Style of Philosophizing*

Arjan van Dixhoorn

Abstract

This chapter argues that Isaac Beeckman's early *Loci communes* reveal two major fields of philosophical action. The most well-known is the academic world of learning shown in his life as a teacher and in his network of correspondents (such as Descartes and Mersenne). Less well-known is the other world, which, inspired by Edgar Zilsel's thesis on the role of 'superior artisans' in the making of the mechanical science, has often but mistakenly been identified with the world of 'Beeckman the artisan'. This second philosophical world is the world of *consten*-culture; a literary world of the Dutch vernacular, in which explicit, bookish, academic, theoretical learning and tacit, bodily, artisanal, practical, experience-oriented knowledge had already been 'interpenetrating' for two centuries.

Keywords: Isaac Beeckman, *consten*-culture, rhetoricians, recipes, philosophical exercises, virtuosi

It is because we supply these living conversational contexts that we understand the directions of philosophic writing. Philosophic writing begins in conversation, and it returns to it. Because the talk that surrounds the writing of philosophy is not mere talk, philosophy is dangerous and subversive. Socrates embraced and wrestled with his friends in talk, not only for love and wisdom but also for the sake of Athens.

– Amelie Oksenberg Rorty, 'Experiments in Philosophic Genre: Descartes' "Meditations"'[1]

* This article was largely written during a research stay at the Freie Universität Berlin, graciously funded by a travel grant from the collaborative research centre SFB 980 Episteme in Motion, in the summer of 2019.

Van Berkel, Klaas, Albert Clement, and Arjan van Dixhoorn (eds.), *Knowledge and Culture in the Early Dutch Republic: Isaac Beeckman in Context*. Taylor & Francis Group 2022
DOI: 10.5117/9789463722537_CH13

1 Introduction

> Yesterday at the chamber of rhetoric of the Vreuchdendal [Valley of Joy] in Breda, I heard someone say that a ship in Harlem's Lake [Haerlemmermeer] will sink deeper than in other waters that are less silty. I will try and come up with the reason, to establish if this is true.[2]

Sometime in the winter of 1618-1619, the 30-year-old philosopher Isaac Beeckman, who was staying in Breda to help an uncle and court a young woman, turned this topic of conversation into a 'problem' of natural philosophy by logging it in his *Loci communes* and reflecting upon its causes.[3] This notebook, which according to humanist pedagogical practice he had been keeping since the age of sixteen, had by then developed from a compilation of reading notes (commonplaces) into a set of comments on problems that he encountered while reading, engaging in conversations or in observations of all sorts. Regularly, when he theorized about causes, the note taking became fully-fledged philosophizing in the sense of developing a philosophical theory on causes of natural phenomena. Such active note taking combined with the production of philosophical explanations may have been a peculiarity of Beeckman, the result of his training at the Latin School and, in theology, mathematics, and medicine, at university. The notes, however, also show how in his social environment many of those with whom he interacted were interested in and marvelled at natural and/or artificial phenomena. 'I heard someone say' references a conversation which he seems to have overheard; this gave him a 'problem', which then became the object of a philosophical inquiry: 'I will try and come up with the reason, to establish if this is true.'

This is how Isaac Beeckman philosophized, engaging with 'natural particulars' based on his own observation or that of others who gifted him their observation in conversations. While he may have been the only one from his immediate social environment to move beyond the finding,

1 Amelie Oksenberg Rorty, 'Experiments in Philosophic Genre: Descartes' "Meditations"', *Critical Inquiry* 9 (1983), pp. 545-564, p. 562.

2 *Journal tenu par Isaac Beeckman de 1604 à 1634*, publié avec une introduction et des notes par C. de Waard, 4 vols. (The Hague: Martinus Nijhoff, 1939-1953) [henceforth *JIB*], I, pp. 257-258: 'Gisteren hoorde ic in de rethoryckcamer, genaemt Vreuchdendal, te Breda seggen, dat een schip int Haerlemmermeer dieper sinct dan in ander water, dat so troublich niet en is. Hiervan soeck ic de reden, en oft ooc waer is, dat daer een schip dieper sinct.'

3 On the significance of his stay in Breda, see: Klaas van Berkel, *Isaac Beeckman on Matter and Motion: Mechanical Philosophy in the Making* (Baltimore: Johns Hopkins University Press, 2013), pp. 19-27.

collecting and contemplation of 'particulars', turning them into philosophical truth-problems by contemplating causes, his philosophizing was deeply rooted in such conversational exchanges with a multitude of people who normally we do not consider as part of the realm of philosophical enquiry; among them, artisans and rhetoricians. Despite this, the *Loci* show that Beeckman's interlocutors were far from passive in these crucial stages of his philosophical inquiries and the *Loci* also suggest that their interests predated their encounter with Beeckman, even though, without doubt, his presence must have heightened these interests. After all, the *Loci* also suggest that, wherever he went, he would serve as the local 'master of nature', to whom the virtuosi would turn with questions regarding natural phenomena that they had learned to marvel at, since these masters were known to have expert understanding of nature's secrets.

The crucial role of both these natural particulars and his many interlocutors in the making of Beeckman's so-called 'mechanical philosophy' raises questions about the role of particulars and problems of nature in the natural philosophical traditions around 1600, and about the reach of the interest in such matters beyond the more narrow confines of Beeckman's philosophical correspondents, that is, those who, like him, also engaged in contemplating natural causes and in sharing the resulting philosophical ideas. These questions are old, but far from answered yet.[4]

Indeed, the conversation on the effects of the silty nature of the lake on the draught of a ship sailing its waters also took place at what at first sight may seem to be an unlikely event: a meeting of a chamber of rhetoric, an institution not commonly known as a 'site of knowledge' of natural (or any) philosophy. Instead, the chamber of rhetoric used to be known in Flemish-Dutch *literary* history as the essentially 'late medieval' type of literary society in the Dutch-speaking Low Countries, engaging in writing, reciting, and performing poems, songs, and plays and the staging of ceremonies in the urban public sphere. It has become increasingly evident in recent years,

4 The problem of the role of artisans in the making of the new science is also known as the Zilsel thesis. On the ideas of Edgar Zilsel and on Beeckman as an ideal case, see: H. Floris Cohen, *The Scientific Revolution: A Historiographical Inquiry* (Chicago: University of Chicago Press, 1994), pp. 336-351. See also: Edgar Zilsel, *The Social Origins of Modern Science*, ed. Diederick Raven, Wolfgang Krohn, and Robert S. Cohen (Dordrecht: Kluwer Academic Publishers, 2000 [a collection of essays from the 1930s and 1940s]; and also: Paolo Rossi, *Philosophy, Technology, and the Arts in the Early Modern Era*, trans. by Salvatore Attanasio, ed. by Benjamin Nelson (New York: Harper & Row, 1970); Pamela O. Long, *Openness, Secrecy, Authorship: Technical Arts and the Culture of Knowledge from Antiquity to the Renaissance* (Baltimore: Johns Hopkins University Press, 2001); Pamela H. Smith, *The Body of the Artisan: Art and Experience in the Scientific Revolution* (Chicago: University of Chicago Press, 2004).

however, that these chambers of rhetoric were conceived by their members, the rhetoricians, *not* as a literary institution (like 'science', 'literature' is a modern, not an early modern category), but rather as the centre of a larger culture of *consten*, the Dutch equivalent of the Latin *artes*. They can be considered the first institutions of '*artes* in society' in the Low Countries and their members, the rhetoricians, were major agents in a movement aimed at accomplishing the domestication of the *artes* in daily life building on early traditions from the fourteenth century.[5]

This chapter argues, then, that Beeckman was not only the product of the Latin culture of learning of the Latin School and the university, but also of this vernacular, practice-oriented, culture of knowledge. The chapter aims to show how key features of *consten*-culture can be identified in the typically Beeckmanian style of philosophizing as recorded in his *Loci*, at least in the early years, before he entered the context of the Latin School as a teacher.[6] These features are essential if we want to solve the problem of his 'hybridity', his mediation between (artisanal) 'practice' and (philosophical) 'theory', and how that 'hybridity' itself influenced his philosophizing.[7] He diverged *from* other virtuosi with whom he engaged by shaping his own

5 On the rhetoricians, see the contributions in: Arjan van Dixhoorn, Samuel Mareel, and Bart Ramakers, eds., *The Knowledge Culture of the Netherlandish Rhetoricians*, special issue of *Renaissance Studies* 32:1 (2018).

6 The focus of this chapter is on those early, formative years of his philosophizing. A next step would be to consider how *consten*-culture developed as an element in his style of philosophizing as it matured, in particular regarding the role of his Collegium Mechanicum. No study deals with *consten*-culture comprehensively. The notion is introduced in: Arjan van Dixhoorn 'Recreating Man's Cunning Virtues: The Philosophical Project of the Netherlandish Culture of Arts', in: Van Dixhoorn, Mareel, and Ramakers, eds., *The Knowledge Culture of the Netherlandish Rhetoricians*, pp. 23-42. For earlier work on (late medieval) Dutch (liberal and mechanical) arts (or *consten*) literature, see: Orlanda Lie and Lenny Veltman, *Kennis-maken: Bloemlezing uit de Middelnederlandse Artesliteratuur* (Hilversum: Verloren, 2008); Frits van Oostrom, *Wereld in woorden. Geschiedenis van de Nederlandse literatuur 1300-1400* (Amsterdam: Bert Bakker, 2013), pp. 79-114. The most comprehensive interpretation of Dutch vernacular culture and the role of *consten* to date in: Herman Pleij, *De sneeuwpoppen van 1511: Literatuur en stadscultuur tussen middeleeuwen en moderne tijd* (Amsterdam: Meulenhoff, 1988). On the epistemology of Dutch *consten*-culture, see also: Jeroen Vandommele, 'Mirroring God, Reflecting Man: Shaping Identity through Knowledge in the Antwerp Plays of 1561', in: Bart Ramakers, ed., *Understanding Art in Antwerp: Classicizing the Popular, Popularising the Classic (1540-1580)* (Leuven: Peeters, 2011), pp. 173-196; Jeroen Vandommele, *Als in een spiegel. Vrede, kennis en gemeenschap op het Antwerpse Landjuweel van 1561* (Hilversum: Verloren, 2011).

7 On Beeckman's hybridity, see: Van Berkel, *Isaac Beeckman on Matter and Motion*, pp. 136-140. The notion itself is taken from: Peter Burke, *What Is the History of Knowledge?* (Cambridge: Polity Press, 2016), p. 39. Van Berkel has pointed to Ramism as another attempt to bridge the divide between artisans and philosophers. This path will not be investigated here.

style of philosophizing *among them*, which he then, through exchange with expert philosophers, thickened into the mechanical philosophical method. The existence of an *artes*- or *consten*-culture as Beeckman's first and daily 'field of action', calls into question the usefulness of 'hybridity' as a category of analysis, since *consten*-culture was *neither* artisanal nor scholarly, it had *already* progressively mediated between the two spheres since at least the fourteenth century, and by the fifteenth century had generated its own (vernacular) sphere of exchange. It is not, then, artisanal practice that came into play in Beeckman's activity as a philosopher, but the exercises of this highly literary *consten*-culture.[8]

Here, Beeckman's notebook is particularly interesting. In his *Loci communes* his 'hybridity' as both a 'mechanical man' and 'rational man' clearly shows; but these *Loci* also show how this 'hybrid' developed a new method, that would come to define the new natural philosophy, and establish a new hierarchy that would come to distinguish the world of philosophizing and the world of practice again.[9] After all, Beeckman co-created the new natural philosophy, but when he and his contemporaries did so, this philosophy with its new theories, new rules of method, and new practices obviously did not yet exist. This, too, raises questions about the nature of the 'hybrids', since, as John Schuster points out in this volume, the leap from the 'revolutionary' world of knowledge with its strong focus on practice but weakness in theory to the new theoretically strong philosophy was not the result of organic growth. How did its founders acquire access to both practical and theoretical knowledges, but, even more importantly, how were they able to turn the multiplicity of knowledges to which they had access into a new, successful, philosophy? Instead of focusing on the moment of the leap itself (mainly the story of how Beeckman and others found their new method to discern between 'true' and 'false' opinion), this chapter adds more precision to our understanding of Beeckman's field of action as a philosopher just before and during his invention of a

8 On the notion of *consten* in the work of Simon Stevin, see: Karel Davids et al., eds., *Rethinking Stevin, Stevin Rethinking: Constructions of a Dutch Polymath* (Leiden: Brill, 2021), in particular the discussion of Stevin's 'program for the *consten*': Davids et al., 'Introduction: Simon Stevin, Polymaths and Polymathy in the Early Modern Period', pp. 1-24, esp. pp. 14-20. The *consten*-culture, here identified as Beeckman's 'field of action', was also the world from which the vernacular learning of Stevin emerged. I am working on a monograph provisionally entitled *Virtuoso Culture* that will focus on the making, mainstreaming and disintegration of *consten*-culture (1400-1700).

9 See also the argument in: Bert De Munck and Arjan van Dixhoorn, 'Working Bodies, Matter and the Performance of Knowledge: The Mechanical and Liberal Arts in the Civic Community', in: Sven Dupré et al., eds., *Embattled Territory: The Circulation of Knowledge in the Spanish Netherlands* (Ghent: Academia Press, 2015), pp. 255-278.

new method. It could be assumed that his new method was the result of his exchanges within that field of action and the problems that it posed. It cannot be understood as the field of natural philosophy *strictu sensu*. The new natural philosophy, after all, grew out of a new re-configuration of the older theoretical discipline *and something else*.

2 The *Loci Communes* and Beeckman's Style of Philosophizing

Due to his style of philosophizing, for which they are our main source, Beeckman's *Loci communes* also provide a unique insight into an everyday philosophical culture that otherwise remains largely invisible.[10] Records of early modern conversations are already rare, but records of philosophical conversations taking place outside the world of expert philosophers are probably even more unique. Beeckman's *Loci* do provide access to the philosophical conversations of people that had no formal philosophical training, but were highly interested in philosophical issues, and loved to engage in philosophical talk. Beeckman was both a member of that group and a philosophical expert who continued to philosophize by entering the topics of conversations in his *Loci* and expanding on them. Since the *Loci* are so intimately related to his philosophizing with others and in writing, their status as a key witness must be investigated. In order to establish with more precision what exactly can be said about his exchanges and his writing exercises based on this unique document, its features must be briefly inspected here.[11]

While one tradition in the history of early modern science, following Edgar Zilsel and Erwin Panofsky, has stressed the importance of hands-on knowledge in the observation and representation of and experimentation with nature,[12] others, critically building on Elizabeth Eisenstein's arguments on the shift from manuscript to print culture, have focused on the role of

10 The use of this notion is partly inspired by Lawrence M. Hinman, 'Philosophy and Style', in: *Philosophy as Style and Literature as Philosophy*, special issue of *The Monist* 63:4 (1980), pp. 512-529, esp. p. 514: 'a change in philosophical position may demand a corresponding change in one's style of philosophizing', 'method is only sedimented style', and p. 527: 'philosophy begins with a discourse of method'.

11 *JIB*, I, pp. xxv-xxxiv, 'Note sur le manuscrit', and pp. xxxv-xxxix, 'Avertissement au premier volume'; Van Berkel, *Isaac Beeckman on Matter and Motion*, pp. 5-6, 13, 17, 19-20.

12 Zilsel, *Social Origins*; Erwin Panofsky, 'Artist, Scientist, Genius: Notes on the "Renaissance-Dämmerung"', in: Wallace K. Fergusan et al., *The Renaissance* (New York: Harper Torchbooks, 1962), pp. 123-182.

'paper technologies' and printed books in the making of the new science.[13] They have uncovered a complex relationship between hands-on knowledge and bookish learning and learned skills and techniques in the cultures of observation. Observations of nature in the work of early modern natural philosophers, well into the eighteenth century, could come from at least three sources: first-hand observations made by the philosopher, second-hand observations from others, and book-transmitted observations that could have a very long tradition well into antiquity.[14] In fact, these observations were part of an old tradition, of natural *problemata* and *quaestiones*, collections of natural particulars, secrets, and problems of nature that provided commonplaces for philosophical or dialectical inquiries.[15] These provided the structure for Beeckman's *Loci*, with several originating from his readings, but many others coming from his own observation or the observation of his interlocutors. One type of natural particulars that was crucial in Beeckman's philosophizing were the *secreta*, the secrets of nature, derived from the same type of sources: first- and second-hand observation and textual tradition. *Secreta*, a paper form of how-to knowledge, linked the knowledge of natural particulars with a quest for rules for manipulating nature's gifts into arts for improving human life, comfort, health, wealth, and power.[16]

Loci communes had been promoted by Erasmus as part of a humanist pedagogy, and had become popular throughout Western Europe. By 1600, these collections of reading notes to be used in rhetorical and dialectical practice, had developed into a more complex paper tool, which was used for the collection not only of reading notes, but also of things said in conversation or observations made. Beeckman's *Loci* also followed this

13 See, for example: Ann M. Blair, *Too Much to Know: Managing Scholarly Information before the Modern Age* (New Haven: Yale University Press, 2010); David Cowling and Mette B. Bruun, eds., *Commonplace Culture in Western Europe in the Early Modern Period* (Leuven: Peeters, 2011).

14 See: Volker Hess and J. Andrew Mendelsohn, 'Sauvages' Paperwork: How Disease Classification Arose from Scholarly Note-taking', in: *A Natural History of Early Modern Writing Technologies*, special issue of *Early Science and Medicine* 19:5 (2014), pp. 471-503; Michael Stolberg, 'John Locke's "New Method of Making Common-Place-Books": Tradition, Innovation and Epistemic Effects', *Early Science and Medicine* 19 (2014), pp. 448-470; Fabian Kraemer, 'Ulisse Aldrovandi's "Pandechion Epistemonicon" and the Use of Paper Technology in Renaissance Natural History', *Early Science and Medicine* 19 (2014), pp. 398-423.

15 See, for example: Jaap Mansfeld and Philippe Rousseau, 'Physikai doxai et Problemata physica d'Aristote à Aétius (et au-delà)', *Revue de Métaphysique et de Morale* 97 (1992), pp. 327-363; Paul J.J.M. Bakker, 'Les Palastrae de Jean de Spello: Exercices scolaires d'un maître en médecine à Pérouse au XIVe siècle', *Early Science and Medicine* 3 (1998), pp. 289-322; Ann Blair, 'Authorship in the Popular "Problemata Aristotelis"', *Early Science and Medicine* 4 (1999), pp. 189-227.

16 See, for example: Eamon, *Science and the Secrets of Nature*; Eaine Leong and Alisha Rankin, eds., *Secrets and Knowledge in Medicine and Science, 1500-1800* (Farnham: Ashgate, 2011).

practice of note taking that was common among in particular medical men and natural philosophers. His *Loci* were initially inscribed on sheets of paper that were kept in a roughly chronological order, and often were explicitly dated; only in a few specific cases did Beeckman use the term *journael* to refer to his notes.[17] They were then later copied into a fair copy of paper sheets kept in two columns with references to the topic in the margin, after which he burnt the draft notes. Later these sheets were collected in a volume of folio paper beautifully bound in a sturdy calfskin binder, with board covered with leather, a double copper clasp and several copper ornaments. The book cover was lost in the Middelburg fires of 17 May 1940, but the pages survived with water damage.[18] Beeckman commented a few times on his note-taking practices. Sometime between March 1615 and February 1616 he wrote: 'Here should be some more observations and meditations that I burned by accident with (the notes) that I already had turned into a fair copy.'[19] In December 1616 he added a few more thoughts on his note taking that show that he approached the practice of note taking as a tool to separate his own 'act of thinking' from the act of reading or engaging in conversations, which he all recognized as the sources of his 'book':

> What I have written in this book is what I have neither read nor heard unless it is indicated. And if since I have read or heard about it, I don't change the note, in order not to blot the page and show how sad it is to lack good masters, unless someone believes that the mind, because of the zeal which is ignited because of the act of thinking, will be sharpened more than it is hurt by the lack of books or masters.[20]

These two fragments also reveal something about the role Beeckman assigned to his 'book' in his thinking or philosophizing. The notes are termed

17 *JIB*, I, pp. xxxvi. The term 'journael' only refers to his meteorological observations from 1612 to 1615.

18 *JIB*, II, [p. 457], 'Post-scriptum', additional notes on the state of the manuscript.

19 *JIB*, I, p. 87: 'Hier moesten noch eenighe observatien ende meditatien staen, die ick onvoorsiens verbrandt hebbe, met degene, die ick al uytgeschreven hadde.'

20 *JIB*, I, p. 112 [23 December 1616-16 March 1618]: 'Hetgene ick in desen boeck scrive, is hetgene ick niet gelesen noch gehoert en hebbe oft staet er by. Ende al hebbe ick wel naederhant veel dingen daervan gelesen oft gehoret, soo laet ick het nochtans staen om niet te kladden ende te toenen, wat een agterdeel dat het is goede meesters te missen, tensy dat iemant achte, dat het verstant door den iver die men door het bedencken krygt, soo veel te meer gescarpt wort, als het missen van boeken ende meesters scadet. Dan alle boecken gelesen hebbende, ende alle meesters gehoret, daer blijft noch sooveel te bedencken, dat men den iver niet sal missen.'

'observations and meditations' regardless of the source, whether they are the result of reading or hearing (the 'books and masters') or of the 'act of thinking' independently of books or masters. These three sources: books, masters and the act of thinking, potentially sharpen the mind, the result of a 'zeal' that they ignite. Beeckman apparently believed the great 'zeal' and the more sharpened mind to be the result of the act of his own thinking more than of the act of reading or hearing others. The paper technology of the *Loci* played a role in the exercise of his mind, by preserving his observations and meditations in a particular order, and opening up the possibility of a comparison between his own thinking, his own past thinking, and his engagement with the thinking of others.

The observations and meditations were used to provide further stimulus for his own 'act of thinking'. This is why, on the first page, he not only refers to his 'book' as *Loci communes*, but also as *Meditata mea*. Judging from Beeckman's own reflections in the *Loci* they were not intended to be a manuscript in preparation for the printing press. Rather, they served a highly personal function in what can be called Beeckman's 'cognitive exercises'.[21] Additional proof for this interpretation comes from 1644, when Beeckman's brother Abraham published a selection from the *Loci* in a topical, non-chronological order in *Mathematico-physicarum meditationum, quaestionum, solutionum centuria*. Here, the *problemata* or *quaestiones* are again referred to as meditations.[22] The dedication to the curators of the school of Dordrecht, referencing Socrates, also refers to contemplations. The notion of meditation could also be a subtle reference to Descartes' *Meditationes de prima philosophia*, published in Paris in 1641, in Leiden in 1642, and in Amsterdam in 1644.[23] Since the 1950s, this title caused much debate about Descartes' use of the Jesuit or even Stoic devotional or spiritual exercises in his style of philosophizing.[24] After a

21 On this notion, see: Pierre Hadot, *Philosophy as a Way of Life: Spiritual Exercises from Socrates to Foucault* (Malden: Blackwell, 1995); Matthew L. Jones, *The Good Life in the Scientific Revolution: Descartes, Pascal, Leibniz, and the Cultivation of Virtue* (Chicago: University of Chicago Press, 2006).

22 *Mathematico-physicarum meditationum, quaestionum, solutionum centuria* (Utrecht: Pieter Daniels Sloot, 1644). I have seen the digitized copy of Leiden University Library, once owned by Isaac Vossius.

23 A subtle reference then since the book does not mention Descartes once.

24 Oksenberg Rorty, 'Experiments in Philosophic Genre'; Gary Hatfield, 'Descartes' Meditations as Cognitive Exercises', *Philosophy and Literature* 9 (1985), pp. 41-58. Interestingly, Marin Mersenne's first publication, *L'Usage de la raison* (1623), was a devotional work with 'spiritual exercises', according to: Peter Dear, *Mersenne and the Learning of the Schools* (Ithaca: Cornell University Press, 1988), p. 5.

sceptical investigation of these claims, Bradley Rubidge concluded that Descartes might have meant his book to be studied as if the reader were performing a 'cognitive exercise', but that this conclusion did not require scholars to change their interpretations of the argument of the book.[25] The relationship of the *Loci* with 'cognitive exercises' is more evident. After all, they do not present a polished 'philosophical argument', but rather give an intimate view into Beeckman's everyday way of philosophizing. Just reconstructing his 'mechanical philosophy' as a set of ideas and related practices does not do full justice to the *Loci* or Beeckman's philosophizing and limits our understanding of how and where the new philosophy came about.

3 Rhetoric and Beeckman's 'Field of Action'

Following his own genealogical notes in the *Loci* and additional information uncovered by his first modern biographer Cornelis de Waard, Isaac Beeckman was a descendent of well-to-do and well-connected immigrant families from Brabant and Flanders. His great-grandfather Gerard Beeckman was a shopkeeper and chandler in Turnhout near Antwerp. Grandfather Hendrick (born 1520) was *hofmeester* (major-domo) at the court of Andrea Doria in the Republic of Genova, where he became friends with Giovan Luigi Vitelli, known as Chiappino, an Italian nobleman and military genius, who later became a general in the Army of Flanders and Tuscan ambassador at the court of Elizabeth I. He married Mariette, born on the island of Kos but educated at the Doria court, and returned with her to Turnhout around 1555, taking over his father's business. Having moved to London in 1566 in the wake of the Iconoclast Fury, he became a member of the Italian church, and befriended Johan Radermacher, the virtuoso-merchant from Antwerp. From his mother's side Beeckman was a descendent of a rich Flemish refugee family which had established itself in Middelburg in the late 1570s. The social status of the family is reflected in the fact that Isabeau van der Varent, wife of Pieter de Rycke, substitute of the First Noble (the Prince of Orange as the representative of the nobility in the States of Zeeland) was a witness at the baptism of Hester, daughter of Pieter Janssen van Rhee and Janneken van Rentergem in 1585. Their daughter Suzanna married Abraham Beeckman in 1588 and

25 Bradley Rubidge, 'Descartes' Meditations and Devotional Meditations', *Journal of the History of Ideas* 51 (1990), pp. 27-49.

gave birth to Isaac Beeckman in December. Beeckman recorded these social facts in the *Loci*.[26]

Clearly, his ancestors and family members were by no means ordinary artisans. Instead, the evidence published by De Waard shows that they belonged to the world of vernacular virtuosi, many of whom were or had been members of a chamber of rhetoric and actively engaged with a vernacular culture of *consten*.[27] The close relationship between the culture of literary exercises of the chambers of rhetoric and a wider *consten*-culture is evident from several 'observations' in Beeckman's *Loci* and from evidence about his youth and the sociocultural environment in which he grew up and with which he actively engaged in the 1610s when he was formulating the principles of his new philosophy. The first set of traces are references to the chambers of rhetoric and their practices and members; and a second set are references to the interest in natural particulars and natural philosophy in these circles. Beeckman's reference to the conversation on the particular nature of the Haarlemmermeer discussed before suggests *how* these traces of a culture of literary exercises and an interest in nature were combined; and it is through these references that the *Loci* also are a major source for the study of vernacular *consten*-culture around 1600. The surviving evidence in archival documents, manuscripts and printed texts, artefacts and visual objects for both sets of traces throughout the Dutch-speaking Low Countries, both north and south, is extraordinarily rich (also for Beeckman's home town Middelburg) and remains largely unexplored. Despite destruction and other forms of loss, this is also true for the isle of Walcheren. Beeckman engaged with many aspects of its rich *consten*-culture from a young age, and had become an active participant by the 1610s.[28]

Apart from the reference to the meeting of the chamber of rhetoric of Breda in 1618, the *Loci* contain more evidence of Beeckman's active engagement with rhetorical culture throughout his life, from a very young age onwards. Combined with traces from other sources, the evidence from the *Loci* implies first that Beeckman grew up among rhetoricians, and secondly, that the rhetoricians were an important constituency of a wider *consten*-culture, which would be in line with their traditional ethos. Youngsters who from a young age witnessed family members writing, reciting and

26 *JIB*, I, pp. i-xxiv, 'Vie de l'auteur'. See also Isaac Beeckman, original notebook 'Loci communes', 48r second column (Zeeuwse Bibliotheek, manuscript no. 6471) for the term 'hofmeester'.

27 *JIB*, I, p. iii.

28 On the world of rhetoricians of Holland and Zeeland, see: Arjan van Dixhoorn, *Lustige geesten. Rederijkers in de Noordelijke Nederlanden (1480-1650)* (Amsterdam: Amsterdam University Press, 2009).

discussing poems, songs and plays, sometimes mimicked their behaviour and this may have happened in the Beeckman family, too. In 1628, Beeckman observed that at the age of eleven, he composed several poems and also embellished, from memory, one of the histories he had been re-reading, turning it into comedy of four characters of about 500 lines in rhyme, which was performed in the presence of friends and neighbours.[29] Also, works by rhetoricians was read at elementary school, such as *Den wtersten wille* (The last will) of Louis Porquin (a mid-sixteenth-century poetical testament by a Lombard banker active in the Scheldt region, the geographical heart of *consten*-culture). Beeckman later also cited the book.[30]

Due to the fact that for the early seventeenth century we only know the names of a few (board) members of the Middelburg chamber of rhetoric Het Bloemken Jesse (The Flowering Jesse) we will never know if Beeckman ever was a member or a regular guest. It is quite possible though, given his interest in and expert knowledge of several technical aspects of rhetorician culture, and the engagement of many members from his family and acquaintances.[31] The fact that at an early age he wrote poetry and staged his own play implies that the young Beeckman was somehow trained in rhetorician culture. The *Loci* also provide much evidence for an active reflection on several of the principles of its tradition throughout his life. His observations and opinions on the lyrical features of the literature of the *rethoryckers* (rhetoricians) indicate that he actively engaged in debates in those circles that post-1600 focused on the use of metre and the use of a purified Dutch. While Beeckman expresses no interest in the second debate,[32] he clearly was interested in the first, joining the traditionalists among the rhetoricians against the 'new poets' such as Jacob Cats, Daniel Heinsius and Philips van Marnix, Lord of Aldegonde. Thus, in late 1612 or early 1613 he already observed that given their use of metre rhetoricians made two types of *refereynen*, and in 1619 he returned to his ideas after having studied the issue again, as is shown in his use of the phrase: 'if you want to investigate the division of the beat more closely'.[33]

29 See note 142 in the chapter by Zuidervaart in this volume for a reference to the *Loci*. See also: *JIB*, I, p. iv.

30 *JIB*, I, p. 350 [22 November 1619]: 'Leest de rethoryck veersen van Lowys Porquyn.'

31 On Beeckman and the rhetoricians, see also: Van Dixhoorn, *Lustige geesten*.

32 See also the contribution by Semra Meray in this volume. On the language debates in the Low Countries, see most recently: Alisa van de Haar, *The Golden Mean of Languages: Forging Dutch and French in the Early Modern Low Countries (1540-1620)* (Leiden: Brill, 2019).

33 *JIB*, I, p. 18 [November 1612-March 1613]; *JIB*, I, p. 349 [20 November 1619]: 'Wilt ghy vorders de kracht ondersoecken van deese bedeelinge der maten.'

The *refereyn* was a typical Dutch ballade with three or more stanzas each ending with the identical refrain line and with its final stanza addressing a 'prince' (the ceremonial head of the community). This address of the prince indicates that rhetorician literature was meant to be read or performed in the community. Since its genres were often composed in response to a set question or proverb in the context of a competition, rhetoricians had to work on the quality of the sound of their texts as well as the quality of their argumentation. The first type of metre (and *refereyn*), as Beeckman's explanation indicates, used the musical cadences of the Dutch language, stressing three or four words or syllables each line; the second type used the more rigid stressed and unstressed syllables, a system imported into Dutch from French examples at the time. In this context Beeckman was also reflecting on the introduction of a new diacritical mark showing which syllables should be stressed. Another related issue seems his interest in ways to organize a text as 'een oratie oft eenighe andere schrifte' ('an oration or another kind of writing'), so that it could easily be read out loud without making mistakes by forgetting lines.[34] Among the rhetoricians, these issues of sound and correct recitation were known as 'pronunciation'. Good 'pronunciation', like the correct answer to a question, was a traditional prize category.

Indeed, speaking to his interest in and expert knowledge of rhetorician poetry, several of Beeckman's observations in the *Loci* over the years concern the issue of metre in songs and *refereynen*. The fact that in the late 1620s he again returns to these notes and comes to a more explicit rejection in 1630 of the use of the French metre as introduced by the 'new poets' such as Cats, Heinsius, and Aldegonde, shows a concern for the natural qualities of the Dutch language as it was used in everyday practice; a use that Beeckman promoted as the foundation for a melodious language, too, because of its proximity to the natural sounds of Dutch.[35] It is obvious from the observations in his *Loci* that his knowledge stemmed from a highly active engagement with rhetorician culture both as a witness, a student of rhetorician literature in its written form, and experiments with song, melody and musical annotation. His deep understanding of the technical problems relating to rhetorician literature make it rather likely that not only in his teens but also during his adult life he tried to write in rhetorician genres. Clearly, too, his interest in metre and the problem of the natural beat in language was linked to his interest in the (mechanical) aspects of the art of

34 *JIB*, I, pp. 19-20 [November 1612-March 1613].
35 *JIB*, III, p. 173 [17 November-1 December 1630].

singing and musical performance in particular on the lute, the clavichord and the organ. The *Loci* show that these interests intensified when in 1618 Beeckman was living in Breda, which, in turn, suggests that his presence at a meeting of the chamber of rhetoric was not an accident but part of a regular pattern in his social life.[36]

Once this relationship is established more evidence in the *Loci* for Beeckman's interest in a social life edified with literary and artful exercises can be seen as part of this broader pattern. An example is his rule, based on observation, dated 7 March 1619 that, when in company, one should not immediately add another *cluchtken* (joke) to the one just told, in order to allow for a good laugh in honour of the previous jester.[37] In the circles frequented by Beeckman, *cluchtkens* of all kind, including all sorts of witty and wonderful stories, tricks, riddles, experiments, and discussions of natural and moral *problemata*, as the *Loci* show abundantly, were part and parcel of 'joyful companies'. There is plenty of evidence that such circles permeated society in the early modern Low Countries, and also that they included the likes of Beeckman, the lawyer Jacob Cats, or the Dordrecht medical doctor Johan van Beverwijck, but also merchants and artisans practising all sorts of crafts.[38] The *Loci* also show that these circles were intertwined, both in the public sphere and in the family sphere, through friendships and marriage, precisely in those circles that were engaged with the world of vernacular rhetoric. The literary record from the fifteenth century onwards implies that *cluchtkens*, either as natural or moral observations or *problemata*, were promoted among vernacular audiences of the type the Beeckman family was rooted in, as key ingredients of a philosophical and more virtuous – and thus more noble – life.[39]

When Beeckman temporarily joined the chamber of rhetoric of Breda, he mingled with men who came from the same type of families, such as the butcher Dingman Beens, whose manuscript of *refereynen*, ballades, and letters of invitation documents the literary life that Beeckman encountered. Beens was the descendent of a well-to-do family of butchers that had been intermarrying with patricians and even nobility since the sixteenth century

36 *JIB*, I, p. 226 [16-28 October 1618]. See also: K. Bostoen, 'Dingman Beens en de kamer van Vreugdendal', *Jaarboek Oranjeboom* 34 (1981), pp. 134-163; Van Dixhoorn, *Lustige geesten*, pp. 11-12.

37 *JIB*, I, p. 280 [Middelburg, 17 March 1619].

38 *JIB*, III, p. 317 [Dordrecht, October-November 1633].

39 See: A. van Dixhoorn, 'Nature, Play and the Middle Dutch Knowledge Community of Brussels in the Late Fifteenth and Early Sixteenth Centuries', in: B. Noak, ed., *Wissenstransfer und Auctoritas in der frühneuzeitlichen niederländischsprachigen Literatur* (Göttingen: Vandenhoeck & Ruprecht unipress, 2014), pp. 99-122.

at least. Beens's great-uncle Willem Beens, mayor of Breda, also was a rhetorician. Like the Beeckman family, the Beens family had converted to Calvinism. The manuscript shows the interesting combination of a butcher who also was an active rhetorician, a deeply convinced Calvinist and anti-Catholic and an admirer of Maurice of Nassau. This type of behaviour was typical of many members of the well-to-do and well-connected socially upwardly mobile families in the Low Countries at the time: the scions of these families were domesticated partly by actively engaging in the literary exercises of the rhetoricians; that is, through practising the art of rhetoric, which often aligned with religious, social, and political activism.[40] The Beens manuscript shows the young butcher taking part in rhetorical contests, initially in those organized during the rhetorical season of his own chamber, and then, representing his chamber, in contests elsewhere, and not just in Calvinist territory. The Breda rhetoricians took part in contests in neighbouring 's-Hertogenbosch, and corresponded with rhetoricians in Antwerp, both (at that time) in Catholic territory. In 1616, Beens submitted two *refereynen*, in response to a set question and a set proverb, to a contest organized by the Het Bloemken Jesse, the chamber of rhetoric of Middelburg. As was customary in these circles, most of his poetical work was religious, moral, or political in nature.[41]

Given Beeckman's involvement with rhetorician culture, Beens may have met the young philosopher during the Middelburg contest in 1616. If Beens was indeed present in person, as was customary, he will certainly have met the Middelburg baker and leading member of the local chamber of rhetoric, Hendrick van Cannenburch. Van Cannenburch is another example of a deeply convinced Calvinist and admirer of Maurice of Nassau. He was both actively engaged in rhetorician life and, while being a baker (and proud of it, too), associated with learned circles. In 1604, Paschasie van Isenhoudt, wife of Herman Faukelius, one of Middelburg's ministers, was a witness at the baptism of Van Cannenburch's daughter. A Calvinist partisan using his rhetorical skills he entered into the pamphlet war between the Antwerp Jesuit Joannes David and the Middelburg minister Willem Teellinck in 1611.[42] Unlike Beens's work, which only survives in his personal manuscript, Van Cannenburch's rhetorical work only survives in print,

40 R. Tempelaars, 'Op de drempel van de Renaissance. De rederijker Dingman Beens (1595-1623) en zijn handschriftelijke bundel Refereynen en baladen (1613-1623)' (MA thesis, Leiden, 1982), part 1, pp. 11-17. I have been able to consult a digital copy of the thesis, with thanks to Bart Ramakers.

41 Tempelaars, 'Op de drempel', part 3. The manuscript is in the Leiden University Library.

42 On Van Cannenburch: Van Dixhoorn, *Lustige geesten*.

except for a pro-war song dating from 1609, collected in the *Album Rotarii*, the *adversaria* of Johan Radermacher (also known as Johannes Rotarius), the Middelburg virtuoso merchant.[43] The fact that this song was added to the album's section with work by rhetoricians such as Lucas D'Heere, a friend of Radermacher's, suggests that Van Cannenburch belonged to the Radermacher circle, just like the Beeckman family.[44] Other members of the Middelburg chamber of rhetoric from this circle include Beeckman's uncle, the carpenter Hans Coene, the physician Jacob Lansbergen (son of the astronomer), and the teacher, pedagogue, and pamphleteer Johannes de Swaef. De Swaef was the son of the baker Hans de Swaef who actively supported Isaac's father in a theological debate with the Middelburg ministers, including Faukelius. The lawyer Pieter de Rycke, a Beeckman connection, was a friend of rhetorician Lucas D'Heere, and his (likely) brother, the physician Lieven Pietersz de Rycke, was a member of the board of the Middelburg chamber of rhetoric in 1580. Radermacher, who wrote and collected rhetorician poetry and was a friend of prominent rhetoricians, may well once have been a rhetorician, too.[45]

With its focus on literary exercises the chamber of rhetoric promoted the idea of engaging with the (liberal) arts as a way of life in accordance with (if not inspired by) ancient practices of 'spiritual' or rather 'cognitive' exercises. Only a few years after Beeckman's death, in his 1644 chronicle of Zeeland, the Leiden professor of rhetoric Marcus Zuërius Boxhorn referred to the Middelburg chamber of rhetoric as a company of 'mannen van lustighen en poëtischen gheest' ('men of joyful and poetical wit'), who used to compose vernacular verse in response to edifying questions, to the 'great pleasure and enjoyment of their audiences and an exercise of the mind', which type of meetings and practices, he explained, 'were also observed in all the cities of ingenious Italy, by their most noble, dignified

43 K. Bostoen, ed., *Het album J. Rotarii. Tekstuitgave van het werk van Johan Radermacher de Oude (1538-1617) in het Album Rotarii, Handschrift 2465 van de Centrale Bibliotheek van de Rijksuniversiteit te Gent* (Hilversum: Verloren, 1999). See the online edition of the Album at: dbnl.org, based on the transcription by K. Bostoen, C.A. Binnerts-Kluyver, C.J.E.J. Hattink, and A.M. van Lynden-de Bruïne.
44 For a different view of the relations of the Beeckman family with Radermacher see the chapter by Huib Zuidervaart in this volume.
45 Based on: *JIB*, I, pp. iii-iv, 'Vie de l'auteur'. Further evidence for a rich rhetorician culture on the isle of Walcheren can be found in the fully unexplored account books of the chamber of rhetoric of Veere (running from 1590 to 1795). Zeeuws Archief Middelburg [henceforth ZAM], Gemeentearchief Veere [henceforth GAV], archive no. 2515, Kamer van Retorica 'Missus scholieren' te Veere, 1530, 1590-1794 (with thanks to the members of the working group Rhetoricians in Zeeland for their transcriptions of these sources).

and witty inhabitants'.[46] Just two years before, in 1642, Jacob Lansbergen (someone who Boxhorn certainly would have met) had been the chamber's bookkeeper. The idea that Beeckman was domesticated from an early age onwards in such a culture known for practising 'cognitive exercises', and was still critically investigating the literary forms in which the rhetoricians did so, aligns well with his philosophical style which he explicitly grounded in 'cognitive exercises' meant to sharpen the mind. The questions, proverbs, and *cluchtkens* rhetorical culture used to exercise and edify the mind are reminiscent, in form though less in content, of the questions and problems collected in Beeckman's *Loci*, which he used to exercise his mind and philosophize. The fact that he overheard a conversation on the silted nature of the Haarlemmermeer at the Breda chamber of rhetoric, however, indicates that the literary exercises of the chambers were not isolated from a broader interest in arts and nature, which more narrowly defined Beeckman's 'observations' and 'meditations'.[47]

4 Natural Particulars, Problems of Nature, and Play

According to the *Loci*, sometime between 30 April and 25 June 1618, in Brussels, Beeckman encountered revolving doors similar to the ones his uncle Hans Coene had once shown and explained to him.[48] Beeckman turned this observation of a mechanical phenomenon that he did not yet fully understand into a philosophical problem, in an attempt to make it intelligible. In the same year, 1618, Coene was bookkeeper of the chamber of rhetoric. Some evidence suggests that carpenters were a rather strong presence among Middelburg's rhetoricians; the register of the carpenters' guild even opened with a few stanzas by rhetorician and author of a manuscript continuation of the first printed chronicle of Zeeland, the carpenter Pieter Joossen. Thus, the mechanical art of carpentry was associated with the liberal art of rhetoric. The association of mechanical and liberal arts had been a tradition in rhetorician culture since the mid-fifteenth century. Since at least the end of the fifteenth century, the poetical form of the rhetoricians was also used to identify and address vernacular audiences interested in particulars and problems of nature. The 'joyful company', also at the heart

46 Marcus Zuerius Boxhorn, *Chroniick van Zeelandt* (Middelburg: Zacharias Roman, 1644), I, pp. 176-177. See: Van Dixhoorn, *Lustige geesten*, pp. 157, 300.

47 See also the arguments in: Van Dixhoorn, *Lustige geesten*.

48 *JIB*, I, p. 181 [30 April-25 June 1618].

of rhetorician culture, became the privileged social sphere where hosts and guests were expected to provide witty entertainment through marvellous performances and ingenious inventions in the form of rhetorical and natural *cluchtkens*, as Beeckman would call them. Such a culture in which these two types of ingenious performances were very much enjoyed as stimulating friendship and edifying wit, was loved by Beeckman, who also continued to study its rules.

The *Loci* show that sometime in 1614 or in 1615, Beeckman began to read *De subtilitate rerum* published in Paris in 1550 by the Italian physician and mathematician Geralomo Cardano. Since then Cardano became one of the main authors with whom Beeckman engaged in his philosophizing. Like Levinus Lemnius's *Occultis naturae miracula* (Antwerp, 1559), Giovanni Baptista della Porta's *Magiae naturalis, sive de miraculis rerum naturalium* (Naples, 1558), Pedro Mexia's *Silva de varia lecion* (Seville, 1540), Henricus Cornelius Agrippa's *De occulta philosophia* (Antwerp, 1531), or Paracelsus's work published from the 1520s onwards, Cardano's book investigates the wonderful secrets of nature and shows the human ability to understand and use that knowledge to manipulate nature.[49] These books gained great popularity throughout early modern Europe, also in the Low Countries. Beeckman became an avid reader of Cardano, Lemnius, Della Porta, and Mexia, as well as Simon Stevin, who built on the same tradition. Based on his reading alone Beeckman can already be situated in this tradition of polymath collectors of secrets and students of nature and arts who aimed to reform or perfect human life by spreading or perfecting knowledge of these secrets of nature and their uses. Art was understood to be the skilful use of the secrets of nature, and science the knowledge of principles that explain these secrets. Both arts (providing particulars and problems) and sciences (investigating causes) were needed in the act of philosophizing. In the vernacular philosophical lifestyle that was centred around Dutch rhetoric, the focus would largely be on the arts (that is, particulars and problems of nature combined with how-to knowledge), but it seems that while this culture matured in the sixteenth century, an interest in causes, and, closer to practice, in rules and principles derived from investigating natural particulars grew with the interest in perfecting available how-to knowledge and improving results. Beeckman's style of philosophizing in the *Loci* seems to be the fruit of this tension between philosophical *lifestyle* and *results*.[50]

49 On the tradition, see, for example: Eamon, *Science and the Secrets of Nature*.
50 See also the argument in: Van Dixhoorn, 'Recreating Man's Cunning Virtues'.

Collecting descriptions of wonderful and useful phenomena and recipes for the creation of wonderful and useful effects became a rage in the cultivated circles of early modern Europe; it was not only fashion among the rich and powerful or among Latinized scholars, in some regions of Europe the fascination with the wonders and secrets of nature deeply penetrated the world of daily practice, from lawyers and government officials, to merchants, artisans and artists. The result of this deepening interest was that from the early 1400s onwards at least making (and doing) things and making knowledge became more closely associated than before. Depending on local and regional circumstances, the way in which the association between making things and making knowledge was socially and culturally organized differed across Europe. All over the continent, knowledge communities came into being outside of the universities that shaped new philosophical cultures more dedicated to perfecting practice through theory and theory through practice. These communities shared an interest in the wonders and secrets of nature, in collecting marvellous objects and special recipes, prescriptions for how to act and how to manipulate nature. In the Dutch-speaking world, rhetorician culture became instrumental in shaping this audience, joining artists, artisans, printers, merchants, priests, and men of government into one community.

Several other manuscripts, apart from Beeckman's *Loci*, that were created in this social world confirm that an interest in practical arts was associated with a philosophical way of life. This could happen implicitly through the use of 'rhetorical' forms often in introductory sections or explicitly by comparing the life of the virtuoso practitioner with that of the philosopher. An example among many is the manuscript of a French translation of Bernardino de Escalante's *Dialogos del arte militar* originally printed in Seville in 1583 and later also in Brussels in 1588 and 1595.[51] A note at the beginning of the manuscript indicates that the *Cinq dialogues de l'art militaire* was gifted by Martin Drooghe to the library of Johan Huijssen, Lord of Kattendijke. Drooghe was a well-known high commander of the fleet of Zeeland in the 1580s. Huijssen was a high official from Middelburg (and a member of the Admiralty board of Zeeland). He combined antiquarian interests (in 1614 he acquired a unique manuscript chronicle of Holland and Zeeland which became known as the Huijssen van Kattendijke chronicle) with an interest in botany, like many in this world. He was the dedicatee of 'exercises in my office', the eighth part of 'Dapes inemptae', the poem in praise of a

51 ZAM, archive no. 157: Schorer family archives, inv. no. 1205; 'Cinq dialogues de l'art militaire de Bernardino d'Escalente' (dated c. 1600).

contemplative life in the countryside written by his friend, the botanist Petrus Hondius, who was minister of Terneuzen.[52] Since Huijssen was not a military man himself, the fact that he owned the manual rather suggests a more 'philosophical' interest in the art of war. The text introduces itself with a French sonnet explaining that becoming a good soldier requires knowledge of how to behave in war (i.e. it requires mastering the art of war):

> Pour estre bon soldat, et scavoir quil faut faire,
> Quand sera en la guerre, pour bien s'entretenir,
> et pour en son debvoir tousiours se maintenir.

The dedication of the book to the 'tresillustres seigneurs de l'infanterie d'espaigne' explicitly links the military art with philosophy, contrasting the ancient commander Hannibal in his army with the ancient philosopher Phormio in his academy.[53]

A bookish interest, comparable to the one expressed in this manuscript, in all sorts of arts had been permeating Flemish-Dutch culture since the fourteenth century already (developing in close conjunction with readership and translations and adaptations of French and Latin texts). Increasingly, many aspects of life became scripted in the vernacular through rules/recipes passed on in manuscript or printed books of arts, some for practical use but most also for pleasure, and often both for joyful entertainment and serious perfection of practices. Another local example of the rise of books of arts, also from a military context, again shows how verse was used to express the love of the liberal and mechanical arts and sciences, and address the so-called 'lovers of arts', the intended readership of those books. This manuscript dates from the year of Beeckman's death, 1637, and contains two books.[54] The first is an art of fortification and the second an art of the production and

52 Petrus Hondius, *Dapes inemptae of De Moufe-schans, dat is de soeticheyt des buyten-levens vergheselschapt met de boucken* (Leiden: Daniel Roels, 1621), 'Den achsten ganck. Ouffenijngen op t'cantoor'. On Hondius, also see the article by Zuidervaart in this volume.

53 'Il est a craindre qu'aucuns diront de moy ce qu'Hannibal disoit du philozophe formion quand il y oyoit traiter en son academie les qualitez que doibt avoir le Cappitaine general et plusieurs autres officiers de la guerre pource il ma semble bon de vous faire entendre l'occasion que i ay eue d escrire ces dialogues de l art militaire.' The encounter of Phormio, 'the well-known Peripatetic', and Hannibal is recounted in Cicero's *De oratore*, bk II.

54 ZAM, GAV, collection no. 2854, Verzameling handschriften (manuscripts), no. 9, 'Vestingbouwkundige tekeningen en voorschriften tot vervaardiging van saltpeter'. The manuscript treatise on salpetre making has been dated 1637 in: M. de Jong, *'Staat van oorlog'. Wapenbedrijf en militaire hervorming in de Republiek der Verenigde Nederlanden 1585-1621* (Hilversum: Verloren, 2005), p. 235.

use of gunpowder. Both are written in Dutch with some traces of German. It is unknown who wrote these books or created the manuscript and why; but it may very well be associated with the world of gunpowder makers active on the isle of Walcheren, one of the centres of Dutch mercantile and military power at the time. The first book, on fortification, presents drawings of various types of fortifications, one of them accompanied by a brief poem expressing some of the core notions of the discourse of this arts culture, with its stress on labour and diligence in the study of arts that will bring about inventions and new forms of ingenuity:

Where Labour is not spared, Nor diligence, Art can be seen to give birth to Many inventions.[55]	Daermen geen Arbeyt en spaart noch Diligentie, Sietmen dat Conste baert, menige inventie.

The second book, on the art of gunpowder making and its use, provides a series of recipes for the making of saltpetre, according to the well-known recipe formula (my italics added):

> *If you want to* boil good saltpetre, *you will need* four standing tubs.
> *In what manner this matter can be* separated and reduced to saltpetre.
> *Take* a thin and long rod ...[56]

The evidence that locally rooted and regionally and internationally networked communities in the Low Countries (in particular the Dutch-speaking coastal regions) were creating a society in which philosophizing (that is, engaging with the arts and sciences) was in the process of being domesticated, deeply permeating social relations, has been piling up for decades now. Lime Street and other communities such as those identified by Deborah Harkness in London around 1600 (which by the way was strongly focused on Antwerp *and* Middelburg) had been multiplying for at least a century, with the earliest traces dating back to around 1400.[57] Just one example of how such networks operated comes from the recipe book of the Rotterdam surgeon Barent Hovius (1642-1696). It contains a copy of a 'secret' for the making of gold gifted to Hovius by his neighbour, pharmacist

55 ZAM, GAV, 2854, 9, 'Vestingbouwkundige tekeningen'.

56 ZAM, GAV, 2854, 9, 'Vestingbouwkundige tekeningen', saltpetre recipes.

57 Deborah E. Harkness, *The Jewel House: Elizabethan London and the Scientific Revolution* (New Haven: Yale University Press, 2007). Harkness does not once mention Middelburg.

Pieter Doelman (d. 1696). The recipe recalls an event on 3 March 1604 in Middelburg, in the house of the goldsmith Fredrick Muntynchx, at the corner of the Groenmarkt. That day, a Scottish nobleman made 1¾ ounce of gold from 4 ounces of lead that Muntynchx at his behest melted in a mug with some sulphur. When the sulphur was glowing he asked Muntynchx to add some heavy red powder, and after heating the mixture for half an hour, and pouring it out, it had become gold. The mint essayer Herman Cluyssen tested the substance, and confirmed it was over 23 carats of gold. Muntynchx gave his siblings a piece of gold in memory of the event, and a document was made that confirmed the veracity of the transmutation.[58]

These examples attest to a world in which an audience of artisans, experts, officials, and virtuoso aristocrats joined in 'experimenting'. In Beeckman's critical assessment, this playing with nature, acted out in public, often was nothing more than mere dissimulation. In any case, it had become an important constituent element of social life for a significant substratum of society, for a community that, at least in the Dutch-speaking Low Countries, was open to the active engagement of artisans as well as noblemen. This is the type of social life, with its forms of gift-giving, that underpins most of Beeckman's *Loci*, providing a significant amount of the particulars and problems, the 'observations' that he used for his 'meditations'. Its existence is documented in manuscript evidence that is close to actual practices, expertise, or experimenting, but its results are also recorded in printed books. In many cases, these texts are embedded in this world of virtuosi through rhetorical tools such as little verses that address the 'lover of arts'. In the Dutch-speaking world, the discourse of the 'lover of arts' acquired its form from the styles and language of the rhetoricians, to such an extent, that the presence of those forms and this language alone must have been an indication that 'lovers of arts' were being addressed.

The (often) versified discourse addressing these virtuosi that connects so many of the *artes* texts in the vernacular, associates the exercise of arts with the joy and pleasure of knowing and mastering nature. Indeed, the joyful mind, which in the arts culture of the Dutch-speaking Low Countries was believed to be engendered through the exercise of rhetoric, was the object of intensive discursive practices. The discourse embedding books of arts produced in Dutch in a culture of philosophical 'experimenting' in the 'joyful company' explicitly associates using tricks of nature with *cluchtkens* or comical play. Indeed, many of the recipes for pleasure are meant for dissimulative performances, to use tricks of nature to deceive

58 F. Wiersum, *Alchemie te Middelburg in 1604* (Middelburg: J.C./W. Altorffer, 1918).

friends, for example, at a dinner table. Such recipes can also be found in Beeckman's *Loci*, and, reading these 'joyful recipes', in conjunction with the other recipes in Beeckman's manuscript, shows that the *style of thinking* used to create comical tricks of nature must have sprung from the same mindset used in conjuring up useful arts. Thus, techniques used to create comical texts and literary performances can immediately be related to the playful way with which Beeckman and his associates approach nature. The *Loci* clearly register this playful, witty mind aligned with a playful, witty nature *in action*. What is more, they show that Beeckman was an active observant of these practices, and, through his 'meditations', came up with a rule to perfect the practice based on his fundamental principle in line with Stevin's personal motto 'Wonder en is gheen wonder' ('Magic is no magic'): in nature nothing exists without a cause that can ultimately be understood.[59] The 'natural master' then, was allowed by Beeckman to use tricks to create magical effects but not to leave people stuck in awe. Instead, the trick should be used to edify by revealing its causes.[60]

5 Questions, Recipes and Marvels as Playful Exercises

The *Loci* are partly structured around four types of 'exercises' or 'observations' that are affiliated with the tradition of secrets and marvels that had become deeply ingrained in Dutch vernacular culture by 1600, that is, in the socially open world of the 'lovers of arts'.

The first type are anonymous questions that ask: why does this happen? For example, in late 1612 or early 1613, Beeckman noted, possibly in Zierikzee:

> Why does it often happen that one side of a street is more often used than the other side? Is it because streets are curved and people look for a shortcut? Or is it because on one side the streets or markets are better than on the other side?[61]

59 One is reminded of Stevin's motto in the motto of two members of the Breda chamber 'Niet sonder oorsaeck' ('Nothing without a cause') and 'Niet sonder dat' ('Nothing without something') in a letter sent by thirteen Breda rhetoricians to the Antwerp teacher R. Lenaerts dated between 1614 and 1621. Tempelaars, *Op de drempel van de Renaissance*, p. 181. Beeckman may in fact have been one of them (the problem is that they only sign with their mottos and Beeckman's motto, if he ever had one, is unknown), as he was in Breda in 1618/1619.

60 *JIB*, III, p. 65 [19 June-7 July 1628].

61 *JIB*, I, p. 18 [November 1612-March 1613]. 'Waerom gebeurt het dickwils, dat de een syde van sommighe straten meer begaen wort dan dander syde? Ist omdat de straten krom loopen ende

The second type are observations of the kind: 'they say', 'someone says', or 'I have seen', such as the one Beeckman noted in March 1615, probably while he was in Antwerp:

> Carps in Antwerp (they say) are kept alive the entire winter in fresh-fish ships without additional feeding, but after Easter they are given bread.[62]

The third type are recipes or prescriptions of the kind 'if you want to get this effect, do this', such as the 'playful' one he noted in 1615 or 1616, possibly in Zierikzee:

> If you want something printed on paper, such as little figurines, invert it and copy it to another paper, cover the little figurine with Spanish soap until it is quite slick. Then take another clean piece of paper and make it wet, and press it strongly on the lathered figurine or rub it with your nail or else, and the little figurine will be copied inverted on the clean piece of paper with all that is inscribed on it.[63]

The fourth type are questions, comments, or recipes garnered from acquaintances or from books. For example, in 1615 or 1616, possibly in Zierikzee, Beeckman 'observed':

> The felt used for hats, says Mr. Leil, who is a hatter, does not shrink in hot water, unlike leather, so it is waterproof, which makes it suitable for making valves used by brewers to pump hot beer.[64]

dat de luyden den kortsten wech soecken? Ofte ist omdat aen deen syde beter straten ligghen ofte marckten dan aen dander syde?'

62 *JIB*, I, p. 62 [March 1615]. 'De carpels (seggen se) houdt men t'Antwerpen den heelen winter in de vischschepen levendich sonder eten te geven, maer na Paesschen moet men se broot geven.'

63 *JIB*, I, p. 69 [March 1615-6 February 1616]. 'Als ghy yet, dat geprendt is op pampier, als figuerkens etc., op een ander pampier brenghen wilt, maer averechs staende, so bestryckt het figuerken met spaensche seepe totdat het figuerken wel glat is. Neempt dan een ander schoon pampierken ende maecket wat nat ende lecht het over het bestreken figuerken ende drucket styf daerop, ofte wryvet met u naghele) ofte andersins, so sal het figuerken op het schoon pampierken averechs staen met al datter in geschreven staet.'

64 *JIB*, I, p. 80 [March 1615-6 February 1616]. 'De vilten van de hoeden, seght Mr. Lejl die eenen hoetmaker is, en krimpen in het heet water niet, gelyck het leer doet, ende syn dichte, also datse water houden. So syn se dan goet om clappen te maken voor de brouwers om heet bier te pompen.'

In the *Loci*, observations from anonymous sources or a Mr. Leil are awarded the same epistemic status as an observation acquired from a bookish 'authority':

> If you would like to say something that could be heard far away, you could make something to stand in which would amplify the sound, or create the same effect through the reflection of the echo, or by making an artificial man, such as was made by Albertus Magnus, that could speak.[65]

A related observation, but not a full recipe either, was registered by Beeckman during his stay in Breda in November or December 1618, as the result of a conversation with a local official:

> Today, in Breda, I counselled Bergainde, a steward at Breda, how to listen in from his office to what was being said in the rest of his house, and how to be heard in the rest of the house, from his office. That is, to lay conduits of tubes from the office over the ceilings of all rooms.[66]

The affinity of these 'observations' with the more playful recipes from books of arts printed in the Low Countries from at least the early sixteenth century onwards, can easily be noticed if they are compared with one of the joyful recipes from the so-called *Tbouck van wondre*, printed in 1517 in Brussels by Thomas van der Noot to whom printing itself was a philosophical art. This type of joyful recipe is also known as 'performative magic', and it clearly is of the type that Beeckman enjoyed, and recommended, but only to be used for the purpose of enlightenment. The brief introduction of the book explicitly joins 'joyful arts' and 'useful arts', including the very mechanical arts of painting and of making iron or steel, into one 'culture of marvellous arts':

> So that everyone understands what is useful and profitable in man, beautiful and diverse arts will follow here, including many joyful arts, which is why this book is called Book of Wonder, in which you will find

65 *JIB*, I, p. 83 [March 1615-6 February 1616]. 'Als men yet segghen wilt, dat men verre hoore, so soude men moghen yet maken om in te staen, twelck het geluydt vergroot; ofte tselve deur de reflectie van de echo teweghe te brenghen, ofte deur eenen gemaeckten man, gelyck van Albertus Magnus gemaeckt is geweest, die spreken konde.'

66 *JIB*, I, p. 261 [23 November-26 December 1618]: 'Ic gaf vandaech Bergajnde, een rentmeester van Breda, te Breda, raet, om uyt syn cantoor al te hooren, wat men in syn heel huys sprack, en gehoort te werden, waer men is int huys, uyt syn cantoor. Te weten met buysen te leyden uyt het cantoor tot deur de solders van alle camers etc.'

> many truly marvellous and many profitable arts. First of all, the art of painting all sorts of substances, then the art of making iron and steel hard or soft, and many more.[67]

The recipes in the book are reminiscent of Beeckman's recipes and, very much in the same spirit, entice their audience into 'exercises' that stimulate the study of natural particulars. They explicitly call upon readers to engage in 'experimenting':

> If you want to see the stars during the day, take a clean and clear bowl and put in the middle a clean, clear mirror, then add clean water, so that the mirror is well-covered, and you will see the stars.[68]

> How to make two little men – one who will light a candle and the other who will put it out. Take a cabbage and make on one side two little men with faces. Put in the mouth of the first a bit of gunpowder, and in the mouth of the second some sulphur. Then take and lit a candle until its wick burns well. Move the candle to the little man with the gunpowder in his mouth and he will put it out. Then move it to the other man with the sulphur while your wick is still glowing and he will light your candle again.[69]

6 Epilogue and Conclusion

Not long after the death in 1617 of the virtuoso merchant Johan Radermacher a family member recalled his character in a document entitled *Loci communes*. It chronicles how Johan and his family moved from Aachen to Antwerp, from there settling in London after the Iconoclast Fury, returning to Antwerp in the 1580s, where Johan became a leading member in the Calvinist republican government, and after the seizure of Antwerp by Farnese in 1585 back to Aachen. From there the family came to Middelburg in the 1590s, settling there with many members of the former Calvinist

67 *Tbouck van wondre* (Brussels: Thomas van der Noot, 1517), prologue. On this book and its tradition, see: W.L. Braekman, *Den Sack der Consten. Een Vlaams volksboek gereproduceerd naar de Antwerpse druk van Jacob van Liesvelt uit 1528* (Bruges: M. Van de Wiele, [1989]).

68 *Tbouck van wondre*, 'Wilt gy in de schone dag de sterren sien'.

69 *Tbouck van wondre*, 'Om te maken twee mannekens'.

regimes in the south.[70] Radermacher was an expert in astronomy, but also had a great interest in promoting the study of the arts and sciences in the Dutch language, to which his grammar of Dutch, the oldest one to have survived, in manuscript, attests. The description of his character gives a deeper insight into the reason why sharpening wits was considered a crucial first step in the shaping of a life of learning. The Radermacher family *Loci* associated perfect memory and wit, with perfect quality of the senses of hearing, vision, smell and taste:

> He was a wise and expert, learned, and godly man, who had a reputation in all virtuous sciences, was not arrogant but humble, which made him beloved by the little and the great, having great understanding of all things, was able to live among people of all abilities, so that everyone liked to be with him, he always lived a frugal life and though not very strong physically speaking, the lord blessed him with the ripe old age of 79 years, and at that time he still had such strong memory and wit, hearing, vision, smell, and taste, that it was a marvel.[71]

Reviewing the various types of 'rhetorical' and 'natural' exercises that can be associated with the philosophical life of Dutch *consten*-culture as it had developed until the seventeenth century, and can be witnessed in Beeckman's *Loci*, with the expert in the virtuous sciences invoked by the Radermacher family *Loci*, it can be concluded that the joyful exercises of a pedagogical nature, along the lines of Beeckman's prescriptions, were intended to further memory and wit (the rhetorical exercises) and hearing, vision, smell, taste and touch (the natural exercises). It has also been shown that Beeckman's *Loci*, apart from developing philosophical ideas, were used by him to improve his thinking capacities, that is, his capacities to

70 On the Radermacher family and Johan Radermacher's learning. see: Marisa A. Bass, *Insect Artifice: Nature and Art in the Dutch Revolt* (Princeton: Princeton University Press), pp. 42-48 and pp. 106-108.

71 ZAM, 157, Schorer family archives, 37, *Loci communes* by Johanna Radermacher, 45. 'Hij was een seer wijs ende ervaeren, geleert, ende godsaelich man, bij allen daer voor erkent in alle eerlycke wetenschappen, waeren niet hooggevoelende van hem selven maer nederich en clijn gevoelende hem bemindt maekende so wel bij clyne als by groote hebbende verstant van alle dingen connende leven onder alle menschen van wadt faculteijt sy waeren so dat een iegelyck gerne by hem was, leefden altyt seer sober, en maetich door dien hij niet sterck van naturen scheen te sijn evenwel segende hem de heere toto inden hoogen ouderdom van 79 jaeren ende was nog met sulcken stercken memorie en verstandt, gehoor, gesicht, rueck, en smaek begaeft dat wonder was.' See also Abraham Beeckman's description of his deceased brother: *JIB*, I, p. xxiii, esp. notes 5 and 6.

philosophize. Even though, in his brief description of his brother after his death, Abraham Beeckman did not mention the senses, the *Loci* show many exercises to use and improve them. The *Loci* then, were an instrument deeply integrated with Beeckman's life as a philosopher, with writing becoming part of a series of stimuli that were meant to help him turn his entire life into a philosophical project, the object of philosophical reflection from a perfected philosophical mind. The Radermacher *Loci* also reveal another feature of the virtuoso: the ability to engage with people of all sorts. They thus identify 'conversation' and 'use of the senses' as the main characteristics of a man of learning such as Radermacher. Abraham Beeckman on the other hand reduced his brother's style of philosophizing to the two qualities of *converseren* and *speculeren*.

In an article on possible influences on John Locke's *Some Thoughts Concerning Education* (London, 1693) which was conceived while the philosopher was living in Rotterdam, Brita Rang has pointed to a Dutch tradition of schoolmaster-pedagogues whose arguments on the importance of teaching experience and useful knowledge in the schools preceded Locke's. Schoolmaster-pedagogues such as Johannes de Swaef combined a 'strong, critically reflective, humanist-theoretical tradition' with a 'tradition of experience-oriented craftsmanship'. As described by Edgar Zilsel 'for the process of early modern development of the sciences', Rang agrees, these two traditions began to 'interpenetrate one another, their languages and methods becoming mutually related and partially intermingling'.[72] Yet, the process by which these traditions were mingled had begun much earlier in the world of *consten*-culture. In its vernacular spheres, since the fifteenth century, the two traditions of book-centred learning on the one hand and experience-centred practice on the other, had already been mingled with great success. In this world, the chamber of rhetoric had become the core institution. Playful exercise of the mind and playful experimentations with nature had become the style of philosophizing of this world. The *Loci* show how certain key aspects of Beeckman's unique style of philosophizing were deeply influenced by and intertwined with this world. Thus, Isaac Beeckman's *Loci* reveal two major fields of philosophical action. The most well-known is the academic world of learning shown in his lifeworld as a teacher and the network of his correspondents (such as Descartes and Mersenne). Less well-known is the other world, which has often mistakenly

72 B. Rang, 'Letters across the North Sea: A Dutch Source of John Locke's Letters Concerning Education', in: Juliette Roding and Lex Heerma van Voss, eds., *The North Sea and Culture (1550-1800)* (Hilversum: Verloren, 1996), pp. 378-395, esp. p. 394.

been identified with the world of 'Beeckman the artisan'. This second philosophical world is the world of *consten*-culture. This chapter has argued that Beeckman should be considered as a major witness for this world, in which explicit, bookish, academic, theoretical learning and tacit, bodily, artisanal, practical, experience-oriented knowledge had already been 'interpenetrating' for two centuries.

Beeckman's *Loci* were deeply embedded in this *consten*-culture. It provided much of the content of the book, which in turn reflects on its rules. Indeed, the *Loci* provide a unique insight into this culture and its participants. Beeckman's style of philosophizing shows a deep affinity with *consten*-culture in his observations, his joyfulness, his conversations with an unrestricted circle of virtuosi and experts, and also, in his use of cognitive exercises instead of, for example, publication as a preferred philosophical tool. The *Loci* show, in fact, that this movement between two philosophical cultures (high and low, theoretically oriented and practically oriented, Latin and vernacular) drove Beeckman's philosophical productivity. Moving between the two, from *converseren* to *speculeren*, (the first with all sorts of virtuosi and experts in arts and crafts, the other in his 'book' and in letter-writing with philosophers), using (and exercising) his wit and senses like Radermacher did, is *how* Isaac Beeckman philosophized.

About the Author

Arjan van Dixhoorn is Hurgronje Professor by special appointment for the History of Zeeland in the World at Utrecht University and teaches history at its International Honours College, University College Roosevelt, in Middelburg. He was FWO-research fellow at the universities of Antwerp and Ghent from 2005 until 2014. His publications investigate the early modern history of public opinion and the social history of knowledge, with a focus on the role of vernacular literary cultures in Europe.

14 Harnessing the Elements

Beeckman and Atmospheric Instruments

Fokko Jan Dijksterhuis

Abstract

Isaac Beeckman's investigations of atmospheric instruments shed new light on his epistemic and natural philosophical ideas and practices. Atmospheric experiences in the late sixteenth and early seventeenth century are closely linked to the tradition of natural magic and the critique of Aristotelian meteorology. Beeckman's dealings with the thermoscope and other apparatus fit this picture, too, and add a new dimension to his pioneering in mechanical philosophy. Central to this chapter are his responses and eventual understanding of the perpetuum mobile of Cornelis Drebbel, a figure who appears intellectually more close to Beeckman than is often assumed.

Keywords: Isaac Beeckman, Cornelis Drebbel, atmospheric instruments, natural magic, perpetuum mobile

In September 1618, when he was almost 30 years old, Isaac Beeckman made some biographical entries in his notebook. From the age of 21, he recollected, he had devoted himself 'more than a little' to mechanical operations.[1] During the 1610s, having set up shop as a candle maker in Middelburg and Zierikzee, Beeckman was engaged in all kinds of projects involving conduits, pumps, and fountains. First assisting his father, he soon took on work of his own. These activities turn up in his notebook in reflections upon hydrological machines, but also chimneys and stoves. After he moved to Utrecht and

1 *Journal tenu par Isaac Beeckman de 1604 à 1634*, publié avec une introduction et des notes par C. de Waard, 4 vols. (The Hague: Martinus Nijhoff, 1939-1953) [henceforth *JIB*], I, pp. 217-219, esp. p. 218: 'Post annum 21um negotia mechanica non parum me confecerunt, fistulas et curas paternas tractantem.'

Van Berkel, Klaas, Albert Clement, and Arjan van Dixhoorn (eds.), *Knowledge and Culture in the Early Dutch Republic: Isaac Beeckman in Context*. Taylor & Francis Group 2022
DOI: 10.5117/9789463722537_CH14

Rotterdam and became a schoolmaster, the waterworks did not disappear from the notebook, but the emphasis shifted from technical installations to curious apparatus like thermoscopes and perpetual motion machines.[2]

This chapter focuses on Beeckman's engagement with such atmospheric instruments in his Rotterdam period (1620-1627). Both the contents and the epistemic significance of his notes on these topics have not been studied in much detail. They shed an interesting new light on his learned *persona* and put him in a new historical context related to late-sixteenth-century meteorology. The notes on atmospheric instruments show a kind of reasoning that I call 'artefactual': getting a conceptual grip on matters by tinkering with and reflecting upon artefacts like thermoscopes and stoves. This aspect of Beeckman adds, I maintain, to our understanding of his learning and his place in the early-seventeenth-century history of science.

Beeckman's engagement in waterworks is well known but does not figure prominently in historiography. Besides the introductions and comments of Cornelis de Waard in the *Journal*, only Klaas van Berkel discusses them and mainly in general terms. Focusing on Beeckman's natural philosophy, Van Berkel emphasizes ontological and methodological aspects of his notes on flows and pressure. John Schuster has discussed Beeckman's ideas on hydrostatics with respect to the connection between Simon Stevin and René Descartes regarding the development of the latter's mechanistic philosophy. I believe that Beeckman's dealings with atmospheric instruments deserve further scrutiny as they reveal new aspects of his learning.[3]

This chapter discusses phenomena and apparatus that involve the dynamics of air, water and fire: pressure, weight, flow, balance, expansion, condensation, and so on. I call these atmospheric because this comes closest to the way they were understood at that time. It is a broad category that relates to conduits and pumps, stoves and chimneys, thermoscopes and perpetuums. They employ the hidden powers of the elements as they were contrasted to the manifest forces of mechanical actions. The phenomena involved are hydraulic, thermodynamic, and pneumatic, but such designations

2 I will use the shorthand 'perpetuum'.

3 Klaas van Berkel, *Isaac Beeckman on Matter and Motion: Mechanical Philosophy in the Making* (Baltimore: Johns Hopkins University Press, 2013), pp. 16-19; Klaas van Berkel, 'Vruchtbaar isolement. Isaac Beeckman in Zierikzee (1611-1616)', *Kroniek van het Land van de Zeemeermin (Schouwen-Duiveland)* 11 (1986), pp. 59-74; Klaas van Berkel, 'Descartes' Debt to Beeckman: Inspiration, Cooperation, Conflict', in: Stephen Gaukroger, John Schuster, and John Sutton, eds., *Descartes' Natural Philosophy* (London: Routledge, 2000), pp. 46-59; John Schuster, *Descartes-Agonistes. Physico-mathematics, Method & Corpuscular-Mechanism 1618-33* (Dordrecht: Springer, 2013), as well as his chapter in this volume.

are from nineteenth-century physics and do not express the particular way they were understood in the early modern period. Hero's classic works were a main point of reference at that time. In the early modern period these were referred to with the Latin term *spiritalium* rather than with a derivation from the Greek *pneuma*.[4] 'Meteorological' also suggests modern understandings of such instruments and is too confined to measuring in my view. The link with Hero also points towards a specific interpretation of atmospheric phenomena and apparatus: as ways to get a practical as well as a conceptual grip upon natural phenomena. This combination of utility and inquiry puts instruments such as thermoscopes and perpetuums in a similar category as the waterworks of pumps and conduits. Not surprisingly, many atmospheric contemplators had a water engineering background. In this chapter I will follow Beeckman and use this case to shed light on how matters of heat, flow, and pressure were handled and interpreted in the early seventeenth century.

The Delft Weatherglass

On the 9 November 1621, Isaac Beeckman visited the City Hall of Delft. In the 'best room' he saw a glass, given to the gentlemen (the regents of Delft) by some people from Bohemia.[5] It was constructed (*van fatsoen*) as follows (fig. 14.1): the upper bulb *c* was empty, tube *bd* was filled with water and it was inserted in a container with a wooden cover *a*.[6]

> When I put my hand on top of the bulb *c*, then the water sank down and came to *e*, so that *be* was also empty. But as soon as I took my hand away from it, it gently rose again upward until *b*.[7]

4 The English edition of Hero's *Pneumatics* (1851) introduced the term 'pneumatics'. Before that, 'pneuma' was mainly used in theological and philosophical contexts. Giambattista della Porta's use of the term 'pneumatics' in his *Magia naturalis* (1558) and other publications seems to be an exception.

5 *JIB*, II, pp. 186-187: '(Vitrum aeris calorem indicans primò a me visum) Den 9en Novemb. was ick te Delft, ende sach daer in de beste kamer van het stadthuys een gelas, twelck de heeren gegeven was van eenighe persoonen uyt Bohemen.'

6 *JIB*, II, p. 187: 'ende was van fatsoen als men hier siet. *c* was de opperste rondicheyt ende was ledich; van *b* tot *d* toe wast vol water, maer men konde het water maer tot *d* sien, want het onderste deel vant glas was met het houte baxken *a* bedeckt.'

7 *JIB*, II, p. 187: 'Als ick myn handt leyde boven op de rondicheyt van *c*, so sonck het water neder ende quam tot *e*, also dat *be* oock leegh was. Maer so haest als ick myn handt daeraf dede, so reest wederom al soetkens opwaerts tot aen *b*.'

Figure 14.1 Delft weather glass, as depicted by Beeckman

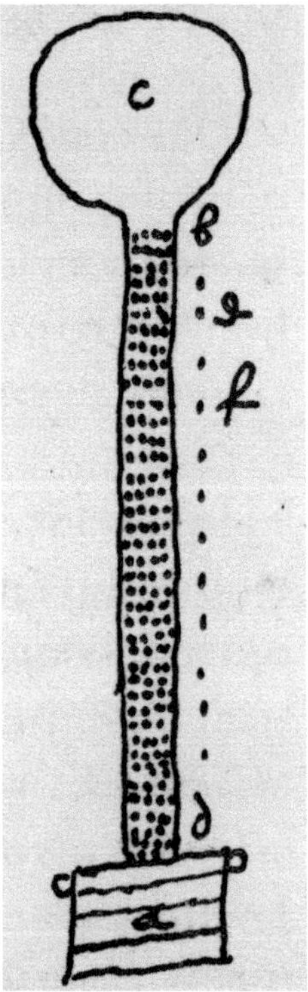

From: *JIB*, II, p. 187 [9 November 1621]

The chamberlain explained:

> in the evening the gentlemen stick a pin in *b* or *e* or *f*, as high as the water then is; and they see the next day whether it is higher or lower. If it is higher, it has been colder in the night; if it is lower, warmer. He also said that it can come into the middle of *c*; and when it is bad weather at sea, the water in *c* also goes up and down with billows.[8]

For Beeckman the glass and its uses were an occasion to raise further questions:

8 *JIB*, II, p. 187: 'De knecht van de kamer seyde dat de heeren savons eenen pinneken steken in *b* of *e* of *f*, so hoghe alst water dan is ende sien sanderdaegh oft hoogher of leegher is. Ist hoogher, so heeft het snachs kouders geweest; ist leegher, warmer. Seyde oock dat het wel kompt tot int midden van *c*; ende alst dan quaet weder in see is, so gaet het water in *c* oock op ende neder met baren.' Beeckman used the Dutch word for waves (*baren*), which coincidentally corresponds with the current measure of air pressure, the meteorological bar (from the Greek *baros* = weight).

> it is surprising that the water goes *down* rather then up with the heat, as heat expands the oil etc. Secondly, how it can go upwards and downwards, as *c* has to be full of air, that can go out nowhere when it comes upwards; and when the water sinks down, there ought to come some air from outside, which cannot happen in this glass. The reason for this I may consider another time.[9]

We see how Beeckman linked what he saw to his understanding of the dynamics of elements. Expecting heat to cause expansion of fluids, he was surprised to see the water level drop and wondered where the surplus air would come from. The entry in the notebook is characteristic of Beeckman in several ways. It shows his insatiable curiosity and interest in the workings of the world. Any observation would induce questions about the properties and reasons of phenomena. He would connect his observations with earlier ones, with readings, and so on, looking for relationships, contradictions, generalities. In this curiosity there was no real distinction between the natural and the artificial: contraptions like the one in Delft offered clues to heat and temperature of bodies and skies, and vice versa. Machines were a way to think about the world. With his background in waterworks it does not come as a surprise that the Delft glass captured Beeckman's attention. Still, any intriguing phenomenon would do, and could be subject to Beeckman's perceptive considerations.

The Delft glass, the phenomena it displayed, and its uses were quite common in learned circles at that time. It is a thermoscope, an instrument that had emerged during the previous decade and was discussed in several places. Well known is the correspondence between Galileo Galilei and Giovanni Sagredo on the latter's trials, inspired by the 1612 account of Santorio Santori.[10] Such instruments were around in many places and appeared in publications. Beeckman knew some of these publications but I have not found a direct reference to an account of the thermoscope. He did, however, frequently refer to Hero of Alexandria, a main figure in early-seventeenth-century discussions of atmospheric instruments. Hero had been on Beeckman's university reading list and

9 *JIB*, II, p. 187: 'Hier is te verwonderen dat het water met de warmte nedergaet ende niet liever opwaerts, dewyle de warmte de oly etc. vermeerdert. Ten tweeden, hoet kan opwaerts of nederwaerts gaen, nadien *c* vol lochs moet wesen, dewelcke nergens uyt en kan alst opwaerts kompt; ende alst water nedersinckt, so behoorder eenighe locht van buyten in te kommen, twelck niet geschieden en kan in dit gelas. De reden hiervan mach ick op een ander tyt bedyncken.'

10 W.E. Knowles Middleton, *A History of the Thermometer and Its Use in Meteorology* (Baltimore: Johns Hopkins University Press, 1966), pp. 3-14; Matteo Valleriani, *Galileo Engineer* (Dordrecht: Springer, 2010), pp. 165-169 (the letters are reproduced in the appendix).

was mentioned regularly in his considerations of waterworks, returning now in his reflections on thermoscopes and other instruments. The thermoscope responds to changes in both temperature and pressure. These two properties were not yet distinguished at that time; atmospheric pressure would gradually be recognized in the course of the seventeenth century. Beeckman also talked about atmospheric changes in a general sense, phrasing it in terms of temperature and with only an implicit indication of pressure.

In historiography, the thermoscope is treated as a precursor of the thermometer and the barometer.[11] In his classic accounts of these instruments, W.E. Knowles Middleton focuses on demarcating them as meteorological instruments. As a result, he excludes earlier atmospheric experiments of Hero and late-sixteenth-century authors on meteorology. Recently Arianna Borelli and Fabrizio Bigotti have convincingly argued that it is necessary to place the early atmospheric instruments in the wider context of late-sixteenth-century meteorology and medicine.[12] As regards the physical and philosophical background of these instruments, historiography has been dominated by the question of vacuum and its repercussions on natural philosophy, isolating Torricelli's experiment from this wider meteorological context.[13] Cesare Maffioli has placed those experiments in the context of hydraulic engineering and recently Matteo Valleriani has studied how Hero was read in late-sixteenth-century engineering circles.[14] In this chapter I adopt the perspective of recent historiography of meteorology, putting Beeckman's reflections on atmospheric instruments in the context of Heronic experiments and discussions of the dynamics of the elements.

11 Knowles Middleton, *A History of the Thermometer*, pp. 3-26; J.A. Chaldecott, 'Bartolomeo Telioux and the Early History of the Thermometer', *Annals of Science* 8 (1952), pp. 195-201. Valleriani, *Galileo Engineer*, 155-186, largely adopts the chronology and categorizations of Knowles Middleton. See also the chapter on Cornelis Drebbel by Vera Keller in this volume.

12 Arianna Borrelli, 'The Weatherglass and Its Observers in the Early Seventeenth Century', in: Claus Zittel et al., eds., *Philosophies of Technology: Francis Bacon and His Contemporaries* (Leiden: Brill, 2008), pp. 67-130; Fabrizio Bigotti, 'Weight of the Air: Santorio's Thermometers and the Early History of Medical Quantification Reconsidered', *Journal of Early Modern Studies* 7 (2018), pp. 73-103.

13 Van Berkel, *Isaac Beeckman on Matter and Motion*; Schuster, *Descartes-Agonistes*; Alan Chalmers, *One Hundred Years of Pressure: Hydrostatics from Stevin to Newton* (Dordrecht: Springer, 2017), pp. 60-68, 99-106.

14 Cesare Maffioli, *Out of Galileo: The Science of Waters 1628-1718* (Rotterdam: Erasmus Publishing, 1994); Matteo Valleriani, 'From Condensation to Compression: How Renaissance Italian Engineers Approached Hero's Pneumatics', in: Harmut Böhme et al., eds., *Übersetzung und Transformation* (Berlin: Walter de Gruyter, 2007), pp. 333-353.

Atmosphere and Perpetuums

Towards the summer following his visit to Delft, Beeckman returned to his note on the Delft glass. He had seen Willem Jansz. Blaeu (1571-1638), the Amsterdam publisher with whom he maintained close relations. Blaeu had told him about an Amsterdam doctor who could compare the 'temperament' of each room 'and that in the way I told him I had seen at the Delft city hall'.[15] The encounter gave rise to a lengthy series of notes on atmospheric contraptions in the summer of 1622.

Beeckman first gave the Delft glass a second consideration, providing a detailed description of its construction and working (fig. 14.2).

> Now *a* is full of air, that is slightly thinned and thickened by the heat. If it is thickened by the cold, then it takes up less place, by which the liquor has to come up from below to *d*, in order to fill the space made empty by the cold.[16]

The emphasis of his interpretation of the phenomena had shifted from the water in the tube to the air in the bulb.[17] In his first note on the Delft glass he had expressed his surprise that the water level dropped with rising temperature because he expected expansion of a fluid when the heat went up. Now he zoomed in on the air above and concluded that it indeed 'thickened' when temperatures went down. This solved the problem he had raised in his first note on the issue, that the water level rose without air going out. This re-interpretation of the phenomena displayed by the Delft glass is a fine example of the way the *Journal* documents Beeckman's inquisitive thinking and developing understanding.

15 Beeckman used the word 'temperament' in Dutch, disposition, adding 'calor aeris' in the margin. *JIB*, II, p. 199: '(Vitri quo calor aeris examinatur ratio.) Also my Wilem Jansen verhaelde datter een doctoor was t'Amsterdam, die het temperament van elcke camer int bysonder weten conde, hoeveel deen van dander in hitte ende coude verschilde, ende dat op de manniere, die ick hem verhaelde gesien te hebben te Delft opt stadthuys, daer ick vooren wat van geschreven hebbe.' In a footnote, Beeckman's editor, Cornelis de Waard, identifies this 'doctor from Amsterdam' as either Pieter Janszoon Hooft or Nicolaes van Wassenaer. Vera Keller discusses this issue in more detail in her contribution to this volume.

16 *JIB*, II, p. 199: 'Nu *a* is vol lochts, dewelcke van de hitte ende koude lichtelick verdunt ende verdickt wort. Alse verdickt wort door de koude, so beslaet se min plaetse, waerdoor het liqueur van onder opkommen moet tot aen *d*, om die ledich gemaeckte plaetse door de koude te vervullen. Maer als de locht in *a* door de wermte verdunt wort, so beslaetse meer plaetse ende stoot het water na beneden toe, tot aen *b*, twelck lichtelick op ende neer gaen kan, dewyle het bacxken *c* open is, ende wort also deen tyt volder ende dander tyt legher.'

17 An anonymous reviewer points out that it is surprising that Beeckman first focused on the fluid, as Hero already discussed the compression and expansion of the air.

Figure 14.2 New sketch of the Delft weather glass by Beeckman

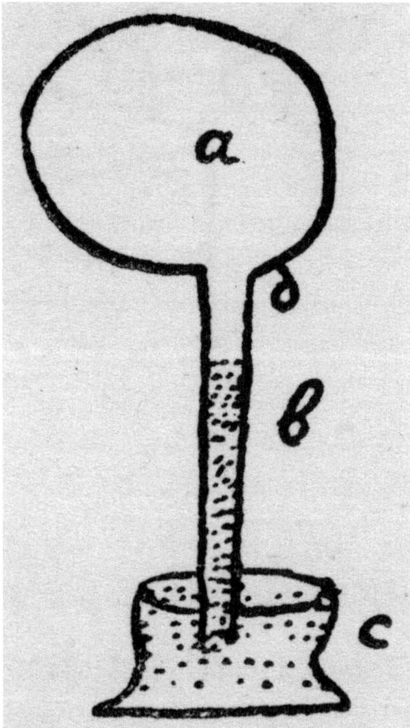

From: *JIB*, II, p. 199 [May-July 1622]

Beeckman immediately thought of a useful application of this insight: 'By this one could make a motum perpetuum.'[18] The idea was to equip the glass with a valve *t* so that the water cannot descend when the air warms up. The water is guided to a basin, also equipped with a valve, that has a hole to release water steadily. The drops from the basin set in motion a 'wheel of fortune [with] kings, emperors, popes, princes, gentlemen, noblemen, merchants, artisans, beggars' (fig. 14.3). And Beeckman concluded: 'Now as the day is always warmer than the night, so the water will not stop with going up and down.'[19] The wheel of fortune is certainly

18 *JIB*, II, p. 199: '(Motus perpetuus in rotâ inventus.) Hierdoor soude men een motum perpetuum <konnen> maken, te weten: [...].'

19 *JIB*, II, p. 200: 'Neempt dat *A* sy een radt van avontueren ende dat *a, b, c, d, e, f, g* persoonen syn, te weten, koningen, keysers, pausen, princen, heeren, edelen, cooplien, ambachslien, bedelaers. Ick segghe datmen sal konnen maken dat dit radt altyt drayen sal, sonder eenighe nieuwe hulpe. Want het liquer deur de koude int glas *p* opgetrocken synde, wort door de klappe *t* opgehouden, also dat het weder werm wordende, het water niet wederom in denselfden back en kan persen, want de klappe *t* gaet dan toe, ergo het water moet door de klappe *r* in den anderen back gaen. Also wort desen back altyt volder ende volder. Maer indien men in dien backs bodem by *s* een

Figure 14.3 Beeckman's wheel of fortune, driven by a perpetuum mobile machine

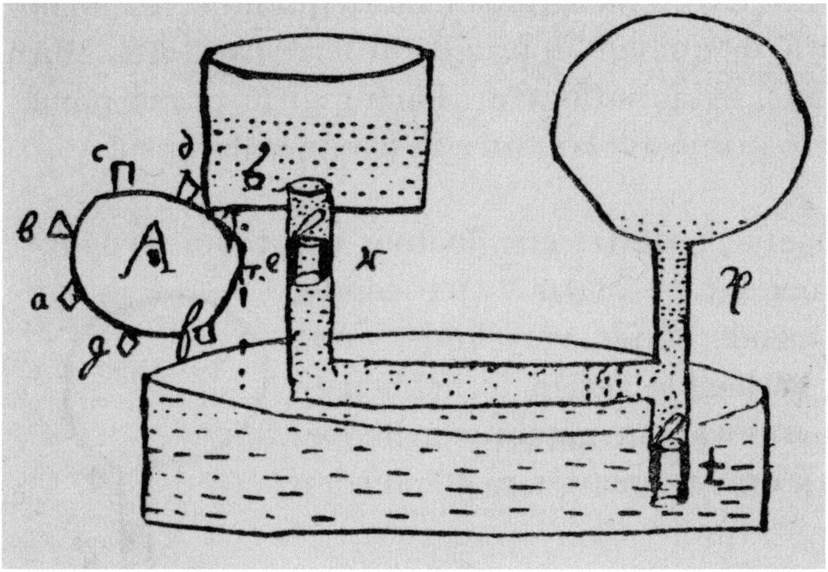

From: *JIB*, II, p. 200 [May-July 1622]

symbolic of the idea of a perpetuum mobile, but the key is Beeckman's realization that the daily cycle of heat and cold could be employed to generate motion.

The idea of a perpetuum was not new to Beeckman. Right from the start of his notebook we find references to perpetual motion machines. These always concern 'atmospheric' apparatus, entailing what we would call effects of fluid mechanics and thermodynamics: fountains, pumps, chimneys, and so on. Like the wheel of fortune, these machines employed the daily changes of the atmosphere to create perpetual motion. A 'timepiece [...] that go[es] faster or slower by the changes of the weather', he aptly called it at some point.[20] These atmospheric apparatus employed the hidden powers of the elements rather than the manifest forces of mechanical actions, a distinction that

gaetken maeckt, daerdoor het water trachelick drupt op het rat, so sal het rat, licht synde, van den val van die druppelkens omdrayen met persoonen ende al. Nu doordien den dach altyt warmer is dan den nacht, so en sal t'water niet ophouden van op ende neder te gaen. Detur nunc opera ut hinc in hoc motu aliqua aequalitas possit reperiri per aequalem casum guttarum; id enim horologium perpetuum efficiet.'

20 *JIB*, II, p. 366: 'Dit uerwerck is gelyck alle andere, die door veranderinghe des weders rasscher of tragher gaen, also dat de tyden niet recht ende effen gelyck en konnen syn.'

was regularly made at that time.[21] Beeckman used the phrase 'perpetuum mobile' in the sense that was common at that time: a machine that employed the forces of nature, an automaton. The perpetuum in this passage employs the dynamics of the atmosphere, changes in pressure and temperature, like a windmill employs the force of wind to create controlled movement.

From the viewpoint of Beeckman it is not surprising that the thought of a perpetuum came up when he was considering atmospheric apparatus like the Delft glass and the Amsterdam thermoscope. Moreover, this connection was quite common at the time. After the consideration of his wheel of fortune, he turned to one of the most famous perpetuums of his day, the one of Cornelis Drebbel (c. 1572-1633). It had drawn attention in learned circles all over Europe since its inception around 1604. In 1612, for example, Galileo received a report on a Drebbel instrument seen in Brussels.[22] Beeckman had not seen the instrument himself but wrote that those who had seen it said that it looked like this (fig. 14.4).[23] In Beeckman's understanding it was a variation of the Delft glass, only with the tube being split into two circular sections that converge again at the top. He assumed the advantage of the construction was that 'the curvature gives lightness so that the liquid is not so heavy to pull up'.[24] Beeckman was apparently thinking that the circular tube supported the liquid, but it is not entirely clear what he had in mind with this remark.

Beeckman went on to discuss the use of the glass and its workings: it could be seen as an indicator of the tides or the weather, the water waxing and waning accordingly.[25] He noted the similarities between the instrument

21 See, for example, D. Antonini to Galileo in a 1612 letter (reproduced in: Valleriani, *Galileo Engineer*, pp. 227-228), discussing a Drebbelian perpetuum: '[T]here is no difference between this motion and that of a water mill apart from the cause of motion, which is seen by everybody, whereas in this case it is not.'

22 The letter of Antonini reproduced in: Valleriani, *Galileo Engineer*, pp. 227-228. See also: Vera Keller, *Cornelis Drebbel (1572-1633): Fame and the Making of Modernity* (PhD diss., Princeton University, 2008).

23 *JIB*, II, p. 202: 'Die het motum perpetuum van Drebbel gesien hebben, segghen, dat het twee glase halve rynghen syn, tegen malkanderen kommende, waerin een liqeur is, twelck met het getye op ende neer gaet in de rynghen, also dat het van beyde syden ontrent *a* byeen komt, ende dan na *b* toe wederom afwyckt. Segghen daerenboven, dat men daerin oock siet wat weer dattet in see maeckt.'

24 *JIB*, II, p. 202: 'Ja de rondicheyt geeft lichticheyt, also dat het liqeur so swaer niet op te trecken en is, omdatter veel plaetse verandert, weynich verhooght synde.'

25 *JIB*, II, p. 202: 'Wat aengaet het wassen van het water, dat is misschien geseyd per similitudinem, te weten, gelyck het water wast ende daelt, also ryst dit oock ende daelt, twelck de lieden hoorende, kunnen gedocht hebben, dat men de getyen daerdoor weten konde als per signum.' ('Regarding the rising [waxing] of the water, that maybe said *per similitudinem*, that is: as the

Figure 14.4 Beeckman's sketch of Drebbel's instrument, on display in Brussels

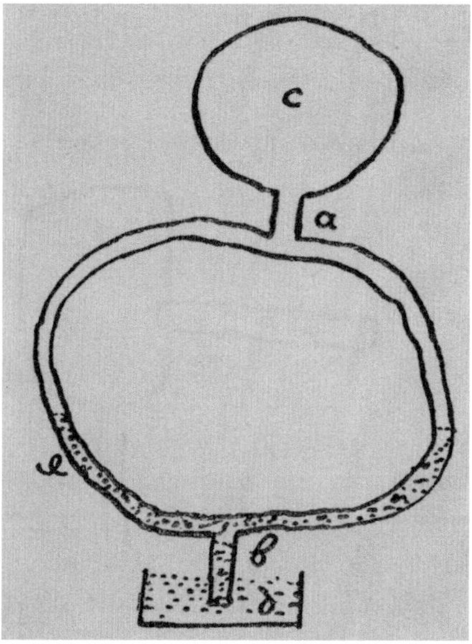

From: *JIB*, II, p. 202 [May-July 1622]

and the sea, implying it could be considered as a model of the atmosphere. The water below the air, he remarked, should be expected to 'shoot its vapours' that cannot escape because the glass above is closed.[26] Whether the atmosphere was closed off in a similar manner, he was not certain yet, but this would suggest an interpretation of the perpetuum. He did consider, however, whether this atmospheric effect could offer a more probable explanation of the relationship between the Moon and the tides: 'Why should the operation of the Moon not have more force in the air, that one sees can so easily stretch and thicken, than in the water?'[27]

water waxes and wanes, so this also rises and decends, [...] suggesting that one could know the tides as *per signum*.')

26 *JIB*, II, p. 202: 'Dan wat dat de veranderinghe des weers voor vapeuren of anders int glas verwecken sal, dewyle de locht int glas behoort verandert te worden gelyck de locht buyten, nadien datter onder oock water is, | also wel als in de see, die also wel hier als daer syn dampen behoort te schieten, welcke oock opkommende niet en konnen verdwynen dewyle het glas boven toe is, de locht van eenderley natuere synde als buyten – dat sal den tyt leeren.'

27 *JIB*, II, p. 203: 'Ende wat aengaet het hoogh ende leegh water, dewyle de reden, waerdoor dat de Mane dat veroorsaeckt, noch niet ter deghen bekendt en is, waerom en soude de operatie van de Mane niet meer kracht hebben in de locht, die men siet ende so gemackelick recken ende dicken kan, dan int water? De locht, dan verdickt synde door eenighe kracht der Mane, maeckt ontrent het water een ydelheydt, alwaer de Mane gaet; om welcke ydelheydt te vullen, so volcht haer het water van den Oceaen, ende steutende teghen America, maeckt daer seer hoogh water,

After these reflections, Beeckman returned to thermoscopes like the Delft glass. He surmised that the beaker of the Delft glass was covered to avoid evaporation, considered the relation between the volume of air in the bulb and that of the column of water, and the particular nature of the 'stretching of air' the force it exerted. Finally, he brought up the idea that this atmospheric phenomenon could be used to make a 'horologium perpetuum', referring to a design of Hero.[28] Later that summer, Beeckman made a final note on thermoscopes, reflecting upon the dilatation of the air and proposing another, weather glass arrangement.

From what Beeckman knew about Drebbel's perpetuum in 1622, he assumed it was a kind of thermoscope similar to the Delft glass, except for the double, circular tube suspending the water between the air bulb and the container. This seemingly unnecessary bifold connection apparently provided some kind of support for the water column. Beeckman's assumption was wrong, and he would find out a few years later. In the intervening time, thermoscopes and other atmospheric apparatus are largely absent from his notebook. In the spring of 1623, Beeckman made another couple of notes on the Delft glass, for the time being the last ones. They contain an interesting observation: the glass 'had the fault of not maintaining proportion in the lifting of the humour, because when it is high it cannot pull equal change as strong as down'.[29] That is: the rises and declines of the fluid in the tube are not uniform. After a corpuscular account of hydrostatic pressure, he once again reconsidered the design of the glass and ways to trace the phases of the Moon.[30]

The Secrets of Drebbel

In the summer of 1626 Beeckman's interest in atmospheric instruments was rekindled in discussions with Jan Jansz Stampioen Sr. (before 1610-after 1660), a well-known mathematical practitioner in Rotterdam. In the notebook,

gelyck vooren ergens van de manniere daervan geseydt is.' ('The air being thickened by some force of the Moon, causes idleness around the water, where the Moon goes; to fill the idleness, so the water of the ocean follows here, and pushing against America makes very high water.')

28 Beeckman regularly referred to Hero, as well as Cardano, in this context. It concerned Commandino's 1575 edition *Heronis Alexandrini spiritualium*.

29 *JIB*, II, p. 234: 'Het instrument, daermen mede siet hoe koudt ende heet het is, heeft die foute dat het geen proportie en houdt int optrecken van het humeur, want alst hooghe is en kant evenveel veranderinghe so sterck niet trecken als omleeghe, omdat het humeur, hoe hoogher het hanght, hoe het meer teghen treckt.'

30 *JIB*, II, pp. 236-238. See also: Van Berkel, *Isaac Beeckman on Matter and Motion*, pp. 134-136. In 1626 he used the word 'eenparichlick' explicitly (*JIB*, II, p. 361).

Figure 14.5 The 'diarium Drebbelii' according to Beeckman

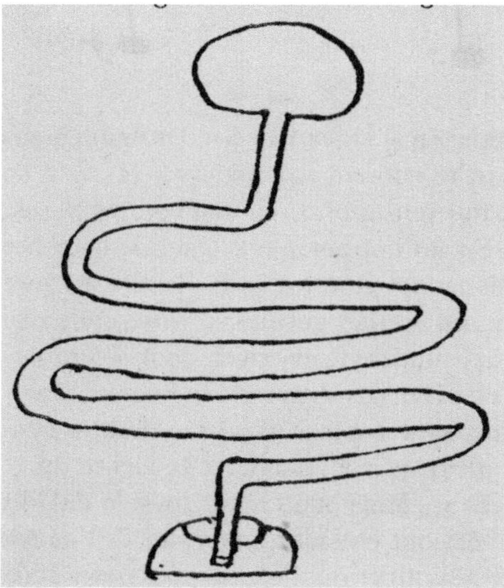

From: *JIB*, II, p. 366 [August-September 1626]

the entries on thermoscopes are preceded by Beeckman's account of his involvement in the debate over a perpetual motion machine. This was a mechanical apparatus and he made clear to the inventor and the projected investors that it is was ill-conceived.[31] According to Van Berkel, this affair was the occasion to establish the Collegium Mechanicum, to discuss topics of the mechanical arts and natural philosophy.[32] That summer Stampioen began to figure prominently in the notebooks and he would be the most important member of the Collegium when it started at the end of August.

On 15 August, Beeckman and Stampioen had discussed the thermoscope – the 'Drebbelian instrument', as he now called it (fig. 14.5).[33] He had proposed to Stampioen to figure out how to measure heat and cold by making the rise and decline of the water column behave uniformly, returning to his final reflection of the 1621-1623 notes on atmospheric apparatus.[34] To that end he proposed to figure out how exactly air expanded. After a consideration of the effects of accumulated pressure – not unlike Beeckman's analysis of acceleration – he described an experimental set-up to test it. He immediately took a step further

31 *JIB*, II, pp. 350-356.

32 Van Berkel, *Isaac Beeckman on Matter and Motion*, pp. 35-37.

33 *JIB*, II, pp. 361-367.

34 Albrecht Heeffer discusses this theme in more detail in his recent article, 'Quantifications of the Secondary Qualities Heat and Cold on the Earliest Scales of Thermoscopes', *Early Science and Medicine* 25 (2020), pp. 562-593.

Figure 14.6 A thermoscope inside a thermoscope, design by Beeckman

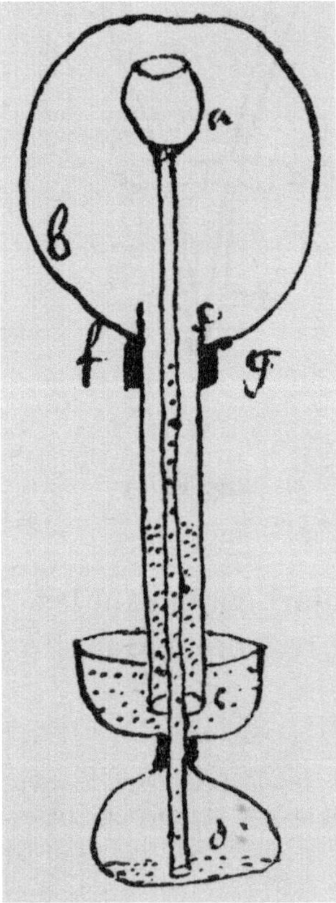

From: *JIB*, II, p. 370 [September-November 1626]

by projecting how one could 'calculate the force of the cold' by registering the proportion of cold over extended periods. Keeping track of fortune and ill could do great benefit for prognostication. With this idea he turned to his wheel of fortune, projecting to use the rotating wheel as a recording device. To this end the 'diarium Drebbelij' could also be used, especially when it's construction was so altered as to make the passage of times visible.

A little while later, early November 1626, Beeckman had yet another idea of constructing a weather clock. This time he conceived of a thermoscope within another thermoscope, mounted in a kind of double container. He noted that the container at the top of *da* prevented spilling of the water in the tube (fig. 14.6). Beeckman had gotten the idea of a nested thermoscope from 'the pedagogue of the children of Pensionary Pauw'.[35] The teacher in

35 *JIB*, II, pp. 370-371: 'Dit bacxken boven aen te setten hebbe ick gesien by den paedagoge van den pensionaris Pauws kinderen, doch in alio casu niet verschillende vant gene vooren geseyt is, ergens op een ander plaetse.'

question was Henricus Reneri (1593-1639), at that time student of medicine at Leiden and a protégé of Beeckman's acquaintance André Rivet (1572-1651).[36] Reneri also knew the secret of the perpetuum of Drebbel. He said he knew someone 'who was as familiar with Drebbel as with his own brother'.[37] As it turned out, Drebbel's instrument was an ordinary thermoscope, only arranged differently. The tube was not double but bent into a circle, with some liquid in the lower part (fig. 14.7). The circular tube only connects at one side to the bulb, while the other is open to the air. The liquid in the tube is thus pushed left and right depending upon the temperature of the bulb and the atmospheric pressure. Rather than the equilibrium Beeckman had assumed at first, the secret of the instrument consists in the way the liquid is suspended in the tube. Drebbel's perpetuum thus functioned in the same way as the Delft glass; with only a different layout of the parts – air bulb, water container, tube. Drebbel had made an engaging design of his instrument, concealing the actual construction and placing the air bulb in the centre. After hearing the secret of Drebbel's perpetuum from Reneri, Beeckman immediately proposed an adjusted design to make the working more transparent.

Beeckman had no direct knowledge of Drebbel's perpetuum, despite the fact that he had relatively close connections to the circles around him (as Vera Keller explains in her contribution to this volume).[38] His father was probably acquainted with Drebbel when the latter constructed a fountain in Middelburg in 1601 and was kept informed of his doings.[39] On 10 November 1619, Beeckman wrote that he had had an occasion to read Drebbel's treatise on the nature of the elements, a very rare publication that the author had distributed only among good friends.[40] When he discussed Drebbel's perpetuum in the summer of 1622, Beeckman mentioned a Haarlem patrician alchemist who

36 Robin Buning, *Henricus Reneri (1593-1639): Descartes' Quartermaster in Aristotelian Territory* (PhD diss., Utrecht University, 2013), pp. 76-79, 116-121, 168-171.

37 *JIB*, II, p. 372: 'dewelcke seght kennisse te hebben met een, die so familiaer met Drebbel is als met syn eyghen broeder'. Vera Keller also discusses this entry in her contribution to this volume.

38 Van Berkel, *Isaac Beeckman on Matter and Motion*, pp. 49-50.

39 On 15 March 1620 Beeckman writes about a letter from his father, discussing Drebbel's submarine. *JIB*, II, p. 25.

40 It may have been a copy of the 1604 first edition and it may have belonged to his father or Philippus Lansbergen; Van Berkel, *Isaac Beeckman on Matter and Motion*, p. 50. *Tractaet van de Natuere der Elementen* has a complicated history of editions and translations, but was to become quite popular and widely distributed. See: Keller, *Cornelis Drebbel*, pp. 109-110: the 1604 edition may be the Dutch edition which Johann Ernst Burggrav mentioned in his own 1628 edition of Drebbel's *On the Nature of the Elements*. According to Burggrav, Drebbel had a few copies of his natural philosophy printed. He shared these copies only with 'good friends and philosophers'. See also the chapter on Drebbel by Vera Keller in this volume.

Figure 14.7 Drebbel's weather glass and Beeckman's idea for improvement

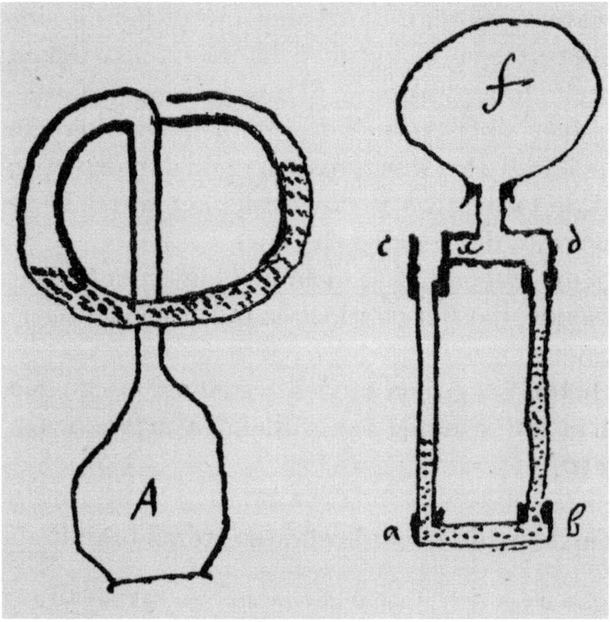

From: *JIB*, II, p. 372 [September-November 1626]

had worked closely with Drebbel.[41] In 1626 Reneri turned out to be a direct informant on Drebbel's inventions. Still later, Beeckman met Sibertus Küffler, one of Drebbel's sons-in-law and an important promoter of his legacy.[42] In the meantime, on 15 March 1631 he had made his famous copy of Drebbel's 1613 letter to the English king and the drawings of the microscope.[43]

Engineering Magic

In Beeckman's considerations of the Delft glass, Drebbel's perpetuum, and other atmospheric apparatus, we see a particular way of knowing consisting of handling artefacts both cognitively and manually. They offer

41 *JIB*, II, p. 201. He heard it from Jacobus Bernardi, who had apparently blown glasses for the man. Beeckman thought Bernardi's own inventions were nonsense.
42 *JIB*, III, p. 367 [15 October 1634].
43 *JIB*, III, pp. 203-204, 438-442. See also: Fokko Jan Dijksterhuis, 'Magi from the North: Instruments of Fire and Light in the Early Seventeenth Century', in: Arianna Borelli, Giora Hon, and Yaakov Zik, eds., *The Optics of Giambattista Della Porta (ca. 1535-1615): A Reassessment* (Cham: Springer, 2017), pp. 125-143.

a fine example of what Bertoloni Meli has called 'thinking with objects' and that I denote as 'artefactual knowing'.[44] Observing and reflecting upon the workings of thermoscopes and weather glasses, Beeckman raised questions, connected phenomena, instruments, constructions, remodelling the things he understood into new objects of consideration. I now want to see how Beeckman's dealings with atmospheric instruments and this artefactual way of knowing fit in our understanding of his oeuvre as a whole and how they position historically in early-seventeenth-century learning.

Beeckman's notebook entries from the 1620s on thermoscopes, perpetuums, and other atmospheric instruments were part of his life-long engagement with hydrology. He had worked with his father in waterworks and in the 1610s, when he had set up as a candle maker, he undertook numerous projects and business trips concerning pumps, fountains, conduits, valves and so on.[45] In April 1618, for example, he went to Brussels with his cousin Andries Lambrechts to inspect some of their fountains.[46] These can be considered engineering activities, although the word 'engineer' was reserved for the military before the twentieth century. Hydrological 'engineering' also differed in kind: rather than the geometry of defence constructions and surveying operations, it concerned the dynamics of the elements. Chimneys, pumps, and conduits were means of managing fluids, flows, heat and pressure. Beeckman's interest in perpetuums fitted these activities. Such atmospheric automatons were ways to harness the elements in what we would call an engineering way.

In typical Beeckman fashion, these dealings with waterworks were occasion to reflect upon technical, physical, and even geological aspects. A recurrent theme was the issue of water shock in pipes, thinking about the reason of its destructive effects as well as ways to avoid it by proper bending. But he also thought along grander lines, comparing pumping of water to both blood circulation and subterranean flows, and considering issues of density, rarefaction, and vacuum. In these 'philosophical' reflections, Beeckman appears as a 'learned practitioner', the kind of sophisticated engineer that was emerging all over Europe in the fifteenth and sixteenth centuries.[47] Part of these learned activities were dealings with books: reading, compiling, commenting. In this

44 Domeni Bertoloni Meli, *Thinking with Objects: The Transformation of Mechanics in the Seventeenth Century* (Baltimore: Johns Hopkins University Press, 2006).

45 Van Berkel, *Isaac Beeckman on Matter and Motion*, p. 18.

46 *JIB*, I, pp. 173-174 (with further reflections on pp. 175-176, 181, 187-188).

47 See, for example: Pamela O. Long, *Artist/Practitioners and the Rise of the New Sciences, 1400-1600* (Corvallis: Oregon State University Press, 2011); Ursula Klein, *Nützliches Wissen. Die Erfindung der Technikwissenschaften* (Göttingen: Wallstein, 2016); Lilian Hilaire-Pérez, Fabien

way the ideas of Hero entered Beeckman's reflections on chimneys, fountains, as well as thermoscopes and perpetuums. Matteo Valleriani has shown how the work of Hero – in the circulation of manuscripts and the preparation of editions – was instrumental in the emergence of a group of learned engineers in late-sixteenth-century Italian states.[48] In northern Europe, men like Drebbel, Salomon de Caus, Andries Vierlingh and – in his own way – Simon Stevin, fit this profile, and we may count the Beeckman of the 1610s among them.

Still, Beeckman does not completely fit such a profile of 'learned practitioner'. He had received academic training and his notebook shows that he also had philosophical inclinations towards inquiring and developing formal principles for understanding nature. When, in 1618, he re-entered university and delivered a dissertation, it contained five *corollaria* and two *quodlibeta* on natural philosophy, including an argument against 'horror vacui'.[49] Beeckman's promotion was the prelude to his switch from practitioner in Zeeland to schoolmaster in Rotterdam. There his engineering work shifted more towards advising and assessing and his engagement with waterworks to curious atmospheric instruments.[50] These instrument display, I have argued, an artefactual way of knowing that differs from the philosophical pursuit of developing general principles. After his move to Dordrecht in 1627, Beeckman was immersed in the scholarly circles of René Descartes, Pierre Gassendi, and Marin Mersenne, and we can see him increasingly engaging with issues of natural philosophy. Still, in Dordrecht he also fully entered into lens grinding and his notes display a similar artefactual way of thinking that we encountered in the Rotterdam dealings with atmospheric instruments.[51] In Beeckman and his notebooks we encounter both philosophical and artefactual ways of knowing – and often in very interlinked manners. This Beeckman may well be compared to someone like Drebbel, dealing with practical things but with a bent for philosophizing.

In his monograph, Van Berkel strongly contrasts Beeckman and Drebbel; too strongly in my view. There are undeniable differences between the two

Simon, and Marie Thébaud, eds., *L'Europe des sciences et des techniques. Un dialogue des savoirs, xve-xviie siècle* (Rennes: Presses Universitaires de Rennes, 2016).

48 Valleriani, *Galileo Engineer*, pp. 172-178; Matteo Valleriani, 'The Garden of Pratolino: Ancient Technology Breaks through the Barriers of Modern Iconology', in: Natascha Adamowsky, Harmut Böhme, and Robert Felfe, eds., *Ludi naturae. Spiele der Natur in Kunst und Wissenschaft* (Paderborn: Fink, 2011); Valleriani, 'Condensation to Compression'.

49 Van Berkel, *Isaac Beeckman on Matter and Motion*, pp. 20-22.

50 Van Berkel, *Isaac Beeckman on Matter and Motion*, pp. 34-37.

51 Fokko Jan Dijksterhuis, 'Labour on Lenses: Isaac Beeckman's Notes on Lens Making', in: Albert Van Helden et al., eds., *The Origins of the Telescope* (Amsterdam: KNAW Press, 2010), pp. 275-270.

men, but these are mainly in style and context – the lucid (private) reasoning of a practitioner and teacher versus the suggestive spectacle of a projector. As regards the content of their ideas there was much common ground between the two. Van Berkel contrasts Beeckman's 'mechanical' approach to Drebbel's 'mystical' and 'vitalistic' ideas.[52] Drebbel's writings are scarce and not easy to interpret, but I do not find many 'vitalistic elements' there. There are alchemistic ideas in the sense of transformations of substances and elements, but these were also not alien to Beeckman.[53] In my view, terms such as 'vitalistic' and 'mechanical' miss the heart of the matter, namely the dynamic nature of their ideas. Both regarded the atmosphere and the elements in terms of active flows of heat and fluids. Beeckman had an expressly corpuscular conception but his physics exceeds the motions and impact of mechanics; it is active and dynamical. Drebbel is more difficult to figure out because of the paucity of writings from his hand and because of his rather hermetic style. The intellectual ties between Beeckman and men like Drebbel do not mean that he was not critical of such ideas: Beeckman could be dismissive of perpetuum builders like Nicolaas Puyck, and what he considered to be alchemical nonsense and geological fantasies.[54]

The dynamical approaches of atmospheric processes we encounter in Beeckman and Drebbel can be linked to a tradition in late-sixteenth-century meteorology.[55] Originating in a critique of Aristotle's *Meteorologica*, a materialistic interpretation emerged of atmospheric processes, in particular regarding the nature and origin of winds. In the views of people like Giambattista della Porta and Giambattista Benedetti, the atmosphere presented a dynamic realm where heat and transformations created flows and meteorological phenomena. This new meteorology employed various experimental and instrumental accounts, most notably the so-called inverted glass experiment. This consists of an inverted glass alembic in which the level of the water in the neck varies with temperature and weather. Della Porta described it in his 1589 *Pneumaticorum*, reinterpreting it in the 1606 Italian edition.[56] The

52 Van Berkel, *Isaac Beeckman on Matter and Motion*, pp. 50-51.

53 See also: Dijksterhuis, 'Magi from the North'; Fokko Jan Dijksterhuis, 'Understandings of Colors: Varieties of Theories in the Color Worlds of the Early Seventeenth Century', in: Tawrin Baker et al., eds., *Early Modern Color Worlds* (Leiden: Brill, 2016), pp. 225-247; Keller, *Cornelis Drebbel*; Borelli, 'Weatherglass'.

54 On Puyck: *JIB*, II, pp. 352-353, 355-356; Van Berkel, *Isaac Beeckman on Matter and Motion*, pp. 35-39. On alchemy, see above, footnote 41. On earth fantasy: *JIB*, II, pp. 388-389.

55 Borelli, 'Weatherglass'.

56 Borelli, 'Weatherglass', pp. 75-78. Beeckman's *Journal* does not reveal whether he knew the 1604 Italian edition of Della Porta and in what sense he knew the inverted glass experiment.

Delft glass and Drebbel's perpetuum were variations of this experiment. Beeckman was well acquainted with this meteorological tradition, referring to Della Porta, Benedetti, and others, and could have heard about the weather glass through all kinds of routes.

The experimental and materialistic approach to meteorology was closely connected to the tradition of natural magic, epitomized by Della Porta's voluminous *Magia naturalis*. This popular book circulated widely and had been translated into Dutch as well. Arianna Borrelli has lucidly applied Meli's concept of thinking with objects to both the meteorology and the optics of Della Porta to show the epistemic features of the tradition of natural magic.[57] In this tradition, nature was thought to be governed by forces that could be harnessed by apparatus to serve human needs.[58] Instruments like the weather glass and the perpetuum had a double meaning: besides being automata, they were also seen as analogons to nature. By imitating the dynamics of the atmosphere and recreating natural phenomena, such instruments offered clues to the secrets of nature and to the causes of phenomena. This is precisely how we saw Beeckman interpreting Drebbel's perpetuum, as an analogon to the way the atmosphere acted on the oceans to create ebb and flow. Telescope and microscope, thermometer and barometer, later followed by the air pump and the electrical engine, represent a novel type of instruments in the seventeenth century that aided philosophical investigation.

Beeckman and Drebbel inhabited a world of natural magic and atmospheric engineering that was quite common in those days. The link between engineering, magic, and alchemy had ancient roots and can also be found in the writings of someone like Stevin. In a reflection on the age of sages, he placed the mathematical sciences alongside alchemy (*Stofscheyding*, separation of substances/matter) and magic (*Gheesthandel*, operation of spirits) as instances of original wisdom, the operative learning of the *consten*.[59] The epitome of this ingenious knowing was Archimedes and in this regard Stevin reflected a generally shared conviction. In late-sixteenth-century conceptions, engineering and magic were closely connected and we can see a direct line from Archimedes

57 Borelli, 'Weatherglass'; Arianna Borelli, 'Thinking with Optical Objects: Glass Spheres, Lenses, and Refraction in Giovan Battista Della Porta's Optical Writings', *Journal of Early Modern Studies* 3 (2014), pp. 39-61; Vera Keller, 'Drebbel's Living Instruments, Hartmann's Microcosm, and Libavius's Thelemos: Epistemic Machines before Descartes', *History of Science* 48 (2010), pp. 39-74.

58 William Eamon, *Science and the Secrets of Nature: Books of Secrets in Medieval and Early Modern Culture* (Princeton: Princeton University Press, 1996).

59 Fokko Jan Dijksterhuis, 'The Wise Origins of the *Consten*: Stevin and Sixteenth-Century Debates on Arts, Mathematics and Language', in: Karel Davids et al., eds., *Rethinking Stevin, Stevin Rethinking: Constructions of a Dutch Polymath* (Leiden: Brill, 2020), pp. 182-205.

and Hero, to Stevin, Drebbel, someone like De Caus, and Beeckman.[60] Stevin's hydrostatics was also more concerned with pipes and flows than with the classical topic of floating bodies.[61] When Beeckman went through the manuscripts of Stevin he particularly picked out topics of atmospheric contraptions such as the ventilation of latrines.[62] When Stevin discussed a sort of weather glass, Beeckman responded to his ideas about density.

Stevin was prominent in Beeckman's work and the latter had access to Stevin's unpublished papers. The way his reading of Stevin's hydrostatics contributed to the development of Descartes' natural philosophy has been documented very well.[63] My interpretation of Beeckman's considerations of atmospheric instruments raises new issues regarding the relationship to Stevin. His artefactual way of knowing recalls Stevin's ideas about the *consten*. Stevin's oeuvre can be read as a programme for a new science, the operative way of knowing of the *consten*.[64] Central to his works – from the early treatises on statics to the late memoirs on mathematics – was the question what kind of knowledge is acquired through practices and artefacts. Elsewhere in this volume, Arjan van Dixhoorn discusses a 'culture of *consten*' and he shows how ideas about practical understanding were linked to the urban circles of chambers of rhetoric. Despite his deep involvement with Stevin's writings, Beeckman hardly pursued themes regarding the epistemic implications of artefactual practices. He did not reflect upon the kind of knowledge acquired in artisanal work or how it combined with philosophical thinking.

Conclusion

However we may want to interpret Beeckman's work, the fact is that we are able to do so on the basis of the rich and detailed sources found in his unique notebook. If he is exceptional in any way, it is in the first place

60 In the Leiden reading list that Rudolph Snellius compiled for Beeckman, Hero figured prominently in the mechanical arts (rather than pneumatic). *JIB*, IV, pp. 17-19

61 Fokko Jan Dijksterhuis, 'Principles of Weighing Water', *Metascience* 28 (2019), pp. 181-186.

62 *JIB*, II, p. 291. See: Charles van den Heuvel, *'De Huysbou': A Reconstruction of an Unfinished Treatise on Architecture, Town Planning and Civil Engineering by Simon Stevin* (Amsterdam: KNAW Edita, 2005), pp. 91, 228.

63 Van Berkel, *Isaac Beeckman on Matter and Motion*; Schuster, *Descartes-Agonistes*.

64 Karel Davids, Rienk Vermij, and Fokko Jan Dijksterhuis, 'Introduction: Simon Stevin, Polymaths and Polymathy in the Early Modern Period', in: Davids et al., eds., *Rethinking Stevin*, pp. 1-24. See also: Hélène Vérin, 'Rediger et reduire en art. Un projet de rationalisation des pratiques', in: Pascal Dubourg Glatigny and Hélène Vérin, eds., *Réduire en art. La Technologie de la Renaissance aux Lumières* (Paris: Éditions de la Maison des science de l'homme, 2014), pp. 17-59.

because his daily notes have been preserved. In this chapter I have tried to contribute to the Beeckman historiography by highlighting an aspect of learning that has been little discussed, the artefactual way of knowing manifest in his dealings with atmospheric instruments. In conclusion, I would like to consider how this perspective may add to our understanding of Beeckman and his position in early modern history of knowledge. In current historiography Beeckman's dealings with hydrology and meteorology are not central and are mainly considered regarding themes of matter theory and vacuum. Historiography tends to focus on the mechanics of matter and motion and the ensuing mechanistic philosophy.[65] The artisanal background then resulted in a practical take on philosophizing, comparable to the way Galileo read scholastic philosophy from his engineering context.[66] Still, in this way the focus remains on philosophical systems – causal accounts in an ontological and cosmological framework. Beeckman's invention of corpuscular-mechanical philosophizing – cogently argued by Schuster elsewhere in this volume – may have supplementary roots in his dealings with atmospheric apparatus. Rather than driven by theories or principles or aimed at explanatory systems, we see him thinking with the objects and phenomena he encountered everywhere. Some underlying natural philosophical system may be reconstructed from this, but that may not be exactly getting the point of what Beeckman was doing. He was engaging intellectually – and quite ingenuously – with atmospheric effects and dynamical apparatus. We may turn this around and ask how this artefactual way of knowing shaped his corpuscular-mechanical philosophizing. That would emphasize the machine-like approach of nature in mechanistic philosophy rather than its mechanical principles. It would also emphasize the hydrological features of the world machine – the dynamical interactions of the elements that may, or may not, be harnessed for human benefit.

About the Author

Fokko Jan Dijksterhuis is Associate Professor in the History of Science at the University of Twente and Louise Thijssen-Schoute Professor in Early Modern History of Knowledge at the Vrije Universiteit Amsterdam. Originally specializing in seventeenth-century optics, his interests have broadened to

65 Schuster, *Descartes-Agonistes*; Van Berkel, *Isaac Beeckman on Matter and Motion*; Van Berkel, 'Vruchtbaar isolement'.

66 Valleriani, *Galileo Engineer*; Vérin, 'Reduire en art'.

encompass the cultural history of knowledge in the early modern period. He is particularly interested in practical, useful, and artefactual ways of knowing. He is the author of 'The Wise Origins of the *Consten*: Stevin and Sixteenth-Century Debates on Arts, Mathematics and Language', in Karel Davids et al., eds., *Rethinking Stevin, Stevin Rethinking: Constructions of a Dutch Polymath* (Leiden: Brill, 2020), pp. 182-205.

15 'Communicated Only to Good Friends and Philosophers'

Isaac Beeckman, Cornelis Drebbel, and the Circulation of Artisanal Philosophy

Vera Keller

Abstract

This chapter explores Beeckman's communicative strategies through comparison with the case of Cornelis Drebbel, a figure in whom Beeckman was highly interested. Both Beeckman and Drebbel prioritized in-person communication and tended to avoid print communication. This essay discusses some of the concerns that motivated this targeted communication as well as the conditions that made it possible and effective. A transnational network that linked Beeckman and Drebbel circulated knowledge of their works on their behalf in ways that rendered print publication unnecessary.

Keywords: Isaac Beeckman, Cornelis Drebbel, post-Reformation networks, artisanal knowledge, thermometer

In 1634, the Polish naturalist Jon Jonston (1603-1675) complained to the London-based intelligencer from Elbląg, Samuel Hartlib (c. 1600-1662), that 'the rector of Dordrecht Beeckman has something like a thousand experiments and sounds like a great philosopher, but he is morose and incommunicative'.[1] Indeed, the only text that Isaac Beeckman (1588-1637) published during his lifetime was his 1618 dissertation for a medical degree

1 M. Greengrass, M. Leslie, and M. Hannon, *The Hartlib Papers* (Sheffield: The Digital Humanities Institute, University of Sheffield, 2013), https://www.dhi.ac.uk/hartlib, 29/2/43A [henceforth HP]: 'Rector zu Dort Becmannus habet aliquot millia Experimentorum et audit egregius philosophus, sed morosus et non communicativus.'

Van Berkel, Klaas, Albert Clement, and Arjan van Dixhoorn (eds.), *Knowledge and Culture in the Early Dutch Republic: Isaac Beeckman in Context*. Taylor & Francis Group 2022
DOI: 10.5117/9789463722537_CH15

at Caen. This survives in only a single, incomplete imprint.[2] Beeckman kept his extraordinary body of scientific work confined to his manuscript daily notes, the *Loci communes*. Yet, in Hartlib's next and only other discussion of Beeckman in his own copious daily notes, the *Ephemerides*, Hartlib noted Beeckman's desire to establish a 'college of inventions', hardly the mark of an incommunicative individual.[3] How might we square Hartlib's two remarks and make sense of Beeckman's communicative strategies?

In his chapter in this volume, Arjan van Dixhoorn argues that Beeckman should be understood through the lens of a very long-standing culture centred on sociable communication: the *consten*-culture of the chambers of rhetoric, which prized joking, ingenuity, and cognitive exercises through engagement with the liberal arts. Van Dixhoorn argues that through the *consten*-culture, 'explicit, bookish, academic, theoretical learning and tacit, bodily, artisanal, practical, experience-oriented knowledge had already been "interpenetrating" for two centuries'. Van Dixhoorn's criticisms of a misleading dichotomy between artisanal and textual approaches and identities are well taken. However, waves of emigration cutting across national and linguistic boundaries brought varied and sometimes competing forms of knowledge into conversation and spurred dynamic experiments in how to communicate and collaborate. Beeckman's 'college of inventions', which he called the Collegium Mechanicum, was one such experiment that distinguished itself from already extant forms of sociability, such as the joyful companies gathered by chambers of rhetoric.

As a counterpoint to Dutch *consten*-culture, this chapter places Beeckman amid a transnational mercantile and religious émigré network that was attempting to negotiate varied and sometimes conflicting cultural values and forms of knowledge.[4] At stake was how to communicate and collaborate while safeguarding incipient and experimental forms of knowledge that both depended upon material resources and that bore immediate financial repercussions. A comparison of Beeckman to Cornelis Drebbel (1572-1633), whose many philosophical similarities to Beeckman are further explored by Fokko Jan Dijksterhuis in this volume, will shed light on the issues faced and choices made in their targeted communicative strategies. Beeckman's ability

2 *Journal tenu par Isaac Beeckman de 1604 à 1634*, publié avec une introduction et des notes par C. de Waard, 4 vols. (The Hague: Martinus Nijhoff, 1939-1953) [henceforth *JIB*], IV, pp. 42-44; Klaas van Berkel, *Isaac Beeckman on Matter and Motion: Mechanical Philosophy in the Making* (Baltimore: Johns Hopkins University Press, 2013), p. 20.

3 HP 29/3/34B: 'Rector of Dort voluit Collegium Inventionum instituere.'

4 On Beeckman and immigration see: Van Berkel, *Isaac Beeckman on Matter and Motion*, pp. 8-10.

to gain knowledge about Drebbel, who did not communicate his knowledge widely and whom Beeckman did not meet in person, offers evidence of the importance of this network in the communication of artisanal philosophy.

Drebbel's Turn against the Liberal Arts

In his early career, Drebbel was surrounded by the *consten*-culture Van Dixhoorn describes. His engravings of the seven liberal arts after the elegant designs of his teacher and brother-in-law, Hendrick Goltzius, are his best known. Drebbel also engraved images after designs by Karel van Mander, with Latin mottoes by Cornelis Schonaeus. Van Mander was a member of a chamber of rhetoric, Goltzius produced artwork for the Pellicanists, a chamber in Haarlem, and both he and Schonaeus participated in its public events.[5] Such intersections with the world of the rhetoricians, however, did not preclude an engagement with other, possibly rival, cultural streams. Goltzius also practised alchemy alongside Drebbel at the same time that the two of them produced the series on the liberal arts. Drawing on motifs from that series, Goltzius put the liberal arts in conversation with alchemy in his grand painting, the *Allegory of the (Alchemical) Arts*, a work that offered an open-ended investigation of how the liberal arts and alchemy related to painting.[6]

While Goltzius blended the liberal arts with his alchemical interests, Drebbel disengaged from the liberal arts. In his texts, Drebbel presented himself as aggressively opposed to rhetorical compositions and linguistic facility. Drebbel called his *On the Nature of the Elements* a 'little book' and proved critical of those who wrote 'fat books'.[7] He argued that he would rather demonstrate his ideas manually through his instruments than describe them at length in words.[8]

5 Arjan van Dixhoorn, 'Chambers of Rhetoric: Performative Culture and Literary Sociability in the Early Modern Northern Netherlands', in: Arjan van Dixhoorn and Susie Speakman Sutch, eds., *The Reach of the Republic of Letters: Literary and Learned Societies in Late Medieval and Early Modern Europe*, 2 vols. (Leiden: Brill, 2008), I, pp. 119-157, esp. p. 121.

6 Christine Göttler, 'Tales of Transformation: Hendrick Goltzius's *Allegory of the (Alchemical) Arts* in the Kunstmuseum Basel', *21: Inquiries into Art, History and the Visual/Beiträge zur Kunstgeschichte und Visuellen Kultur* 1 (2020), pp. 403-446.

7 Cornelis Drebbel, *Ein kurtzer Tractat von der Natur der Elementen* (Leiden: Haestens, 1608), sig. A2^{v}: 'sollen wir grosse Bücher schreiben, Gott dar mit zu loben? Ist es nicht eittelheit?'

8 Vera Keller, 'Re-entangling the Temperature Concept: Cornelis Drebbel's Description of his Self-regulating Oven, the Regiment of Fire, and the Early History of Temperature', *Nuncius: Journal of the Material and Visual History of Science* 28 (2013), pp. 243-275, esp. pp. 253-254.

Like Beeckman, Drebbel published little during his lifetime. He had his short *On the Nature of the Elements* printed in Dutch in 1604 in only a few copies, 'which he communicated only to good friends and philosophers'[9]; no copies of this original imprint survive, and Isaac Beeckman is the best attested reader of it. This was the only publication of his that Drebbel ever saw to press. His few other, short published texts were printed on the initiative of others. For instance, Drebbel's friend Gerrit Pietersz Schagen (1573-1642) recalled how Drebbel gave him a copy of a text addressed to King James I to accompany the celebrated perpetual motion device presented to the English monarch in 1607.[10] It was Schagen, rather than Drebbel, who then chose to publish this dedication, the *Wonder-vondt*, appending his own dedication to the cartographer, engineer, mathematician and mayor of Alkmaar, Adriaen Anthoniszoon (1527-1607).

In his dedication of the Dutch translation of Drebbel's letter to King James I, Schagen noted that while it is 'apparent that the seven liberal arts tend towards the great advantage of mankind, they should hardly be considered the first principles of the perpetual motion with which the Philosopher Cornelis Jacobszoon Drebbel of Alkmaar has honoured the mighty King James of Britain'.[11] This was based rather, on the investigation of nature, allowing the 'Alkmaarian philosopher' to be able to demonstrate the structure of the universe, 'not only with reasoning/speaking (*reden*), but with living instruments'.[12]

The Lost Edition of 1604

As a result of his aversion to textuality, Drebbel has been criticized as secretive, and his identity as the author of philosophical texts minimized. He

9 Johann Ernst Buggrav, 'Vorrede', in: Cornelis Drebbel, *Ein Kurtzer Tractat von der Natur der Elementen* (Frankfurt: Fitzer, 1628), sig. Aii^{v}: 'etlich wenig Exemplaria fur sich drucken lassen / und allein guten Freunden unnd Philosophis mitgetheilet.'

10 G.P. Schagen, ed., *Wonder-vondt van de eeuwighe beweginghe* (Alkmaar: Jacob de Meester, 1607), sig. A3^{r}: 'Naedien de voorsz. Drebbel de Copy van de Dedicatie oft tooeygheninghe van de eewigh beweginghe / aen Coningh Jacob my te handen bestelt heeft / om te lesen. [...] Soo heeft my goet ghedocht die selfde Dedicatie ofte toeeygheningh / in Druck te laten uytgaen.'

11 Schagen, ed., *Wonder-vondt*, sig. A3^{r}: 'T is oock open-baer / dat de seven vrije Consten tot grootê voordeel der Menschen zijn streckende / maer mogen naulijcx de be-ginselen gheacht worden / van de eeuwigh beweginghe / daer den Philosooph Cornelis Jacobsz. Drebbel van Alckmaer / den grootmach-tighen Coningh Jacob van groot Brittangen heeft mede vereert.'

12 Schagen, ed., *Wonder-vondt*, sig. A3^{r}: 'Maer desen Alckmaersche Philosooph can 't selfde niet alleen met reden / maer oock met levendige Instrumenten bewijsen.'

has been called an 'eccentric Dutchman' and a 'Dutch machine maker' who 'kept silent about the principle' on the basis of which his devices worked.[13] Lynn Thorndike singled Drebbel out as 'probably the most pretentious, secretive and magical figure in the scientific and technical world of the early seventeenth century'.[14] Even historians of alchemy, well-versed in practices of secretive communication, have described Drebbel as a 'shadowy figure' and an 'eccentric mechanic'.[15]

Drebbel was not silent, but he preferred speech to text, and manuscript authorship to print. His account of his self-regulating oven, extant in two German translations, has recently been rediscovered and published.[16] Other reported manuscripts include treatises on marine salt, vitriol, and mercury, a collection of optical experiments, a treatise on the element of fire, and a collection of images of his inventions.[17] A large group of Drebbelian manuscripts, according to Samuel Hartlib, were kept by one 'Host' or 'Hoft' in Amsterdam, whom Hartlib suspected of claiming Drebbel's works as his own.[18] This would have been Pieter Janszoon Hooft (1576-1636), who, together with Jacob de Graeff, according to several sources, once worked alongside Drebbel and later was connected with disputes over credit for the perpetual motion machine.[19] Drebbel moved to Ipswich, England, in 1604, where he worked on developing his perpetual motion machine for several years before presenting it at the court of King

13 David Freedberg, *Eye of the Lynx: Galileo, His Friends, and the Beginnings of Modern Natural History* (Chicago: University of Chicago Press, 2002), p. 151; Matteo Valleriani, *Galileo Engineer* (Dordrecht: Springer, 2010), pp. 162-163.

14 Lynn Thorndike, *A History of Magic and Experimental Science* (New York: Macmillan, 1958), VII, p. 492.

15 For Drebbel as a 'shadowy figure' see: William Newman and Lawrence Principe, 'Alchemy and the Changing Significance of Analysis', in: Jed Z. Buchwald, ed., *Wrong for the Right Reasons* (Dordrecht: Springer, 2005), p. 86; 'rather eccentric inventor': Lawrence Principe, *Aspiring Adept: Robert Boyle and His Alchemical Quest* (Princeton: Princeton University Press, 1998), p. 86.

16 Keller, 'Re-entangling'.

17 Joachim Morsius noted around Drebbel's inscription in Morsius's *Stammbuch* unpublished manuscripts on marine salt, mercury, vitriol, and other topics. Morsius, Lübeck, 4a 25, 2, 344. Staats-und Universitätsbibliothek Hamburg cod. alchim. 668, 202v. In his preface to his 1628 German edition, Burggrav mentioned that he was in possession of other unnamed manuscripts by Drebbel. Sir Christopher Gardiner was said to have inherited 'all Drebbel MS. and Arcana.' HP 29/5/102B. In 1649, William Petty praised Drebbel's 'Optical manuscripts'. HP 28/1/5A.

18 HP 29/3/55B. These included a 'Treatise on the Element of Fire', optical experiments and images.

19 Ole Borch, *Itinerarium 1660-1665, Vol. II: Oct. 1661-May 1663* (London: Brill, 1983), p. 143. Geraardt Brandt, ''t Leeven van den Weleedelen, gestrengen, grootachtbaaren Heere, Pieter Corneliszoon Hooft', in: P.C. Hooft, *Nederlandsche Historien* (Amsterdam: Wetstein, 1703), p. 2.

James. In Ipswich, he lived with a fellow Dutchman, 'Peter Hoste'.[20] In 1622, Isaac Beeckman appeared to refer to a dispute over credit for the perpetual motion between Drebbel and a rich doctor of Amsterdam who, as Beeckman's editor Cornelis de Waard pointed out, fits Hooft's description.[21]

Before he moved to England, it appears, Drebbel published a small run of his major work, *On the Nature of the Elements*, in Dutch. This edition is no longer extant, and its existence has been doubted.[22] Several reasons exist to accept its existence, however. In a 1628 German edition of *On the Nature of the Elements*, Drebbel's translator and editor, Johann Ernst Burggrav, recalled how he first encountered Drebbel's short work. While abroad during his academic peregrination 20 years previously,

> I became, through a trusted friend, a sharer in this treatise *On the Nature of the Elements*, which Cornelis Drebbel had written in Dutch and had printed for himself in only a few exemplars which he communicated only to good friends and philosophers, which little book I then translated into German and had printed, and it has since become a pleasing little book to many understanding philosophers who have been able to read it.[23]

After he had translated and published Drebbel's work, claimed Burggrav, he met Drebbel in England and was taken into his confidence.[24]

20 Ipswich archives, c2843, Headborough's Court 1-9 Jas I, p. 17, 3 May 1604, 'Defective paving. Cornelius Dreble & Peter Hoste, Dutchmen', p. 29, 3 July 1604, 'Cornelius Dreble & Peter Hoste next Jn. Herne, innholder'. Thanks to the late John Blatchly for pointing me toward these sources.

21 *JIB*, II, pp. 199-203.

22 Volker Fritz Brüning, *Bibliographie der alchemistischen Literatur. Band 1: Die alchemistischen Druckwerke von der Erfindung der Buchdruckerkunst bis zum Jahr 1690* (Munich: K.G. Saur, 2004); Klaas Hoogendoorn, *Bibliography of the Exact Sciences in the Low Countries from ca. 1470 to the Golden Age (1700)* (Leiden: Brill, 2018), p. 305.

23 Johann Ernst Buggrav, 'Vorrede', in: Cornelis Drebbel, *Ein Kurtzer Tractat von der Natur der Elementen* (Frankfurt: Fitzer, 1628), sig. Aii^v: 'bin ich damaln durch einen vertrauweten Freundt dieses Tractats, von der Natur der Elementen/welchen Cornelius Drebbel damaln in Niderteutscher Sprach verfertiget / und etlich wenig Exemplaria fur sich drucken lassen / und allein guten Freunden unnd Philosophis mitgetheilet / theilhafftig worden / welches Büchlein ich hernacher in die hochteutsche Sprach ubersetzt / und in Druck damals befördert: Ist auch bey vielen verständigen Philosophis, so es zu lesen bekommen / ein angenemmes Büchlein gewesen.'

24 Burggrav, 'Vorrede', in: Drebbel, *Von der Natur der Elementen,* sig. A2^v: 'mit welchem ich hernacher in Anglia, in vertrauwliche Freundt: und Kunstschafft gerahten'.

In 1868, Fritz Burckhardt, a teacher in Basel, reported receiving a 1604 Dutch edition of Drebbel's *On the Nature of the Elements* from an engineer, Th. van Doesburgh of Rotterdam, which included a woodblock print entitled 'Cornelius Drebbel, Alcmariensis, 1604'.[25] As Burckhardt later noted in 1903, this edition was printed by 'Gillis Roomann' in the 'vergulde Parsse' in Haarlem.[26] Rooman's publications have a low rate of survival.[27] The Short Title Catalogue Netherlands lists only one 1604 publication from Rooman's press.[28] Rooman would have been a likely candidate to print such an edition; Karel van Mander and Cornelius Schonaeus also employed him.[29] Goltzius's estate owed debts to the estate of Rooman.[30] It was in Rooman's shop that Jacob de Meester, who would later print Drebbel's 1607 *Wonder-vondt*, first worked when he emigrated from Bruges.[31] It seems unlikely that Burckhardt or Doesburgh would bother to falsify the detail of Rooman as the printer or even be aware of Rooman's correct address in Haarlem. A portrait of Drebbel, attributed to Christoffel van Sichem I and dated 1604, is still extant (fig. 15.1).[32] It is the basis of all subsequent author portraits in editions of Drebbel, including the 1608 German translation.[33] It is reasonable to think that it was originally produced as an author portrait.

The testimony of Beeckman offers another compelling piece of evidence. In 1604, Beeckman was only sixteen and perhaps just beginning to record notes in what would become his *Loci communes*.[34] However, by 1619, he was reading and experimenting from a Dutch edition printed in Haarlem of Drebbel's works

25 Fritz Burckhardt, 'Historisch Notizen vom Professor Burckhardt', *Annalen der Physik und Chemie* 133:1 (1868), pp. 681-682.

26 Fritz Burckhardt, 'Zur Geschichte des Thermometers', *Verhandlungen der Naturforschenden Gesellschaft in Basel* (Basel: Georg & Co., 1903), XVI, p. 3.

27 H.J. Laceulle-van de Kerk, *De Haarlemse drukkers en boekverkopers van 1540 tot 1600* (The Hague: Nijhoff, 1951), p. 105.

28 Leenaert Clock, *Het groote liedeboeck* (Haarlem: Rooman, 1604). See: www.stcn.nl.

29 Cornelius Schonaeus, *Comediarum* [...] *altera pars* (Haarlem: G. Rooman, 1599). See: Laceulle-van de Kerk, *De Haarlemse drukkers*, pp. 87, 93-94.

30 Lawrence W. Nichols, 'Hendrick Goltzius – Documents and Printed Literature Concerning His Life', in: *Goltzius-Studies, Hendrick Goltzius (1558-1617), Nederlands Kunsthistorisch Jaarboek* 42-43 (1991-1992), pp. 77-120, 111.

31 Boukje Thijs, *De hoefslag van Pegasus: Een cultuurhistorisch onderzoek naar Den Nederduytschen Helicon (1610)* (Hilversum: Verloren, 2004), p. 28.

32 See: Dieuwke de Hoop Scheffer and George S. Keys, *C.V. Sichem I* (Hollstein's Dutch and Flemish Etchings, Engravings and Woodcuts, c. 1450-1700, vol. XXVII), ed. K.G. Boon (Amsterdam: Van Gendt & Co., 1983), p. 33; H.F. Wijnman, 'De Van Sichem-puzzle', *Oud-Holland* 46 (1929), p. 233.

33 De Hoop Scheffer, *C.V. Sichem I*, p. 33.

34 *JIB*, I, p. xxv.

Figure 15.1 ***Cornelis Drebbel*** **by Christoffel van Sichem (I)**

Rijksmuseum Amsterdam RP-P-1913-2257

that predates the first Dutch edition of 1621.[35] Finally, Peter Lauremberg, a professor at the Hamburg Gymnasium, in his 1621 Latin translation of *On the Nature of the Elements*, criticized the previous German translation and claimed to translate from the original Dutch. It is unlikely that Lauremberg based his translation on the Dutch edition printed that same year in 1621, as his editor, Joachim Morsius, described Lauremberg as having produced the translation many years prior.[36] Cumulatively, these offer reasons to accept the existence of a lost first edition, and thus to situate Beeckman among those few friends and philosophers to whom Burggrav says this text was communicated.

35 *JIB*, I, p. 346: 'Den 10 November de Middelb., occassionem praebente cap. 6 libri Drebbelii Alcmariensis, gedruckt to Haerlem, *Van den natuyre de Elementen,* int Duytsch.'
36 Joachim Morsius, dedicatory letter to Georg Schumacher, in: Drebbel, *Tractaus duo*, sig. A2^{v}: 'primum a multis jam annis a praeclarissimo ejus interprete P. Laurembergio nostro'.

Face-to-Face Sociability and the Transnational Reformed Network

Following this 1604 edition, Drebbel communicated mainly verbally with individuals who gained his trust. He became great friends with Constantijn Huygens Sr.[37] The Bohemian alchemist Daniel Stolz von Stoltzenberg recalled how on his 1623 trip to England he met many famous men, chief among whom were Fludd and Drebbel. Drebbel received him warmly.[38] Drebbel signed Stolzenberg's *album amicorum*.[39] The Herborn philosopher Johann Bisterfeld complained that, when in London in 1625, he couldn't visit Drebbel because of the plague, Bacon was leading a 'private life', and there was nobody else in London worth talking to.[40] Jon Jonston, who had characterized Beeckman as morose, had better luck with Drebbel. He fondly recalled examining peacock-coloured flies through the microscope alongside him.[41] Jonston also admired Drebbel's *On the Nature of the Elements*.[42] In 1629, Abram Booth, an employee of the VOC, noted in his travel diary that at Stratford-Langton, three miles from London, one could find 'the world famous Cornelis Drebber [sic], a Hollander, who was hired by the current King for his wonderful and almost unbelievable inventions'.[43]

Drebbel, however, turned a cold shoulder to Paul Marquard Slegel (1605-1653). Slegel, who had already studied for ten years at the universities of Rostock, Altdorf, Wittenberg, Jena and Leiden, went to London for five months to learn English from October 1631 to February 1632. There he conversed much with Theodore Mayerne, William Harvey, and Robert Fludd. By contrast, he found Drebbel 'a jealous hider of his secrets' ('occultatore secretorum suorum invido').[44]

37 Rosalie Colie, 'Cornelis Drebbel and Salomon de Caus: Two Jacobean Models for Salomon's House', *Huntington Library Quarterly* 18 (1954), pp. 245-269; Rosalie Colie, *'Some Thankfulnesse to Constantine': A Study of English Influence upon the Early Works of Constantijn Huygens* (The Hague: Nijhoff, 1956).

38 Johann Daniel Horst, ed., 'Epistola M. Danielis Stolcii de Stolzenberg [...] ad Gregorium Horstium', *Gregorii Horstii operum medicorum tomus secundus* (Nürnberg: Endter, 1660), pp. 500-501.

39 Uppsala University Library, Y 132 d, https://www.alvin-portal.org/alvin/imageViewer.jsf?dsId=ATTACHMENT-0056&pid=alvin-record:103741.

40 Howard Hotson, *Johann Heinrich Alsted, 1588-1638: Between Renaissance, Reformation, and Universal Reform* (New York: Oxford University Press, 2000), p. 231.

41 Jon Jonston, *Historiae naturalis de insectis libri III* (Frankfurt: Merian, 1653), p. 67.

42 Jon Jonston, *De constantia naturae* (Amsterdam: John Jansson, 1634), pp. 68-69.

43 A. Merens, ed., *Een Dienaer der Oost-Indische Compagnie te London in 1629. Journael van Abram Booth en zijn descriptie van Engelandt* (The Hague: Stols, 1932), pp. 196-197.

44 Johann Moller, *Cimbria Literata* (Copenhagen: Gottmann and Kisel, 1744), p. 637.

The Küffler brothers of Cologne, who attached themselves to Drebbel, also had great difficulty extracting secrets. In 1623, Abraham Küffler (1598-1657) married Drebbel's daughter Anna and Dr. Johann Sibertus Küffler (1595-1677), who held a medical degree from Padua, married Drebbel's daughter Catharina in 1627. In 1624, Abraham and another Küffler brother, Gilles (1596-1658), were interviewed in Paris by Nicholas Claude Fabri de Peiresc extensively about Drebbel.[45] The Küffler brothers were in Paris to market Drebbel's microscopes in the French court. They also managed this international trade in Drebbel's optical devices via Cologne, as Peiresc would report to a friend in 1625.[46]

In their 1624 interview with Peiresc, the Küffler brothers reported how Drebbel told them that he had more than 200 inventions and more than a thousand secrets that he did not wish to teach to anyone.[47] He scorned formal learning, and 'growing in age, he continually grew in inventions, which came from the vivacity of his spirit, without the help of the reading of books which he always disdained, holding it as a maxim that the excellence of sciences lay in the knowledge of the secrets of nature, of which they are completely composed'.[48] Drebbel was very choosy when it came to revealing his knowledge, but there was one way to get him to do it.

> Drebbel claims simplicity and ignorance; if someone asks him if he knows how to do something, he says no, and he will only reveal himself to those whom he believes to be intelligent or desirous of becoming so. About three or four years ago, he started taking tobacco, which he used to hate. He became so addicted to it, that he would spend whole days and nights taking it, and he believed that those who didn't take it were not smart. When he found someone who took a lot of tobacco, he admired and loved

45 Nicholas-Claude Fabri de Peiresc, 'Relation de ce que j'ai appris de la vie et des inventions de Corneille Drebbel', Bibliothèque Municipale Inguimbertine, Carpentras Ms. 1776.

46 Nicholas-Claude Fabri de Peiresc, *Lettres de Peiresc aux Frères Dupuy*, ed. Philippe Tamizey de Larroque (Paris: Imprimerie Nationale, 1888), I, pp. 485-486: 'Cologne, où resident les parents de Corn. Drebels, qui en font profession.'

47 Peiresc, in: Carpentras Ms. 1776, fol. 411^{r}: 'Il y en a plus de deux cens de choses qui n'ont iamais esté faictes. Et dict qu'en mourant il en enterrera avec luy plus de mille secretz qu'il ne veut enseigner a personne.'

48 Peisresc, in: Carpentras Ms. 1776 fol. 408^{v}: 'en croissant d'aage [sic] il aloit tousiours croissant d'inventions, qui procedoient de la vivacité de son esprit, sans ayde ny lecture de livres qu'il a tousiours mesprisé, tenant pour maxime que la verité et l'excellence des sciences consiste en la cognoissance des secrets de la nature dans laquelle elles sont tout comp[osées]'.

> him, and in this case, he might tell that person his secrets, but otherwise he was very ill at ease.[49]

Such an attitude explains the widely differing receptions various individuals experienced when they sought to make Drebbel's acquaintance and why he might have reacted poorly to the very academic Slegel.

Despite the limited reach that such selective, in-person sociability could foster, knowledge of Drebbel and his works spread far and wide. It did so in part via a transnational Reformed network with nodes in cities such as Cologne, Middelburg, Amsterdam, and London.[50] This web connects Beeckman to Drebbel via the intermediaries of Beeckman's brother-in-law Justinus van Assche, a preacher in Cologne, his friend, Johann Moriaen, and their acquaintances from Cologne, Drebbel's associates and sons-in-law, the Küfflers. Even after the Küfflers moved to London to work with Drebbel, they travelled back to Cologne to sell Drebbel's optical devices. There, they would have known the Pergens, the Van Zevels, the Velthuysens and other Reformed families who moved between Cologne, the Netherlands, and London.[51] The Küfflers were also in-laws of the Vernattis, originally of Chieri in Savoy, via the two daughters of Pieter van Gheel, a wealthy Amsterdam merchant who lived in Cologne from 1585 to 1592.[52] Members of this network shared spiritual ideals, collaborated in business, married and met in Calvinist places of learning such as Herborn, where Johann Moriaen, Jacob and Johann Pergens, Jan Amos Comenius, Adam von Zevel,

49 Peiresc, in: Carpentras Ms. 1176 fol. 408^{v}: 'Derbbel [sic!] fait profession de simplicité et d'ignorance; si on luy va demander s'il ne scait point faire telle chose, il dict que non, et ne se descouvre qu'à ceux qu'il croit estre intelligens ou desireux de l'estre. Despuis trois ou quatre ans il s'est mis à boire du tabac, qu'il haissoit auparavant. Il s'y est tellement adonné, qu'il passe les jours et les nuitz entieres à en boire, et tient que ceux qui n'en boivent poinct, n'ont pas bonne teste. Quand il trouve quelqu'un qui en boit bien, il l'estime et aime grandement; et en ce rencontre luy pourroit declarer de ses secretz; hors de là il est bien malaisé.'

50 John Young, *Faith, Medical Alchemy and Natural Philosophy: Johann Moriaen, Reformed Intelligencer, and the Hartlib Circle* (Brookfield: Ashgate, 1998), p. 122; Ole Peter Grell, *Brethren in Christ: A Calvinist Network in Reformation Europe* (Cambridge: Cambridge University Press, 2011).

51 Rudolf Löhr, *Protokolle der Hochdeutsch-Reformierten Gemeinde in Köln von 1599-1794*, vol. 1 (Köln: Rheinland-Verl., 1976).

52 H.F.Wijnman, 'Vernatti, Sir Philibert (1)', in: *Nieuw Nederlandsch Biografisch Woordenboek*, 10 vols. (Leiden: Sijthoff, 1911-1937), IX, cols. 1201-1203. In 1620, Philibert Vernatti's brother Johann Vernatti (1595-1637) married Sara van Gheel, whose sister Margaretha married Gilles Küffler. On Pieter van Gheel, see: Jessica Roitman, *The Same but Different? Intercultural Trade and the Sepharadim* (Leiden: Brill, 2011), p. 173.

Johann Sibertus Küffler, and Philibert, Gabriel and Pieter Vernatti all studied between 1595 and 1616.[53]

In 1626, Beeckman knew both Abraham and his uncle Sir Philibert Vernatti; De Waard suggests that the young Abraham may have been his student.[54] Beeckman visited Sir Philibert's house three times in hopes of testing the knowledge of Johannes Torrentius or Jan Symonsz. van der Beeck (1589-1644), who, according to Sir Philibert, 'knew everything in philosophy'.[55] Sir Philibert would move to England in 1628, taking out numerous patents and working on the drainage of the Fens with Cornelius Vermuyden, Matthias van Valkenburg (1588-1644) and later, Abraham Vernatti.[56] These social, mercantile and kinship relationships are captured in the *album amicorum* of Abraham Velthuysen of Cologne, who partnered with the Küfflers and Johann Moriaen in developing an industry based on Drebbel's scarlet dye.[57] Velthuysen's friendship book was signed by several Küfflers, Abraham Vernatti, Matthias van Valkenburg, and many other foreign projectors resident in England.[58]

Johann Moriaen, who later came to possess a great deal of Drebbel's inventions, was summoned from Frankfurt to serve as preacher to the Cologne Reformed community in 1619, where he married Peter von Zevel's daughter, Odilia, in 1633.[59] Johann and Jacob Pergens and Adam and Peter von Zevel, who were in-laws, would collectively serve as the dedicatees of Johann Ernst Burggrav's 1628 Latin edition of Drebbel's works (colour ill. 16). Burggrav called them 'careful and industrious investigators of nature and of the mysteries of true and more secret philosophy'. They, who had often requested the work, zealously pursued the study of 'purer philosophy'

53 Gottfried Zedler and Hans Sommer, eds., *Die Matrikel der Hohen Schule und des Paedagogiums zu Herborn* (Wiesbaden: Bermann, 1908), pp. 21, 56, 65, 59, 241, 243, 260.

54 *JIB*, II, pp. 364-366.

55 *JIB*, II, p. 365, 'dat hy alles wist in philosophie'.

56 Jill Turnbull, *The Scottish Glass Industry, 1610-1750: 'To Serve the Whole Nation with Glass'* (Edinbugh: Society of Antiquaries of Scotland, 2001), pp. 96-102; Gerard Doorman, *Octrooien voor uitvindingen in de Nederlanden uit de 16e-18e eeuw: met bespreking van enkele onderwerpen uit de geschiedenis der techniek* (The Hague: Nijhoff, 1940), p. 197.

57 Augustinus Petraeus to John Winthrop Jr., *Winthrop Papers, Vol. 4: 1638-1644* (Boston: Massachusetts Historical Society, 1944), pp. 368-369; Frans Maurits Jaeger, *Cornelis Drebbel en zijne tijdgenooten* (Groningen: Noordhoff, 1922), p. 47; Paulus Nijhoff, *Inventaris van het Oud Archief der Gemeente Arnhem* (Arnhem: Nijhoff, 1865), p. 456.

58 British Library MS Additional 19882.

59 Johann Pergens and Adam von Zevel married the Von Driesch sisters, Magdalena and Elisabeth. See: Hermann Friedrich Macco, *Geschichte und Genealogie der Familie Pastor* (Aachen: Aachener, 1905); Borch, *Itinerarium 1660-1665*, II, p. 165.

and would be able to judge the 'high mysteries of nature' contained within Drebbel's text.[60]

While no others sources attest to these philosophical interests, we do know that this circle invested in practical inventions. In 1626, Adam van Zevel acquired a 15-year patent for a water-draining engine.[61] In 1638, Abraham Küffler witnessed an account of Jacob Pergens's successful use of raincoats.[62] In Dordrecht in 1649, Benjamin Worsley consulted with Pergens at length about land drainage. Pergens described the negative experience of many of his friends who had previously invested in the English drainage of the Fens and 'beene themselves dreyned'. Nevertheless, he communicated a 'secret' about draining to Worsley.[63] In 1643, Pergens had become director of the Amsterdam chamber of the West India Company and eventually one of the wealthiest figures of the Dutch Golden Age.[64]

Beeckman could utilize this network to find information concerning a wide array of individuals. Over the course of many years, he investigated Drebbel and his works.[65] Johann Moriaen kept Justinus van Assche, and Van Assche and Küffler kept Beeckman informed about Drebbel.[66] Beeckman also heard about Drebbel via Henri Reneri (1593-1639), who was connected to the Cologne network.[67] Beeckman himself had relatives in London, and his own father, Abraham, wrote to him in 1620 with the news of the submarine that Drebbel had demonstrated.[68] Drebbel and Beeckman never directly interacted, although the Dutch Reformed Church of Dordrecht made Drebbel a payment for unknown reasons in 1633.[69]

60 Johann Ernst Burggrav, 'Dedicatio', in: Cornelis Drebbel, *Tractatus de natura elementorum* (Frankfurt: Fitzer, 1628), pp. 3-4. 'Vobis Dn. Cognati atque amici honorandi libellum hunc quod à me petijstis saepius, inscribere atque dedicare volui, cùm quòd purioris Philosophiae studium vos mirum in modum teneat, atque eo quoque non leviter tincti ac imbuti, de hisce alijsque summis Naturae Mysterijs dextrè iudicare valeatis [...].'

61 Doorman, *Octrooien*, p. 167.

62 HP 67/16/1B.

63 HP 26/33/1A-1B.

64 Adolf von den Velden, 'Die Pergens, niederländische Reformierte in Köln', *Familiengeschichtliche Blätter* 14 (1916), cols. 353-358; Young, *Moriaen*, p. 10; Kees Zandvliet, *De 250 rijksten van de Gouden Eeuw: kapitaal, macht, familie* (Amsterdam: Rijksmuseum, 2006), pp. 310-311.

65 *JIB*, II, pp. 2, 201, 202, 363, 372 and *JIB*, III, pp. 203-204, 302-304, 358, 367.

66 *JIB*, III, p. 302.

67 *JIB*, II, p. 272. Robin Buning, 'A Circle of Former Reformed Ministers in Cologne', in: Buning, *Henricus Reneri (1593-1639): Descartes' Quartermaster in Aristotelian Territory* (PhD diss., Utrecht University, 2013), pp. 194-199.

68 *JIB*, II, p. 25. Van Berkel, *Isaac Beeckman on Matter and Motion*, p. 50.

69 Gilles Dionysius Jacobus Schotel, *Kerkelijk Dordrecht*, 2 vols. (Utrecht: N. van der Monde, 1841-1845), I, p. 341.

This network travelled constantly between its branches. As Pieter Boudaen Courten recorded in his diary, when he married Catherina Fourmenois (a relative) of Cologne in 1618, his mother, Margaret Courten, his uncle William Courten and his sister Anna came from London for the wedding. When the next year his sister Anna married Jacob Pergens (originally from Cologne) in London, he and Catherina attended the wedding there. When Pieter and Catherina's first son, Johan (1620-1634), was born in Middelburg in 1620, Catherina's mother (Hortensia Courten, née Del Prato) and Pieter's stepfather, John de Moncij (Moncy) of London, stood witness. Johan Courten attended 'de schole van Isac Beeckman' ('the school of Isaac Beeckman') in Dordrecht.[70] Jacob Pergens, travelling between the Netherlands and London, mediated the correspondence of Johann Moriaen, Justinus van Assche, and Samuel Hartlib.[71] Travelling so often, such figures built the infrastructure of transnational communication.

Drebbel, the 'Thermometer', Beeckman, and Transnational Circulation

Analyzing how knowledge moves across this network can shed light on old questions, such as the invention of the thermometer over which fierce battles for priority have been waged.[72] In the 1860s, Emil Wohlwill and Fritz Burckhardt pointed out that a description of the thermometer as the 'Drebbelian instrument' was first made three years after Drebbel's death and thus was at best a second-hand report and not an acceptable claim to invention.[73] Other historians celebrated the revelation of this 'blunder'.[74] The discovery seemed to explode the 'Drebbel legend'.[75] The 'darkness, which

70 Unfortunately, he drowned alongside other students who were fleeing plague in 1634. Pieter Boudaen Courten, 'Familjie boeckje van dheer Pieter Boudaen Courten en sijn huijsvrouw Catharina Fourmenois', *Bulletin van het Rijksmuseum* 53 (2005), pp. 52-61, esp. p. 53. See also *JIB*, III, p. 369.

71 HP 37/50A. See also 37/40A-B, 37/76A, 37/80A-B, and 37/116A-B, and Library of the University of Amsterdam, Letters of Justinus van Assche, 65.e and 65.b.

72 Emil Wohlwill, 'Zur Geschichte der Erfindung und Verbreitung des Thermometers', *Poggendorff's Annalen* 124 (1865), pp. 163-178; Fritz Burckhardt, *Die Erfindung des Thermometers und seine Gestaltung im XVII. Jahrhundert* (Basel: Schultz, 1867); Henry Carrington Bolton, *Evolution of the Thermometer, 1592-1743* (Easton: Chemical Publishing, 1900).

73 Caspar Ens, *Thaumaturgus mathematicus* (Köln: Constantinus Münich, 1636), pp. 125-128.

74 Wohlwill, 'Zur Geschichte' (1865), p. 164; Burckhardt, *Erfindung* (1867), p. 7. For 'blunder', see: Bolton, *Evolution* (1900), p. 6.

75 Siegmund Günther, *Handbuch der Geophysik* (Stuttgard: Enke, 1899), II, p. 48.

for so long lay upon the invention history of the thermometer, has been illuminated', proclaimed one professor of physics.[76] The term 'Drebbelian instrument', however, is not a legend in need of explosion, but evidence of the transnational circulation of information concerning Drebbel that occurred at a temporal and geographic remove from Drebbel himself.

As Dijksterhuis discusses in this volume, the part of Drebbel's devices that eventually became a candidate for being a thermometer began in the perpetual motion device Drebbel built between 1604 and 1607 for King James I. Drebbel claimed to King James that he could apply it to uses. One such application was Drebbel's self-regulating oven, which used the oven's heat to move a liquid and thereby regulate an air hole supplying oxygen to the oven's fire in a feedback loop. According to the surviving German translations of Drebbel's description of his self-regulating oven (one of which is associated with Moriaen), Drebbel's oven included two glass parts, one called the *regiment*, which used the moving liquid to control the air supply, and a second, called the *judicium*, marked with seven degrees, indicating at what degree of heat the fire was set. Today we would call the *regiment* a thermostat and the *judicium* a thermometer.[77]

In 1624, the same year that the Küfflers described Drebbel's self-regulating oven to Peiresc in Paris, the term *thermomètre* first appeared in print in a French publication at Pont à Mousson ascribed to Jean Leurechon, S.J.[78]

In 1626, Beeckman referred in Latin and Dutch to the 'Drebbelian instrument' ('instrumento Drebbeliano', 'Drebbelianum instrumentum quo aestum maris imitatur', 'Drebbeliaensche instrument', 'Drebbeliaens instrument', and 'het instrument van Drebbel'), sometimes to the 'Drebbelian glass' ('Drebbeliaens glas'), or, because it could be used to reckon years by the change in seasons, the 'diarium Drebbelii'.[79] He also used the term 'Drebbelian instrument' in his correspondence with Mersenne in 1629-1630.[80] In 1631, Beeckman copied a circa 1613 letter from Drebbel to King James I in which Drebbel described several of his earlier inventions, including

76 E. Gerland, *Bericht über die Wissenschaftlichen Apparate auf der Londoner Internationalen Ausstellung im Jahre 1876* (Braunschweig: Vieweg, 1878), p. 69. 'Das Dunkel, welches lange Zeit auf der Erfindungsgeschichte des Thermometers lag, ist erst in neuerer Zeit vollständig aufgehellt worden.'

77 Keller, 'Re-entangling'.

78 Arianna Borrelli, 'The Weatherglass and Its Observers in the Early Seventeenth Century', in: Claus Zittel, Gisela Engel, Nicole C. Karafyllis, and Romano Nanni, eds., *Philosophies of Technology: Francis Bacon and His Contemporaries* (Leiden: Brill, 2008), pp. 67-132, esp. p. 119.

79 *JIB*, II, p. 346, 361-383.

80 *JIB*, IV, p. 147, 186.

his perpetual motion device. On the verso, Beeckman offered sketches of three of these inventions, which he cumulatively entitled 'Instrumenta Drebbeliana'. Interestingly, for his image of the 'motus perpetuus fluxus et refluxus' he did not sketch an image of the perpetual motion device as he knew it to have appeared circa 1607, that is, as a round tube showing the tides. Rather, he sketched a metric device measuring seven degrees, much as the *judicium* of Drebbel's oven in the 1620s did (see fig. 17.9); the thermometer pictured by Leurechon had nine degrees.[81]

In a 1636 Latin translation of Leurechon's text produced by Caspar Ens (1569-c. 1642), the image which had previously been labelled a 'thermomètre', gained the description 'Instrumentum Drebilianum'.[82] Peter Burke has characterized Caspar Ens as a cultural go-between who became a translator in part due to his own displacement.[83] Ens, born in Lorch (Württemberg), worked in both the Netherlands, where he was conrector of the Latin School of Delft from 1594 to 1596, and Cologne.[84] He dedicated a work to Johann Pergens in 1620 (whose brother Jacob, he mentioned in the dedication, was in England at the time).[85] Thus, Ens was likely to be familiar with Drebbel's device from acquaintances in Cologne and perhaps elsewhere.

Among the many uses the original author Leurechon had offered for the thermometer in the 1620s was keeping 'a room, a furnace, a stove, in a constant heat by making it so that the water in the thermometer always stays on the same degree'.[86] The thermometer served not only a diagnostic, but a regulatory function even in this early description. When Ens described Leurechon's 'thermometer' as a Drebbelian instrument, he may well have been referring not only to the thermometer itself, but to Drebbel's application of it to control heat. By 1636, Beeckman had already used the term 'Drebbelian instrument' for a decade. Ens's 'Drebbelian instrument' thus represents less a moment of creation of the 'Drebbel legend', as Wohlwill

81 *JIB*, III, p. 442.

82 *JIB*, II, pp. 346, 361-366 and: Ens, *Thaumaturgus mathematicus*, pp. 126-127.

83 Peter Burke, 'The Renaissance Translater as Go-Between', in: Andreas Höfele and Werner von Koppenfels, eds., *Renaissance Go-Betweens: Cultural Exchange in Early Modern Europe* (Berlin: De Gruyter, 2005), p. 22.

84 Wilhelm Kühlmann, 'Ens, Caspar', in: W. Killy, red., *Literaturlexikon: Autoren und Werke deutschsprachiger kulturraum* (2008), III, p. 266.

85 Caspar Ens, *Mantissa apothegmatum* (Köln: Lutzenkirchen, 1620).

86 I was unable to consult the rare 1624 first edition. Jean Leurechon, *La Récréation mathématique* (Pont à Mousson: N.A., 1626), p. 77: 'On peut entretenir une chambre, un fourneau, une estuue, en chaleur tousiours égale faisant en sorte que l'eau du thermometre demeure tousiours sur un mesme degré.'

and Burckhardt supposed in the 1860s, than a trace left by a largely oral and scribal transnational network already deploying this term.

Conclusion: The Collegium Mechanicum, the Vauxhall Operatory, and Intellectual Theft

As Dijksterhuis discusses in this volume, both Beeckman and Drebbel deployed an 'artefactual' form of reasoning that relied upon the manipulation of material resources. Moreover, their researches often bore useful applications, thus setting a great deal of profit at stake in communicating their knowledge, especially when that knowledge was still in development. As Pamela Long has discussed, many practitioners of technical arts elected to become published authors in order to gain epistemic authority, thus trading the protection of their knowledge for status. Such traditions of authorship 'present visible manifestations of collaboration and communication between practitioners, learned humanists, other university-educated men, and ruling elites'.[87] However, not everyone followed this model. Artisanal philosophers might see little to gain, and much to lose, by partnering with more textually oriented individuals or attempting to advertise themselves broadly in print. In part by drawing on resources supplied by their transnational friends and relatives, figures like Drebbel and Beeckman did not need to rely upon print celebrity nor participation in the Republic of Letters.

Concerns about intellectual theft were well justified in such stories as the one Beeckman recounted in 1622 concerning a dispute for credit over Drebbel's perpetual motion device. Beeckman had his own experience with theft, after Descartes viciously attacked him for his efforts to defend the originality of his ideas. It was ridiculous of Beeckman, Descartes claimed, to attempt to distinguish so carefully between his ideas and those of another, as though knowledge were a possession like that of land or money.[88] Pace Descartes, not only was money at stake, but artefactual reasoning could not be carried out without it.

In his Collegium Mechanicum, Beeckman participated in experimenting in a new form of sociability that was not geared towards the publication

87 Pamela Long, *Openess, Secrecy, Authorship: Technical Arts and the Culture of Knowledge* (Baltimore: Johns Hopkins University Press, 2001), p. 246.

88 Descartes to Beeckman, 17 Oct. 1630, in: *JIB*, IV, p. 197: 'et ridiculum est tam accuratè, ut facis, in scientiarum, tanquam in agrorum vel pecuniae possessione, inter tuum alienumque distinguere'.

of texts, but rather the production of useful inventions and projects, and which did not ignore financial issues. Among the eight members, only one other besides Beeckman had a university degree.[89] Beeckman's own notes recorded the conversations held there, the college had a collective book,[90] and Beeckman described his plan to bring in a folio writing book for drawings.[91] These were not intended for publication.

Beeckman took the issue of finances seriously. When the college discussed plans to make models, Beeckman asked if everyone wished to share in the expenses. One member did not wish to, nor to join in potential profits.[92] Financial models were at the top of Beeckman's mind, both in plans for how to keep the Collegium Mechanicum afloat (after his departure for Dordrecht), and in his later plans for a Collegium Physico-Mathematicum teaching 'natural sciences and mathematical arts'.[93] These were institutions that made the resouces required by artefactual reasoning available to a group of inventors.

Van Berkel compares Beeckman's Collegium Mechanicum to Interregnum English plans such as the Gymnasium Mechanicum proposed by William Petty in 1648.[94] There was another realized institution that was even more closely related: the Vauxhall Operatory established by Charles I beginning in 1629.[95] Stretching over eight riverside acres, the Operatory held nine forges, thirteen vices, seven furnaces, and ten workbenches, a mill for boring guns and a brass foundry.[96] In a memorandum supporting the preservation of the royal Vauxhall Operatory for Interregnum use, Hartlib's interlocutor Cressy Dymock noted that it would provide 'a place of Resort, wherunto Artists and Ingeneres from abroad and at home may repaire to meete with one another, to conferre together, and improoue many Way's their abilities'.[97] This was the reason the late king had founded it.[98] In 1654, as Hartlib reported to Robert Boyle, Edward Somerset, Marquis of Worcester had purchased

89 Van Berkel, *Isaac Beeckman on Matter and Motion*, p. 37.
90 *JIB*, II, p. 440: 'ons gemeyn boeck'.
91 *JIB*, II, p. 441: 'een schryfboek in fol'.
92 *JIB*, II, p. 447.
93 *JIB*, III, pp. 61-62.
94 Van Berkel, *Isaac Beeckman on Matter and Motion*, p. 200.
95 Guy Wilson, *The Vauxhall Operatory: A Century of Inventions before the Scientific Revolution* (Leeds: Basiliscoe, 2009), p. 32.
96 Rhys Jenkins, 'The Vauxhall Ordnance Factory of Charles I', in: *Links in the History of Engineering and Technology from Tudor Times: The Collected Papers of Rhys Jenkins* (Cambridge: Newcomen Society, 1936), p. 28.
97 HP 8/64/3B.
98 HP 8/64/4A.

Vauxhall in order to bring Caspar Kalthoff and his son back to England, 'for he intends to make a college of artisans'.[99]

Caspar Kalthoff the Elder (1606-1664) was a central figure at the royal Vauxhall and its later resurrection. Born in Westphalia to a gunsmithing family, Kalthoff had many relatives pursuing careers at courts across Europe. His son, also Caspar Kalthoff, lived in Dordrecht and worked as a lens grinder, where he was known to Beeckman's friend Andreas Colvius (1594-1671).[100] Kalthoff was even less literary than Drebbel. Joachim Hübner lamented to Hartlib in 1640 that 'Kalthoffs several Inventions are not brought to paper'.[101] Another correspondent noted that Kalthoff 'doth not care for writing or cannot write'.[102] Nevertheless, figures such as Benjamin Worsley, who visited Kalthoff the Elder when he retreated to Dordrecht during the Interregnum, characterized Kalthoff as a fascinating interlocutor 'about Glasses Telescopes Things in Natural Philosophy and Chimistry'. His home was always full of gentlemen dropping in 'to speake with him'. Worsley wished to unite William Petty with Kalthoff so they could collaborate on perpetual motion and achieve 'wonders.'[103]

Sir John Heydon, Lieutenant-General of the Ordnance Office and Drebbel's patron and supervisor, recruited Kalthoff in 1633, seemingly to replace Drebbel, who had died that year; steam-driven perpetual motion became one of Kalthoff's main investigations.[104] Although, as was the case also for Drebbel, Kalthoff's inventions were kept secure by royal command, he also partnered with transnational collaborators on the side. In particular, Kalthoff worked with the roving Moravian nobleman Johann Christoph Berger von Berg, aka Jan Kryštof Pergar z Pergu, signing a contract for inventions with him in Hartlib's home in 1640.[105]

Berger travelled frequently between the Netherlands and England, raising funds to support Moravian victims of the Thirty Years' War and taking out patents for a perpetual motion and an invention for raising shipwrecks.[106] He

99 Royal Society MS, Boyle Letters 7.2 [1A-2B], via HP.

100 Huib J. Zuidervaart, 'The "Invisible Technician" Made Visible: Telescope Making in the Seventeenth and Early Eighteenth-Century Dutch Republic', in: Alison D. Morisson-Low, Sven Dupré, Stephen Johnston, and Giorgio Strano, eds., *From Earth-Bound to Satellite: Telescopes, Skills and Networks* (Leiden: Brill, 2011), pp. 41-102, esp. p. 80.

101 HP 30/4/50B.

102 HP Royal Society MS, Boyle Letters, 7.2 1A-2B.

103 HP 8/50/1A-2B.

104 John Bruce, ed., *Calendar of State Papers Domestic: Charles I, 1631-1633* (London: Her Majesty's Stationery Office, 1862), p. 554; Jenkins, 'Vauxhall', p. 30.

105 HP 4/1A-2B.

106 Doorman, *Octrooien*, pp. 176-177. See also 23 October 1631.

imitated inventions by Simon Stevin, Drebbel, and others.[107] He collaborated with fellow Moravian Jan Amos Comenius on perpetual motion, informing him in 1632 that Drebbel was still sweating after the perpetual motion, but in vain, for all of his 'boasts'. According to Comenius, Berger reported that he heard Drebbel say that 'either he would find it or nobody would'.[108] In 1635, Hartlib described Berger as 'very intimate' with 'Kuffler successor Drebbeli' ('Küffler, the successor of Drebbel').[109] Eventually tiring of Berger, Moriaen criticized Comenius's attempt to rival Drebbel's perpetual motion.[110] Berger's attempts at perpetual motion were dismissed by Hübner as 'for the most part nothing else but new Applications vpon old Principles or Inventions'.[111]

It was Berger who informed Hartlib of Beeckman's 'college of inventions', by which he most likely meant the Collegium Mechanicum. Unsurprisingly, given his imitative tendencies, Berger had his own 'description of a Colledge for divers Inventions'.[112] Berger's constant preying upon the international network he managed to access illustrates some of the issues faced by artisanal philosophers that more formalized institutions were designed to ameliorate.

Research aiming at ambitious goals required both funding and collaboration between individuals with varying forms of knowledge. Relying on kinship networks and vetted face-to-face sociability was one way to protect knowledge still in development. That network of supporters, as the term 'Drebbelian instrument' attests, would associate knowledge with an individual even when that individual did not fight for credit. Yet, this network, as the predation of Berger upon it illustrates, offered an imperfect solution. More formal groups, such as Beeckman's Collegium Mechanicum or the Vauxhall 'college of artisans', offered the means to combine intellectual and material resources needed for artefactual reasoning without recourse to print publication. Thus could Beeckman be described simultaneously as incommunicative and as someone full of plans for a 'college of inventions.' He was wary of communicating in print, yet, within the curated space of the Collegium Mechanicum, he was a key interlocutor.

107 HP 8/63/3A-4B. HP 29/2/32A.

108 Jan Amos Comenius, 'De Arte Spontanei Motus', *Opera omnia* (Prague: Czech National Academy of Sciences, 1978), XII, p. 315: 'ex ore eius auditam sibi referre vocem: / se reperturum aut neminem'. Berger told Hartlib, 'hee made cum Comenio Primum Mobile'. HP 29/2/23B. See also HP 29/3/47B.

109 HP 29/3/44A-B.

110 HP 37/54A and HP 42/2/26B.

111 HP 30/4/46B and HP 29/3/41B.

112 HP 29/2/28A.

About the Author

VERA KELLER is an Associate Professor of History at the University of Oregon. She is a historian of science and of knowledge in early modern Europe and is the author of *Knowledge and the Public Interest, 1575-1725* (Cambridge: Cambridge University Press, 2015).

16 What's in a Language?

Dutch and Latin in Isaac Beeckman's *Journal*

Semra Meray

Abstract

This chapter aims to explore Beeckman's use of Dutch and Latin in combination with his thoughts on the purification of language by using his *Journal*. When studying Dutch in the Low Countries Simon Stevin (1548-1620) cannot be overlooked. This essay firstly investigates specifically which words Beeckman used to document his observations by studying whether he used any of Simon Stevin's neologisms; however, only seven were found in the manuscript. Secondly, it examines Beeckman's code switching and finds that Beeckman uses Dutch for artisan-related entries. Lastly, it searches the manuscript for entries on languages but merely three related entries were found. It concludes with the statement that Beeckman did not show active involvement in the purification of the Dutch language and mostly preferred to use the Latin language.

Keywords: Isaac Beeckman, code switching, Latin versus Dutch, Simon Stevin, purification, vernacular

Introduction

Throughout the Middle Ages, Latin had been the dominant language of learning and knowledge in Europe. Any conversation between educated men, and occasionally women, about learning or about science would be conducted in Latin. Its dominance remained the standard in the sixteenth and seventeenth century, although it gradually received more and more attention. Around the beginning of the sixteenth century, the functions of language in general and the relationship between Latin and the vernacular languages in particular developed into a significant and critical subject of

Van Berkel, Klaas, Albert Clement, and Arjan van Dixhoorn (eds.), *Knowledge and Culture in the Early Dutch Republic: Isaac Beeckman in Context*. Taylor & Francis Group 2022
DOI: 10.5117/9789463722537_CH16

discussion for religious and social reformers. At this time, multilingualism was not a new concept; Europeans were accustomed to the many different dialects and languages within and between the countries of Europe.[1] Around the second half of the sixteenth century, however, these encounters with various languages sparked something described as a 'fascination with language'.[2] The Latin language, in particular, received increasing attention in these changing views. It became a popular idea to look at the riches and poverty of a certain language and to compare these to those of other languages. The comparison between a vernacular language and Latin generally resulted in an emphasis on the poverty found in the former in terms of vocabulary. In order to tackle this, some proposed to borrow words from different languages, while others preferred to invent new words.

These ideas were all part of the general debate in Europe about the mingling and mixing of different languages and whether purification of the vernacular was necessary.[3] This debate instigated a new preference towards the vernacular languages over Latin and this change of preference, in turn, played a role in the increasing number of publications in vernacular languages. Slowly, Latin was pushed more to the background, which was a process predominantly seen in English, French, German, Italian, Spanish, and Dutch. The effects of these reflections on the vernacular languages can be seen in the writings of influential natural philosophers of these times. For instance, the Italian Galileo Galilei (1564-1642), the Frenchman René Descartes (1596-1650) and Jan Baptista van Helmont (1577-1644) of Brussels have been studied for their use of code-switching in their works. These natural philosophers showed a switch from Latin to their vernacular languages in their writings even though they were educated in Latin and accustomed to this language as the primary one to be used for science.[4] In the case of the Dutch language in the Low Countries, these new ideas and changing habits related to languages co-occurred during a particularly tumultuous time. The political, social and religious turmoil of the Reformation and the Dutch Revolt resulted in migrations which increased the impact of encounters with different

1 Peter Burke, *Languages and Communities in Early Modern Europe* (Cambridge: Cambridge University Press, 2004), pp. 18-19.

2 Alisa van de Haar, *The Golden Mean of Languages: Forging Dutch and French in the Early Modern Low Countries (1540-1620)* (Leiden: Brill, 2019), p. 42.

3 Burke, *Languages and Communities*, pp. 18-19.

4 Sietske Fransen, 'Latin in a Time of Change: The Choice of Language as Signifier of a New Science?', *Isis* 108 (2017), pp. 629-635, esp. p. 630.

languages. This in turn further instigated people to reflect on the relationship between their language or dialect and those of others.[5] For example, in 1584 the important writer and rhetorician Hendrik Spieghel from Amsterdam published his work *Twe-spraack vande Nederduitsche Letterkunst* (Dialogue of Dutch grammar) in which Dutch was said to be richer than other languages.[6]

Particularly interesting in this conflict between Latin and the Dutch language in the Low Countries are the natural philosopher Isaac Beeckman (1588-1637) and his *Journal*. From 1604 until 1634, Beeckman closely kept a notebook, in which he collected his observations and reflections on diverse topics of which alchemy, astronomy, medicine, music, mechanics and logic are a few examples. Besides these scientific topics, he also wrote about personal matters and about his family. In terms of a 'fascination with language', Beeckman's notebook is especially interesting because of its multilingual nature. Beeckman wrote some parts of his manuscript in Latin and other parts in the Dutch vernacular. Surely, these two languages were not the only important ones in the Low Countries at the time. In addition to Dutch and Latin, the French language played a great role as well. It was a prominent language that was rarely viewed as foreign by the native Dutch speakers. Most people were able to speak or understand the language, and both French and Dutch were generally used in domains such as administration, commerce, and jurisdiction.[7] Beeckman also wrote letters in French, which shows he was able to speak this particular language as well, but he did not, however, use it to write sections in his manuscript.[8] This code-switching between Latin and Dutch in Beeckman's notebook is what makes it particularly valuable in the study of the switch to vernacular languages and provides intriguing material to analyse, yet this topic is surprisingly underexplored. Therefore, this contribution will specifically focus on Beeckman's use of Latin and Dutch and on his reflections on the vernacular language by analysing his language use in his manuscripts. The argument is divided into three parts that each discuss different language-related aspects of Beeckman's *Journal*. Firstly, it examines how Beeckman documented his scientific ideas in terms of language and topic. Secondly, it investigates specifically which words Beeckman used to document his

5 Van de Haar, *The Golden Mean*, pp. 49, 79-80.

6 Marijke J. van der Wal, 'Early Language Typology Attitudes towards Languages in the 16th and 17th Centuries', in: Klaus D. Dutz and Kjell-Åke Forsgren, eds., *History and Rationality: The Skövde Papers in the Historiography of Linguistics* (Münster: Nodus Publikationen, 1995), p. 97.

7 Van de Haar, *The Golden Mean*, pp. 30, 42.

8 Van Berkel, *Isaac Beeckman on Matter and Motion*, p. 25.

observations by studying whether he used any of Simon Stevin's neologisms. Lastly, it aims to explain Beeckman's choice of language or words by searching the manuscript for entries on the topic of foreign languages or language use.

Simon Stevin's Neologisms

When specifically looking at the Dutch vernacular in the Low Countries, the important figure Simon Stevin (1548-1620) cannot be overlooked. Stevin was a mathematician, physicist and engineer from Bruges. He was convinced that Dutch was the most appropriate language to use for scientific thought and decided to only write his works in Dutch, or *Nederduytsch* as he called it.[9] He was not the only one with these views. De Eglantier, a chamber of rhetoric in Amsterdam, also played an important role in the development of Dutch. Its members supported, for example, the use of Dutch for educational purposes and longed for a replacement of Latin, which was very much in line with Stevin's views on foreign languages.[10] Stevin believed that *Nederduytsch* had the ability to more concisely and clearly express and bring across ideas than Latin or Greek because of the fact that Dutch had more monosyllabic words. These short words made it possible to briefly but coherently express thoughts and were easier to use for compound words.[11] To further optimize the Dutch language, Stevin set out to popularize the use of uncustomary Dutch terms as opposed to their more popularly used foreign variants and at times created new terms. For instance, it is believed that he created the terms *wiskunde* (mathematics), *evenaar* (equator), *middelpunt* (centre point) and *meetkunde* (geometry), which are still used in contemporary Dutch.[12] His ideas on the vernacular spread to some extent and were adopted by pastors, statesmen, jurists, and mathematicians of his time.[13]

9 E.J. Dijksterhuis, *Simon Stevin* ('s-Gravenhage: Martinus Nijhoff, 1943), p. 304.

10 Dijksterhuis, *Simon Stevin*, pp. 303-304; Harm Klifman, 'Dutch Language Study and the Trivium', in: Jan Noordegraaf, Kees Versteegh, and E.F.K. Koerner, eds., *The History of Linguistics in the Low Countries* (Amsterdam: John Benjamins, 1992), p. 77.

11 Van der Wal, 'Early Language Typology', p. 99; Klifman, 'Dutch Language Study', pp. 76-77.

12 Dijksterhuis, *Simon Stevin*, p. 306; Marijke J. van der Wal, 'Simon Stevin', in: Wim Van Anrooij, Ingrid Biesheuvel, Karina van Dalen-Oskam, and Jan Noordegraaf, eds., *Bio- en bibliografisch lexicon van de Neerlandistiek* (Leiden: Stichting DBNL, 2003), p. 328.

13 Van der Wal, 'Simon Stevin', p. 328.

Moreover, the Dutch language was slowly more accepted as a language for educational and scientific purposes.

While Stevin was working on his ideas for the purification of Dutch, Beeckman was not yet born. Stevin lived from 1548 until 1620 while Beeckman lived from 1588 until 1637, which means there was a 32-year overlap between them. Yet, Stevin and Beeckman were not close friends and we do not know whether they ever met each other. It is certain, however, that Beeckman at least knew about Stevin since Beeckman frequently mentions Stevin and his works in his notebook. Moreover, Beeckman was allowed to read Stevin's unpublished manuscripts a few years after his death.[14] This information leads to the possibility that Beeckman used some of Stevin's Dutch neologisms in his manuscript, which is a reason to analyse Beeckman's *Journal* in order to find out whether Stevin's terms appear in it.

Although Stevin is often discussed for his influence on the Dutch language, it is unclear to what extent Stevin's work has influenced natural philosophers, such as Beeckman, and their ideas about multilingualism and the purification of vernacular languages. Therefore, the first aspect that will be studied here is Beeckman's use of the Dutch language and, more specifically, Beeckman's use of Stevin's neologisms. The second aspect examined in the *Journal* is Beeckman's use of both Latin and Dutch. The three philosophers mentioned earlier have been studied for their switch between Latin and the vernacular. For Galileo, this was Italian, for Descartes, this was French, and for Van Helmont, it was Dutch.[15] Sietske Franssen, who studied these three philosophers, claims that the increasing use of the vernacular can be explained mainly by a rising literacy in these vernaculars. This gave multilingual writers the ability to reach various audiences with their publications.[16] Elaine Limbrick comes to the same conclusion in her study focused solely on Descartes.[17]

Apart from his M.D. thesis, which was in Latin, Beeckman never published any works. Nevertheless, the explanations given by Franssen and Limbrick could explain Beeckman's switch between languages as well. In order to learn more about Beeckman's code-switching, this chapter analyses the languages used by Beeckman in his manuscript in detail, attempting to see whether a certain relationship can be found between the topic of a section and its language.

14 Van Berkel, *Isaac Beeckman on Matter and Motion*, p. 32.

15 Fransen, 'Latin in a Time of Change', p. 632.

16 Fransen, 'Latin in a Time of Change', p. 635.

17 Elaine Limbrick, 'To Write in Latin or in the Vernacular: The Intellectual Dilemma in an Age of Transition: The Case of Descartes', *History of European Ideas* 16 (1993), pp. 75-80, esp. p. 79.

A Few Notes on Method

In order to find an answer to these three questions, three different approaches have been used. In the first part, Stevin's influence on Beeckman's use of Dutch has been examined. Several authors have studied Stevin's influence on the Dutch language, and they have published lists of words that can most likely be attributed to Stevin. K.W. de Groot's article 'Het purisme van Simon Stevin' from 1919 contains a list of terms.[18] E.J Dijksterhuis's monograph *Simon Stevin*, published in 1943, also contains a list.[19] Marjolein Kool published a list in her 1992 article 'Rekenkundige termen van Simon Stevin'.[20]

For the present study, these lists have been critically assessed, after which they have been combined into one database of terms that can most likely be attributed to Stevin. This critical assessment was deemed necessary since the most difficult aspect of Stevin's contribution to the Dutch language has been to distinguish between the terms Stevin truly created and the terms that have been incorrectly attributed to him. Stevin came up with several neologisms, but he also popularized the use of already-existing words.[21] The latter resulted in false attributions of those words to Stevin. Up to a certain degree this has been studied before, but it remains unclear to what extent the terms ascribed to Stevin are truly created by him. Marjolein Kool is the only one who has taken a closer look at Stevin's arithmetical terms by looking at arithmetic books from before Stevin's work. This research allowed her to exclude several terms that were previously ascribed to Stevin because they had already been in use before him.[22] K.W. de Groot and E.J. Dijksterhuis took a different approach. Whenever one of Stevin's terms also appeared in Christoffel Plantijn's (Christophe Plantin) dictionary *Thesaurus Theutonicae Linguae. Schat der Neder-duytscher spraken* from 1573, or in Cornelis Kiliaan's (Cornelis van Kiel) *Dictionarium Teutonico-Latinum* from 1574 with the same meaning, they concluded that Stevin did not come up with it. If it appeared in one of these dictionaries with a different meaning, they concluded that Stevin had created a semantic neologism. However, Frans Claes found that Kiliaan had probably used Stevin as inspiration when creating his

18 K.W. de Groot, 'Het purisme van Simon Stevin', *De Nieuwe Taalgids* 13 (1919), pp. 168-181.

19 Dijksterhuis, *Simon Stevin*, pp. 307-315.

20 Marjolein Kool, 'De rekenkundige termen van Simon Stevin', *Scientiarum Historia* 18 (1992), pp. 96-107.

21 Jozef T. Devreese and Guido Vanden Berghe, *'Magic Is No Magic': The Wonderful World of Simon Stevin* (Southampton: WIT Press, 2008), p. 205.

22 Kool, 'De rekenkundige termen van Simon Stevin', pp. 96-107.

dictionary.[23] Subsequently, Claes created a list of words that do appear in Kiliaan's dictionary but that are in fact taken from Stevin. Therefore, the method that De Groot and Dijksterhuis take is not completely infallible. For this chapter, the methods adopted by De Groot and Dijksterhuis on the one hand and Claes on the other have been combined. The words that, according to De Groot and Dijksterhuis, were not created by Stevin because they appeared in Kiliaan or Plantijn are not included, except for those that are in fact taken from Stevin according to Claes. Ultimately, this resulted in a list of 419 terms (see Appendix A).

The next step was to search the entire *Journal* for these 419 terms using the software AntConc.[24] The list of hits resulting from this search has been analysed in detail in order to give a reliable image of Beeckman's use of Stevin's terms. This analysis shows, firstly, which of the 419 terms occur in the manuscript and, secondly, whether Beeckman used the newer Dutch version or the previous original variant of this word. In the attempt to further determine for the hits whether these terms have been created by Stevin or not, the online historical dictionary *De Geïntegreerde Taalbank* has been used.[25] In some cases, this dictionary gives information about the first use of a certain term. If this given year of first use precedes Stevin, it can be concluded that Stevin did not create the term. In addition, Ruud Ryckaert's *De Antwerpse spelen van 1561 naar de editie Silvius (Antwerpen 1562) uitgegeven met inleiding, annotaties en register* has been used in this chapter to determine the originality of Stevin's terms as well.[26] This work has an index of terms that were used in texts for a literary competition of chambers of rhetoric in Antwerp in 1561. These texts discuss many of the arts and sciences and therefore mention terms that were common for these subjects in sixteenth-century Dutch. In order to create a more reliable and accurate list of Stevin's terms, Ryckaert's index has been used. If the term appeared in the register, it was concluded that the term had been in use before and was therefore not a neologism made up by Stevin.

This method is only effective, however, when Stevin's term and the identical term also have the same meaning. Otherwise, it is possible that Stevin

23 Frans Claes S.J, 'Simon Stevin als bron voor Kiliaan', *Tijdschrift voor Nederlandse Taal- en Letterkunde* 111 (1995), pp. 55-63.

24 Anthony Lawrence, AntConc, computer software, version 3.5.6 (2018).

25 *De Geïntegreerde Taalbank*, Instituut voor de Nederlandse taal (2018), http://gtb.inl.nl/search/ (accessed 1 November 2021).

26 Ruud Ryckaert, *De Antwerpse spelen van 1561 naar de editie Silvius (Antwerpen 1562) uitgegeven met inleiding, annotaties en register* (Gent: Koninklijke Academie voor Nederlandse Taal- en Letterkunde, 2007).

created a semantic neologism, which entails that he gave an existing word a different meaning for a different context. Therefore, it is important to know exactly which meaning Stevin had in mind when creating his terms. Although this is unfortunately not always clear, for the current research it will suffice to know whether Beeckman used the original meaning or Stevin's meaning. In order to determine whether Beeckman used Stevin's version or the original version of these semantic neologisms, the original meaning of the specific word has been determined by using Ryckaert's work and the historical dictionary. Then, by looking at how and in which context Beeckman used the word, its meaning has been determined as well. If this meaning does not correspond to the meaning found in Ryckaert's work or the historical dictionary, the possibility exists that Beeckman used Stevin's meaning for the word. However, if the meanings do concur, it can be concluded that Beeckman did not use the word with Stevin's intended meaning. Additionally, to give a comparison of Beeckman's use of the Dutch and the Latin language, his manuscript has also been searched for the (Latin) equivalents of Stevin's terms. This can reveal whether Beeckman had a preference for either the Latin terms or the new terms.

Then, in order to answer the second question of this chapter and to gain more knowledge about Beeckman's behaviour of using Latin and Dutch in relation to the subjects he wrote about, the first volume of De Waard's edition of the *Journal* has been examined. This volume consists of 351 pages, excluding the appendix. Of these 351 pages, the first 60 and the last 60 have been analysed. This gives an overview of Beeckman's use of language in his manuscript from 1604 until January/February 1615 and from April 1619 until November 1619. First, the topic of each entry written on these pages has been listed. The website created by Ad Davidse and Cathie Schrier gives Dutch translations of certain Latin parts of the manuscript.[27] These have been used to determine the subject of the sections written in Latin. Secondly, for each entry it has been noted which language Beeckman used for it. Finally, for some entries Beeckman mentioned in which city or village he wrote them, which has been noted as well. This ultimately resulted in a table that contains a list of topics with their corresponding language, date and, if provided, the name of the city. In order to identify a possible relationship between topic and language, the topics have been reordered into a list of categories. The following categories have been used: Candles; Earth and the Universe; Human Body; Light; Mathematics; Meteorology;

27 Ad Davidse and Cathie Schrier, 'Isack Beeckman' (2018), https://adcs.home.xs4all.nl/beeckman/ (accessed 1 November 2021).

Mirrors and Lenses; Practical and Personal; Pumps, Pipes, and Waterworks; Vacuum and Pressure; and Water, Wells, and Caves. These categories might reveal a specific connection between a certain topic and a certain language.

Finally, the third part focuses on any reflections on language that Beeckman mentioned in his manuscript. In order to find these in the Dutch and Latin sections, the website created by Ad Davidse and Cathie Schrier has been used. Their website contains an index of terms or subjects that appear in the entire manuscript. In addition, AntConc has also been used to search the manuscript for certain key words related to the topic of language or linguistics. These key words are in Dutch and in Latin. Examples are *taal* and *tael* (language), *spraecke* (language, speech), *Duytsch* (Dutch), *woorden* (words), *lingua* (language), *grammatica* (grammar), and *verbum* (word).

Beeckman's Use of Stevin's Terms

In total, 180 terms from the list of Stevin's words have been found by AntConc in Beeckman's entire *Journal*. Many of these terms were excluded from the list because they were not used with the same meaning or in the same context as Stevin had intended for them. An example is the noun *val*. This word appears in the *Journal*, but Beeckman used it to describe the 'fall' of certain objects while Stevin intended this word to describe a 'cadence' in the field of music. Another example is the term *derde*. Stevin turned this word into a semantic neologism by giving it the specific meaning in arithmetic context of 'a thousandth of one'. Beeckman used this term many times but in its more common meaning 'third', as in, for instance, *het derde getal* (the third number).

Also, many of Stevin's terms found in the manuscript by AntConc were used by Beeckman but only as a reference to Stevin's works. For instance, Stevin created the term *weereltschrift* (cosmography). He used this term as the title for the first part of his *Wisconstige Gedachtenissen*. AntConc gave two hits for this term in the manuscript, but these two hits were Stevin's work mentioned by Beeckman. Any other term similarly used has been removed from the list of final results. Ultimately, only seven of Stevin's terms have been found in the manuscript. For these words it can be said that they have most likely been created by Stevin and that they have been used by Beeckman in his *Journal*. Table 1 shows the number of hits for the Dutch term and its Latin equivalent, as well as an English translation. The term *glasgront* lacks its Latin equivalent in the table because it is not noted by Dijksterhuis. Moreover, the 74 hits of the term *parallelus* also includes variants such as *parallelum*, *parallel*, and *parallele*.

Table 1 Stevin's Terms Used by Beeckman

Stevin's term	Dutch count	Latin term	Latin count	English translation
Evenwydich	6	parallelus	74	parallel
Glasgront	1	–	–	floor
Sne	2	nodus	–	the spot in which certain lines or surfaces intersect
Soetluydicheyt	1	harmonia	16	harmony
swaerheytsmid-delpunt, swaerheyts middelpunt	18	centrum gravitatis	15	gravity centre point
Waterwicht	4	hydrostatica	–	hydrostatics
Sichteynder	5	horizon(t)	5	horizon

The small number of hits comes unexpected, since Beeckman mentioned Stevin many times in his manuscript. Although spelling variations have been taken into account for most terms, it remains a possibility that some terms were missed due to Beeckman using a different spelling. Another explanation for the small number of hits can be the fact that Stevin actually did not create that many terms. Out of the 180 terms initially found by AntConc, many terms were excluded because they had been used before Stevin's time. This relates to the major challenge of determining whether a certain term can truly be attributed to Stevin or if this has incorrectly been done so in the past. If Stevin did not create as many neologisms as has been believed, it can still be said that Beeckman used very few of his terms. A possible explanation for this could be that most of Stevin's terms are very discipline specific. For example, many of his terms are only applicable to arithmetic and since Beeckman does not focus on this field as much, most of Stevin's terms do not appear in the manuscript. On the other hand, the term *swaerheytsmiddelpunt*, which occurs eighteen times in the manuscript, belongs to a discipline Beeckman did focus on. This is also the only term he used more than the Latin equivalent.

Another option is that Beeckman simply did not feel comfortable using Stevin's terms because he was more familiar with the Latin terms. Table 1 shows that in most cases Beeckman used the Latin equivalent more often than Stevin's term. The terms Beeckman used most are *parallelus* and its variants. The Latin *harmonia* is used sixteen times more than Stevin's equivalent. It seems that he preferred to use the Latin terms over the Dutch ones.

Despite the possibility that Stevin did not create as many terms as believed or the possibility that some terms were missed in the search due to spelling variations, the small number of Stevin's (semantic) neologisms in Beeckman's *Journal* can be explained by his familiarity with and his habit of writing in Latin. Beeckman had always used Latin in the context of knowledge and learning and was therefore accustomed to these Latin terms. It was simply quicker and easier to express his thoughts using the words that were already present in his vocabulary. Besides, he might not have come across many terms he thought were important or convenient enough to adopt because of how discipline specific they were. In the end, the option remains that Beeckman did not think it necessary to use Dutch terms, which says something about his level of purism.

Beeckman's Use of Latin and Dutch

For the second parts of this study, which examines the relationship between Beeckman's choice of language and the topic he wrote about, the first volume of De Waard's edition of the *Journal* has been used. Altogether, the first 60 and the last 60 pages of this volume contain 236 sections. Of these sections, 150 sections (63.6 per cent) were written in Latin and 86 sections (36.4 per cent) were written in Dutch. Of the 150 sections in Latin, five sections contained Dutch words or phrases in the text or in the margins. From the 86 Dutch entries, one contains a French quote, and six entries contain Latin words or phrases. From these results it seems that Beeckman had certain short phases in which he would write using the same language for consecutive days. For example, on page 17, between November 1612 and March 1613, he starts with three entries in Dutch, after which he writes one in Latin, five in Dutch, four in Latin, seven in Dutch, eighteen entries in Latin, one in Dutch, three in Latin, one in Dutch, one in Latin. He keeps alternating between the two languages and mostly uses the language for more than one entry before he switches to the other. These phases can be linked to the topics he writes about, which can be seen in Table 2. The number of Latin entries and the number of Dutch entries within these themes have been noted. The separate tables for each category can be found in Appendix B.

Table 2 confirms the fact that Beeckman primarily used Latin. When writing about the themes Earth and the Universe (Table 5 in Appendix B); Human Body (Table 6); Meteorology (Table 10); Music and Sound (Table 11); and Vacuum and Pressure (Table 14), Beeckman used Latin for the majority

Table 2 Overarching Themes

Theme of topics	Number of entries in Latin	Number of entries in Dutch	Latin and Dutch	Total
Candles	2	5	1	8
Earth and the Universe	9	2	1	12
Human body	10	3	2	15
Light	3	2	–	5
Mathematics	4	3	2	9
Meteorology	9	2	–	11
Mirrors and Lenses	3	–	–	3
Music and Sound	33	7	–	40
Practical and Personal	10	17	2	29
Pumps, Pipes, and Waterworks	5	15	–	20
Vacuum and Pressure	8	1	–	9
Water, Wells, and Caves	2	4	1	7

Source: *Journal tenu par Isaac Beeckman de 1604 à 1634*, publié avec une introduction et des notes par C. de Waard, 4 vols. (The Hague: Martinus Nijhoff, 1939-1953), vol. 1.

of the entries. The vernacular language is used mostly for the themes Candles (Table 4); Mirrors and Lenses (Table 9); Practical and Personal (Table 12); Pumps, Pipes, and Waterworks (Table 13), and Water, Wells, and Caves (Table 15). The difference between Latin and Dutch entries within the category Light (Table 7) is small given that, of the five entries, two entries are written in Dutch and three in Latin. The same counts for the category Mathematics (Table 8) in which four topics are written in Latin and three topics are written in Dutch. For two sections in this category, however, both languages are used. This is also the case for one section in Candles, one section in Earth and the Universe, two sections in Human Body, two sections in Practical and Personal, and one section in Water, Wells, and Caves.

Looking at these various categories can help explain Beeckman's phases of using a language. It seems that he sometimes writes about similar topics for several entries in a row and therefore keeps using the same language for that specific topic. This is the case, for example, on pages 36 and 37 of the first volume of the *Journal*. Beeckman writes an entry about vacuum and pressure, one about pressure and magnets, and the next one about suction cups. These entries about similar topics are all written in Latin.

However, these consecutive topic and language phases do not always occur. For example, on pages 49 to 58, Beeckman writes a short entry on the F-note in Dutch, after which he continues writing about music for eleven entries in Latin and then writes about the size and weight of water in Dutch. These eleven entries on music in Latin followed by a switch to another language and another topic show that he does at times stay with the same language when talking about the same topic. Yet, the first entry on the F-note in Dutch, which is also linked to music, shows that he is not necessarily always consistent in his language and topic combination. These inconsistencies can be found within many categories. The entries on meteorology, for example, are mostly in Latin. Two entries that are linked by the topic of the altitude of clouds are written in Dutch. Yet, something that intervenes with this is the entry written on low-hanging clouds in Latin. Although this seems to be related to the altitude of clouds, Beeckman uses a different language to discuss a similar topic in this instance.

Yet, it is still possible to detect consistency from a larger perspective. As Van Berkel puts it, 'Beeckman was both a craftsman and a scholar.'[28] This seems to be the perfect division in terms of language. The craftsman uses Dutch and the scholar uses Latin. The category Pumps, Pipes, and Waterworks is one of the three categories dominated by Dutch. Five of the 20 entries have been written in Latin, but a relationship between the reason for writing in Latin and the topic of the entry is difficult to find in this case. It seems that Beeckman had a preference for Dutch when writing about topics related to his craftsmanship but at times decided to use Latin, nevertheless. It is visible as well in the two other Dutch-dominated categories Candles, and Practical and Personal. These two categories are part of Beeckman's background in the world of artisans; he learned these crafts in Dutch and therefore continued to write about them in the same language.

The other category dominated by Dutch, Practical and Personal, mostly includes entries that are relatively practical and not so much focused on explaining abstract theories. These are entries in which Beeckman discusses, for example, the fact that a certain year had been a good year for cherries, advice about friendships, the changing value of money, and his reflections on the weight difference between an unboiled egg and a boiled egg. Apparently, whenever Beeckman wrote about these practical topics that did not involve complicated theories, he preferred Dutch. Nevertheless, it cannot be said that all practical entries in this category have been written in Dutch. For example, on page 311 of the first volume, Beeckman writes about a brook near Renesse on the isle of Schouwen using Latin. This does not seem to be a very theoretically abstract subject, yet he does not write about it in Dutch.

Beeckman also seems to prefer Dutch when explaining how to carry out certain activities, as if he is writing a manual. For instance, an entry

28 Van Berkel, *Isaac Beeckman on Matter and Motion*, p. 4.

can be found in which Beeckman discusses how to make glue. In the category Human Body, he writes about how to avoid a painful spleen. Similarly, in the category Universe and the Earth, he describes how to make a compass and he writes about making a portable sundial. The category Pumps, Pipes, and Waterworks contains three more entries in which Beeckman explains how to use unevenly laid pipes, how to keep pumps wet, and how to make a fountain pump. These seven entries have all been written in the style of a manual, and they have all been written in Dutch. Again, these entries are concrete and practical. As Peter Burke explains, 'to write in Latin was to cut themselves off from ordinary people, but to write in a vernacular was to cut themselves off from the rest of Europe'.[29] If Beeckman had intended his manual-like entries to be read by ordinary people around him at some point, it made sense to write them in the language they understood best.

Something else is going on in the eight entries in which Beeckman uses both Latin and Dutch. In these entries, Beeckman starts writing in one of the two languages and then either completely switches to the other language or keeps switching between the two throughout the entry. For example, in the category Water, Wells, and Caves, Beeckman discusses the depth of water wells in the following entry on page 14 of the first volume:

> Jan oom seyde, dat hy een kunste wist, die schier niemant meer en wiste ende was, dat hy eenen eemer waters op konde halen met half soveel touwe op te trecken als den put diep was ende en wilde dat my niet leeren. Ego verò quaesitum excogitans, hoc theorema condidi: Omnia instrumenta, quae motum facilitando, tempus motûs producunt, eadem è contrario possunt motum, gravando tempus motûs, minuere; si subjectum loces ubi erat vis, et vim ubi subjectum, ut patet in figurâ.[30]

He starts this entry in Dutch and explains something his uncle Jan Pieters van Rhee told him about the depth of water wells, but then switches to Latin for the remaining part of the entry after this first sentence to explain his own views on the matter after having given it some thought himself.

29 Peter Burke, 'Heu Domine, Adsunt Turcae: A Sketch for a Social History of Post-Medieval Latin', in Roy Porter and Peter Burke, eds., *Language, Self, and Society: A Social History of Language* (Cambridge: Polity Press, 1991), p. 27.

30 *Journal tenu par Isaac Beeckman de 1604 à 1634*, publié avec une introduction et des notes par C. de Waard, 4 vols. (The Hague: Martinus Nijhoff, 1939-1953) [henceforth *JIB*], I, p. 14 (in the margin: 'Chordâ multo minore quàm est puteus, aquam haurire' = 'to draw water from a well with a rope that is much shorter than the well').

Another example of this code-switching can be found in an entry on the smallness of things on page 17:

> Miranda est subtilitas rerum in rebus. Want een dinck in een man, dat maer een aesken en weecht, en weecht qualick het 2413de deel van een aesken in een puero primum conformato. Denckt hoe kleyn dat moet syn, ende denckt, hoe kleyn dat is datter alle ure aenwast totdattet so groot is als in een volwassen man. Ita ut Anaxagorae sententia: 'Omnia sunt in omnibus' mirabilior non esse videatur.[31]

In this entry, Beeckman discusses his amazement of the fact that adults are so much smaller in the early stages of their lives and that they grow from something so small to something much bigger. In this case, Beeckman does not switch between the languages once but keeps switching between them within his sentences.

In these entries in which Beeckman using both Latin and Dutch, the combination of the two can be clarified using Peter Burke's explanation. He states that Latin texts had the tendency to slip into the vernacular when a certain untranslatable term was needed; however, 'the reverse process was also a common one. In both speech and writing there was a tendency to slip into Latin at certain points.'[32] Whenever Beeckman was writing something in Dutch, he could have had a certain word or phrase in Latin in mind. If he was accustomed to use this specific word in Latin and he could not immediately find a satisfying Dutch translation for it, it is possible he opted to use the Latin term instead. Of course, this can then also work the other way around.

All in all, looking at the entries chronologically reveals that Beeckman had phases in which he used the same language for multiple topics before switching to the other. In this case, the language is not linked to topic per se, since Beeckman sometimes switched to another topic in the same language and at other times wrote about the same topic in a different language. Apart from this, the most important observation is the fact that the majority of the entries were written in Latin. This majority focuses on the more abstract theories on, for example, vacuum, pressure, sound or the human body. Beeckman shows a preference for the Dutch language in the three categories Candles; Practical and Personal; and Pumps, Pipes, and Waterworks. Moreover, he used the vernacular for manual-like entries

31 *JIB*, I, p. 17 (in margin: 'Parvitas rerum quae sit' = 'how small things are').

32 Burke, 'Heu Domine', p. 39.

in which he explains how to make certain objects and prevent or solve certain issues. This creates a consistency between language and Beeckman's scholarly side on the one hand and his craftsmanship on the other hand. Lastly, the combination of Dutch and Latin that appears in some entries has probably been done out of convenience and habit.

Beeckman's Reflections on Language

The entries in the manuscript in which Beeckman wrote something related to foreign languages have been found by using the search words in AntConc and by looking through the index on Ad Davidse's website. This resulted in only three entries, all in the first volume. Two of these are in Dutch and the other one is written in Latin.

The first Dutch entry appears on the second page of the first volume, written in 1608 or 1610.

> Cal. aux *Cor. 14, 5*: 'Quelle sottise seroit-ce de prononcer une mesme chose en plusieurs langages sans necessité quelconque? Mais il peut souvent advenir qu'il y aura bonne occasion d'user de langue estrange.'
> Welck occasie ick misschien t'Amsterdam gesien hebbe, als den predicant op my begeerde, dat ick int latyn soude propheteren in de kercke, naerdien dat ick het Engels wel verstondt, maer niet spreken en konde; ende hy beloofde dat het volck int Engels te segghen. – Vide vers 13.[33]

The French quote by John Calvin translates to the following: 'How foolish would it be to express the same matter in different languages without any cause? But often occasions can occur in which it is suitable to use a foreign language.' In Dutch, Beeckman explains that he possibly saw such an occasion in Amsterdam. The preacher asked Beeckman to prophesy in Latin because Beeckman could understand but not speak English. The preacher promised him that he would explain it to the people in English.

The next entry can be found on page 19 of the first volume of the edition.

> Daer is een gebreck in de talen, dat sy de mannieren van spreken int schryven niet uytdrucken en konnen, sodat men niet en kan weten met wat een affectie dat een dynck gesproken geweest is. Twelck men, naer

33 *JIB*, I, p. 2 (in the margin: 'Linguarum in Ecclesia non intellectarum usus' = 'the use of languages in the church that are not comprehended').

> myn oordeel, wel soude eenichsins konnen beteren, somen op die syllaben, die met een ἔμφασις, dat is die styf gepronuncieert werden ende die men int spreken meest gehoort begeert te syn, dat men daerop sedt sulck een streepken: /, gelyck alsmen seght: 'Kondt ghy dat doen?' ende dat men op die syllaben, die clemachtich gesproken worden, sedt: ~, gelyck als men seght: '*Men moet niet Claeỹs Claeỹs eten, maer Wouter Wouter*', dat is niet traech, maer rasch.[34]

In this section, Beeckman expresses his opinion about a certain imperfection of written language. He explains that this imperfection consists of the problem that one cannot express the way of speaking or pronunciation in written language and therefore one will not know the intended tone of something that is written. Beeckman thinks this can be solved by using the / symbol on the syllable that needs to be said the loudest if the sentence was to be spoken. The ~ symbol is used to indicate which syllable should be emphasized. He gives the example of a sentence in which the word 'Claeys' is pronounced twice with emphasis because of the tilde (~) placed on the y (ỹ). Moreover, Beeckman uses the Greek spelling 'ἐμφασις' for the term 'emphasis'.

The last entry, which has been written in Latin, can be found on page 147. On Ad Davidse's website, this entry has been named 'Jong geleerd, oud gedaan', which is a proverb that translates to 'What is learned in the cradle is carried to the tomb.' The margins belonging to this section are 'Linguae extraneae pronunciatus difficiles. Cur.' which addresses the difficulties of pronouncing foreign languages. This entry is interesting because it shows that Beeckman spent some thought on the aspects of learning foreign languages.

The entry on foreign languages in church shows that Beeckman spent time thinking about the use or effect of a multilingual environment. The section on page 19 says something about Beeckman's interest in improving the way spoken language was written down. Noteworthy is the fact that nothing has been found in the other two volumes of the manuscript. The three entries are all found in the first volume, within the short time period between the years 1608 and 1618. This can say something about a shift in Beeckman's interests. Possibly, he had been interested in languages when he started his manuscript; the three entries show that he was thinking about foreign languages and that he was aware of the importance of certain

34 *JIB*, I, p. 19 (in the margin: 'Scriptura belgica etc. accentus desiderat' = 'the Dutch language requires an accent').

aspects of language at some point in his life. Three entries are relatively insignificant, however, in comparison to the total number of topics in the entire manuscript. Possibly, Beeckman was not very interested in the topic of foreign languages besides these three entries.

Since the search using the key terms and the index on Ad Davidse's website have not given many results, it is safe to say that Beeckman's interest in languages was very short-lived. Beeckman's manuscript does not supply much information on his ideas and reflections on language. The limited number of three entries found shows that he was aware of the differences or situations that occurred in a multilingual environment, but simultaneously shows that he only paid attention to them three times in his manuscript.

Conclusion

It has become obvious that Beeckman used Latin more than he used the vernacular language in his manuscript. Important to note here, however, is that the Latin language was still a relatively more important language than any vernacular until the late seventeenth century. It was very common for works of learning to be written and published in Latin.[35] This is the main explanation for why only seven of Stevin's terms were found and why the majority of the sections studied were written in Latin. Although spelling variations might play a minor role in the number of hits of Stevin's terms and the majority of these terms might have been too discipline specific, the small number of hits shows that Beeckman did not actively pursue or advertise purification of the vernacular. The fact that only three entries in the manuscript have been found in which Beeckman discusses topics related to languages confirms this. Moreover, his use of Latin and Dutch was at times arbitrary and incidental, which is probably due to the fact that he felt more comfortable using Latin terms at times but preferred Dutch terms at other moments.

On top of this, this study has revealed new knowledge specific to Beeckman's manuscript writing. Interestingly, he showed a clear preference for the Dutch language when discussing topics in the three categories Candles; Practical and Personal; and Pumps, Pipes, and Waterworks, as well as in his manual-like entries. These more practical, artisan entries were written in Dutch because he had learned them in this language and possibly to make sure that ordinary people interested in these topics would be able

35 Burke, 'Heu Domine', p. 29.

to read them. Most of the other, more abstract and scholarly entries have been written in Latin, most likely because Beeckman was accustomed to reading and learning about these topics in this language.

Concerning limitations of the study, the uncertainties surrounding the details of Stevin's terms is the most significant one. If a complete list with meanings could be constructed, the manuscript could be searched for these terms. This would give a precise overview of the presence and number of Stevin's terms, either the Dutch ones or the equivalent terms, in Beeckman's *Journal*. Being able to compare the number of occurrences of Stevin's Dutch terms to the number of occurrences of the original equivalents of these terms would reveal more accurate details on Beeckman's choice and use of language. Moreover, the complete manuscript should be included in the analysis of the relationship between topic and language. This would result in a more thorough picture of Beeckman's manuscript behaviour and it could reveal interesting changes over time. Related to this change is the fact that Beeckman starts to note the name of the city or village in which he wrote a section more often over time. It could certainly be interesting to incorporate this information while studying later years of the manuscript in order to find a possible relationship between language, topic and place. In addition, the reasons for the lack of French in the manuscript could also be a topic of future research. These ideas would be a valuable addition to the study of Beeckman's multilingual manuscript behaviour.

About the Author

Semra Meray completed her bachelor's degree in linguistics and literature at University College Roosevelt (Middelburg) in 2018 and afterwards specialized in forensic linguistics in the research master's programme at the Vrije Universiteit Amsterdam.

Appendices

Appendix A

Table 3 Stevin's Terms

Stevin's term	Latin translation or other equivalent
aertconst	physike
aertist	physicien
afcomst	–
afdaking	Abdachung
afwesen	privatio
afwijcking	declinatio
almeter	holometrum
anvangtijt	aera, radix annorum, apocha
aspunt	polus
balck	–
banierleyer	colonel, collonel
bebardering	lamberceering, lambercering
bedeckt graven	zapperen
bedeckte wech	strada coperta, corridot, chemin couver
begheerde	quasitum
beghin	principium
beghinsel	–
bepaeld, bepaelt	
bepalinghe Definitio	limitatus
bereytsel	praeparatio
beschoeying	scarpa, talud, escarpe, Boschung
beschryvelick	inscriptibilis
bevechtcrijch	bellum offensivum
bewijsconst, bewysconst	dialectica
bewyser	dialecticien
bewysreden, bewijsreden	syllogismus
bijl	(trapezium)
bol	convexus
boumeester	ingenieur
boveplaets	place supérieure
brander	conoides
brantsne	parabola
breetheyt	latitudo
breuk	–
burgherlickheydt	politia
buyteschoeysel	contrescarpe
byclank	accentus

Stevin's term	Latin translation or other equivalent
cabeschoeysel	contrescarpe
constwoorden	vocabula artium
crans	ghirlanda
crijchstuych	amonitie van oorlogh
cromstreeck	loxodromia
dachschrift	journael
dachtafels	ephemerides
daelick maecksel des sancx	compositio cantus
daellini	–
daelwicht	–
damphooghde	altitudo vaporum
derde	thousandth part of one
deursichtighe	perspectiva
dobbeling	duplicatio
doende	activum
drieroe	triquetrum
duysteraer, duystering	zodiac, ecliptica
dwaelder	planeta
dwaelderloopen	motus planetarum
ebront	–
eenheyt	unitas
eenichvorst, eenig-vorst	monarch
eenichvorstheydt	monarchia
eerlyck vertreck	honourable retraicte
eerste	tenth part of one
eertcloot	globus terrae, globus terrestris
eertclootschrift	geographia
eintlicke oirsaeck	causa finalis
erm	–
evebeenich	isosceles
evenwydich	parallelus
everedelick	proportionaliter
everedenheyt	proportio
everedich	proportionalis
evesijdich	aequilaterus
evestaltwichtich	–
evestofswaer	–
evewijdighe vierhouck	parallelogrammum
eysch	problema
gaslagher, gaslager	observator
gaslaghing, gaslaging	observatio
geesthandel, gheesthandel	magia
geeygenwetticht	ghepreviligeert
geleyen	convoyen

Stevin's term	Latin translation or other equivalent
geloofbrieven	credentiebrieven
geometrice	meetconstlick
gesangkmaecksel	compositio cantus
getalt	genombreert
getuychnissen	attestatiën
ghebreeckende	–
ghecort	truncatus
ghedachtenissen	commentaria
ghedeurighe everedenheyt	continua proportio
ghegheven	datum
gheley	convoy
gheloofbrief	credentiebrief
ghemeene rechtsraet, gemeene rechtsraet	hoff provinciael
ghemeene reghel	axioma
ghemeenheyt, gemeenheyt	democratia
ghemeenschool	academia, universitas
gheneesstof	medicamentum
gheneser, geneser	medicus
gheschickt	regularis
ghestelde	hypothesis
gheweldighe	provoost generael
glas	(tafereel)
glasgront	–
goddist	theologus
godheyt, godtheyt	theologia
gordine	cortina
grachtcant	contrescarpe
grijphouck	diameter visualis
grofschutweer	baterie
gront	basis
grontteyckening	ichnographia, plan semiarcus
halfmiddellijndeel	sector
halfmiddellini	semiarcus
halfpeze	semichordum
handthaef	–
hanghende lini	perpendicularis
hangsnoer, hanghende	perpendicularis
heflini	–
hefwicht	–
hemelclootsch, clootsch	sphaericus
hemellooptuych	instrumenta astronomica
herfstsne	sectio autumnalis
heymelicke laech	embuscade
heymelicke uytgang	sortita

Stevin's term	Latin translation or other equivalent
holgraven	mineren
holgraver	mineur
hondertleyer	centurio
hoochmetingleer	scala alimetra
houckmaet	sinus
houckmaetpijl	sinus versus
hulpe	assistant
huysbou	architectura
huysgenoot	domestijcq
inront	epicyclus
inscrivelick	inscriptibilis
juweelkamer	tablinum
keer	coude
keghelhandel	tractatus conicorum
keghelsne, kegelsne	conisectio
lanckrondt	Ellipisis
langde, lanckheyt	longitudo
langhworpigh	–
leege walganck	terreplein
leeghste punt	nadir
lentsne	sectio vernalis
letteraer	grammaticus
leyer	dux
leytsman	conducteur
lichtplaets	atrium
lijck-spreuck	metaphora
lijckstandig	homologus
lijckstandighe sijden	homologa
lijkspreuckelick	metaphorice
logierder	fourier, quartiermeester
loofweerdicheyt	authoritas
machtelick	potentiâ
Maechtsare	Spica Virginis
mael	quotiens
maenduystering	eclipsis lunaris
maniere	modus
meetconst	geometria
meetconstich	geometricus, geometrice
meetconstich viercant	quadrans geometricus
meetdaet	praxis geometriae
meetstrael	radius geometricus
meter	geometra, geometricus
middachront	meridianus circulius
middelgracht	

Stevin's term	Latin translation or other equivalent
middellinie	fossetta, contrefossé
diameter	
middelplaets	
middelpunt	centrum, place moyenne
middelront	aequator
misschaeuwing	refractio
moortcuyl	casematte
naelde	pyramis
naeldwijsing	–
naerdering	aprochering
naestepunt, naeste punt	perigeum
nagevolck, nagesleyp	treyn
naghesteld	postposé
natuerlick	physicus, physice
natuerlick jaer	annus naturalis
noemer	nominator
oirde	methodus
oirdensch	ordinaris
oirspronck des naems	etymologia
omstreck	periferie
omtreck	peripheria
onderanderden	subalterni
onderandert	subalterne
onderplaets	place inférieure
onderschrijver	clerck, klerck
onghemiddelt	immediate
onparich	oneven
opclimming	ascensio
opklimming	ascensus
opperschool, hoochschool	academia, universitas
opsiender	controlleur
opsnijding	anatomia
overst	generael
pael	limes
pael	terminus, elementum
pael	terminus, elementum
parich	even
pees	chorda
peesdeel	segmentum
perck	–
pijl	sagitta
pilaer	columna
platcloot	astrolabium, planisphaerium
platmeter	planimetrum

Stevin's term	Latin translation or other equivalent
platte form	piatta forma
plomphouck	angulus obtusus
plucken	peculeren
plucker	peculator
raecklijn	tangenus, tangens
raet van de ghemeensaecx rekening	raet van finance
rechte streeck	(orthodromia)
rechthoeckigh	–
rechtlinich	rectilineus
reden	ratio
redenaer	orator, rhetoricus
redenstryt	disputatio
reetschap	instrumenta
rekenbaer	contabel
rekeninghe	–
reysen	op marsch zijn
riem	zona
ront	circulus
saempunt	–
saming	conjunctio
sanglini	regula haromices, monocordus
sangmaker	componista
sangsnaer	monochordum
schaeu	imago, simulacrum
scheefrondt	ellipsis
scherphouck	angulus, acutus
scheydoven	furnus chimicus
schietweer	batterie
schijn	aspectus
schiksel	fatum
schilbooch	arcus complementi
schilhouck	complementum
schoensche sijde	hypotenusa
schrijfcamer	contoir
schult en hebbing	debet en credit
schutsel	orecchione, oreillon
schuyfcruys	Radius, baculus astronomicus
selfslachtich	eiusdem generis
selfwoordicheyt	anaphora
sichteynder, sichteinder	horizon
sichteyndersch	horizontalis
sichtgaetkens	dioptra
sichtlijn	spira
sichtrye	linea fiducialis

Stevin's term	Latin translation or other equivalent
sienderlijn	–
siendermaet	–
sijde	root of a number
singconst	musica
singconstich	musicalis
singleer	scala musica
singteijcken, singteijckens	notula, notulae
sne	nodus
snijlijn	secans linea
soetluydicheyt	harmonia
soomenichmael	quotiens
soorte	figura
soppich vierendeel ronts	quadratum azimuthale
soppunt	punctum verticuli
speeltuyghen	instrumenta musica
spieghelconst	catoptrica
spieghelschaeu	–
spijswaert (cleene)	gardemenge
spijswaert (groote)	magasyn
stadthouder	lieutenant
stadtwetten	leges civiles
staltwicht	–
stantteyckening	orthographia, profil
stelreghel	algebra
stelreghelsch	algebraicus
stelreghelstal	numerus algebraicus
stercktebewaring	garnisoen
stercktebou	fortification
sterctebou	fortificatie
steroirdeelen	iudiciariae astrologicae
steunstijl	contrefort
stijflichaem	–
stoffering	hyperbole
stoffscheyder	alchimus
stoffscheyding	alchimie
stoflichter	–
stofscheyder	chimicus, alchimicus
stofscheyding	alchimia
stofswaerder	–
stomphouck	angulus obtusus
stootvougen	joinctures
strijckelick	nettoyable
strijcken	nettare, strisciare, nettoyer, streychen
strijckhoucken	fianchi, flancs
strijckingh	nettoyement, flancquement

Stevin's term	Latin translation or other equivalent
strijding	argumentatio
strijen	argumentari
strijtreden	argumentum
stuck	volumen
swaerheydts middellini	–
swaerheydtsmiddelpunt, swaerheyts middelpunt	centrum gravitatis
swijchteijckens	pausae
tant	merlone
teerlinck	cube root
teerlinckxwortel	cube root
teghenpunt, teghepunt	contrapunctum
teghenschoe, teghenverbeschoeysel	contrescarpe
teghestant	oppositio
telconst	arithmetica
teldaet	praxis arithmetica
telder	rekenkundige, arithmeticus
thiende	–
thiendetal	–
thienich	decimalis
thienleyer	decurio
toomprang	–
topbooch	arcus verticalis
toppunt	punctum verticale
trecklini	–
treflickheden	facultates
tuych	instrumenta
tuychwerckelick	mechanice
tweeclieving, tweescheyding, tweespalting	dichotomia
tweede	hundredth part of one
tweeich liet, tweeick	duo
twelfgrondich lichaem	dodecahedrum
uyrront	circulus horarius
uytbreng	productum
uytbreng	outcome of calculation
uytmiddelpuntich	excentricus
uytoirdensch	extraordinaris
uytwetticheyt	dispensatio
vaenleyer	capiteyn, capitein
val	cadens
veelhouck	polygonus, polygonum
veelsijdich	multilaterus
vergaren	addere
vergaring	college
verheffing des aspunts	elevatio poli

Stevin's term	Latin translation or other equivalent
verkeerde reden	inversa ratio
verschaeuwing	scenographia, sciographia
verscheensicht	parallaxis
verscheenslachtich	heterogeneus
versielt	animatus
versoeckbrief	–
versoucker	suppliant
verste punt	apogeum
verstijving	consolidatio
vertooch	theorema
vertreckwal	retranchement
vervlieting	resolutio
verwachtbrief	expectabrief
verwachter	expectant
viercanten wortel	–
viercanting	quadratura
viercantsijde	radix quadrata
viergrondich lichaem	tetraedrum
vlackvat	–
vliet	studium
vloettop	–
volmachtichvorst, volmachtig vorst	absoluyt monarch
voor rechte crijgh	justum bellum
voorderbetrecking	appel, appellatie
voornaemlickheydt, voornamelickheydt	aristocratia
voorstel, stelling	propositio
voortschuppen	zapperen
voortslag	propositie
voorttreden	marcheren
vorstraet	cancelry, secretenraet, priveenraet
vorstschrijver	secretaris
vrying, vrijing	affranchissement
waerder	commis
waerschouwinghe voor rechte crijc	clarigatio ante justem bellum
wanschaeuwing, misschaeuwing	refractio
wantij	–
wassende sne	hyperbola
waterhol	–
watersch	aquaticus
waterwicht	–
waterwichtdaet	–
wechmaker	pionier
weeghconst	ars ponderaria
weeghdaet	praxis artis ponderariae

Stevin's term	Latin translation or other equivalent
weercrijch	bellum defensivum
weereltschrift	cosmographia
werck	constructio
werck	besoigne
wercking	operatio
werckstuck	problema
werckstuck	vraagstuk
werf	quotiens
wijsheyt, wijsheit	philosophia
wijsreghel, wijsrye	alidada
wisconst	ars mathematica
wisconstenaer	mathematicus
wisconsttuych	instrumenta mathematica
wisconsttuyghen	instrumenta mathematica
worteltrecking	extracio radicum
zeeraet	admiraliteyt
zeesaken	admiraliteyt, navigatie
zeesaken	navigatie
zeeschrift	hydrographia
zeylstreeck	–
zuydlantd	terra Australis

Appendix B

Table 4 Candles

Page	Date	Topic	Language
6	18 July 1612	candles, determining the proportion of diameter of candles using their weight	Latin
7	18 July 1612	burning time of candles	Dutch
7	18 July 1612	length of candles	Dutch and Latin
8	18 July 1612	weight of wick of a candle	Dutch
21	June 1613-April 1614	candle wax	Latin
35	April 1614-January 1615	(melted) candle wax floating in melted candle wax	Dutch
38-39	April 1614-January 1615	why a candle goes out when it is in a glass, how to make a lantern, improvements of smoke in chimney/fireplace	Dutch

Table 5 Earth and the Universe

Page	Date	Topic	Language
1	1604-1608	stars are spheres	Latin
1	1604-1608	argument in favour of Copernicus	Dutch and Latin
10-11	18 July 1612	sundial	Latin
12	18 July 1612	meridian	Latin
13-14	18 July 1612	portable sundial	Dutch
14	November 1612	spheres	Latin
21	June 1613-April 1614	third Earth movement explained	Latin
24	June 1613-April 1614	movement of the Sun	Latin
26	June 1613-April 1614	the Earth in the middle	Latin
27	June 1613-April 1614	lune of Hippocrates	Latin
60	January and February 1615	making a compass	Dutch
326-327	11-23 July 1619	the Earth is growing	Latin

Table 6 Human Body

Page	Date	Topic	Place	Language
15	November 1612	stopping a nosebleed	–	Dutch and Latin
17	November 1612-March 1613	the smallness of things, how something so small can grow into an adult man	–	Latin and Dutch
22	June 1613-April 1614	help/cure for kidney stones	–	Dutch
34	12 April 1614	Beeckman's heartbeat	–	Dutch
59	January and February 1615	how to not get *Milte*, how to avoid a hurting milt	–	Dutch
296-297	14 May 1619	the elderly and nutrition	–	Latin
297	14-23 May 1619	variation in diet	–	Latin
307	4 June 1619	peritoneum	–	Latin
308	4-10 June 1619	pain in the limbs	–	Latin
310	4-10 June 1619	human state and size	–	Dutch
315-317	18 June 1619	estimating distance with one eye	–	Latin
339	16 September 1619	pain	Middelburg	Latin
339	16 September 1619	pain (*venijn)* and nerves	–	Latin
340	1 October 1619	blood	Middelburg	Latin
342	19 October 1619	illness and fluids	Veere	Latin
344	19 October 1619	unconsciousness	–	Latin
344-345	19 October 1619-7 November 1619	inflamed intestines and boiling water	Veere	Latin
348	16 November 1619	nutrition	Veere	Latin

Table 7 Light

Page	Date	Topic	Place	Language
6	18 July 1612	shining light through a small hole into a dark room	–	Dutch
10	18 July 1612	light	–	Latin
28	June 1613-April 1614	light reflected	–	Latin
327	11-23 July 1619	light and colours	–	Latin
329	23 July 1619	small hole and light beam	Middelburg	Dutch

Table 8 Mathematics

Page	Date	Topic	Language
4	18 July 1612	spherical trigonometry,	Latin
4-5	18 July 1612	(*swaerheyts middelpunt*) gravity centre point of a block	Dutch
5	18 July 1612	sinus	Dutch and Latin
5	18 July 1612	truncated pyramid	Dutch and Latin
9	18 July 1612	miles per degree	Latin
18	November 1612-March 1613	geometry and diameter	Latin
26	June 1613-April 1614	quadrature of a circle	Latin
34	12 April 1614	making big circles	Dutch
303-304	23 May 1619	flywheel ratio	Dutch

Table 9 Mirrors and Lenses

Page	Date	Topic	Language
12-13	18 July 1612	principal of the telescope	Latin
30	June 1613-April 1614	two parallel mirrors show more than just one	Latin
35	April 1614-January 1615	hollow paintings	Latin

Table 10 Meteorology

Page	Date	Topic	Place	Language
10	18 July 1612	(weight of the) sky	–	Latin
15	November 1612	how to find the source of the wind	–	Latin
15	November 1612	sources of rain	–	Latin
16	November 1612-March 1613	height/altitude of the clouds	–	Dutch
17-18	November 1612-March 1613	altitude of the clouds	–	Dutch
32	April 1614	wind breaking (*windbreken*)	–	Latin
291	29 April-2 May 1619	mist and clouds	–	Latin
313	10-15 June 1619	wind	–	Latin
315	17 June 1619	clouds	Noordgouwe	Latin
321	18-30 June 1619	low-hanging clouds	–	Latin
346	10 November 1619	direction of the wind	Middelburg	Latin

Table 11 Music and Sound

Page	Date	Topic	Place	Language
12	18 July 1612	canon, about French song	–	Dutch
18-19	November 1612-March 1613	two different kinds of chorus	–	Dutch
29	June 1613-April 1614	geometric proportion of tones	–	Dutch
48-50	April 1614-January 1615	music, fa	–	Dutch
50-51	April 1614-January 1615	notes in music	–	Latin
52	April 1614-January 1615	four music notes is enough	–	Latin
52	April 1614-January 1615	octaves	–	Latin
52	April 1614-January 1615	variation in songs	–	Latin
53	April 1614-January 1615	dividing sound into more strokes	–	Latin
53	April 1614-January 1615	harmony/consonance of the octave is best	–	Latin
54	April 1614-January 1615	fifth quint (music)	–	Latin
54	April 1614-January 1615	string vibrations (music)	–	Latin
55-56	April 1614-January 1615	major third (music)	–	Latin
56	April 1614-January 1615	psalms	–	Latin
58	April 1614-January 1615	harmonically better than perfect (music)	–	Latin
292	29 April-2 May 1619	tone system	–	Latin
293-294	2-14 May 1619	inharmonious psalms	–	Latin
305	2 June 1619	long tubes and low tones	–	Latin
307	2-4 June 1619	human voice	–	Latin
311	10-15 June 1619	mixed tonality	–	Latin
312	10-15 June 1619	flute	–	Latin
312	10-15 June 1619	flute holes	–	Latin
319	18-30 June 1619	singing	–	Latin
319-321	18-30 June 1619	music and fugue	–	Latin
321	18-30 June 1619	long tubes and sounds	–	Latin
323	11 July 1619	Gioseffo Zarlino	Middelburg	Latin
323-324	11-23 July 1619	music: fourth and third	–	Latin
324	11-23 July 1619	octave jump, even/uneven octaves	–	Latin
327	11-23 July 1619	music: half-tone	–	Latin
328	11-23 July 1619	music, sequence of notes	–	Latin
329	11-23 July 1619	limit of audibility with high and low tones	–	Latin
335	11 August 1619	mixed tonality	Rotterdam	Latin
337	28 August 1619	music and singing	Breda	Latin
337	29 August 1619	music, same tone	Breda	Latin
338	10 September 1619	music, psalms, cadence	Veere	Latin
340	6 October 1619	high voice	Serooskerke	Latin

Page	Date	Topic	Place	Language
343	19 October 1619	high-pitched voices carry further	–	Latin
348	20 November 1619	singing and percussion	Utrecht	Dutch
349	20-22 November 1619	fascination with language: notes and words	–	Dutch
350	22 November 1619	hymns and long notes	–	Dutch

Table 12 Practical and Personal

Page	Date	Topic	Place	Language
1	1604-1608	how ordinary people can judge about the meaning of the H. Scripture	–	Latin
1	1604-1608	how to stir up the study of sciences	–	Latin
4	18 July 1612	practical, about friendship and how to become better friends, life advice	–	Dutch
4	18 July 1612	habit can turn something bad into something not bad anymore	–	Latin
6	18 July 1612	practical, life advice about friendship	–	Dutch
10	18 July 1612	tolling bells, the biggest clock of Roanen	–	Dutch
15	November 1612	extinguish fire with salt	–	Dutch and Latin
15	November 1612	how to make glue	–	Dutch
18	November 1612-March 1613	change of value of money	–	Dutch
19	March-July 1613	written accents to signal emphasized syllable	–	Dutch
19-20	March-July 1613	useful division or organization of writings	–	Dutch
21	June 1613-April 1614	good year for cherries	–	Dutch
22	June 1613-April 1614	unboiled egg is lighter	–	Dutch
22	June 1613-April 1614	using a candle stub as bullet	–	Dutch
22	June 1613-April 1614	waking up at the same time in the morning	–	Dutch
40-41	April 1614-January 1615	dice	–	Dutch
299-302	14-23 May 1619	jammed door	Veere	Latin
308	4-10 June 1619	washing with salt or fresh water	–	Latin

Page	Date	Topic	Place	Language
311	10 June 1619	brook near Renesse	–	Latin
313	10-15 June 1619	finding rabbits	–	Latin
344-345	19 October 1619-7 November 1619	inflamed intestines and boiling water	Veere	Latin
322	30 June-3 July 1619	fascination with language: why is it that you can sing together but not speak together	–	Dutch
343	19 October 1619	spasms and a jammed door	–	Latin

Table 13 Pumps, Pipes, and Waterworks

Page	Date	Topic	Language
37	April 1614-January 1615	fountains, waterworks, improving pumps	Dutch
41-43	April 1614-January 1615	water pipes/waterworks	Dutch
45	April 1614-January 1615	airflows through pipes	Dutch
46	April 1614-January 1615	secretly speaking/writing through pipes/telescope	Dutch
46	April 1614-January 1615	rising water around a tower using pipes	Dutch
46-48	April 1614-January 1615	system for pumps	Latin
48	April 1614-January 1615	siphon	Latin
61	February-March 1615	unclogging water pipe	Latin
62	March 1615	attaching pipes to a wall	Dutch
64	March 1615-February 1616	long pumps, mistakes in pumps	Dutch
65	March 1615-February 1616	how to use uneven laid pipes	Dutch
65	March 1615-February 1616	wide pumps and pipes	Dutch
66	March 1615-February 1616	how to keep pumps wet	Dutch
66	March 1615-February 1616	improving a piston pump	Dutch
66	March 1615-February 1616	pipes of stone are better than pipes of led in soft earth	Dutch
67	March 1615-February 1616	making pipes with different substances/matters	Dutch
293	2-14 May 1619	fountain using a pump activated by people walking by	Dutch
298	14-23 May 1619	making a fountain pump	Dutch
327	11-23 July 1619	water pipes	Latin
325-327	11-23 July 1619	slope and water pipes	Latin

Table 14 Vacuum and Pressure

Page	Date	Topic	Language
22	June 1613-April 1614	bleeding (*aderlaten*) with cupping glass	Latin
22	June 1613-April 1614	heat in cupping glass	Latin
23	June 1613-April 1614	vacuum exists, it is proven with motion, vacuum because of air streams	Latin
24-25	June 1613-April 1614	stone in vacuum	Latin
26	June 1613-April 1614	movement and vacuum	Latin
36	April 1614-January 1615	vacuum and pressure, how do things move in nature without vacuum?	Latin
36	April 1614-January 1615	magnet and pressure, how does a magnet pull iron?	Latin
37	April 1614-January 1615	suction cups, something hollow can stick to the wall/ceiling	Latin
39	April 1614-January 1615	little ball on airflow	Dutch

Table 15 Water, Wells, and Caves

Page	Date	Topic	Place	Language
6	18 July 1612	the tide in a barrel of water	–	Dutch
8	18 July 1612	water in caves	–	Latin
14	November 1612	water wells	–	Dutch and Latin
58	1615	size and weight of water, the course of water through a hole	–	Dutch
60	January and February 1615	warmth in a well	–	Dutch
329	25 July 1619	water wells (2)	–	Latin
338	8 September 1619	water wells (3)	Middelburg	Dutch

17 'Ut patet in figura'

On the Use of Images in Beeckman's *Journal**

Klaas van Berkel

Abstract
Isaac Beeckman's notebook includes some 240 drawings made by Beeckman himself, but thus far no attention has been paid to them. Most of these drawings are indeed illustrations that support the argument in the text of the notebook, but in a couple of cases the drawings do more than that and replace the argument; they do not illustrate the text, but are the argument itself, the text serving as illustration to the picture. These pictures document Beeckman's visual way of thinking and reveal that his mechanical philosophy is in part the product of a realistic interpretation of illustrations found in the work of Simon Stevin, especially his picture of the *clootcrans*, or wreath of spheres.

Keywords: Isaac Beeckman, drawings, pictorial argument, Simon Stevin, wreath of spheres

Isaac Beeckman included around 240 images of his own making in his *Journal*. In addition, there are some musical scores and images taken from books studied or referred to by Beeckman and inserted in the *Journal* by its editor, Cornelis de Waard.[1] In itself, the number of 240 images is not exces-

* This chapter is a revised and annotated version of a paper written for the conference, 'How Science Makes Sense', held on 17 November 2005 in Amsterdam on the occasion of the presentation of the 2005 Erasmus Prize to Simon Schaffer and Steven Shapin. I thank Fokko Jan Dijksterhuis, Arjan van Dixhoorn and Eric Jorink for their comments on an earlier draft.

1 De Waard also redrafted several figures, in most cases because the original drawings were too vague or too crude. In his 'Note sur le manuscrit', he does not discuss the images in the notebook. *Journal tenu par Isaac Beeckman de 1604 à 1634, publié avec une introduction et des notes par C. de Waard*, 4 vols. (The Hague: Martinus Nijhoff, 1939-1953) [henceforth *JIB*], I, p. xxxiii. Whether the images in Beeckman's manuscript were really his own or were redrawn

Van Berkel, Klaas, Albert Clement, and Arjan van Dixhoorn (eds.), *Knowledge and Culture in the Early Dutch Republic: Isaac Beeckman in Context*. Taylor & Francis Group 2022
DOI: 10.5117/9789463722537_CH17

sively high. The printed *Journal* has some 1,270 pages, so, on average, there is an image on every fifth page of the book. Nevertheless, it is a substantial number, and it is therefore strange that Beeckman's images have not been the subject of scholarly study before.[2] All the more so since we know that Beeckman himself had a strong preference for *Anschaulichkeit* (picturability), both in the context of discovery and in the context of the dissemination of natural philosophical ideas. Beeckman only accepted explanations that could be represented by a real or mental image. Until the 1980s, historians of science in general were inclined to overlook the presence of images in texts and manuscripts. Since then, however, the importance of these visual tools has become widely recognized.[3] It is therefore about time to ask what kind of images Beeckman used in his notebook, what these images were used for and, most importantly of all, what they tell us about the development of his mechanical philosophy. I will argue that these images are not merely illustrations of the text, but form an integral part of the argument that Beeckman wishes to make.

Picturing the New Science

Today, it is hard to imagine science without the use of images. Computer simulations, mathematical diagrams, three-dimensional models, photographs, graphs – there is no field of science that can do without these visual resources, which have become indispensable both in teaching and research. We live in a visual culture and science is no exception.[4]

by a draftsman or by the same secretary that transcribed Beeckman's rough notes in the neat version of the manuscript remains to be studied. This makes it difficult to establish whether the images show some development in Beeckman's way of picturing his ideas. See also footnote 12.

2 The cover of my Dutch dissertation, *Isaac Beeckman (1588-1637) en de mechanisering van het wereldbeeld* (Amsterdam: Rodopi, 1983), features one of Beeckman's images of a perpetuum mobile, but I treated this picture simply as an illustration to decorate the cover of the book and not as a part of an argument Beeckman was making.

3 Brian S. Baigrie, ed., *Picturing Knowledge: Historical and Philosophical Problems Concerning the Use of Art in Science* (Toronto: University of Toronto Press, 1996); Wolfgang Lefèvre, Jürgen Renn, and Urs Schoepflin, eds., *The Power of Images in Early Modern Science* (Basel: Birkhäuser, 2003); Sachiko Kusukawa and Ian Maclean, eds., *Transmitting Knowledge: Words, Images, and Instruments in Early Modern Europe* (Oxford: Oxford University Press, 2006); Christoph Lüthy and Alexis Smets, 'Words, Lines, Diagrams, Images: Towards a History of Scientific Images', *Early Modern Science and Medicine* 14 (2009), pp. 398-439.

4 Klaus Hentschel, *Visual Cultures in Science and Technology* (Oxford: Oxford University Press, 2014).

The omnipresence of images in present-day science can easily make us forget that in the past the use of pictures was not always as common as it is today. For centuries, the study of nature could easily do without the use of pictures. The works of Aristotle included no illustrations and even in medieval manuscripts mathematical diagrams were far from common. It was only in Renaissance Europe that pictures became a common element in diverse fields such as anatomy, botany and mechanics. In the field of alchemy or 'chymistry', images also played an increasingly important role. The invention of the printing press, especially, greatly enhanced the possibilities for the dissemination of stable and standardized images. In the sixteenth and seventeenth centuries, mathematicians, physicians and chemists quickly became used to supporting their arguments with visual artefacts. Any reader of Andreas Vesalius's *De humani corporis fabrica* (1543), for example, will look for the argument in the beautiful drawings, not in the accompanying texts.[5]

In natural philosophy, the situation was somewhat different, as this was a discipline that did not focus on the outward appearances of natural objects, but on their essence and on the often hidden causes of natural phenomena. Aristotelian handbooks of natural philosophy could still dispense with images deep into the seventeenth century. In the last quarter of the sixteenth century, Giordano Bruno was perhaps the only natural philosopher who extensively employed images (crude woodcuts that he made himself) to support his ideas about the constitution of the world.[6] Nevertheless, even natural philosophy caught up with the general trend towards more visualization in science and scholarship (including such fields as the study of antiquities and numismatics) and by the mid-seventeenth century, sophisticated images became a normal feature of treatises on non-Aristotelian natural philosophy. The works of Descartes feature dozens of, at times, highly elaborate pictorial representations of natural philosophical concepts and ideas.[7] Both the *Essais* that accompanied his *Discours de la*

5 Sachiko Kusukawa, *Picturing the Book of Nature: Image, Text, and Argument in Sixteenth-Century Human Anatomy and Medical Botany* (Chicago: University of Chicago Press, 2012); Florike Egmond, *Eye for Detail: Images of Plants and Animals in Art and Science, 1500-1630* (London: Reaktion Books, 2017).

6 Christoph Lüthy, 'Centre, Circle, Circumference. Giordano Bruno's Astronomical Woodcuts', *Journal of the History of Astronomy* 41 (2010), pp. 311-327. See also: Christoph Lüthy, 'Bruno's *Area Democriti* and the Origins of Atomist Imagery', *Bruniana et Campanelliana* 4 (1998), pp. 59-92.

7 Klaus Zittel, *Theatrum philosophicum. Descartes und die Rolle ästhetischer Formen in der Wissenschaft* (Berlin: Akademie Verlag, 2009). It is important to note that Descartes was very much stimulated to add illustrations to his printed works by people around him such as Jacob Golius, Constantijn Huygens and Frans van Schooten.

méthode (1637) and his *Principia philosophiae* (1644) are full of carefully executed images of a rainbow, vortices, globules and magnetic particles. As Christoph Lüthy amply demonstrated, these visual images were crucial to convince the reader of the plausibility of concepts and ideas that were in fact contrary to the logic of Descartes' metaphysics.[8] Especially his picture of the mechanism of magnetism, with its fusilli-shaped *particula striatae* (inflexible particles that are impossible in an ontology that defines matter by mere extension), has become an iconic representation of mechanical philosophy. One of the most famous textbooks in the early modern history of science, Richard Westfall's *The Construction of Modern Science* (1971), has this very image on its front cover.

Considering that Descartes, to a certain extent, derived the inspiration for his mechanical philosophy from Beeckman, it might be interesting to understand how Beeckman used images in his notebook. Descartes had been privileged in 1628-1629 to see Beeckman's notebook and he would certainly have seen Beeckman's drawings. At the time, Beeckman was deeply absorbed in the works of Johannes Kepler, especially his *Epitome astronomiae Copernicanae*, which has the same type of illustrations as Beeckman's manuscript.[9] Like Kepler, Beeckman was fascinated with the behaviour of the spinning top, a toy that to him proved that an object (such as planet Earth) could have two or even three motions at the same time.[10] Beeckman may even have told Descartes that he was considering writing a book with more or less the same format as the *Epitome*, which Kepler had set up as a series of questions and answers to introduce his pupils at the *Landschaftsschule* in Linz to his physical cosmology.[11] In this book, which Beeckman was certainly thinking about, he would substitute Kepler's

8 Christoph Lüthy, 'Where Logical Necessity Becomes Visual Persuasion: Descartes' Clear and Distinct Illustrations', in: Kusukawa and Maclean, eds., *Transmitting Knowledge*, pp. 97-133. Eric Jorink has pointed out that the fact – not discussed by Lüthy – that the illustrations in Descartes' works were not put together in a separate section at the end of the book but spread through the actual text is essential. Only in this way could the pictures acquire the argumentative force they now have. See: Eric Jorink, *'Geef zicht aan de blinden'. Constantijn Huygens, René Descartes en het Boek der Natuur* (Voorburg: Huygens Museum Hofwijck, 2008), pp. 31-34.

9 Johannes Kepler, *Epitome astronomiae Copernicanae, usitata forma quaestionum & reponsionum conscripta* (Linz: Johannes Plancus, 1618), in: Johannes Kepler, *Gesammelte Werke*, ed. Walther von Dyck and Max Caspar, 22 vols. (Munich: Beck, 1937-2017), VII (1953). On the illustrations in Kepler's *Epitome*, see: Isabelle Pantin, 'Kepler's *Epitome*: New Images for an Innovative Book', in: Kusukawa and Maclean, eds., *Transmitting Knowledge*, pp. 217-237.

10 *JIB*, III, pp. 115-121, esp. p. 118 'Terrae motus cum turbinis motibus collati.'

11 In his dedication to the higher nobility of Austria, Kepler said the book was meant 'Nobilissimam vestram juventutem ore-tenus [orally] in hac scientiâ institutendi.' Kepler, *Epitome*, p. 8 (= fo. 2^{recto}). Kepler taught mathematics, philosophy and history in Linz from 1611 to 1626.

magnetic explanations of the movements of the Sun and the planets with his own purely corpuscular and mechanical explanations. Even if it cannot be established whether Descartes took an interest in Beeckman's images, it is worthwhile establishing how Beeckman visualized his ideas and concepts, and where he drew his inspiration from. While Beeckman himself did not reflect on his use of images, the way he used them can tell us something about his natural philosophy.

Devising Machines and Instruments

Of the 240 images in Beeckman's notebook, some are very small and some are very elaborate. As is evident from the images reproduced here, Beeckman had no special talent for drawing. As far as we know, he had no formal training in drawing (nor had Descartes, who complained about it), unlike Christiaan Huygens half a century later. Most of Beeckman's drawings are rather crude, and some are even childlike. Beeckman was no Galileo, who, according to Horst Bredekamp, was 'a draftsman of the first rank' and whose drawings can be considered art.[12] Nevertheless, it is instructive to see what kind of images Beeckman used to illustrate, clarify or prove his ideas. It is possible to broadly distinguish between geometrical and physical-mathematical diagrams, pictures of machines (real and imaginary), of hydrological instruments (involving flowing water or dripping quicksilver), pictures illustrating physical principles and, finally, natural philosophical images, visualizing things that cannot be seen but are merely postulated. By discussing these images, it becomes evident that they may have had very different functions: some are demonstrative, others attempt to represent objects that Beeckman saw, heard about or conceived of, while others play a more heuristic role, as if Beeckman was trying to come to grips with new ideas by drawing a picture.

Of all the images, only a handful are purely geometrical diagrams. Beeckman was not really a mathematician and his notes are only seldom devoted to strictly mathematical problems. Much more common are images that are

Beeckman may have seen Kepler as a kindred spirit, both of them teaching at a local secondary school.

12 Horst Bredekamp, *Galileo's Thinking Hand: Mannerism, Anti-Mannerism, and the Virtue of Drawing in the Foundation of Early Modern Science* (Berlin: De Gruyter, 2019), p. vii. Beeckman was also no Antoni van Leeuwenhoek, who made drawings himself but also leaned heavily on the work of several draftsmen. See: Sietske Fransen, 'Antoni van Leeuwenhoek, His Images and his Draughtsmen', *Perspectives on Science* 27 (2019), pp. 485-544.

Figure 17.1 The mechanics of a spinning top

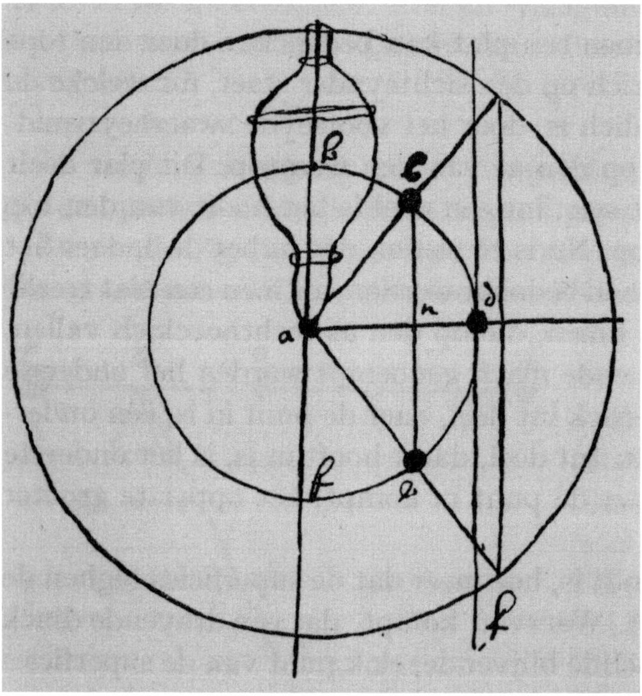

From: *JIB*, I, p. 32 [April 1614]. In this note Beeckman discusses what will happen to the spinning top when its centre of gravitaty (b) is no longer straight above de ground at a.

Figure 17.2 Light of the Sun refracted by clouds around the Earth

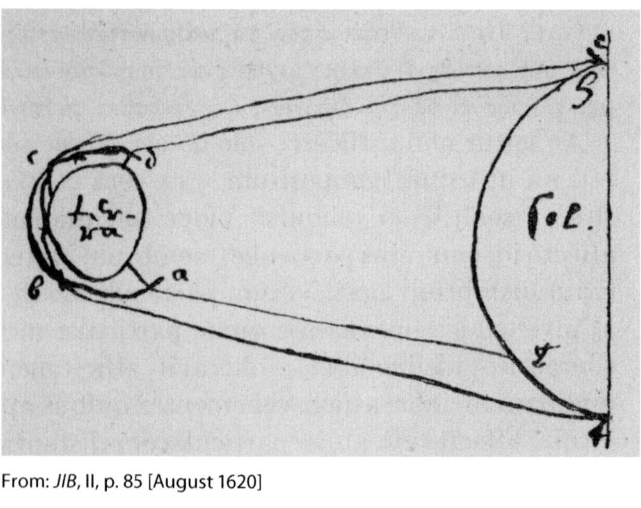

From: *JIB*, II, p. 85 [August 1620]

in themselves not mathematical, but in which the lines and circles directly refer to objects and configurations in the physical world – the mechanical, astronomical, optical or, in general, mathematical-physical figures, such as the images of the movement of a spinning top (fig. 17.1).[13] Another drawing shows how the light of the Sun is refracted in a cloud surrounding the Earth (fig. 17.2),[14] and a third example concerns the convergence of rays of light in a telescope (fig. 5.3).[15]

Even more common than the physical-mathematical figures are the drawings of machines and mechanical devices, real or imaginary. Beeckman's father was a craftsman; he earned his living as a candle maker, but he also constructed conduit pipes in gardens and breweries. His son also did so, at least in the years before he studied medicine and became a teacher at several prestigious grammar schools in Utrecht, Rotterdam and Dordrecht. Even after he had become a schoolmaster, Beeckman was regularly asked to give his advice on matters that related to technical inventions, from ways of deepening a harbour to the construction of a perpetuum mobile machine. A large number of images in the *Journal* testify to this continuing interest in mechanical engineering. On visiting his uncle, Jan Pietersz. van Rhee, in 1612, Beeckman heard about an unspecified device for drawing water from a well with a rope that was half as long as the well was deep, and he immediately thought out a construction that showed how it might be done. He simply made a drawing, formulated the mechanical rules that applied to the case and concluded with the words 'ut patet in figura', that is, 'as is shown in the figure', as if the picture spoke for itself (fig. 17.3).[16] The small hand in the picture that holds the rope is, of course, not Beeckman's own artistic invention, but a visual quotation from the textbooks in mechanics that abound with such hands. They can be seen in the works of Simon Stevin, such as in his *Beghinselen der Weeghconst* (The Art of Weighing) from 1586, with which Beeckman was familiar.[17]

Beeckman was, so it seems, less familiar with the Renaissance tradition of machine albums, in some of which we find, according to Paolo Galluzzi, a 'revival of interest in technical issues of original discussions on the very concept of machine, and of the development of new graphic conventions

13 *JIB*, I, p. 32 [April 1614].

14 *JIB*, II, p. 85 [August 1620].

15 *JIB*, II, p. 209 [July-October 1622].

16 *JIB*, I, p. 14 [October 1612].

17 Simon Stevin, *De Beghinselen der Weeghconst* (Leiden: Chr. Plantijn, 1586), pp. 2, 4-5, 9-13 et passim. For Beeckman's familiarity with this book, see: *JIB*, I, p. 4, 112 et passim.

Figure 17.3 Drawing water from a well with a short rope

From: *JIB*, I, p. 14 [October 1612]

for effectively representing its structure, components and functions'.[18] Beeckman does not refer, for example, to Jacques Besson's *Theatrum instrumentorum et machinarum* (1588), Agostino Ramelli's *Diverse et artificiose machine* (1571-1572) or to books by other authors which often included both very realistic depictions of machines and designs of machine constructions that were patently impossible.

18 Paolo Galluzzi, *The Italian Renaissance of Machines* (Cambridge, Mass.: Harvard University Press, 2020), p. vii.

Figure 17.4 Bucket and vessel with water

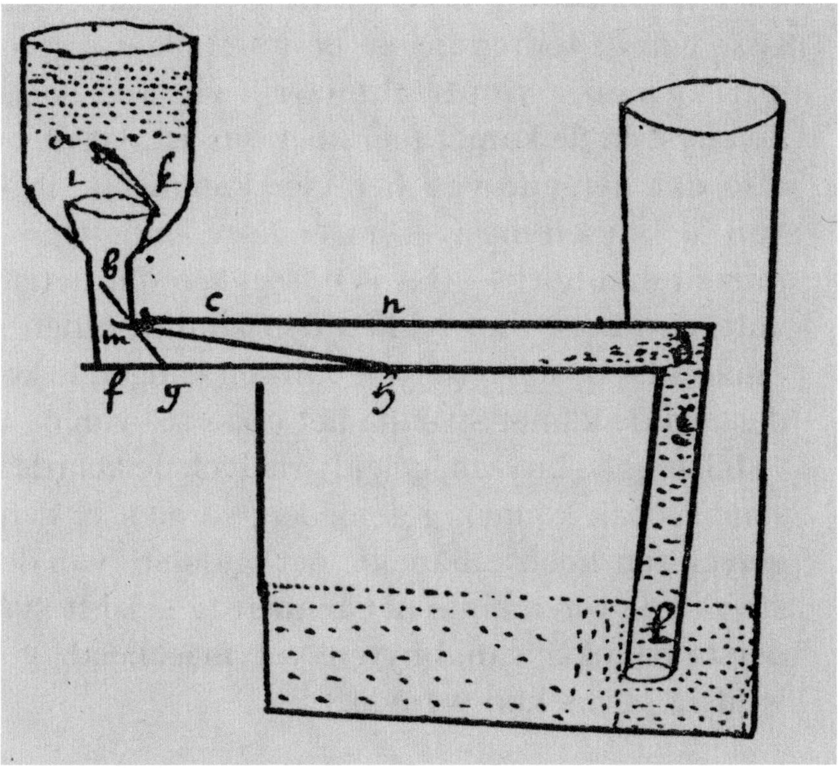

From: *JIB*, I, p. 46 [April 1614-January 1615]

Often, the contrivances are much more complicated. Not surprisingly, many of these devices were equipped with pumps and pistons, like the imaginary vessel or bucket (*haustrum*) that is connected to a pipe and a reservoir full of water (fig. 17.4).[19] Sometimes these machines were purely fictitious, such as the perpetuum mobile constructions he conceived over the years.[20] Although Beeckman believed that a perpetuum mobile was impossible in the real world, he indulged in working out machines that imitated a perpetuum mobile. In 1614, Beeckman was so pleased with the perpetuum mobile machine he had devised that he reserved his notes for

19 *JIB*, I, p. 46 [April 1614-January 1615]. The *Oxford Latin Dictionary* only cites one place in classical Latin for *haustrum* – Lucretius's *De rerum natura*, 5.516, where it means the scoop on a waterwheel. Is this yet another indication of the formative influence of Lucretius on Beeckman?
20 See, for example: *JIB*, I, p. 39 [April 1614]. See also the chapter by Fokko Jan Dijksterhuis in this volume.

Figure 17.5 ***Urelooper*** **or clepsydra with quicksilver, designed by Beeckman**

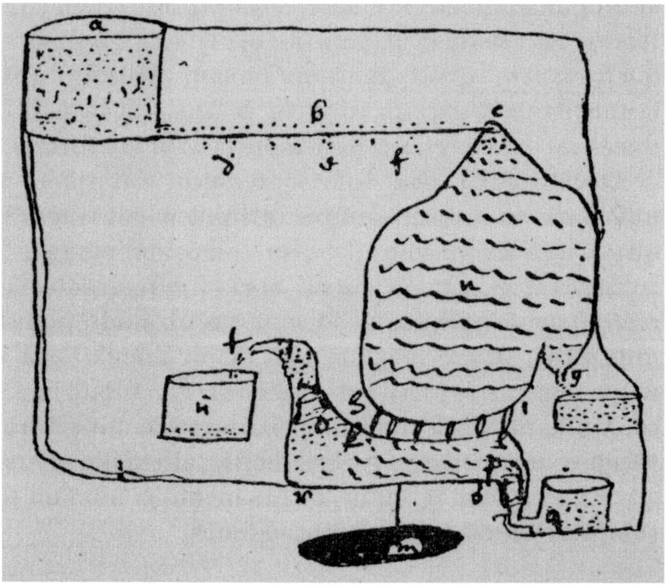

From: *JIB*, I, p. 48 [April 1614-January 1615]. The idea of a clepsydra (water clock) goes back to antiquity, but Beeckman was especially inspired by the quicksilver clepsydra discussed by Tycho Brahe in his *Astronomiae instauratae progymnasmata* (Prague, 1602). See: *JIB*, I, p. 33 [April 1614].

Figure 17.6 Balance to find the maximum speed of a falling object

From: *JIB*, I, p. 268 [26 December 1618]. Beeckman called the point at which an object reaches its maximum speed when falling the *punctum aequalitatis*. He discussed it shortly after discussing with Descartes the law of free falling bodies.

a book he intended to publish around 1630, which was, however, only published posthumously by his brother Abraham: the *Mathematico-physicarum meditationum, quaestionum, solutionum centuria* (1644).[21]

Not all water instruments contained pumps and pistons. There is, for example, an interesting instrument with which Beeckman hoped he could solve the problem of measuring longitude, one of the most challenging nautical problems in a seafaring nation. In 1614, he devised a timepiece (*urelooper*) which would allow the captain of a ship to count the hours that had passed since the ship had left the harbour (fig. 17.5).[22] This was an intricate system of reservoirs filled with quicksilver; however, the accompanying note is barely sufficient to understand how the instrument was supposed to function: actually the volume of quicksilver dripping from *g*, to the right, is the measure of the time that has passed since the ship sailed from the harbour. Either Beeckman drew the picture before making the note (and forgot to be explicit on every detail) or he had the idea that the figure would in some sense be self-explanatory, at least to him.

With this timepiece, we have arrived at another category of images, those of instruments connected with his principles of natural philosophy, either by them demonstrating certain ideas or by illustrating the ideas of others. A simple example is the balance with which Beeckman hoped to be able to determine the point at which an object reached its maximum speed falling towards the Earth (fig. 17.6).[23] Beeckman believed that the acceleration of a falling object would be neutralized at some point by the growing resistance of the air. In order to measure this point, he devised a pair of baskets. The force of one stone falling into one of the baskets would eject a second stone in the other basket. By varying the height from which the first stone was dropped and keeping track of the height the ejected stone reached, until it no longer increased, Beeckman wanted to measure this 'point of equality'.

One man who was very skilled in devising machines and instruments was Cornelis Drebbel (1572-1633), a celebrated figure with whom Beeckman was fascinated throughout his life.[24] While he never met Drebbel, each time he heard something about him, he made some kind of note in his manuscript, and so the *Journal* includes several drawings illustrating the

21 *JIB*, I, p. 39 [April 1614]. The drawing in the manuscript was so crude compared to the figure in the posthumous *Centuria*, in which this note was included, that De Waard reproduced the 1644 figure in his edition of the *Journal*, although it is unknown who drew this figure. It could be Beeckman, his brother Abraham or some unknown artist. See also footnote 1.

22 *JIB*, I, p. 48 [April 1614-January 1615].

23 *JIB*, I, p. 268 [26 December 1618].

24 See also the contributions of Fokko Jan Dijksterhuis and Vera Keller in this volume.

Figure 17.7 Beeckman's idea of Drebbel's submarine

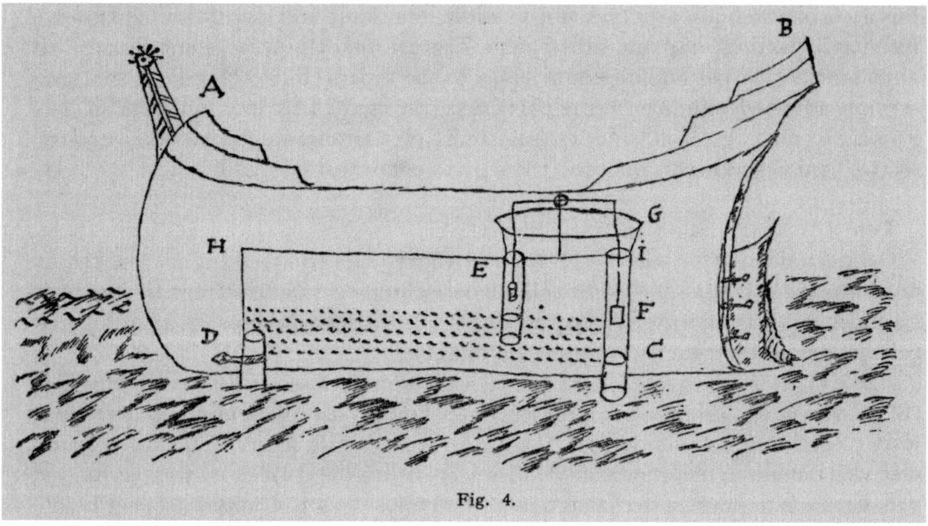

From: *JIB*, II, 26 [March 1620]

Figure 17.8 Drebbel's and Beeckman's design of a weather glass

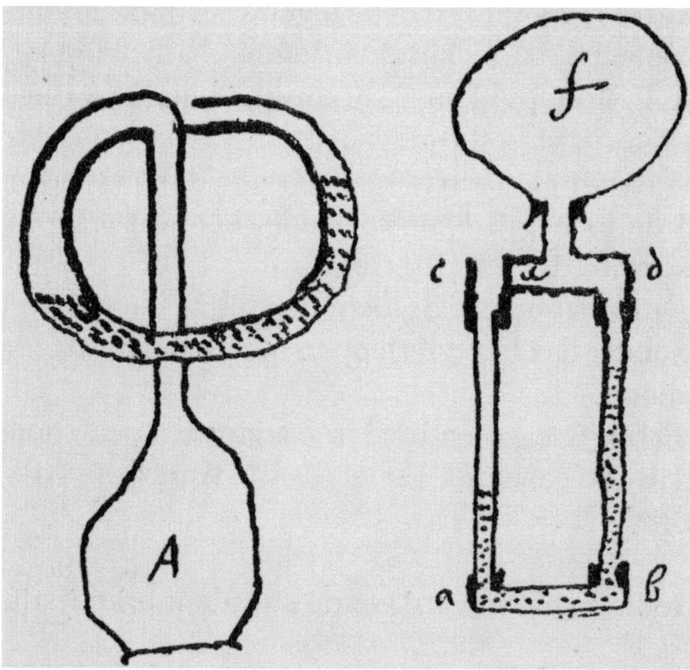

From: *JIB*, II, p. 372 [September-November 1626]. Drebbel's thermoscope is to the left, Beeckman's improved design to the right.

Figure 17.9 Beeckman's drawing of Drebbel's microscope and a calibrated weather glass

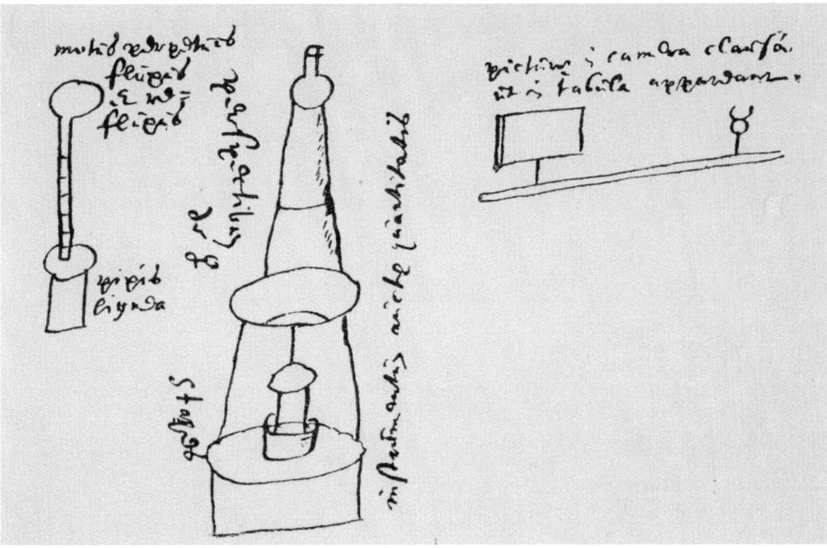

From: *JIB*, III, 442 [March 1631]

bits of news Beeckman had received about Drebbel. There is, for example, a realistic-looking depiction of Drebbel's submarine; at least of the submarine Beeckman imagined Drebbel might have built (fig. 17.7).[25] Beeckman was also very curious about what was called the 'thermoscope', presented by Drebbel as an instrument that imitated the ebb and flow of the sea, but interpreted by Beeckman as an instrument that registered differences in temperature (fig. 17.8).[26] The figure to the left is Beeckman's depiction of what he thought Drebbel's instrument would have looked like, the figure to the right represents how Beeckman thought he could improve the instrument. Finally, we owe the first picture of the microscope (fig. 17.9) to Beeckman's interest in Drebbel's instruments. It was copied (in 1631) from a letter from Drebbel to King James I written in 1613.[27] We should not fail to notice that Beeckman also copied another thermoscope to the left and a camera obscura to the right.

25 *JIB*, II, p. 26 [15 March 1620]. Beeckman received the news about the submarine through a letter from his father, Abraham, in Middelburg, and De Waard suggests that Abraham Beeckman had been informed by some nephews who lived in London.
26 *JIB*, II, p. 372 [September-November 1626].
27 *JIB*, III, p. 442.

Figure 17.10 Van der Veen's theory of the hollow Earth illustrated by Beeckman

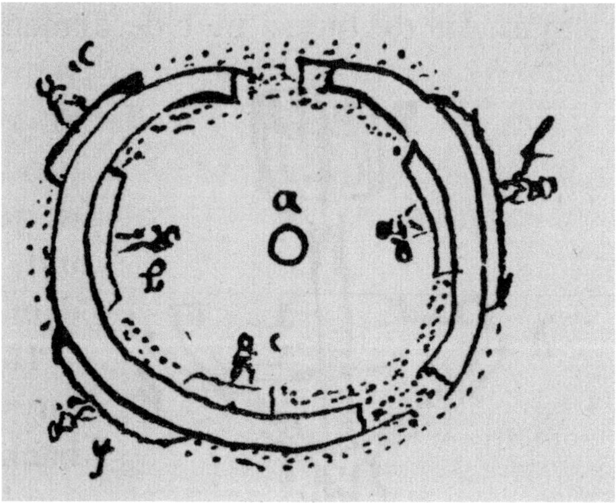

From: *JIB*, II, 389 [4 March 1627]

The images discussed thus far are not in any way exceptional. However, they do testify to Beeckman's need to create a visual representation of the objects he discussed in his notebook. He may not have been particularly talented as an artist, but he certainly liked to make drawings. An example of this is another perpetuum mobile attached to a wheel of fortune (fig. 14.3).[28] The construction itself is hardly new, but his appetite for drawing must have inspired him to create this picture. That he liked drawing can also be inferred from the fact that he regularly did drawings illustrating the strange ideas he had heard about. One example is his rather clumsy sketch dating from March 1627 of the world-picture of one of his acquaintances, the miller Balthasar van der Veen (or Balthasar Jacobs) from the nearby city of Gorcum (fig. 17.10).[29] Van der Veen argued that inside the Earth there is another world, in which people live just like us on the outer surface. He also maintained that our world and this other world are connected through two holes at the bottom of the sea.

In the image of the world according to Van der Veen, several letters can be seen that are not discussed in the accompanying text. For example, there are letters indicating people on the inside and on the outside, while the letter *a* indicates a central Sun in the world that is inside the Earth. The fact that

28 *JIB*, II, p. 200 [27 May 1622].
29 *JIB*, II, p. 389 [4 March 1627].

these letters are not referred to in the text again suggests that Beeckman first made the drawing and afterwards forgot to explain what these letters were meant to indicate. In a sense, therefore, this is not a mere illustration, in the sense of an image summarizing what is said in the text. Rather, in this case, we have a text giving a partial explanation of what can be seen in the picture.

There are more examples of the primacy of the pictures over his text. As a rule, pictures in the *Journal* illustrate a specific idea. In the text, Beeckman then uses phrases such as, 'as you can see in the picture below' or 'as you can see here'; quite often he also uses phrases derived from the reasoning applied in mathematics and mechanics: 'Let *A* be the Earth', or 'Let *ab* be a ray of light'. However, now and then the relationship between the words and the images is reversed; for example, when discussing Drebbel's perpetuum mobile, Beeckman says: 'In the preceding figure it is argued, among other things, that [...]'.[30] According to De Waard, who footnotes this remark, this was simply a slip of the pen: Beeckman did not mean to refer to the 'figure' but to the text. However, it is much more likely that in Beeckman's mind, pictures had acquired a value of their own. *Pictures became arguments; picturing had become a way of reasoning.*

Picturing Natural Philosophy

The most intriguing figures in Beeckman's *Journal* are undoubtedly the images one might categorize as natural philosophical. A particularly interesting note concerns Beeckman's explanation of the phenomenon of the refraction of light. In this note, dating from 1618, Beeckman added a figure in which he represented a ray of light by a chain of small round particles (fig. 17.11).[31]

In this figure, the line *ab* represents the surface of water (or a piece of glass), while *dc* is a ray of light. Beeckman discusses what will happen to the ray of light when it enters (or leaves) the water. From experience, he knew that the ray of light would not continue in a straight line, but he was curious about the causes of this phenomenon. In line with his corpuscular philosophy of nature, according to which light consisted of small particles or *globuli*, he represented the beam of light not by a line, as in geometrical optics, but by a chain of small round circles. These tiny circles were indeed intended as a pictorial representation of the atoms of light and not some abstract or purely symbolic representation, which would have been the

30 *JIB*, II, p. 205 [2 June 1622]: 'In de voorgaende figuere is onder anderen geseydt, dat [..].'
31 *JIB*, I, p. 211 [August-September 1618].

Figure 17.11 Refraction of light according to Beeckman

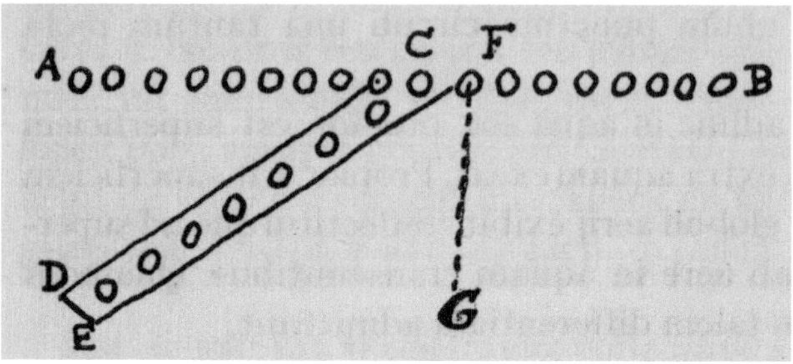

From: *JIB*, I, p. 211 [August-September 1618]

case if he had used simple dots or a line. In this explanation of the cause of refraction, the mechanical properties of these atoms of light were of real significance. The fact that on passing the boundary between water and air, one side of the particle of light touches the new medium earlier than the other side of the particle was, to Beeckman's mind, the most important factor in explaining the behaviour of the beam of light. His image, therefore, was not simply meant as a fictional aid to understand the phenomenon. It was rather a serious representation of what was really going on, although we are unable to actually see the atoms themselves.

In this context, another image is worth considering. It is a picture drawn by Beeckman in the spring of 1623. In the accompanying note, he discusses the pressure exerted by water in a vessel, referring back to the first of the *corrolaria* in his 1618 thesis in Caen on air pressure as the explanation of the *fuga vacui*. When discussing the behaviour of water, he had always illustrated these reflections with figures picturing water as an amorphous mass represented by little dots or short lines or waves (see e.g. fig. 17.4).[32] However, in the figure accompanying the 1623 note, the form of representation is different. Instead of representing water as an amorphous mass, he now draws a picture in which the water directly adjacent to the sides of the vessel is represented by a chain of little round particles (*globuli*) (fig. 17.12).[33] First, he imagines these particles to be connected to each other ('si inter se forent connexi'), but later he states that it is irrelevant whether these particles are really linked to each other or not. The pressure they would exert would be the same even if they were simply lying next to each other.[34]

32 See also figs. 17.5, 17.7 and 17.8.

33 *JIB*, II, p. 236 [February-April 1623].

34 *JIB*, II, p. 236; 'Jam vero solum contigui existentes, non minus afferunt ponderis fundo quam ante.'

Figure 17.12 Water in a vessel represented as a chain of *globuli*

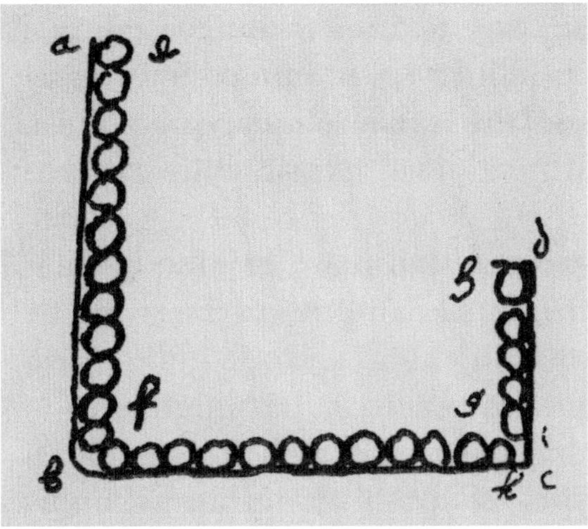

From: *JIB*, II, p. 236 [February-April 1623]

Anyone familiar with the history of sixteenth- and seventeenth-century mechanics will note the parallel here with Simon Stevin's *Weeghconst*; more specifically, with the *clootcrans*, or wreath of spheres. Stevin deals, among other things, with oblique forces and gives his own proof of the law of the inclined plane. He discusses what will happen to a chain of connected equal spheres hanging on a triangle with unequal sides and concludes that since a perpetuum mobile is absurd, this chain will be at rest, even though there are four balls to the left and only two balls on the right side of the upper angle of the triangle. Stevin was so proud of this proof that he used his illustration of it as an emblem in the great majority of his publications, its caption the now famous words: 'Wonder en is gheen wonder' ('Wonder is no wonder') (fig. 17.13).[35] Stevin also used it as a seal for his letters and as a mark on his mechanical inventions. There is perhaps no image in early modern mechanics that has been so widely disseminated as this wreath of spheres.

Beeckman was well acquainted with the work of Stevin. He had read his books, studied the manuscripts left by Stevin after his death, and in one of his notes he uses (without mentioning him) Stevin's motto 'Wonder is no wonder'

35 For a clear exposition of the *clootcrans* proof of the law of the inclined plane (and its inadequacy), see: E.J. Dijksterhuis, *The Mechanization of the World Picture* (Oxford: Clarendon Press, 1961), pp. 326-327.

Figure 17.13 Title page of Stevin's *Beghinselen der Weeghconst*, 1586

DE
BEGHINSELEN
DER WEEGHCONST
BESCHREVEN DVER
SIMON STEVIN
van Brugghe.

Tot Leyden,
Inde Druckerye van Chriſtoffel Plantijn,
By Françoys van Raphelinghen.
cIↃ. IↃ. LXXXVI.

Stevin used the vignet with the wreath of spheres also in his *Wisconstige Gedachtenissen* (Leiden 1608).

explicitly.[36] By picturing light and water as chains of little round particles Beeckman gave a natural philosophical interpretation of a figure that, in Stevin's work, had a purely mechanical meaning. In depicting the little round balls as representing the atoms of water and light, Beeckman crossed the line that separated mechanics from natural philosophy. This is the 'mechanization of the world picture' in a nutshell, to use Eduard Jan Dijksterhuis's famous phrase. Whereas Stevin had worked with balls that could in principle be seen without difficulty, Beeckman's little *globuli* could only be imagined; his atoms were too tiny to be seen and one had simply to accept his assumption that matter consists of little particles with purely mechanical properties. What is more

36 *JIB*, II, p. 375 [November-December 1626]: 'In philosophy, one always has to proceed from wonder to no wonder, that is to say, one has to investigate [a thing] until what was strange to us, is not strange anymore.' In June 1624, Beeckman had consulted Stevin's papers at the home of his widow, who lived in Hazerswoude, a village near Leiden.

important, however, is that Beeckman used these little round balls to *explain* natural phenomena, whereas Stevin only used them to *describe* a phenomenon. Beeckman was an atomist and Stevin probably not, but in their way of picturing the behaviour of material objects the resemblances are obvious.

This is not to suggest that the roots of Beeckman's mechanical philosophy are simply to be found in his reading of sixteenth-century mechanical treatises. His atomism goes back to his reading of the ancient atomists, such as Lucretius and Democritus, while his outspoken preference for concepts that could be visualized (at least in the mind) was very much inspired by the philosophy of Petrus Ramus, the sixteenth-century French mathematician and educational reformer.[37] Furthermore, his experience as a mechanical practitioner, who had to use crude outlines for his constructions of water conduits, may have familiarized him with the use of illustrations and drawings.

Nevertheless, the fact that he could visualize his corpuscular theory of matter with similar imagery to that used by Stevin in his mechanical works a few decades earlier, added to the credibility of his mechanical philosophy. The same step was taken by Descartes in 1637 and 1644, when he also used the convention of visualizing concepts and ideas used in books on mechanics to strengthen the persuasive force of his ideas about the constitution of the material world. Descartes is a good example of how a natural philosopher could impress his readers just as much with his verbal arguments as with his way of visualizing his ideas.[38] This, of course, does not imply that a figure can only have one specific meaning or only serve one purpose; an image, as is shown in the case of Stevin and Beeckman, can acquire new meanings and new purposes when used in a new context. After all, even images do not speak for themselves.

About the Author

KLAAS VAN BERKEL is Emeritus Rudolph Agricola Professor of History at the University of Groningen. He has published widely on cultural history and

37 Klaas van Berkel, *Isaac Beeckman on Matter and Motion: Mechanical Philosophy in the Making* (Baltimore: Johns Hopkins University Press, 2013), esp. chap. 6. For an excellent discussion of the trend towards visualization in the philosophy of Petrus Ramus, see: Howard Hotson, *The Reformation of Learning: Post-Ramist Method and the Reception of the New Philosophy, 1618-1670* (Oxford: Oxford University Press, 2020), pp. 60-66, where the author analyses the work of Walter Ong. This chapter on Ramus is directly followed by a chapter on Beeckman, whose *Journal* is seen, as I previously claimed in my book on Beeckman, as the 'demonstrable link between Ramism and the advent of the mechanical world view' (p. 66).
38 Lüthy, 'Where Logical Necessity Becomes Visual Persuasion'.

history of science and has a special interest in the history of early modern science and the history of scientific institutions. He is the author of *Isaac Beeckman on Matter and Motion: Mechanical Philosophy in the Making* (Baltimore: Johns Hopkins University Press, 2013).

18 Concluding Remarks

Klaas van Berkel and Arjan van Dixhoorn

The study of Isaac Beeckman's mechanical philosophy began in the early twentieth century in the margins of the study of Descartes and Mersenne. It also began in the context of the early professionalization of the history of science. Beeckman's editor, Cornelis de Waard, after all belonged to the first generation of specialized historians of science. In the second half of the twentieth century the study of Beeckman's work was emancipated from the study of Descartes. Beeckman emerged as a pioneering contributor to the new mechanical philosophy in his own right, and this volume shows that the study of his ideas about nature is by no means exhausted. Yet, in the context of the social history of science and the newer history of knowledge, in the late twentieth and early twenty-first century Beeckman has gained more significance as a crucial witness not only to revolutionary developments in natural philosophy. His *Loci communes* have also become testimony for the 'otherness' of the local, regional, and trans-European knowledge cultures that preceded the new science that emerged with Descartes and Newton. His lack of interest in communicating his work in print has long had a marginalizing effect on the role awarded to him in the shaping of the new science, that is, in the lineage from the well-published Descartes to Newton. Ironically, it is precisely because of his manuscript book that Beeckman is emerging as an interesting case in the new history of knowledge. The *Loci* provide great material for those interested in 'knowledge-in-the-making' rather than 'ready-made-knowledge', to quote Bruno Latour. They bristle with notes that allow us to peak into knowledge-making practices, the evolution of rules for such practices, and the interactions of a wide range of 'knowledge-makers' that often remain invisible or difficult to trace. Several contributions to this volume show how the Beeckman studies can be expanded beyond the big questions related to the rise of the new science, using those notes. Building upon those contributions, Beeckman's *Loci* can

Van Berkel, Klaas, Albert Clement, and Arjan van Dixhoorn (eds.), *Knowledge and Culture in the Early Dutch Republic: Isaac Beeckman in Context*. Taylor & Francis Group 2022
DOI: 10.5117/9789463722537_CH18

become crucial evidence in writing cultural and social histories of the world *before* the new science emerged.

Of course, the study of these wider cultures of knowledge *through* Beeckman's testimony should not divert from the study of the man, his work and his legacy, which is also far from exhausted. In the following paragraphs we outline a few aspects of Beeckman's life and work that deserve to be explored in more depth and have only partly been addressed in this volume.

Beeckman and England: in the Beeckman studies so far, the French connection has been dominant, with a minimal interest in the influence of Francis Bacon on Beeckman's philosophizing. Yet, the chapters by Vera Keller and Huib Zuidervaart in this volume already suggest that the relations with England in the world of commerce, the communities of religious refugees, and the world of arts, sciences and technology were intensive around 1600. As could already be glimpsed from Deborah Harkness's study of the 'proto-Baconian' Lime Street immigrant community in London, Middelburg at the time was part of a triangle with Antwerp and London. The traffic in people, goods, and ideas was intense, also across confessional borders. From Middelburg, London was close and could be reached easily. The role of England in the life, work and legacy of Beeckman therefore alone deserves a deeper investigation, with attention paid to his life in Utrecht, Rotterdam and Dordrecht, and not just his life in Middelburg.

An interesting new source in this connection are the Hartlib Papers. A quick look at these papers, prompted by Keller's contribution in this volume, reveals that there is more to be found in the notes scribbled down by Samuel Hartlib, the London-based great intelligencer with a wide network of informants all over Europe, than the 1634 remark by the young Polish naturalist John Jonston quoted by Keller. Jonston also informed Hartlib about the friendship between Descartes and Beeckman, while another informant, the dissident Scottish minister John Dury, a year later added news about their falling out.[1] Also in 1635, a third informant, the exiled Bohemian inventor Christoph van den Berch (Berger), talked to Hartlib about Beeckman's plans to institute a 'collegium inventionum' inspired by a Venetian example (to judge on all sort of inventions and to reward those

1 Hartlib Papers Online, 29/2/42A. Descartes 'colit etiam amicitiam cum Rectore Dordrechtano nî fallor qui plurimum etiam in Physicis excellit, et Colloborationem agere posset'. Hartlib Papers Online, 29/3/15B. 'Iam contentionis serram reciprocat cum Beeck Rectore Dordraceno super Problemata aliquid opticum. Inde invidia et quidem odium exortum inter illos duos. Est enim Beeck ille insignis philosophus inprimis Physicus uti etiam Mathematicus.' Apparently the conflict between Beeckman and Descartes was widely discussed at the time.

who had contrived something useful).[2] The Hartlib Papers, and who knows what other collections, may hide more information that deserves to be explored; and instead of looking directly for clues to Beeckman, taking the detour by studying his connections may also yield further results.

Beeckman's family network: one such detour could lead through the many members of the extended Beeckman family and related families. Huib Zuidervaart's contribution to this volume already underlines the importance of Beeckman's family in his becoming a mechanical natural philosopher. How important family relations were to Beeckman and other members of his family is already evident from the genealogical notes added to the *Loci* by him and his brother Abraham. Further study of his family network could start there, from his own and his brother's conceptions of family and the importance they attached to their descent. What role did family members, such as his father, his uncles, his nephews ('cousins') and others play in Beeckman's upbringing and formative years as a young virtuoso? Did the family have a virtuoso tradition, maybe dating back to his grandfather Hendrick Beeckman and his stay in Italy? How did the social status of his family influence his access to the world of learning? Beeckman has been presented as a hybrid, someone from a world of artisans who also acquired academic learning, but the more we know about Beeckman and his family, the less convincing this simple dichotomy seems helpful in explaining Beeckman's life and work. A better understanding of the social standing of his family could be served by a more in-depth study of the family network, taking into account earlier generations, and the worlds from which they emerged in Flanders and Brabant. This study could be extended by looking into friendships and close relations of the Beeckman family, in particular of Isaac Beeckman's grandfather Hendrick and his father Abraham.

2 Hartlib Papers Online, 29-3-34B. 'Rector of Dort voluit Collegium Inventionum instituere. Est egregius Mathematicus cum Dn. Bergero. Venetiani jam habent tale aliquod Collegium, in quo Examinantur omnes Inventiones et approbaris stipeniis donantur.' Johan Christoph van den Berch had been a provincial governor under the unfortunate Frederic V, King of Bohemia, and in 1619 had to flee with his superior to the Dutch Republic. Before and after his stay in England, where he met Hartlib, Van den Berch tried to make a living in the Netherlands by offering inventions to the authorities. See, for instance: G. Doorman, *Octrooien voor uitvindingen in de Nederlanden uit de 16e-18e eeuw, met bespreking van enkele onderwerpen uit de geschiedenis der techniek* (The Hague: Nijhoff, 1940), G 295 (1629), G 323 (1631), H 44 (1638). See also Keller's contribution elsewhere in this volume, esp. pp. 411-412.

Beeckman's religious affinities: long-term effects of Christianity and short-term effects of Calvinism have been presented by some as crucial in the rise of the new science that evolved in the seventeenth century. In Beeckman's case we are not only presented with a community of religious refugees and the legacy of the Protestant diaspora of the sixteenth century, but evidence about several members of the extended Beeckman family and network and Isaac Beeckman's *Loci* also provide quite a wide spectrum of religious beliefs and affinities. Beeckman's father was a leading figure in a dispute within the Reformed Church of Middelburg. He vehemently opposed baptism of children of citizens that were not members of the Church. As a result, he was not allowed to commune, and the family attended church outside the city. The family became involved with the Brownists and other English non-conformists, who had a strong presence on the isle of Walcheren. The family also had strong connections to ministers in the Church, including the Calvinists Philippus Lansbergen, a close friend, and Willem Teellinck, a prominent Pietist. Isaac Beeckman's own religious beliefs, affinities and sensibilities can be glimpsed from the *Loci* and a few other sources. Like many Calvinists in a multi-confessional society, he easily conversed with people of other convictions, including his in-law Justinus van Assche, who had Remonstrant affinities. It is remarkable how strongly Beeckman remained attached to Van Assche even after he had been refused to the ministry at Veere (he afterwards switched to medicine and settled as a medical doctor in Amsterdam).[3] He also showed curiosity in the ideas of the Haarlem painter and notorious libertine Johannes Torrentius.[4] The *Loci* alone show that a more in-depth study of Beeckman's religious convictions and affinities can bring more precision to our understanding of his social context but also of his philosophizing.

Beeckman and vernacular knowledge: Beeckman's position in the world of natural philosophy has been the focus of most work on him so far.

3 On Van Assche, see: *Journal tenu par Isaac Beeckman de 1604 à 1634*, publié avec une introduction et des notes par C. de Waard, 4 vols. (The Hague: Martinus Nijhoff, 1939-1953) [henceforth *JIB*], I, p. 219, III, 3; E.G.E. van der Wall, *De mystieke chiliast Petrus Serrarius (1600-1669) en zijn wereld* (PhD diss., Leiden University, 1987). From 1623 to 1627 Van Assche had been minister of the Dutch Reformed Church 'under the cross' in Cologne and Frankfurt, where he was influenced by heterodox ideas of (most likely) Johann Moriaen and John Dury. See: John T. Young, *Faith, Medical Alchemy and Natural Philosophy: Johann Moriaen, Reformed Intelligencer and the Hartlib Circle* (Aldershot: Ashgate, 1998).

4 *JIB*, II, pp. 364-365. Beeckman suspected Torrentius to be a follower of the radical reformer David Joris (1501-1556).

Increasingly, it has become clear that Beeckman also participated in another knowledge culture, using the Dutch vernacular, which is characterized by Van Dixhoorn in this volume as *consten*-culture. As Keller argues as well, these worlds are not entirely separated, but they did diverge. The social and cultural reach, the 'tools of knowledge' that were employed, the interests and the questions asked, and the communities, networks and organizational structures in which vernacular knowledge thrived were different from those of the natural philosophers and their Latin-speaking world. Indeed, several contributions to this volume show that Beeckman's role in vernacular knowledge cultures was significant, and certainly cannot be understood as just 'artisanal'. Historians have only just begun to explore the attitudes towards secrecy and publication, scholarly (bookish) learning and practical, expert knowledge, and the deep affinities with artisanal and artistic practices in these vernacular worlds of knowledge. The growing interest among the Latin-speaking scholars in the achievements of vernacular learning is evident from Beeckman's philosophical journey. A better understanding of vernacular knowledge cultures in Europe around 1600 will help us assess with more precision their role in Beeckman's philosophizing and his influence in its developments. Beeckman has emerged as a key witness, but the investigation has just begun.

Acknowledgements

Most of the contributions to this volume originate from the conference 'Beeckman in Context', held in Middelburg in September 2018 on the occasion of the 400th anniversary of Beeckman's dissertation defence in Caen and the meeting of Beeckman and René Descartes in Breda. We are particularly grateful to all those people and organizations that made the event possible. First of all, we acknowledge the hospitality offered by University College Roosevelt at Middelburg, whose premises were used for the sessions of the conference. Second, we are thankful to the Koninklijke Hollandsche Maatschappij der Wetenschappen of Haarlem and the Koninklijk Zeeuwsch Genootschap der Wetenschappen, which together took this conference under their wings. Last, but certainly not least, we acknowledge the support we got from our research assistant, Dániel Moerman, and from Renate Huttinga, who designed the website for the conference and the publicity material.

At a very early stage Amsterdam University Press, represented by Julie Benschop, showed an interest in publishing this collected volume in its series Studies in the History of Knowledge. We thank Julie Benschop and Mike Sanders of AUP for the care they took in transforming a collection of individual texts into a real book.

Finally, we gratefully acknowledge the financial support for both the conference and the publication by the University of Groningen and University College Roosevelt.

Klaas van Berkel, Albert Clement, and Arjan van Dixhoorn

Index

References with Roman numerals refer to the colour section (cs).

For Product Safety Concerns and Information please contact our EU representative GPSR@taylorandfrancis.com Taylor & Francis Verlag GmbH, Kaufingerstraße 24, 80331 München, Germany

Batch number: 10347037

Printed by Printforce, the Netherlands